Organic Synthesis Using Biocatalysis

Organic Synthesis Using Biocatalysis

Edited by

Animesh Goswami
Chemical Development, Bristol-Myers Squibb,
New Brunswick, NJ, USA

Jon D. Stewart
Department of Chemistry, University of Florida,
Gainesville, FL, USA

ELSEVIER

Amsterdam • Boston • Heidelberg • London • New York • Oxford
Paris • San Diego • San Francisco • Singapore • Sydney • Tokyo

Elsevier
Radarweg 29, PO Box 211, 1000 AE Amsterdam, Netherlands
The Boulevard, Langford Lane, Kidlington, Oxford OX5 1GB, UK
225 Wyman Street, Waltham, MA 02451, USA

British Library Cataloguing in Publication Data
A catalogue record for this book is available from the British Library

Library of Congress Cataloging-in-Publication Data
A catalog record for this book is available from the Library of Congress

ISBN: 978-0-12-411518-7

Table of Contents

List of Contributors

Samantha K. Au
School of Chemical and Biomolecular Engineering, Georgia Institute of Technology, Parker H. Petit Institute of Bioengineering and Bioscience, Atlanta, GA, USA

Andreas S. Bommarius
School of Chemical and Biomolecular Engineering, Georgia Institute of Technology, Parker H. Petit Institute of Bioengineering and Bioscience; School of Chemistry and Biochemistry, Georgia Institute of Technology, Atlanta, GA, USA

Chen Cao
Department of Bioengineering, Graduate School of Bioscience and Biotechnology, Tokyo Institute of Technology, Nagatsuta-cho, Midori-ku, Yokohama, Japan

Pere Clapés
Department Química Biológica y Modelización Molecular, Instituto de Química Avanzada de Cataluña, IQAC-CSIC, Barcelona, Spain

Rodrigo O.M.A. de Souza
Biocatalysts and Organic Synthesis Lab, Organic Chemistry Department, Chemistry Institute, Federal University of Rio de Janeiro, Rio de Janeiro, Brazil

Brent D. Feske
Chemistry and Physics Department, Armstrong State University, Savannah, GA, USA

Michael J. Fink
Vienna University of Technology, Institute of Applied Synthetic Chemistry, Vienna, Austria

Animesh Goswami
Chemical Development, Bristol-Myers Squibb, New Brunswick, NJ, USA

Gideon Grogan
Department of Chemistry, University of York, Heslington, York, UK

Harald Gröger
Faculty of Chemistry, Bielefeld University, Universitätsstr, Bielefeld, Germany

Jonathan Groover
Chemistry and Physics Department, Armstrong State University, Savannah, GA, USA

Melissa L.E. Gutarra
Escola de Química, Federal University of Rio de Janeiro, Pólo Xerém, Estrada de Xerém, Xerém, Duque de Caxias, Rio de Janeiro, Brazil

Romas Kazlauskas
Department of Biochemistry, Molecular Biology & Biophysics and The Biotechnology Institute, University of Minnesota, Saint Paul, MN, USA

Tomoko Matsuda
Department of Bioengineering, Graduate School of Bioscience and Biotechnology, Tokyo Institute of Technology, Nagatsuta-cho, Midori-ku, Yokohama, Japan

Marko D. Mihovilovic
Vienna University of Technology, Institute of Applied Synthetic Chemistry, Vienna, Austria

Leandro S.M. Miranda
Biocatalysts and Organic Synthesis Lab, Organic Chemistry Department, Chemistry Institute, Federal University of Rio de Janeiro, Rio de Janeiro, Brazil

Thomas S. Moody
Department of Biocatalysis and Isotope Chemistry, Almac, Craigavon, Northern Ireland, UK

Ramesh N. Patel
SLRP Associates, Consultation in Biotechnology, Bridgewater, NJ, USA

Laila Roper
Department of Chemistry, University of York, Heslington, York, UK

J. David Rozzell
Provivi, Santa Monica, CA, USA

Florian Rudroff
Vienna University of Technology, Institute of Applied Synthetic Chemistry, Vienna, Austria

Jon D. Stewart
Department of Chemistry, University of Florida, Gainesville, FL, USA

CHAPTER 1

Introduction, Types of Reactions, and Sources of Biocatalysts

Animesh Goswami*, Jon D. Stewart†
*Chemical Development, Bristol-Myers Squibb, New Brunswick, NJ, USA
†Department of Chemistry, University of Florida, Gainesville, FL, USA

1 INTRODUCTION

1.1 Enzymes and Their Roles in Nature

Enzymes are nature's catalysts, facilitating the creation, functioning, maintenance, and ultimately the demise of all living cells. Enzymes are proteins composed of 20 natural amino acids joined together by peptide bonds, in some cases augmented with additional organic or inorganic species known as cofactors.[1] In addition to their primary molecular structures dictated by the amino acid sequence, enzyme catalytic function also depends upon subsequent folding into specific three-dimensional shapes that contain a variety of secondary and tertiary structural elements. These architectures determine not only enzyme function but also how they interact with the external solvent medium in which they are dissolved or suspended. This can have important ramifications when enzymes are employed under partially or completely nonaqueous conditions.

Like all catalysts, enzymes increase reaction rates by lowering their activation energies. The most important difference between enzymes and simple catalysts such as a proton or hydroxide is that the former are much more restrictive in the range of acceptable substrates. The three dimensional structure of the enzyme allows binding of only those starting materials (usually referred to as substrates) whose structures are congruent with the size, shape and polarity of the catalytic portion of the enzyme (the "active site"). *Formation of this noncovalent complex prior to chemical conversion is the key to the high selectivity of enzyme-catalyzed reactions since it places the substrate into a specific location where its functional groups are oriented precisely with those on the enzyme.*

[1] Some RNA molecules also possess catalytic abilities; however, their substrate and range of chemical conversions seems rather limited and for this reason, catalytic RNA molecules lie outside the scope of this book.

Organic Synthesis Using Biocatalysis. http://dx.doi.org/10.1016/B978-0-12-411518-7.00001-9

Noncovalent complex formation allows chemical reactions between specific amino acids on the enzyme and functional groups on the substrate to occur in an environment that is kinetically equivalent to a unimolecular process. Transforming what would otherwise be bimolecular reactions into effectively intramolecular conversions is a major reason that enzymes can accelerate reactions by up to 23 orders of magnitude over the background (uncatalyzed) reaction [1].

Selectivity is the second benefit from forming a noncovalent enzyme–substrate complex prior to chemical conversion. By focusing the catalytic attention of the enzyme onto a specific area of the substrate, reactions can be restricted to a single portion of the molecule that may or may not be the most reactive portion of the overall substrate structure. This allows enzyme-catalyzed reactions to be selective in many respects: chemoselective (carrying out only one specific transformation while others are possible), regioselective (transforming only one among several possible sites), and stereoselective (producing and/or consuming one stereoisomer in preference to others).

2 DEFINITION OF BIOCATALYSIS

In nature, enzymes catalyze transformations of metabolites that occur within and/or outside of living cells. Although some enzymes accept only a limited variety of substrates, a large fraction is more tolerant and allows conversions of nonnatural starting materials. The field of biocatalysis rests upon this partial promiscuity. If enzymes were truly selective for only a single substrate, it would be impossible to utilize them for synthesizing new molecules from nonnatural substrates. The goal is to identify or engineer enzymes that are sufficiently *general* to accept a variety of related substrates, but *selective* enough to yield single products or stereoisomers. We use the term "biocatalysis" to describe the use of enzymes (either native or modified) for synthetic transformations of nonnatural starting materials. Enzymes used for *in vitro* synthetic transformations are called biocatalysts, and the processes are called biocatalytic transformations.

3 SCOPE OF THIS BOOK

Some enzymes catalyze the reactions that build up large molecules from simple building blocks, for example, complex carbohydrates from carbon dioxide and water or the synthesis of steroids and terpenoids from acetate. Others are involved in the degradation of large assemblies to small molecules, for example, hydrolysis of proteins to amino acids and the oxidative degradation of lignin. Although some of these native conversions are industrially important and practiced on large scales, the use of enzymes to produce their normal primary and secondary products of cells lies outside the scope of this book. Here, our focus is on preparing *nonnatural* compounds using enzymes since this addresses the need commonly encountered in organic synthesis. However, it should be noted that the native reactions of an enzyme can often be used as starting points for their applications to nonnative reactions.

The field of metabolic engineering also lies outside the scope of this book. These efforts use two or more enzymes to catalyze sequential steps in a pathway that links a simpler starting material such as glucose with a final intracellular target product such as butanol or lysine. In some cases, the complete pathway already exists within a single organism; in others, enzymes from different sources are assembled into an artificial metabolic pathway in a suitable host cell. The key difference between biocatalysis and metabolic engineering is that the molecular skeletons are provided *in vitro* in the former case and *in vivo* in the latter. Although it is economically attractive to produce a target molecule by metabolic engineering, this benefit must be balanced against the (usually) lengthy optimization phase required for efficient production and the restriction that intermediates and the final product should be nontoxic to the host cells.

4 KEY BENEFITS OF EMPLOYING ENZYMES IN SYNTHESIS

Enzymes offer several attractive features as catalysts for organic synthesis. They often show high selectivities, they can operate under mild conditions and they are completely biodegradable catalysts constructed solely from renewable resources. They are thus ideal strategies as chemistry embraces sustainability.

4.1 Selectivity: Chemo-, Regio- and Stereo-

Biocatalytic reactions can show very high selectivities in all respects (Figure 1.1). When similarly-reactive functional groups are present in a molecule, the enzymes often catalyze only the reaction of one while leaving the others intact. For example, nitrile hydratases catalyze the partial hydrolysis of a nitrile group to yield a primary amide without cleaving an ester moiety present in the same molecule or further hydrolyzing the amide product, a property referred to as "chemoselectivity."

"Regioselectivity" is another useful property displayed by many enzymes. This refers to the transformation of one functional group while leaving other identical (or nearly identical) moieties at different locations within the molecule untouched. For example, among three esters in a triacylglyeride, many lipases hydrolyze only one position, and do not catalyze further hydrolysis of the diester product.

All but one of the amino-acid building blocks have at least one chiral center,[2] and for this reason, enzymes are intrinsically asymmetric catalysts. The asymmetric nature of enzymes results in enantioselectivity when biocatalysts convert a prochiral starting material into a single product enantiomer, for example, in ketone or imine reductions. The same asymmetric nature also causes the biocatalysts to preferentially transform only one stereoisomer of a starting material that contains a mixture of diastereomers or enantiomers. Such a process is referred to as a kinetic resolution and the ratio of rate constants for the fast- and slow reacting enantiomers is termed the enantioselectivity (E) ratio. E values higher than 100 are commonly observed for biocatalysts, and values in this range allow both the residual starting material (the slow-reacting enantiomer) and the product

[2] Glycine is the only achiral amino acid normally found in proteins.

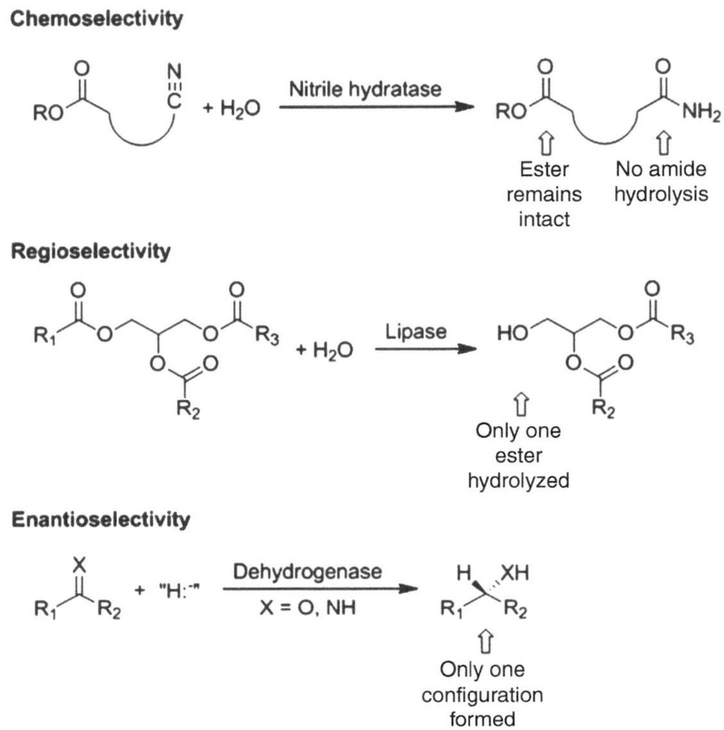

FIGURE 1.1
Types of selectivity exhibited by enzymes. Biocatalytic processes can provide chemo-, regio-, and stereoselective conversions. Representative examples can be observed in reactions catalyzed by nitrile hydratases, lipases and dehydrogenases.

(from the faster-reacting enantiomer) to be obtained with high optical purities from a single reaction run to 50% conversion.

4.2 Biocatalyst Reaction Conditions

Enzymes have generally evolved to function best under the conditions that exist within their respective cellular environments. This usually means ambient temperatures (20–40 °C), near-neutral pH values, and with water as the solvent. Such conditions are particularly appropriate for sensitive starting materials and/or products, and this constitutes an important reason to employ enzymes in organic synthesis. It should be noted, however, that some enzymes have evolved to function under extreme conditions. For example, thermophilic bacteria that thrive at temperatures more than 100 °C have been valuable sources of enzymes with much higher-than-usual thermal stabilities. In addition, enzymes normally found outside cells (extracellular enzymes) are also generally more stable since their operating environment is less predictable and largely uncontrolled. This diversity in thermal stabilities often allows one to choose a reaction temperature that balances good reaction rates with enzyme stability and also maximizes space–time yields.

In nature, most enzyme-catalyzed reactions occur in an aqueous environment and many synthetic applications, therefore, also utilize water as solvent. This is often advantageous with regard to maximizing the sustainability of a chemical process. Moreover, enzymes are usually most stable in water. These benefits, however, must be balanced against two disadvantages of using water as a solvent for biocatalytic processes. Water is a reactant in hydrolytic processes and its concentration must be minimized when such reactions are run in reverse in order to shift the equilibrium, for example, when using enzymes to synthesize esters or amides from carboxylic acids and alcohols or amines, respectively. In such cases, water-organic biphasic systems or completely organic solvents can be used. The second complication associated with aqueous conditions is that many of the starting materials and products of synthetic interest have very limited solubilities in water. This either requires the use of dilute solutions, which lowers space–time yields and also generates large volumes of wastewater that must be treated, or the use of organic solvents as additives or replacements for water.

Biocatalysts are proteins and composed of natural amino acids. In some cases, additional natural ligands (cofactors) are present. This means that biocatalysts are inherently nonhazardous materials, although it should be noted that, because they are proteins, some enzymes may cause allergenic reactions in susceptible individuals. Although such reactions are rare, normal care should be taken when handling solid enzyme powders.

5 MECHANISM AND KINETICS OF ENZYME-CATALYZED REACTIONS

Although it is not necessary to determine kinetic parameters in order to use enzymes for chemical synthesis, this knowledge can often be useful in deciding which avenues offer the best opportunities for process improvements. Likewise, it is not essential - but often highly useful - to understand the mechanism of the enzyme-catalyzed reaction for the same reasons that one often benefits from knowing the mechanisms of any reaction employed in a synthetic route. Although every individual enzymatic reaction has a unique combination of kinetic properties and reaction mechanism, several useful generalizations are summarized in the subsequent section.

5.1 Features of Enzyme Catalyzed Reactions

As noted previously, noncovalent association between the enzyme and its substrate(s) is the essential first step in biocatalytic reactions. Although early theories of enzyme catalysis focused primarily on interactions between the enzyme and substrate, it was later appreciated that maximizing selective, noncovalent interactions between the enzyme and the high-energy transition state(s) that linked enzyme-bound complexes of substrates and products was the key to efficient rate enhancements. A somewhat oversimplified view is that ground-state interactions determine substrate specificity and transition-state interactions yield rate enhancements.

In addition to noncovalent associations (by van der Waals forces, hydrogen bonds, electrostatic and hydrophobic interactions), some enzymes also form covalent bonds between the enzyme and portions of the substrate. Lipases are a well-known example of this phenomenon. These enzymes utilize a specific protein hydroxyl group (most commonly a serine side chain) to form an ester intermediate with the substrate that is subsequently cleaved by an exogenous nucleophile to form the final reaction product and regenerate the free protein hydroxyl group, making the active site suitable for the next catalytic cycle.

5.2 Coenzymes

Although some enzyme-catalyzed reactions utilize only functional groups found in the protein, the limited number and variety of amino-acid side-chain moieties severely limits the range of accessible reactions. For example, there are neither common amino acids with electrophilic side chains nor amino-acid functional groups suitable for redox catalysis.[3] Nature has circumvented this problem by evolving a suite of coenzymes (also known as cofactors) that specialize in particular types of chemical conversions. Table 1.1 lists some common cofactors for biocatalytic processes. In some cases, for example, biotin and some flavins, these cofactors are covalently coupled to the enzyme within the active site. Other cofactors such as nicotinamides are bound reversibly by noncovalent forces during the entire catalytic cycle. Finally, a few cofactors such as pyridoxal phosphate form reversible covalent linkages with the resting form of the enzyme that are cleaved during the catalytic cycle, and then re-formed at the end. When needed by specific enzymes, provision for cofactor supply must also be made. Because of their expense, cofactors are normally supplied in substoichiometric quantities (usually $\ll$ 0.1 mole %). This means that they must be regenerated prior to the start of the next catalytic cycle, and strategies for cofactor regeneration have been developed as an essential adjunct for biocatalytic reactions, particularly for reductions and oxidations.

5.3 Kinetics and Reaction Mechanisms

Reaction mechanisms describe the sequence of bond breaking and bond making steps, whereas kinetics are concerned with the nature and timing of the noncovalent complexes that form and break down during the catalytic cycle. Depending on the number of substrates and products, kinetics can be simple or complex. Single-substrate/single-product reactions are the most straightforward schemes, although there are relatively few examples of such reactions in biocatalysis. More commonly, two or more molecules are bound and/or released. This introduces the question of timing with respect to formation and breakdown of enzyme – ligand complexes. Three common reaction mechanisms for a two-substrate/two-product reaction are illustrated schematically in Figure 1.2. In an ordered mechanism, the enzyme cannot productively bind the second substrate

[3] The only exception is disulfide bond formation between a pair of suitably positioned cysteine side-chains.

Table 1.1 Cofactors Commonly Encountered in Biocatalytic Reactions Applied to Chemical Synthesis

Name	Cofactor Structure	Chemical Function	Enzyme
Nicotinamide adenine dinucleotide, reduced form (NADH)[a]	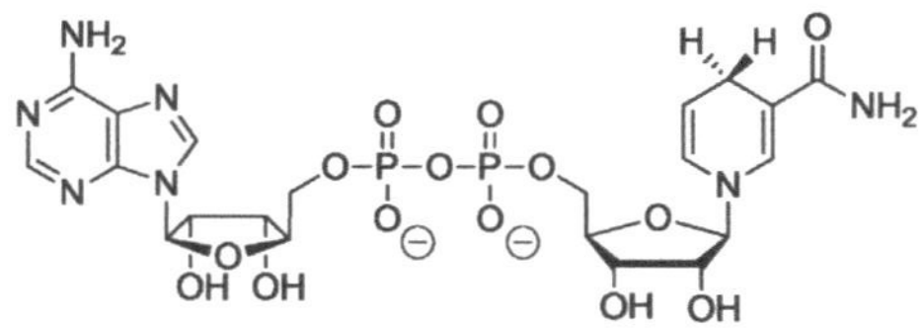	Donates a hydride for polar reductions of functional groups such as C = O and C = N; acts as an electron source for monooxygenases	Alcohol dehydrogenase/ketoreductase; alkene reductase/enoate reductase; amino acid dehydrogenase; monooxygenase; dioxygenase
Nicotinamide adenine dinucleotide phosphate, reduced form (NADPH)[b]	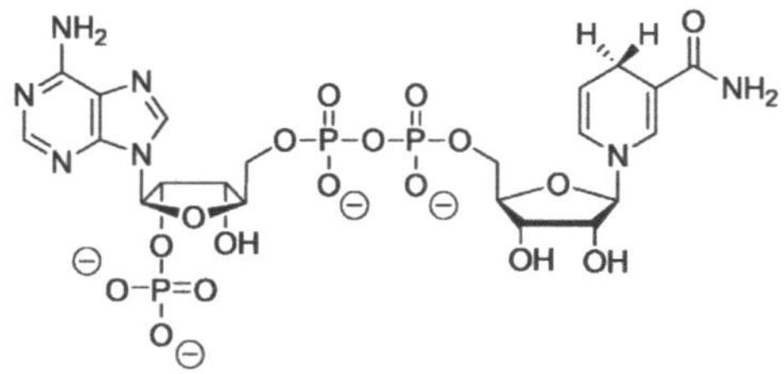	Donates a hydride for polar reductions of functional groups such as C = O and C = N; acts as an electron source for monooxygenases	Alcohol dehydrogenase/ketoreductase; alkene reductase/enoate reductase; amino acid dehydrogenase; monooxygenase; dioxygenase
Flavin mononucleotide (FMN)	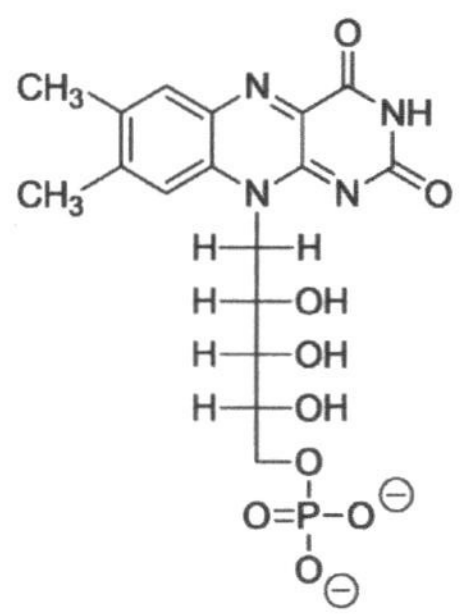	Forms a hydroperoxy intermediate from O_2 that is used by monooxygenases; following 2 electron reduction, donates a hydride for reductions of electron-deficient C = C bonds	Alkene reductase/enoate reductase; amino acid oxidase; Baeyer–Villiger monooxygenase

(Continued)

Table 1.1 Cofactors Commonly Encountered in Biocatalytic Reactions Applied to Chemical Synthesis *(cont.)*

Name	Cofactor Structure	Chemical Function	Enzyme
Flavin adenine dinucleotide (FAD)	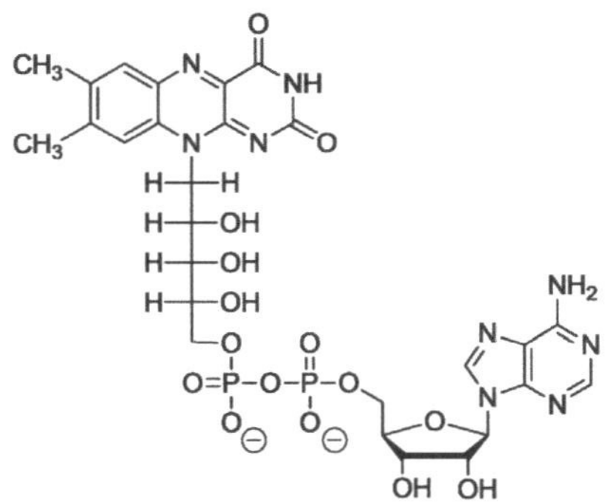	Forms a hydroperoxy intermediate from O_2 that is used by monooxygenases; following 2 electron reduction, donates a hydride for reductions of electron-deficient C = C bonds	Alkene reductase/enoate reductase; amino acid oxidase; Baeyer–Villiger monooxygenase
Thiamine pyrophosphate	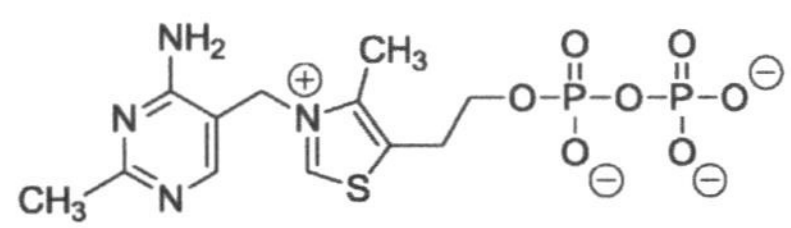	Allows umpolung anion formation by aldehyde carbonyl groups; facilitates α-keto acid decarboxylations	Pyruvate decarboxylase; benzoylformate decarboxylase; phenylpyruvate decarboxylase
Pyridoxal phosphate	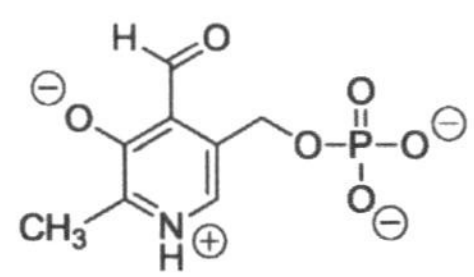	Following reversible Schiff's base formation, the cofactor stabilizes an anion on the carbon adjacent to the amine	Transaminase; threonine aldolase

[a]Known as DPNH in the very old scientific literature
[b]Known as TPNH in the very old scientific literature

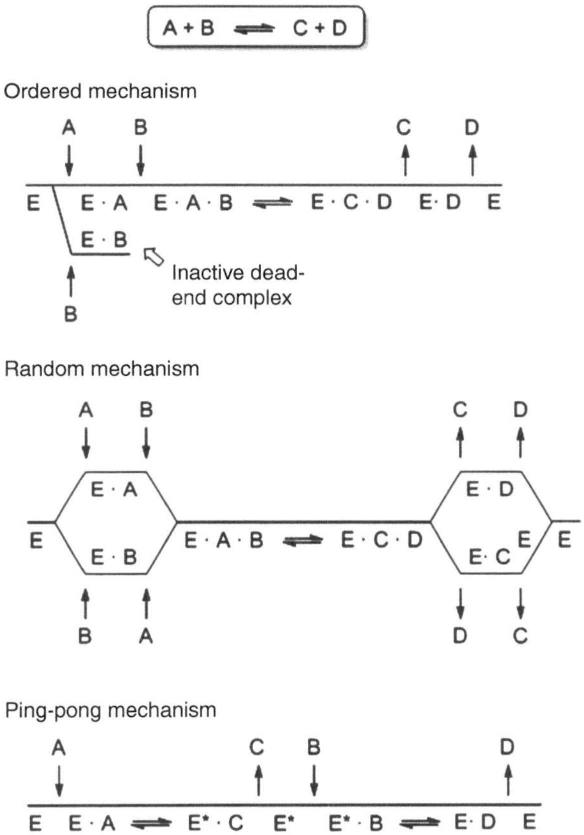

FIGURE 1.2
Three possible kinetic mechanisms for a two-substrate/two-product reaction depicted in Cleland notation. Substrates are designated as "A" and "B", products as "C" and "D" while "E" represents the enzyme and "E*" is a covalently modified enzyme that forms transiently as part of the normal catalytic pathway in the ping-pong mechanism.

(B) until it previously bound the first (A). Enzymes that follow this scheme often show substrate inhibition when the concentration of B is very high (relative to that of A) since this favors formation of a dead-end E · B complex that must dissociate prior to re-joining the productive pathway. A random mechanism removes this restriction and either substrate can be bound first. Product release steps can be similarly described.[4] From a kinetic standpoint, cofactors that do not remain permanently bound to the enzyme are equivalent to substrates and/or products. Cofactors that remain bound to the enzyme during turnover are kinetically treated as part of the enzyme itself. In either an ordered or random

[4] It should be noted that "substrate" and "product" are meaningful only when the direction of the reaction is specified. Like all catalysts, enzymes accelerate both the forward and reverse reactions.

mechanism, the key species is the ternary complex of enzyme with both substrates (E · A · B) since this must be formed in order for the reaction to occur.

Enzymes that form a covalent bond with a part of the substrate often follow a ping-pong mechanism (Figure 1.2). By definition, these must be ordered mechanisms. Ping-pong mechanisms always share two essential features. First, they involve a covalently modified form of the enzyme (or cofactor) as an obligate intermediate.[5] In addition, the first product must be released from the active site *before* the second substrate binds. Together, these two properties of enzymes that follow ping-pong mechanisms open the possibility that the covalent intermediate might be redirected into alternative products by restricting access to the normal second substrate. Employing lipases for synthetic (rather than hydrolytic) acyl transfer reactions is a very common example of this strategy and others can be found in the appropriate sections of later chapters.

Although the methods for determining steady-state kinetic parameters can be complex and the details lie outside the scope of this book, a general understanding of their meaning is useful for applications to biocatalysis. It should be borne in mind that most substrates and products in biocatalytic reactions have limited aqueous solubilities. This complicates kinetic studies since the actual concentration "seen" by the enzyme may not be the same as the concentration added to the reaction mixture unless all of the material is fully dissolved. Therefore, one must be cautious when applying kinetic constants measured under one set of experimental conditions (often dilute aqueous solutions) to reactions run under process conditions that may involve partially dissolved substrates, organic cosolvents, etc.

5.4 Kinetic Constants

Two steady-state kinetic constants are most useful in evaluating biocatalytic reactions. "k_{cat}" is often known as the *turnover number* and higher values indicate more catalytically efficient enzymes. This first-order rate constant describes the speed at which an enzyme converts bound substrates to products and re-forms the free enzyme to prepare for the next round of catalysis. It includes both the "chemical" steps (bond making and bond breaking) as well as the product release step(s). Note that it is not uncommon for product release to be the slowest step. As a practical matter, one normally seeks enzymes with $k_{cat} \geq 1\ s^{-1}$ under the process conditions to ensure reasonable space–time yields along with acceptable catalyst loading levels.

"K_M", also known as the *Michaelis constant,* is a composite of several microscopic rate constants that primarily describes substrate/product binding and has units of concentration (typically *M*). Consistent with normal biochemical convention, the binding interaction is considered in the dissociation direction and smaller numerical values of K_M therefore signal tighter substrate/product binding. When

[5] The covalently modified form of the enzyme has been variously notated as "F," "E,*" or " E′ " in the literature.

the chemical step is slow relative to substrate release,[6] K_M is approximately equal to the actual thermodynamic dissociation constant (denoted as K_D). K_M is also equivalent to the substrate concentration required to give a reaction velocity (V) that is one-half the maximal value (denoted as $V_{max} = k_{cat} \cdot$ [enzyme]). If only a single substrate is involved, only a single K_M value is required to describe the reaction. On the other hand, for enzymes that require multiple substrates, each has an associated K_M value. Because a complete kinetic characterization is laborious in these situations, K_M values are often determined for only the substrate of interest by varying its concentration while holding all others constant. Such studies yield apparent K_M values ($K_{M,app}$). These are very common for nicotinamide-dependent reactions since the cofactors are typically held at a constant value during biocatalytic processes.

Although not a kinetic constant *per se*, the ratio k_{cat}/K_M is often used to characterize enzyme performance in a biocatalytic process. This ratio is often referred as the *specificity constant* and has units of a second-order rate constant ($M^{-1} \cdot s^{-1}$). The k_{cat}/K_M ratio reflects the difference in relative energies between the free enzyme and free substrate compared to the rate-limiting transition state(s). The maximum value is approximately 1×10^8 $M^{-1} \cdot s^{-1}$, which is the diffusion limit of small molecules in aqueous solution. In approximate terms, k_{cat}/K_M describes the performance of an enzyme-catalyzed reaction under nonsaturating conditions whereas k_{cat} reflects its behavior under saturating conditions. The most appropriate descriptor for a given biocatalytic process depends upon the enzyme/substrate affinity and the concentrations actually employed.

5.5 Kinetic Constants and Kinetic Resolutions

When presented with a mixture of substrate isomers, enzymes often show a preference for transforming only one. When the isomers are enantiomers, the process is referred to as a kinetic resolution. Continued conversion of the favored enantiomer increases the optical purity of the residual starting material in favor of the slow-reacting enantiomer whereas the product is drawn primarily from the fast-reacting enantiomer. The enantioselectivity (E) value is defined as the numerical ratio of rate constants for the faster- *versus* the slower-reacting enantiomer. Whether k_{cat} or k_{cat}/K_M values are more appropriate in determining the E ratio for a given reaction depends on the substrate concentration employed and the enzyme/substrate affinity. A simple means to calculate the E value from experimentally observable values are available [2]. As noted previously, one usually targets enzymes with E values ≥ 100 in order to achieve good kinetic resolutions.

It is important to note that racemic mixtures are the thermodynamic minima for all reactions. Because enzymes catalyze reactions in both directions, even enzymes with very high stereoselectivities will ultimately yield racemic mixtures

[6] This is not an uncommon situation in practice, particularly when enzymes catalyze reactions of nonnatural substrates.

if reactions are allowed to reach equilibrium. This can be a problem if extended reaction times and/or very high catalyst loadings are employed. Obtaining racemic products from an enzyme-catalyzed reaction does not necessarily mean that the enzyme has low stereoselectivity; the reaction should be explored further by reducing the reaction time and/or enzyme concentration.[7]

5.6 Kinetic Constants and Temperature

Like all chemical reactions, the rates of enzyme-catalyzed conversions increase with increasing temperature as predicted by the Arrhenius relationship. This relationship only applies to a limited range, however, since enzymes are deactivated by higher temperatures. In addition, substrates and/or products may degrade at elevated temperatures, particularly in water. This means that temperature effects on reaction rate and stability are diametrically opposed and the best choice is necessarily a compromise. In practice, most biocatalytic processes are run at temperatures between room temperature and approximately 50 °C. Somewhat higher optimum reaction temperatures may be achievable with thermophilic enzymes possessing higher deactivation temperature.

6 TYPES OF ENZYME-CATALYZED REACTIONS COMMONLY USED IN BIOCATALYSIS

Though nature uses many different types of enzymes to catalyze a wide variety of different transformations, only a limited number of reaction types have been widely used in biocatalysis. An overview of the most common varieties is provided below, approximately arranged by decreasing number of published examples. Detailed descriptions are available in subsequent chapters dedicated to specific reaction types. Table 1.2 summarizes reaction types, starting materials, products, and enzymes required for the most common types of biocatalytic reactions. This provides a quick reference for retrosynthetic planning and designing syntheses of specific molecules.

6.1 Hydrolysis and Synthesis of Carboxylic Acid Derivatives such as Esters and Amides

This is the most common application of biocatalysis in organic synthesis and represents the majority of published examples. Enzymes that catalyze acyl transfer reactions of esters and amides are widely distributed in nature and belong to the lipase/esterase and protease/amidase families, respectively. They play key roles in the metabolism of lipids and proteins and the choice of names, lipase *versus* esterase, is subject to debate. Normally, acyl transfer occurs almost exclusively to water, resulting in hydrolysis. This is particularly valuable for amide hydrolysis that normally requires forcing conditions and strong acid or

---

[7] Finding products that are completely racemic nearly always indicates that the catalyst concentration and or reaction time is too great. Even poorly stereoselective enzymes typically afford some optical enrichment and only reaching true thermodynamic equilibrium completely erases this preference.

Table 1.2 Summary of Common Biocatalytic Transformations

Reaction Type	Starting Material(s)	Product(s)	Common Uses	Enzyme Types
Hydrolysis	Esters	Alcohols and carboxylic acids	Synthesis of alcohols and carboxylic acids; resolution of esters, alcohols, acids	Lipase, esterase, protease
	Amides	Amines and carboxylic acids	Synthesis of amines and carboxylic acids; resolution of amides, acids and amines	Lipase, esterase, protease
	Nitriles	Amides	Synthesis of primary amides; resolutions of nitriles	Nitrile hydratase
	Nitriles	Carboxylic acids	Synthesis of carboxylic acids; resolutions of acids	Nitrilase
	Epoxides	1,2-Diols	Synthesis of diols; resolution of epoxides	Epoxide hydrolase
Esterification	Carboxylic acid and alcohol	Ester	Synthesis of esters; resolutions of carboxylic acids and alcohols	Lipase, esterase, protease
Amidation	Carboxylic acid and amine	Amide	Synthesis of amides; resolutions of carboxylic acids and amines	Lipase, esterase, protease
Transesterification	Alcohol and ester	Alcohol and ester	Synthesis of esters; resolution of esters, alcohols	Lipase, esterase, protease
Transamination	Ketone and amine	Ketone and amine	Synthesis of amines	Transaminase
	α-Keto acid and α-amino acid	α-Keto acid and amino acid	Synthesis of α-amino acids	Amino acid transaminase
Dehydro-halogenation	Halohydrin	Epoxide	Synthesis of epoxides; resolution of halohydrins and epoxides	Halohydrin dehalogenase
Carbonyl reduction	Ketone	Alcohol	Synthesis of alcohols	Alcohol dehydrogenase/ ketoreductase
	Aldehyde	Alcohol	Synthesis of alcohols	Alcohol dehydrogenase/ ketoreductase

(Continued)

Table 1.2 Summary of Common Biocatalytic Transformations *(cont.)*

Reaction Type	Starting Material(s)	Product(s)	Common Uses	Enzyme Types
Activated alkene reduction	α,β-Unsaturated aldehyde, ketone, ester	Saturated aldehyde, ketone, ester	Asymmetric reduction of C ═ C bonds	Alkene reductase (also known as enoate reductase and ene-reductase)
Reductive amination	Ketone, 2-keto acid	Amine, 2-amino acid	Synthesis of amines	Amino acid dehydrogenase
Alcohol oxidation	Secondary alcohol	Ketone	Synthesis of ketones	Alcohol dehydrogenase, alcohol oxidase
	Primary alcohol	Aldehyde	Synthesis of aldehydes	Alcohol dehydrogenase, alcohol oxidase
Amine oxidation	Amine	Aldehyde, ketone	Synthesis of aldehydes and ketones; resolutions of amines	Amine oxidase
	α-Amino acid	α-Keto acid	Synthesis of α-keto acids; resolutions of amino acids	Amino acid oxidase, amino acid dehydrogenase
Hydroxylation	Saturated carbon–hydrogen bond	Alcohols	Synthesis of alcohols	Monooxygenase, peroxidase
	Aromatic carbon-hydrogen bond	Phenols	Synthesis of phenols	Monooxygenase, dioxygenase, peroxidase
Oxidative dealkylation	Phenyl ethers	Phenol, aldehyde	Synthesis of phenols; dealkylation of methyl or other alkyl ethers	Monooxygenase, peroxidase
	Alkylated anilines	Amine, aldehyde	Synthesis of amines; dealkylation of methyl or other secondary amines	Monooxygenase, Peroxidase
Baeyer–Villiger oxidation	Ketone	Ester	Synthesis of esters and lactones	Baeyer–Villiger monoxygenase
Cyanohydrin formation	Aldehyde or ketone	2-Hydroxy nitrile	Synthesis of 2-hydroxy nitriles	Oxynitrilase (Hydroxynitrile lyase)

Table 1.2 Summary of Common Biocatalytic Transformations *(cont.)*

Reaction Type	Starting Material(s)	Product(s)	Common Uses	Enzyme Types
Aldol condensation	Aldehyde	2-Keto acid	Synthesis of 2-hydroxy ketones	Pyruvate decarboxylase, benzoyl formate decarboxylase, phenylpyruvate decarboxylase
	Aldehyde	2-Hydroxy aldehyde	Synthesis of 3,4-dihydroxy aldehydes	Deoxyribose phosphate aldolase
	Aldehyde, dihydroxyacetone phosphate	1-Phosphorylated 2-keto-3,4-diol	Synthesis of polyhydroxylated 2-ketones	Dihydroxyacetone phosphate aldolase
	Aldehyde	2-amino acids	Synthesis of 3-hydroxy 2-amino acids	Threonine aldolase

The table is organized by reaction type, rather than by enzyme type to facilitate use in synthetic applications.

base. The reverse reaction (ester and amide synthesis) cannot be carried out under aqueous conditions due to the large molar excess of water (whose concentration is 55 *M* in pure water). Enzymes *can* catalyze the reverse reaction – ester and amide synthesis from acids and alcohols or amines – under nonaqueous conditions, although the yields are limited by thermodynamic constraints unless steps are taken to remove the water formed during the reaction. This limitation can be circumvented by using enzymes to catalyze transesterification and transamidation, using an ester or amide as the starting material and relying on Le Chatêlier's principle to shift the equilibrium to the desired product (Figure 1.3).

Acyl transferase enzymes have been widely used to synthesize chiral esters, amides, alcohols, and amines. In many cases, these conversions involve kinetic resolutions of alcohols, acids, esters, amines, and amides. Of course, since each enantiomer makes up half of the racemic mixture, kinetic resolutions can provide a maximum 50% yield. This limitation can be overcome by racemizing or inverting the configuration of the unreacted substrate during the enzymatic reaction. Such a scheme is referred to as a *dynamic kinetic resolution* and theoretically allows complete substrate conversion to product along with 100% chemical yield of a single product enantiomer.

6.2 Carbonyl Reductions to Alcohols

Another common biocatalytic route to alcohols involves enzyme-mediated reductions of the corresponding aldehydes or ketones. Because the starting materials are prochiral, such processes are not kinetic resolutions and are therefore not subject to the 50% yield limitation. Enzymes that catalyze carbonyl

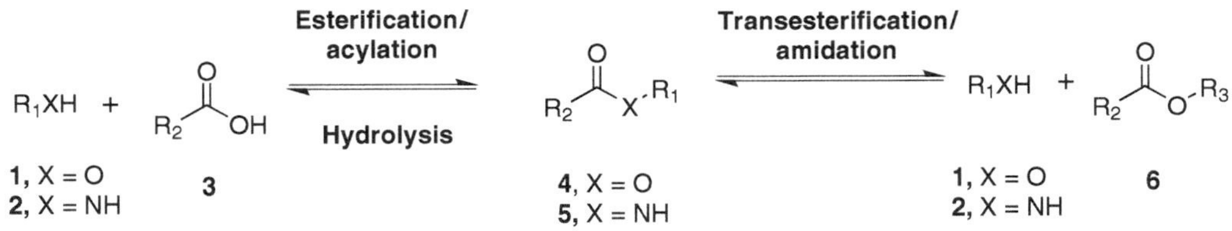

FIGURE 1.3
Hydrolysis and synthesis of carboxylic acid derivatives such as esters and amides.

reductions have been variously named as alcohol dehydrogenases or ketoreductases.[8] Because carbonyl reductions involve the formal addition of H_2 (usually in the form of a hydride ion along with a proton), a second cosubstrate that supplies reducing equivalents must be included along with the aldehyde or ketone of synthetic interest. Nicotinamides are the most common hydride sources for enzymatic reactions and they occur in the form of nicotinamide adenine dinucleotide (NADH) and its phosphorylated analog (NADPH). Some enzymes are highly selective for one cofactor type or the other whereas others accept both NADH and NADPH (a property referred to as *dual specificity*). Carbonyl reduction converts the nicotinamides into their oxidized forms and these must be reduced back to their original oxidation states prior to reuse in subsequent rounds of catalysis. Several methods for *in situ* cofactor regeneration have been developed in response to the high cost of nicotinamides and cofactor turnover numbers of more than 1000 are commonly achievable. This usually makes the cofactor contribution to the overall process costs negligible (Figure 1.4).

Alcohol dehydrogenases/ketoreductases are widely distributed in nature where they play many roles in metabolism. Bakers' yeast is a particularly prolific

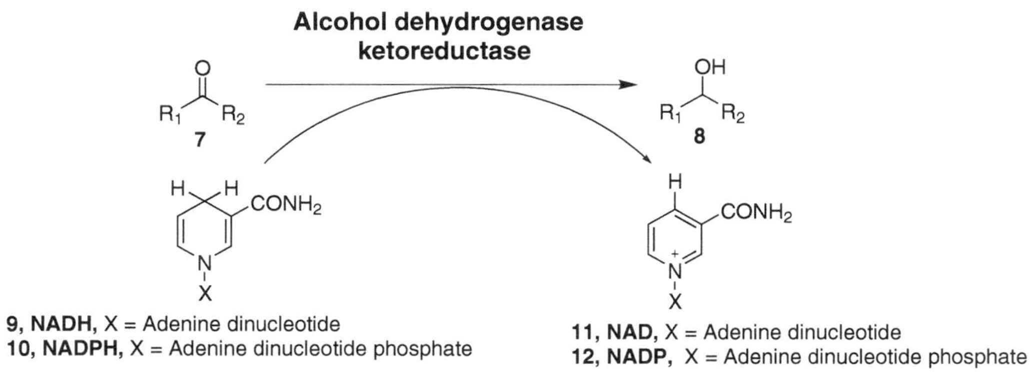

FIGURE 1.4
Carbonyl reductions to alcohols.

[8] The "alcohol dehydrogenase" nomenclature is more common in the biochemical literature whereas "ketoreductase" is typically used in biocatalysis since this is the synthetically more useful direction for the reaction.

producer of carbonyl-reducing enzymes that accept a diverse range of substrates. Moreover, because living yeast cells continuously regenerate nicotinamide cofactors by metabolizing simple sugars such as sucrose or glucose, whole yeast cells purchased from a local grocery store can be used directly as biocatalytic reducing agents in the lab. Many successful examples of this strategy have been disclosed and they constituted one of the earliest widescale applications of biocatalysis to organic synthesis. The major disadvantage of whole yeast cells is the large quantity of extraneous biomass that accompanies the relevant ketoreductase. This often makes scale-up difficult. Several vendors have responded to this problem by making individual ketoreductases available in purified form. In addition to minimizing the amount of added biomass, the use of purified yeast ketoreductases eliminates competition between enzymes with overlapping substrate specificities but divergent stereoselectivities.

Ketoreductases can also be used to catalyze alcohol oxidations to the corresponding aldehydes or ketones when provided with oxidized nicotinamide cofactors. The use of enzymes for this conversion is less commonly employed in synthesis for two reasons. First, alcohol oxidations often involve kinetic resolutions that are subject to the 50% yield limitation. When the same reactions are run in reverse (enzyme-catalyzed reduction of the carbonyl compound), one can obtain a single enantiomer with a theoretical yield of 100%. The second problem with dehydrogenase-catalyzed alcohol oxidations is that provision for regenerating oxidized nicotinamide cofactors must also be made. Until recently, few good methods for this conversion were available, although Bommarius' development of water-producing NADH oxidases has helped to overcome this problem [3]. Another efficient way is coupling with glutamate dehydrogenase enzyme and conversion of glutamate to alpha-keto glutarate [4].

6.3 Transamination

Optically pure amines are very common synthetic targets, either as final products or key intermediates. In fact, many optically pure alcohols were prepared specifically to generate leaving groups for chiral amine synthesis by subsequent S_N2 reactions. Directly converting prochiral carbonyl starting materials to optically pure amines – the synthetic equivalent of a chiral reductive amination – would clearly be much more efficient. Transaminases catalyze these reactions by transferring an amino group from a donor amine to an acceptor ketone or aldehyde. Until recently, virtually all known transaminases required an α-amino-acid donor and showed limited diversity in acceptor ketone structures. Although this allowed synthesis of many amino acids and related analogs, it also limited the range of accessible products. Moreover, because the reaction is reversible, a large excess of the amino acid was usually required to shift the equilibrium toward the desired product and make appreciable quantities of amines from ketones. The last problem has largely been solved by developing transaminases that accept isopropylamine as the amine donor. Not only is the amine donor inexpensive but the acetone by-product can be also removed during the reaction by evaporation that allows the reactions

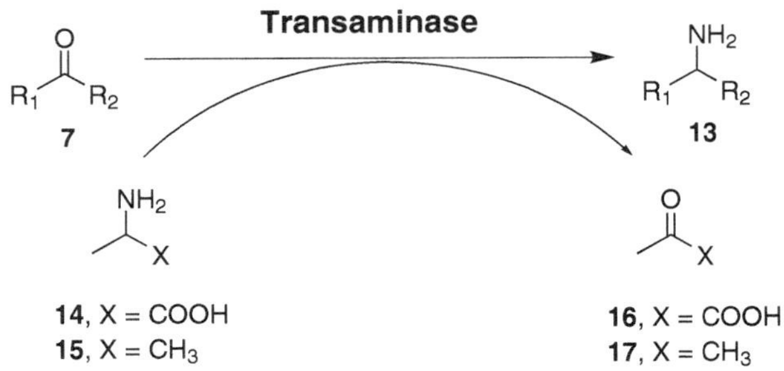

FIGURE 1.5
Transamination.

to be driven to completion without large molar excesses of the amine donor. The same types of protein engineering efforts have also broadened the range of allowable amine acceptors for transaminases to encompass synthetically important structures. The only drawback to transaminases is that relatively fewer are available "off the shelf" with synthetically useful substrate ranges as compared to alcohol dehydrogenases/ketoreductases, although this situation is improving rapidly (Figure 1.5).

6.4 Oxygen-Dependent Oxidations

A variety of enzymes catalyze substrate oxidations that involve atmospheric O_2 as a reactant. Monooxygenases, dioxygenases, peroxidases, and laccases are the most common enzymes in this class and synthetic applications of each type have been developed. Although some enzymes in this class use only organic cofactors, for example, flavins, others contain one or more tightly bound metal ions in the active site. Typical metals are iron, copper, and manganese with the first being most common. Regardless of their identity, the role of the cofactor is to interact directly with O_2 and form the actual oxidant.

6.5 Hydroxylation by C—H Bond Insertion

A variety of enzyme systems catalyze the insertion of oxygen into substrate C—H bonds. Insertions into alkyl C—H bonds yield alcohols as the initial oxidation products while aryl C—H bonds yield phenols. Enzymatic hydroxylation of an otherwise inactive center of an alkane or an aromatic hydrocarbon can provide alcohols or phenols that are often difficult (if not effectively impossible) to obtain by any other means. Moreover, most biological hydroxylations occur with very high regio- and stereocontrol so that only one enantiomer of a single product is typically obtained. In fact, one of the earliest applications of biocatalysis for organic synthesis in pharmaceutical industry was the use of a highly selective enzymatic hydroxylation that greatly facilitated production of corticosteroids [5]. Biocatalytic hydroxylations have been critically important components of the synthetic toolkit for steroid chemistry ever since.

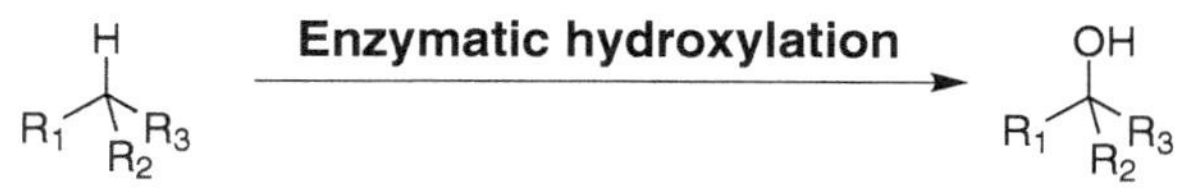

20 21

R_1, R_2, R_3 could be H, alkyl, cycloalkyl, aryl, heteroaryl groups

FIGURE 1.6
Enzymatic hydroxylation by C—H bond insertion.

Cytochrome P-450s are the best-known class of hydroxylation enzyme. Their active sites contain a heme iron that forms a highly activated oxygenating species that reacts by a radical mechanism. In higher animals, they function primarily in metabolite degradation as part of pathways that clear unnatural substances such as toxins and drugs. Hydroxylation increases polarity that facilitates further derivatization by other detoxification enzymes or excretion of the hydroxylated products. Other P-450 family members are involved in secondary metabolite biosynthesis, particularly in plants and microbial cells (Figure 1.6).

With few exceptions, most C—H hydroxylating enzymes are composed of multiple protein subunits and many are membrane bound and unstable in purified form. All of these properties conspire to make such enzymes difficult to handle as isolated proteins. In addition, oxygen activation requires a pair of electrons that are typically supplied by NADH, which introduces cofactor regeneration as an additional complication. For all of these reasons, most preparative biocatalytic hydroxylations have been carried out by mixing intact microbial cells with the substrate of interest (analogous to the way that Bakers' yeast cells have been employed in carbonyl reductions). The main difficulty is that many organisms produce more than one P-450 enzyme and this can lead to multiple products from a single reaction. Several groups have therefore created recombinant strains that express single P-450's in "clean" hosts that minimize side reactions. In addition, by overproducing large quantities of the relevant enzymes, the oxidations are often more efficient with regard to space–time yields.

6.6 O- or *N*-Dealkylations

Enzymatic hydroxylation of a carbon adjacent to an oxygen or nitrogen usually results in dealkylation by spontaneous hydrolysis of the initial hemiacetal or hemiaminal product. These conversions are commonly employed for two synthetic purposes: cleavage of methyl ethers and oxidative deamination of amines. The latter is particularly useful in amino-acid chemistry. These reactions can be catalyzed by P-450 monooxygenases or by flavin-containing monooxygenases (which are typically metal-free enzymes). As in the previous example, these hydroxylations require two electrons that must be supplied by NADH or NADPH, and most synthetic applications have relied on whole microbial cells rather than the isolated enzymes (Figure 1.7).

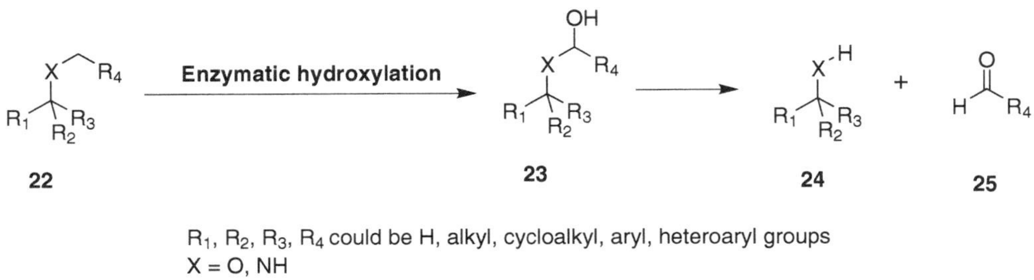

FIGURE 1.7
O- or *N*-dealkylations.

6.7 Oxidative Deamination of Amines to Carbonyl Compounds and the Reverse Reaction

Oxidative deamination by amine oxidases generates ketones or aldehydes. There are two types of amine oxidases. Those of the Type I contain both copper and a covalently attached topaquinone cofactors. In these enzymes, amine oxidation first generates an enzyme bound imine that is subsequently hydrolyzed to a ketone that is finally released from the enzyme. Type II amine oxidases are metal-free enzymes and contain only a flavin cofactor that remains bound to the enzyme throughout the catalytic cycle. Catalysis by these enzymes produces an imine intermediate that is released from the enzyme and subsequently hydrolyzed to the ketone in aqueous medium. To complete the catalytic cycle, the reduced flavin cofactor is oxidized by molecular oxygen generating hydrogen peroxide. In both cases, hydride (or a hydride equivalent) is transferred to the cofactor. Many amine oxidases are highly stereoselective, and this makes them very useful in kinetic resolutions of amines. For example, enantioselective oxidation of one antipode of **26** removes this stereoisomer from the reaction mixture, leaving behind the unreacted enantiomer. As in all kinetic resolutions, however, the maximum yield of the enantiopure product is 50% (Figure 1.8).

Several ingenious solutions to overcome the 50% yield problem inherent in amine kinetic resolutions have been devised. The most practical pairs are an amine oxidase with another enzyme or *in situ* chemical reaction that changes the process into a dynamic kinetic resolution with a 100% theoretical yield. The Turner group developed a very successful strategy in which a biocompatible chemical reducing agent such as amino borane was added along with the racemic starting amine and an enantioselective Type II amine oxidase [6]. Enzymatic oxidation depleted the reactive amine enantiomer; the imine product of this step was reduced in a racemic fashion by the amino-borane reagent. The net result was to racemize the starting amine that yielded 50% of the unreactive (desired) amine enantiomer along with another 50% destined for re-oxidation by the enzyme. After several rounds of oxidation/reduction, essentially all of the "wrong" enantiomer had been converted to the desired product. This is a very powerful strategy, particularly when combined with protein engineering to

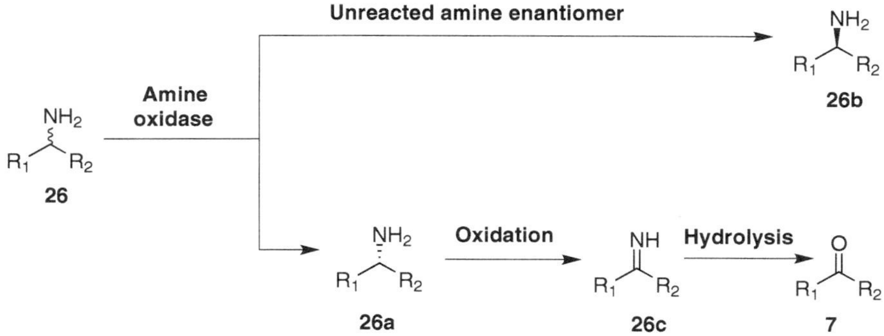

FIGURE 1.8
Amine oxidase – kinetic resolution of amines.

endow amine oxidases with the desired substrate- and stereoselectivities. Since the imine is not released from the Type I amine oxidase enzymes, chemical reduction will reduce the released ketone to the corresponding racemic alcohol. Thus dynamic resolution by combining *in situ* chemical reduction is possible with Type II amine oxidase only. However, both Type I and II amine oxidases can be combined with transaminase (transferring ketone to the desired amine) to effect dynamic resolution (Figure 1.9).

Amino acid oxidase enzymes are similar to the Type II monoamine oxidase and oxidize amino acids to imino acid and then to keto acid. These enzymes can be used for the kinetic resolution of racemic amino acids.

Rather than using O_2 as the ultimate electron acceptor for amine oxidation, amine dehydrogenases transfer hydride from the amine α-carbon to a nicotinamide cofactor. This yields the corresponding imine that undergoes subsequent

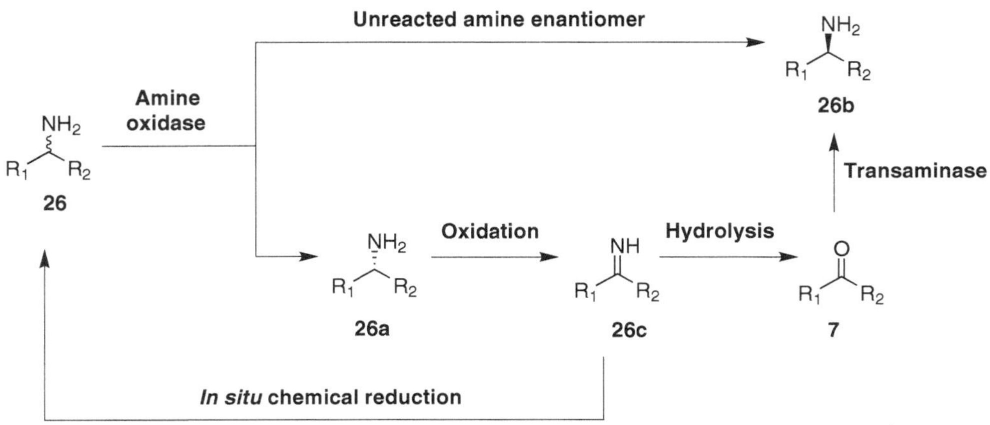

FIGURE 1.9
Chemo-enzymatic dynamic resolution of amines.

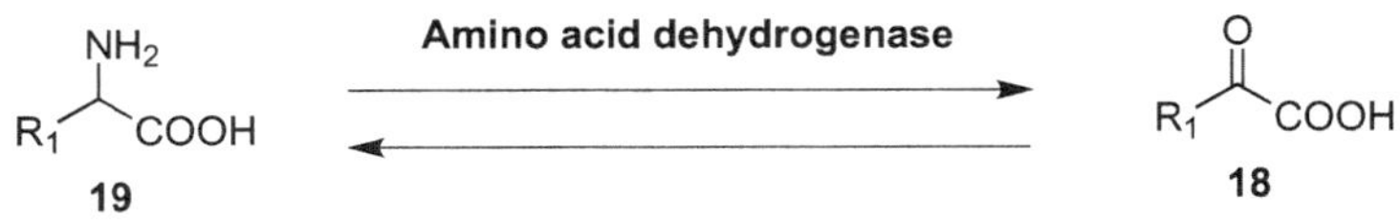

FIGURE 1.10
Conversion of amino acid to keto acid and vice-versa by amino acid dehydrogenase.

hydrolysis. The best-known enzymes in this class accept amino acids as their normal substrates. Amine dehydrogenases can catalyze the reactions in either direction; imine reduction is mechanistically similar to chiral reductive amination and is generally more synthetically useful. This reaction has formed the basis of several commercial processes to produce both natural and unnatural amino acids (Figure 1.10).

From a synthetic standpoint, the major drawback to this class of enzymes is their relatively narrow range of acceptable substrates. With very few exceptions, they require an α-keto acid substrate. Recent protein engineering efforts by Bommarius have overcome this limitation, raising the prospect of tailor-made enzymes for asymmetric reductive amination [7]. This solves a key unmet need in synthesis and details are covered in a later chapter (Figure 1.11).

Like amine oxidases, one can also combine amino acid dehydrogenases with *in situ* chemical reduction, a transaminase, or an amino acid dehydrogenase to effect dynamic kinetic resolutions of amino acids. Details of these more complex processes are described in the appropriate later chapters.

6.8 Nitrile Hydrolysis

Biocatalysis offers two advantages over chemical methodologies for nitrile hydrolysis. First, enzymes operate under mild conditions in the absence of

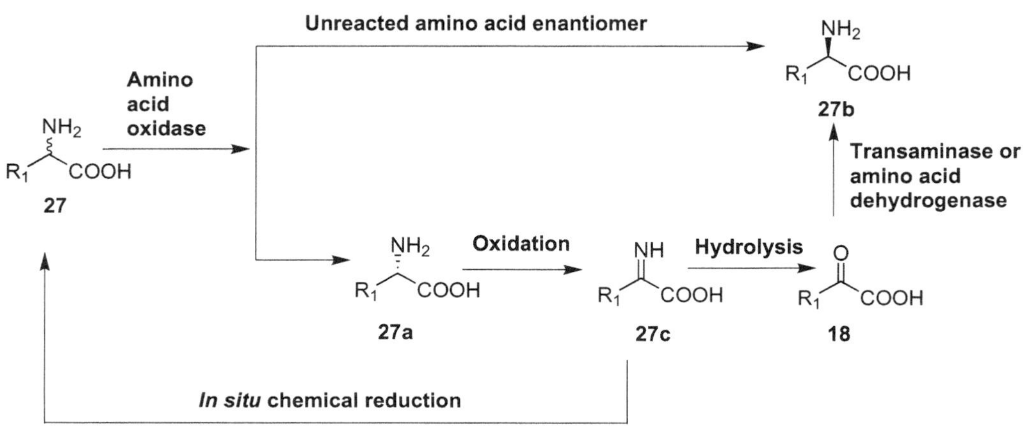

FIGURE 1.11
Chemo-enzymatic dynamic resolution of amino acids.

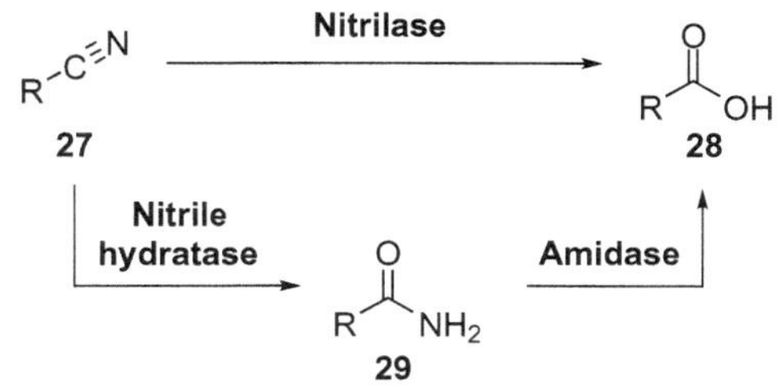

FIGURE 1.12
Hydrolysis of nitrile to amide and acid.

heat, concentrated acid, or base. Their second advantage is that hydrolysis can be halted at the amide intermediate if desired. Two classes of enzymes catalyze nitrile hydrolysis. *Nitrilases* catalyze the exhaustive hydrolysis of nitriles to yield the corresponding carboxylic acids with no release of intermediates. *Nitrile hydratases* catalyze partial hydrolysis of nitriles to the corresponding primary amides. In nature, nitrile hydratases are almost always coproduced with amidases so that the organism can convert the amide intermediate to the corresponding carboxylic acid that is used for catabolism. For synthetic purposes, however, it is more common to eliminate the amidase (either by purifying the nitrile hydratase or employing mutant cells that do not express the amidase) so that the reaction can be halted at the amide stage. Because amides are generally more reactive with respect to hydrolysis than the corresponding nitriles, it is extremely difficult to effect such a chemoselective partial hydrolysis by nonenzymatic means. With enzymes showing chiral discrimination, these enzymes can be used for kinetic resolutions of nitriles, amides, and acids (Figure 1.12).

6.9 Epoxide Hydrolysis

The hydrolysis of epoxides to diols is catalyzed by epoxide hydrolase enzymes. These usually show very high stereoselectivities and have been used for kinetic resolutions of racemic epoxides to yield both chiral epoxides (unreacted enantiomer) and chiral diol products (from reactive enantiomer). For most enzymes, nucleophilic attack by water occurs at the less-substituted end of the epoxide, resulting in retention of configuration at the more-substituted center. Interestingly, for some epoxide hydrolases, the regioselectivity of nucleophilic attack depends on the absolute configuration of the substrate and is opposite for the two antipodes. This makes it possible for some racemic epoxides to be converted to a single diol enantiomer in a process referred to as an *enantioconvergent* process. Some epoxide hydrolases also accept other nucleophiles in addition to water, for example, azide. This significantly increases the range of products that can be formed from epoxide intermediates (Figure 1.13).

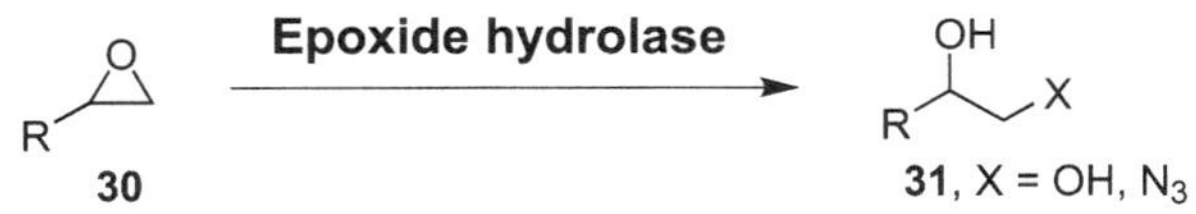

FIGURE 1.13
Hydrolytic/Nucleophilic opening of the epoxide ring.

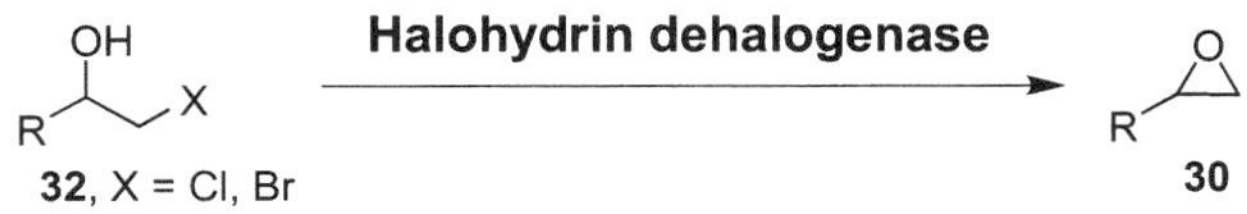

FIGURE 1.14
Synthesis of epoxides from halohydrins.

6.10 Epoxide Synthesis from Halohydrins

Epoxides can be produced by elimination of HCl from vicinal-halohydrins by halohydrin dehalogenase enzymes. These enzymes can be used for the resolution of racemic halohydrins to chiral halohydrins that can be converted easily to chiral epoxides, an important building block for synthesis of variety of compounds (Figure 1.14).

6.11 Baeyer–Villiger Oxidations of Ketones

Baeyer–Villiger monooxygenases catalyze the conversion of ketones to the corresponding esters by oxygen insertion. As the name implies, these enzymes utilize O_2 as a reactant, with one atom incorporated directly into the product and the second lost as water. This process requires a pair of electrons that must be supplied by a nicotinamide (almost always NADPH rather than NADH). The mechanism of the enzymatic Baeyer–Villiger oxidation appears to be identical to that of the usual peracid-mediated reaction; the key difference is that the enzyme forms a hydroxyperoxy-flavin intermediate *in situ* whose reactivity is analogous to that of a peroxycarboxylic acid. The flavin hydroperoxide adds to the substrate carbonyl, forming a tetrahedral Criegee-type intermediate whose breakdown follows the same stereoelectronic rules as the peracid-mediated reaction. This makes it possible to predict the stereochemical outcomes of these biocatalytic oxidations with some confidence. In most cases, migration of the more-substituted (more electron-rich) carbon is observed (as is also true for peracid-mediated oxidations). This intrinsic electronic preference can be overridden in the enzymatic reaction by specific substrate structural features, thereby producing "abnormal" oxidation products. The ability to determine which carbon migrates by correct enzyme choice – along with the mild reaction conditions and nontoxic nature of the oxidant – makes enzymatic Baeyer–Villiger oxidations particularly valuable in synthesis. It is also noteworthy that most Baeyer–Villiger monooxgenases can also catalyze heteroatom oxidations,

FIGURE 1.15
Baeyer–Villiger Oxidation.

converting amines into amine oxides and thioethers into sulfoxides. Such oxidations often occur with high stereoselectivities and constitute useful routes to these chiral building blocks. In this case, the flavin hydroperoxide acts as an electrophilic oxidant, further underscoring the synthetic versatility of these oxidation catalysts (Figure 1.15).

6.12 Alkene Reductions

Biocatalytic alkene reductions have recently grown in popularity. The best substrates for these enzymes involve alkenes conjugated with one or more electron-withdrawing groups, particularly, aldehydes, ketones, esters and, nitro moieties. Reducing equivalents are supplied exogenously by a nicotinamide cofactor. In most cases, the nicotinamide first reduces an active site flavin that subsequently transfers hydride to the electron-deficient alkene β-carbon (a proton is added concomitantly to the α-carbon to complete the reduction process). In nearly all cases, the reaction involves net *trans*-addition of H_2 across the alkene, which makes the biocatalytic strategy nicely complementary to organometallic and organocatalysts for alkene reductions that proceed via *cis*-addition. To date, the major limitations of this methodology have been a lack of stereochemical diversity in the commercially available set of enzymes and the inability to reduce isolated double bonds. The first limitation has been addressed by protein engineering and several successful programs have yielded enantiocomplementary alkene reductase variants (Figure 1.16).

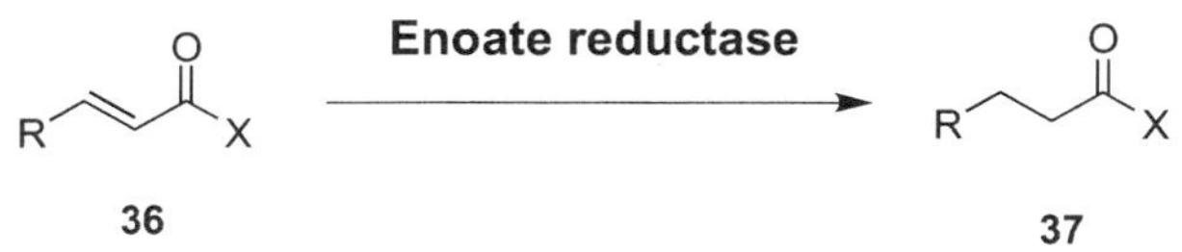

FIGURE 1.16
Reduction of alkene by enoate reductase.

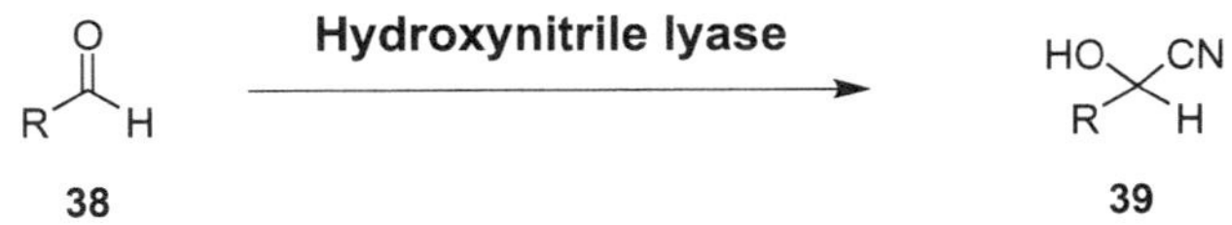

FIGURE 1.17
Addition of HCN to carbonyl compounds.

6.13 Hydrogen Cyanide Addition to Carbonyls

Enzymes of the hydroxynitrilase class catalyze the addition of HCN to aldehydes, producing cyanohydrins. Recently, the reaction has been extended to a few ketones with modified hydroxynitrilase enzymes. In many cases, these are formed with good optical purities and such reactions are the simplest type of enzyme catalyzed carbon–carbon bond formation. By pairing hydroxynitrile lyases with nitrilases or nitrile hydratases, one-pot, multistep conversions become possible, and this also shifts the equilibrium to favor the addition products. Such concerns are particularly important when applying these catalysts to ketones where the equilibrium generally favors the starting carbonyl compound (Figure 1.17).

6.14 Aldol Reactions

The vast majority of carbon–carbon bonds formed in nature occur by either aldol reactions or Claisen condensations. Aldolases catalyze the first type of reaction, and they have proven very useful in synthetic applications. In general, aldolases are highly selective for the enol(ate) partner (the aldol donor), but often tolerate a good deal of structural diversity in the acceptor aldehyde. This tolerance has been exploited in using aldolases to synthesize a wide variety of targets with high stereochemical control. In favorable cases, aldolases with complementary stereoselectivities are commercially available, which enables one to prepare all four possible stereoisomeric products from a single pair of starting materials and no need for protecting groups or chiral auxilliaries.

Aldolases are typically classified by enol(ate) donor, and four basic types are currently known. Pyruvate decarboxylase and deoxyribose phosphate aldolase generate one stereocenter (Figure 1.18).

Dihydroxyacetone phosphate aldolases and threonine aldolases generate two asymetric centers. The former type has been explored extensively and many synthetic examples exist (Figure 1.19).

7 SOURCES OF BIOCATALYSTS

All animals, plants and microorganisms produce enzymes to catalyze the myriad reactions responsible for their survival. It has long been recognized that extracts from natural sources can also catalyze reactions *in vitro* as well. Almost 7000 years ago, ancient cultures discovered that stomach juices could be used to make cheese. Brewing, wine, and bread making using organisms and their native

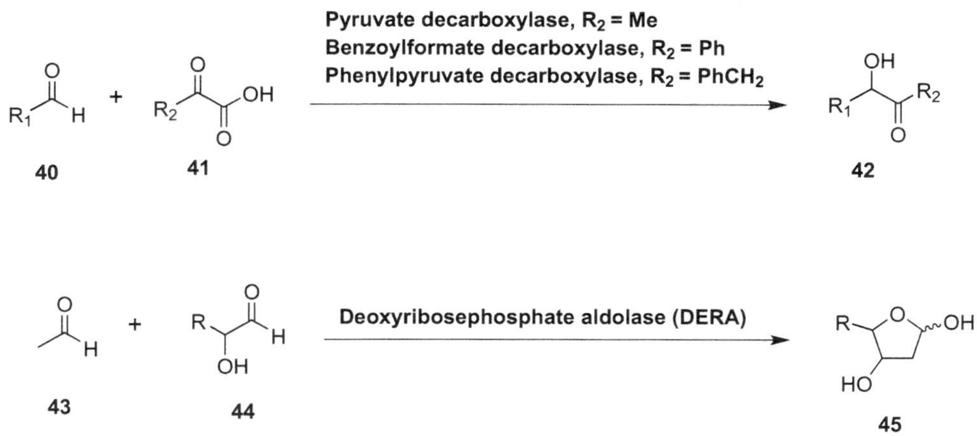

FIGURE 1.18
Aldol condensation reactions generating one new stereocenter.

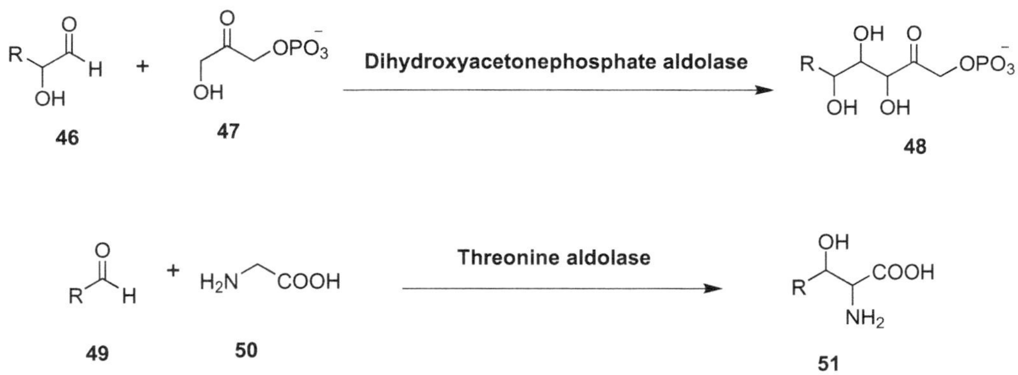

FIGURE 1.19
Aldol condensation reactions generating two new stereocenters.

enzymes have also been known since ancient times. Because enzymes for food and beverage production often involve the same substrates as those found *in vivo,* these industries were among the earliest users of biocatalysis *in vitro.* Until the nineteenth century, enzymes were obtained as mixtures from whole organisms or crude extracts and the nature of the catalyst(s) responsible for a particular reaction were unknown. In 1836, Theodor Schwann discovered that pepsin in stomach caused the breakdown of proteins to peptides. This was followed by Kuhne's first use of the term "enzyme" for this class of catalyst.

The development of "modern" biocatalysis can be traced to Buchner's 1897 study of bakers' yeast extracts. He conclusively demonstrated that a nonliving, cell free mixture retained the intact cell's ability to convert glucose to ethanol and carbon dioxide. This firmly refuted the notion that enzymes contained a mysterious "vital force" that endowed them with the ability to catalyze reactions in living cells. Rather, enzymes are simply another type of catalysts that obey the same laws of chemistry and physics that govern all chemical processes. This development led to the idea that enzymes could be used to catalyze reactions beyond those found in the original organisms, which formed the foundation of biocatalysis as a scientific discipline. Since Buchner's time, identifying and sourcing a suitable enzyme that solves a particular chemical problem has been the start of nearly every project in biocatalysis. In this section, we review some of the common sources of synthetically useful enzymes, which range from microorganisms, plants, and animals to completely synthetic, man-made enzymes that have no known natural counterparts.

7.1 Natural Sources – Animals

Historically, a large number of enzymes used in biocatalysis were obtained from animal sources. Animals are particularly rich sources of hydrolytic enzymes that are used to digest fats, proteins, and complex polysaccharides. In addition, hydrolytic enzymes from animal sources have long been used in the food, beverage, and leather industries, so they are readily available in large quantities. Most animal-derived hydrolytic enzymes are obtained from liver, stomach, pancreas, and other inedible parts of cattle or pig that, in turn, are relatively inexpensive and abundantly available as by-products of the meat industry. Animal-derived lipases, proteases, and esterases have all been widely used for organic synthesis. On the other hand, animal enzymes that hydrolyze complex carbohydrates to simple sugars are important for food industry but have not found significant applications in the synthesis of small organic molecules.

Animal liver is a good source of enzymes that catalyze the *in vitro* hydrolysis of various esters. As early as 1903, Dakin used the esterase from pig liver to resolve racemic mandelate ester [8]. Pig (porcine) liver esterase (often mentioned by its abbreviated name PLE) is commercially available and is a versatile enzyme for the hydrolysis of a variety of esters.

Because of its role in the digestive system, animal pancreas is rich source of secreted proteases and lipases. Pancreatin from porcine pancreas is a mixture of different enzymes capable of hydrolyzing carbohydrates, lipids, esters, amides and proteins. It can be used directly when the presence of competing catalysts are not a problem. Several individual pancreatic enzymes are also readily available. Porcine pancreatic lipase has been widely used in synthesis. Several proteases from this source are sold commercially and used extensively. Porcine pancreatic trypsin is an inexpensive serine protease that hydrolyzes peptide bonds predominantly on the carboxyl side of lysines or arginines in proteins and peptides. Chymotrypsin is another serine protease from the same source that cleaves amide bonds on the carboxyl side of large hydrophobic amino acids such as tyrosine, tryptophan, phenylalanine, and, to a lesser extent, leucine. Several different subtypes of chymotrypsin are known, for example, α-, β-, γ-, with α-chymotrypsin being most commonly used in biocatalysis.[9] The bovine form of α-chymotrypsin is also commercially available.

Animal liver is rich sources of enzymes involved in detoxification and solubilization by introducing more polar functional groups onto hydrophobic starting materials. In particular, the use of horse-liver alcohol dehydrogenase to reduce ketones to alcohols with high stereoselectivity was intensively studied by J. Brian Jones and his group [9]. This early work laid the foundation for subsequent developments in this now well-developed field. Animal livers also produce a variety of cytochrome P-450 monooxygenases that hydroxylate a variety of small organic molecules, and these are key players in drug metabolism and xenobiotic clearance. For this reason, the same liver enzymes have been used *in vitro* to synthesize metabolites of pharmaceutically and agriculturally active components. Because they are often unstable and difficult to isolate, biocatalytic applications of P-450 monooxygenases are usually carried out with liver homogenates or liver acetone powders.

Finally, several other animal tissues yield useful enzymes that have been employed in synthesis. Pepsin is an important digestive protease from animal stomach whose native role is hydrolyzing amide bonds involving hydrophobic, aromatic amino acids, for example, phenylalanine, tyrosine, and tryptophan. Acylase from porcine kidney selectively hydrolyzes *N*-acetyl amino acids and is commercially available. It has long been used for kinetic resolutions of amino acids. In addition to hydrolases, other animal enzymes have found important applications in biocatalysis. Rabbit muscle aldolase is commercially available and was shown to catalyze aldol condensations between dihydroxyacetone phosphate and various nonnatural aldehydes by the Whitesides group in 1989 [10]. This seminal report touched off an avalanche of new applications for this and related enzymes in asymmetric synthesis.

7.2 Natural Sources – Plants

Enzymes from plants have been used since ancient times, even though their nature remained obscure until the nineteenth century. Malting of cereals and the hydrolysis of complex polysaccharides are two of the oldest uses of enzymes in human history. Like their animal counterparts, most commercially available plant-derived enzymes are hydrolases, particularly lipases and proteases. Papain, a cysteine protease from papaya, is the best-known plant-derived hydrolytic enzyme. Bromelain from pineapple and ficain from fig latex are similar cysteine proteases that have also found applications in biocatalysis. Horseradish peroxidase is a versatile oxidative enzyme obtained from its namesake that oxidizes many organic compounds, especially phenols. Hydroxynitrile lyase (often called oxynitrilase) from bitter almond is one of the most important plant-derived

[9] The other forms of chymotrypsin represent alternatively processed forms of the enzyme. Although they generally have similar substrate- and stereoselectivities, this is not always the case and it is safer to employ a single form when possible to avoid side reactions.

enzymes that catalyzes the stereoselective addition of HCN to aldehydes, yielding chiral cyanohydrins. This reaction was first reported more than a century ago by Rosenthaler [11].

7.3 Natural Sources – Microorganisms

The application of microbial enzymes has also been known since antiquity, for example, the use of yeast for brewing and wine making. Whereas animals and plants were originally used as sources of hydrolases, many of the earliest applications of microbial enzymes were for oxidation-reduction reactions. In particular, yeast contains many oxidoreductase enzymes and this organism remains in common use for carbonyl reductions to this day. One of the first large-scale chemical applications of biocatalysis was developed by Reichstein in his 1934 route to vitamin C [12]. The key step of this process utilized *Gluconobacter suboxydans* cells to oxidize l-sorbitol to l-sorbose with essentially complete regioselectivity. Other oxidants gave side products from oxidation of other hydroxyl groups, clearly demonstrating the advantages of biocatalysis in solving this problem. Microorganisms can be used in the form of whole cells for many biocatalytic reactions. This is particularly advantageous for oxidoreductases since these enzymes require cofactors that can be regenerated by normal microbial metabolism of simple carbon sources such as glucose. Novozymes started production of microbial enzymes in 1960, which was followed by several other commercial enzyme producers.

7.4 New Sources – Recombinant Microorganisms

With the increasing need of enzymes for the food, beverage and later, detergent industries, demand rose for enzymes that were inexpensive, easily available and of consistent quality. Animals and plants require relatively long growth times. In addition, the enzymes derived from them are often inconsistent as a result of environmental effects during the growth and development of the organism. Finally, isolation from natural sources does not generally allow for production of mutant enzymes better suited to chemical processes (*vide infra*). By contrast, microorganisms can be grown rapidly, relatively inexpensively, and under controlled and reproducible conditions. These advantages – along with the development of molecular biology in the 1970s and 1980s – drove the development of recombinant microbes that overexpressed exogenous proteins at high levels. Lipases, esterases, and proteases are all produced commercially at high levels using recombinant hosts and this has made these enzymes both accessible and inexpensive. This, in turn, has enabled many applications at both laboratory and industrial scales. One additional advantage of producing enzymes in engineered microbial cells is that they can either be used as sources of purified proteins, or the enzymes can be used within the cells of the producing organism. The latter strategy is particularly useful for oxidoreductases to ensure cofactor supply and regeneration. In some cases, additional cofactor regeneration enzymes have been coexpressed with exogenous oxidoreductases to improve cofactor availability even further.

8 TYPES OF ENZYME PREPARATION

Several strategies have been devised for carrying out biocatalytic reactions. Enzymes can be employed within intact cells that produce them, as purified or partially purified preparations in solution or solid, and immobilized onto a solid support. All have advantages and disadvantages, and these are summarized further.

8.1 Whole Cells

The earliest and simplest application of enzymes to chemical synthesis involved the use of whole cells or tissues. However, directly employing biomass that contains the enzyme of interest does have disadvantages. First, cells need to be grown and/or harvested. Growing microbial cells requires technical expertise in microbiology as well as fermentation equipment. This raises a barrier to employing biocatalysis by an organic chemist in a typically equipped synthesis laboratory. Cells also have to be prepared just in time or immediately prior to carrying out the biocatalytic step, raising further logistical issues of arranging both the fermentation and biocatalytic reaction. It can also be difficult to transport cells over a distance, particularly if they must remain frozen to retain the activity. Although less of an issue for small-scale conversions, as scale increases, a close physical proximity of the cell growth and enzyme-application facilities becomes highly desirable. The last barrier to using whole cells for synthetic applications is the lipid membranes that encase all living cells. This can introduce significant mass transport limitations if the enzyme of interest is located inside the cells (as is usually the case). An intact cell membrane is usually required if native host cell metabolism is used to supply cofactors. This makes it difficult to employ high substrate and product concentrations since small organic molecules usually disrupt the hydrophobic forces that maintain cell membrane integrity. The final disadvantage of using whole cells directly in biocatalysis is that the enzyme of interest makes up only a very small fraction of the total added biomass. This means that the process would have low volumetric productivity and the product of interest must often be separated from large quantities of extraneous biomass. This complicates (and therefore increases the costs of) the biocatalytic reactions carried out in this way.

Despite these disadvantages, there are some specialized cases where whole microbial cells remain the strategy of choice for biocatalysis. As noted previously, bakers' yeast cells produce a wide variety of ketoreductases and alcohol dehydrogenases. Carbonyl moieties can be reduced by mixing commercial yeast cells with the substrate in the presence of sucrose or glucose. Such processes are relatively inefficient on a per-cell basis, but because the biocatalyst is so inexpensive and the cells provide both the required enzyme(s) and a continuing supply of reduced nicotinamide cofactors, the tradeoff is often worthwhile. Cytochrome P-450-catalyzed hydroxylations by whole microbial cells are a second example where native organisms remain a viable choice. These enzymes require several cofactors including heme groups and reduced nicotinamides. They are also often unstable and require several individual proteins. Finally, many are permanently anchored to the cell membranes. All of these properties conspire to make them difficult to handle as isolated proteins and also complicates their overexpression in heterologous hosts. The most common solution is to use these enzymes within whole cells of the native organism to oxidize substrates.

Whole cells of engineered microbes often perform much more efficiently than native cells. *Escherichia coli* is the most common host for these efforts, although other organisms such as *Aspergillus niger*, bakers' yeast and others can be useful in specific cases. The details of how these strains are constructed lies outside the scope of this book. Whole cells of engineered microbes offer two key advantages over their native counterparts. First, if the enzyme of interest can be overproduced at high levels, its catalytic productivity is much improved on a per-cell basis. This directly translates into lower biomass loadings and higher substrate input, which in turns improves volumetric productivity and simplifies workup and product isolation significantly. In the case of cofactor-dependent enzymes, a second enzyme that regenerates the cofactor can be coexpressed, thus creating a two-enzyme metabolic pathway that cycles the cofactor between its two states without relying on any native host enzymes. This strategy has been particularly popular for nicotinamide-dependent reactions.

8.2 Enzymes in Solution

The best way to avoid the mass transport rate penalty associated with substrates and products having to cross the cell membrane is to use the catalyst in aqueous solution. In some favorable cases, the enzymes of interest are secreted by the producing cells and the latter can be removed before substrate addition. Most synthetically useful enzymes, however, are intracellular (produced and retained within the cell). This means that the cell membrane and cell wall must be breached in order to obtain the enzyme of interest. Cells can be broken by a number of methods including high-energy sound waves (sonication; particularly useful for small scales), enzymatic methods (lysozyme) or by passing through a high pressure system (French pressure cell or homogenizer or microfluidizer). These processes release the enzyme into solution, and insoluble debris (most cell membranes and cell wall fragments) are removed by centrifugation or filtration prior to use in chemical conversions. The resulting aqueous-soluble fraction is referred to as a *crude extract*. If desired, the enzyme of interest can be further purified, either partially or completely. The number and type of purification steps is dictated by the presence and number of competing enzymes that convert the substrate of interest to an undesirable product (either an isomer of the desired product or a completely different compound). Although small-scale screening reactions are often carried out with highly purified proteins (to avoid forming by-products that would complicate the screening process), enzymes are more commonly employed in crude or partially purified forms for preparative-scale reactions for economic reasons.

One key advantage of using enzymes in solution is that their concentrations can be relatively high. This improves volumetric productivities and allows significantly higher substrate input. A second advantage to using purified or semi-purified enzymes is that it does not require direct access to fermentation equipment or expertise since cell growth can be decoupled from the biocatalytic reaction. Finally, using cell-free extracts (rather than whole cells) dramatically reduces the amount of biomass added to the reaction, which simplifies workup and product isolation.

Enzyme stability in solution is the major limitation to this strategy. With whole cells under growing conditions, the biocatalyst is automatically regenerated so long as the organism continues to grow; this is not the case when purified or semipurified enzymes are employed. In addition, aside from secreted proteins, nearly all enzymes are much more stable within cells than when placed in solution. Additives are often added to improve protein stability in solution and thereby extending the usable reaction time. Protease inhibitors prevent degradation of the desired enzyme by proteases that are also present in many cell extracts. Water-miscible polyols such as glycerol and ethylene glycol (usually 5–10% v/v) also improve enzyme stability. Because enzymes must maintain specific three-dimensional shapes to retain catalytic activity, solutions are usually stored and transported under cold conditions to minimize thermal unfolding. During the biocatalytic step itself, reaction temperature is a key variable that must be optimized to allow good enzyme longevity (lower is better) along with reasonable reaction rates (higher is better) and minimized substrate and/or product degradation (lower is better). The result must be a compromise between these competing demands.

Many enzyme solutions are commercially available, most notably proteases and lipases, some of which are sold in very large quantities. At small scales, transporting an enzyme solution under cold conditions is not an issue in most cases. On the other hand, transporting large amounts of relatively unstable enzyme solution over long distances can become a challenge when industrial-scale applications are considered.

8.3 Solid Enzymes

Biocatalytic enzymes are usually best stored in the solid form. In favorable cases, they can even be employed in reactions in this form (typically in nonaqueous solvents since the solids normally dissolve in water). Enzyme solids are usually made by removing water from an initial aqueous solution by lyophilization, spray drying, or precipitation. Solid enzymes maximize the catalytic activity per unit weight, which can be important on very large scales. In addition, having enzymes in solid form avoids the pitfalls associated with cell growth and isolation, enzyme preparation, stabilities and transport of enzyme solutions. A large number of enzymes are available in powder form and this is a common form used by suppliers.

Enzyme powders are simply added to reaction mixtures like any other catalyst. The lack of water opens the possibility of carrying out biocatalytic reactions in organic solvents. One caution in using enzyme powders should be observed, however. Like other proteins, enzymes can cause allergic reactions in some individuals. In the early days of adding protease enzyme powders during detergent formulations, respiratory allergies occurred in some workers. This environmental safety issue was solved by developing dust-free product formulations and installing engineering controls. Similar precautions should be observed in biocatalytic processes that use enzyme powders, particularly those carried out on very large scales.

8.4 Immobilized Enzymes

The final common method for employing enzymes in biocatalysis involves their prior immobilization on a solid support by covalent or noncovalent bonds. In many cases, this modification decreases their catalytic efficiencies, and it can also introduce mass transport limitations. Minimizing both of these issues by optimizing both the nature of the solid support and the enzyme/support connection is an area of intensive current research. Some enzymes are available commercially in immobilized form. The best-known example is Novozym 435, which is a trade name for *Candida antarctica* lipase B adsorbed on a polymeric support.

Despite the aforementioned drawbacks, immobilized enzymes have become very popular in biocatalysis for four important reasons. First, they simplify enzyme removal at the end of the reaction. High levels of water-soluble proteins often cause emulsions during solvent extractions, and enzyme removal is therefore often the first step in working up a biocatalytic reaction. Immobilized enzymes can often be recovered by simple filtration and once this has been accomplished, the product(s) can be isolated in the usual way. The second key advantage of immobilized enzymes that they allow re-use of the catalyst in subsequent batches. This is especially critical when high-value catalysts are used to make low-value products and re-use considerably diminishes the enzyme's overall cost contribution to the process. Third, immobilized enzymes often have higher stabilities and longer useful lifetimes than their water-dissolved counterparts. This allows either longer reaction times with correspondingly higher volumetric productivities and/or faster reactions by allowing the use of higher process temperatures. Higher reaction temperatures also increase substrate and product solubilities, which can lead to more intensive processes. The final advantage to immobilized enzymes is that they are very well suited to continuous processes. In a formal sense, a continuous process is a case where the enzyme is recovered and re-used for an infinite number of reaction cycles.

9 AVAILABILITY OF ENZYMES FOR BIOCATALYSIS

9.1 Industrial Enzymes

The food, beverage, detergent, textile, leather, and paper industries are by far the major commercial users of enzymes. In some cases, enzymes appear in the final products, for example, proteases in detergents, β-galactosidase as a digestive aid to mitigate lactose intolerance or as animal feed components. In other cases such as the textile, leather, and paper and pulp industries, enzymes are used as processing aids. In the food industry, large amounts of enzymes are used for baking and fats and oils production. In addition, wine, beer, alcohol, and beverage producers are heavy users of enzymes. For example, the enzymatic production of high fructose corn syrup is one of the largest industrial biocatalytic processes currently practiced. Even today, the use of enzymes for biocatalytic synthesis represents only a very small fraction of the overall enzyme market.

When feasible, using an enzyme for biocatalysis that is already produced for an industrial process is very advantageous. Beyond low cost, these enzymes have been subjected to many years of optimization with respect to stability, catalytic activity and (in some cases) immobilization. An uninterrupted and consistent supply is assured. Among the enzymes used for biocatalysts, trypsin has been available since the 1940s. Novozymes discovered a protease named "Alcalase" from microbial sources in 1958 and started its commercial production in 1960 for the detergent industry. Most commercial enzymes are currently produced by microbial fermentation to meet the very large demands of their major markets. The majority are hydrolytic enzymes, including lipases, proteases, esterases, and glycosidases. Except for the last case, these enzymes have also found wide use in organic synthesis. In recent years, several commercial suppliers have introduced different types of biocatalysts including ketoreductases, transaminases and alkene reductases that are targeted specifically for synthetic chemistry applications. These enzymes are generally available only in relatively small scale ($\leq$ 1 kg), although some are captively produced in large quantities for preparing specific compounds needed in manufacturing. Bakers' yeast is the only microorganism readily available commercially on large scales. The dry powder form sold for baking is used most commonly in organic synthesis since it can be purchased in lab quantities directly from local grocery stores.

Many industrial enzymes are sold as mixtures since additional purification would raise the cost. As a rule, unless the contaminants cause difficulties, they are allowed to remain in the enzyme mixture unless they can be removed very economically. In some cases, slightly different forms of the same enzyme (referred to as *isozymes*) are present. These variants may differ slightly with respect to kinetic properties or stabilities, differences that do not usually cause difficulties for applications in organic synthesis. By contrast, when multiple enzymes with differing substrate- and/or stereoselectivities are present, synthetic applications can be much more challenging. This is particularly important if the composition of the enzyme mixture varies over time and/or with source. Porcine liver esterase provides a well-known example of this phenomenon, as the commercial material contains several enzymes with different specificities that can lead to poor stereoselectivities or even multiple products for some substrates. The fact that the precise composition depends on the animal source makes it very difficult to maintain batch-to-batch consistency [13–15]. The lipase from *Candida rugosa* consists of at least two isozymes that have slightly different selectivities [16]. Finally, it should be noted that commercial enzyme formulations often contain other proteins from microorganisms, processing aids and fillers to aid formulations.

9.2 Recent Advances

With few exceptions, native cells produce enzymes in relatively low amounts (although sufficient for their roles in normal metabolism). By contrast, recombinant cells can overexpress very high levels of target proteins. The need to produce high levels of enzyme and eliminate contaminating enzymes and isozymes have been the key drivers in producing industrial enzymes in heterologous microbial hosts. *E. coli* is the most common host and many enzymes are cloned and expressed in this organism. Unfortunately, *E. coli* secretes proteins relatively inefficiently, and it does not have the biosynthetic capability to covalently attach carbohydrates to specific amino acids in the final protein (a process referred to as *glycosylation*). For this reason, other bacteria such as *Streptomyces lividans*, yeast such as *Pichia pastoris*, *Saccharomyces cerevisae* and fungi such as *Aspergillus* species have also been used as hosts. *C. antarctica* lipase B is now produced in an *Aspergillus* host by Novozymes, and porcine liver esterases (PLE) were cloned and expressed individually in *P. pastoris* [17] and *E. coli* [18]. The recombinant PLE is a nice example that shows how cloning and expressing single genes from the original producing organism completely eliminates the problem of multiple isozymes in the final enzyme stock. In addition, concerns with livestock-associated prion diseases and religious injunctions against certain animal sources from animal-derived proteins can be avoided by using recombinant PLE.

With the explosion of genome sequencing data, one can now identify candidate enzymes in databases on the basis of amino acid sequence alone, create the corresponding gene by solid-phase chemical synthesis and express the protein in a heterologous host without ever handling the original organism. In many cases, one or more properties of a native enzyme, for example, reaction rate, catalyst stability, stereoselectivity, etc., are below desired targets for a given synthetic application. By combining protein engineering strategies with heterologous expression, it has become routine to improve initial enzyme "hits" to the point that they become the most economical choice as a synthetic method. The technology and speed of protein X-ray crystallography has advanced rapidly so that learning the three-dimensional structure of a synthetically useful enzyme is now almost routine. The same techniques can be used to determine the structural perturbations caused by mutations during a protein-engineering program, thereby providing invaluable feedback needed for designing subsequent rounds of mutagenesis. Enabled by advances in molecular biology, rapid automated analysis and informatics, it has now become commonplace to generate and screen large libraries of protein variants to identify changes that yield improved biocatalytic properties, a process known as "directed evolution." This yields improved catalysts much faster than otherwise achievable by traditional natural selection [19–22]. Several techniques have been developed for evolving enzymes. Enzyme activity has been exponentially increased by directed evolution techniques [23–26]. Industrial processes with high productivity have been developed using evolved enzymes. For example, a highly efficient enzymatic process for the synthesis of Sitagliptin was recently developed by using an evolved transaminase enzyme [27].

References

[1] M.G. Snider, B.S. Temple, R. Wolfenden, The path to the transition state in enzyme reactions: a survey of catalytic efficiencies, J. Phys. Org. Chem. 17 (2004) 586–591.

[2] C.-S. Chen, Y. Fujimoto, G. Girdaukas, C.J. Sih, Quantitative analysis of biochemical kinetic resolutions of enantiomers, J. Am. Chem. Soc. 104 (1982) 7294–7299.

[3] B.R. Riebel, P.R. Gibbs, W.B. Wellborn, A.S. Bommarius, Cofactor regeneration of NAD^+ from NADH: novel water-forming NADH oxidases, Adv. Synth. Catal. 344 (2002) 1156–1168.

[4] W. Kroutil, H. Mang, K. Edegger, K. Faber, Biocatalytic oxidation of primary and secondary alcohols, Adv. Synth. Catal. 346 (2004) 125–142.

[5] D.H. Peterson, H.C. Murray, Microbiological oxygenation of steroids at Carbon-11, J. Am. Chem. Soc. 74 (1952) 1871–1872.

[6] F.R. Alexandre, D.P. Pantaleone, P.P. Taylor, I.G. Fotheringham, D.J. Ager, N.J. Turner, Amine-boranes: effective reducing agents for the deracemisation of DL-amino acids using L-amino acid oxidase from *Proteus myxofaciens*, Tetrahedron Lett. 43 (2002) 707–710.

[7] M.J. Abrahamson, E. Vazquez-Figueroa, N.B. Woodall, J.C. Moore, A.S. Bommarius, Development of an amine dehydrogenase for synthesis of chiral amines, Angew. Chem. Int. Ed. 51 (2012) 3969–3972.

[8] H.D. Dakin, The hydrolysis of ethyl mandelate by lipase, Proc. Chem. Soc. London 19 (1903) 161.

[9] J.B. Jones, In: R. Porter, S. Clark (Eds), Enzymes in Organic Synthesis, Pittman, London, 1985; vol. Ciba Symposium 111, pp. 3–21.

[10] M.D. Bednarski, E.S. Simon, N. Bischofberger, W.-D. Fessner, M.-J. Kim, W. Lees, T. Saito, H. Waldmann, G.M. Whitesides, Rabbit muscle aldolase as a catalyst in organic synthesis, J. Am. Chem. Soc. 111 (1989) 627–635.

[11] L. Rosenthaler, Enzyme-effected asymmetric syntheses, Biochem Z. 14 (1908) 238–253.

[12] T. Reichstein, A. Grüssner, Eine ergeibege Synthese der l-Ascorbinsäure (C-Vitamin), Helv. Chim. Acta 17 (1934) 311–328.

[13] D. Farb, W.P. Jencks, Different forms of pig-liver esterase, Arch. Biochem. Biophys. 203 (1980) 214–226.

[14] E. Heymann, W. Junge, Characterization of the isoenzymes of pig-liver esterase. 1. Chemical studies, Eur. J. Biochem. 95 (1979) 509–518.

[15] W. Junge, E. Heymann, Characterization of the isoenzymes of pig-liver esterase. 2. Kinetic studies, Eur. J. Biochem. 95 (1979) 519–525.

[16] C.S. Cobbs, M.J. Barton, L. Peng, A. Goswami, A.P. Malick, J.P. Hamman, G.J. Calton, US Patent 5108916, 1992.

[17] S. Lange, A. Musidlowska, C. Schmidt-Dannert, J. Schmitt, U.T. Bornscheuer, Cloning, functional expression, and characterization of recombinant pig liver esterase, ChemBioChem 2 (2001) 576–582.

[18] D. Bottcher, E. Brusehaber, K. Doderer, U.T. Bornscheuer, Functional expression of the *gamma*-isoenzyme of pig liver carboxyl esterase in *Escherichia coli*, Appl. Microbiol. Biotechnol. 73 (2007) 1282–1289.

[19] K.Q. Chen, F.H. Arnold, Tuning the activity of an enzyme for unusual environments – sequential random mutagenesis of subtilisin-E for catalysis in dimethylformamide, Proc. Natl. Acad. Sci. USA 90 (1993) 5618–5622.

[20] F.H. Arnold, L. Elling, W. Hummel, J.-C. Janson, M. Kataoka, M. Kobayashi, J.C. Moore, J.C. Ogawa, M. Pohl, S. Shimizu, in: F.H. Arnold, L. Elling, W. Hummel, J.-C. Janson, M. Kataoka, M. Kobayashi, J.C. Moore, J.C. Ogawa, M. Pohl (Eds.), Advances in Biochemical Engineering Biotechnology, vol. 58, Springer-Verlag, New York, 1997, pp. 1–14.

[21] M.T. Reetz, Controlling the enantioselectivity of enzymes by directed evolution: practical and theoretical ramifications, Proc. Natl. Acad. Sci. USA 101 (2004) 5716–5722.

[22] J. Sylvestre, H. Chautard, F. Cedrone, M. Delcourt, Directed evolution of biocatalysts, Org. Proc. Res. Develop 10 (2006) 562–571.

[23] S. Lutz, Beyond directed evolution – semi-rational protein engineering and design, Curr. Opin. Biotechnol. 21 (2010) 734–743.

[24] C. Jäckel, D. Hilvert, Biocatalysts by evolution, Curr. Opin. Biotechnol. 21 (2010) 753–759.

[25] A.S. Bommarius, J.K. Blum, M.J. Abrahamson, Status of protein engineering for biocatalysts: how to design an industrially useful biocatalyst, Curr. Opin. Chem. Biol. 15 (2011) 194–200.

[26] E.M. Brustad, F.H. Arnold, Optimizing non-natural protein function with directed evolution, Curr. Opin. Chem. Biol. 15 (2011) 201–210.

[27] C.K. Savile, J.M. Janey, E.C. Mundorff, J.C. Moore, S. Tam, W.R. Jarvis, J.C. Colbeck, A. Krebber, F.J. Fleitz, J. Brands, P.N. Devine, G.W. Huisman, G.J. Hughes, Biocatalytic asymmetric synthesis of chiral amines from ketones applied to sitagliptin manufacture, Science 329 (2010) 305–309.

CHAPTER 2

Practical and Engineering Aspects of Running Enzyme Reactions

Harald Gröger
Faculty of Chemistry, Bielefeld University, Universitätsstr, Bielefeld, Germany

1 INTRODUCTION

The production of fine chemicals and pharmaceutical substances has recently embraced an increasing number of biocatalytic processes [1], particularly for the manufacture of chiral building blocks. At the same time, academic researchers have uncovered novel reactions and reaction sequences that can be performed with enzymes. Thus, for organic chemists, a broad range of biotransformations is available for process-development work [2]. At the same time, a range of challenges still remain when developing an efficient and economically viable organic process, even after a suitable enzyme has been identified. This chapter provides an overview of main challenges and successful case studies in biocatalytic process development with a major focus on biocatalytic aspects from an organic chemist's perspective.

When choosing a biotransformation step from the organic process development perspective, in general, the same requirements as in a "classic chemical" or chemocatalytic process must be considered. Some of the key criteria to run a process efficiently are graphically summarized in Scheme 2.1. One critical issue is catalyst availability for the desired reaction. For industrial use, this typically means either a commercial enzyme or access to an efficient recombinant enzyme. This recombinant enzyme should be produced in a host organism under high-cells density conditions, leading to both high expression of the protein (>20% of total cellular protein) and a high biomass yield in the fermentation process (>200 g/L wet biomass per liter). A further aspect related to the choice of catalyst is its formulation. Several options such as their use in isolated purified form or as a crude or semipurified cell extract, in "free" or immobilized form or alternatively in form of a recombinant whole cell system have to be considered. Each approach has both advantages and disadvantages in synthetic applications, and choices must be made on a case-by-case basis, considering the needs of the particular process (and scale).

Organic Synthesis Using Biocatalysis. http://dx.doi.org/10.1016/B978-0-12-411518-7.00002-0

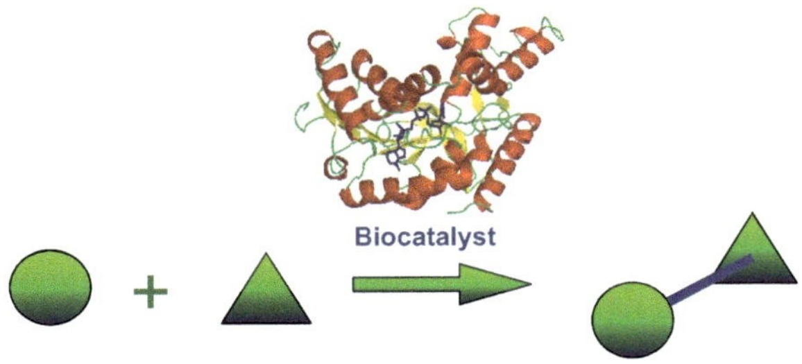

Criteria in process development of **biocatalytic syntheses:**

- Atom economy and sustainability
- Selectivity (enantio-, region-, chemo selectivity)
- Enzyme activity
- Enzyme stability
- Economical access to enzyme (recombinant form?)
- High conversion
- High volumetric productivity
- Simple downstream processing and high yield

SCHEME 2.1

Apart from these catalyst issues, the biotransformation process must also fulfill specific criteria to be economically attractive. A typical key feature of a successful large-scale biotransformation is a high product concentration and space–time yield, which can be reached. This usually means running processes at a high-substrate input (either all at once or by periodic additions over time). High space–time yield together with high conversion and excellent selectivity are key issues to be realized for obtaining a practical and economically attractive process. In terms of selectivity, in most cases of biotransformations this is related to enantioselectivity issues although others such as diastereo- or regioselectivity issues may play a role as well. Finally, downstream processing also represents an important task to be solved. Workup can be the most time-, capacity- and solvent-consuming step as well as the largest waste-producing step. Achieving a high economy for this unit operation is also a high priority for the overall process. Finally, catalyst recycling and a switch from batch to a continuously operating process can also be significant issues to consider.

The numerous key criteria (only some of which are mentioned in Scheme 2.1) illustrate the complexity in designing an efficient and practical synthetic biocatalytic process (ideally with potential for large-scale use) and at the same time also the challenges and the exciting tasks being ahead when conducting organic synthesis with enzymes as catalysts. In the following sections, these key criteria (according to Scheme 2.1) for developing a practical biocatalytic process will be discussed in more detail along with some engineering aspects. These discussions will be illustrated by successful case studies of efficient biotransformations.

2 AVAILABILITY OF AN EFFICIENT ENZYME AS A CATALYST: SCREENING AND PROTEIN ENGINEERING

The first step in development of efficient and practical biotransformations includes the search for suitable enzymes as biocatalysts [3,4]. Because many organic chemists do not have access to microbiological laboratories for preparing their own enzymes, availability of commercial enzymes often is an advantage. For the decision of the "biocatalyst of choice," enzyme properties have to be known. Typical criteria for choosing the "best" enzyme are high specific activity, enantioselectivity, stability, and no inhibition by substrate or product. For preparative utility in organic synthesis, a useful "rule of thumb" is a specific activity of at least 1 U/mg (although numerous exceptions with lower specific-enzyme activities are known). Ideal specific activities are in the range of 100–1000 U/mg.

Today, representatives of many enzyme classes can be prepared in recombinant host organisms (such as *Escherichia coli*) or obtained commercially [5]. The latter is particularly true for hydrolases, however, more and more redox enzymes, such as alcohol dehydrogenases, are becoming commercially available. Hydrolases are especially popular in organic synthesis because they are readily available and important products in the biotech industry with applications in many other industrial segments such as for manufacture of detergents and in the food industry. Furthermore, hydrolases do not require a cofactor. Such enzymes can be directly added to a reaction mixture, which dramatically simplifies their use. Not surprisingly, hydrolases account for about two-thirds of the reported biotransformations in chemical synthesis [6]. Thus, hydrolases are often considered as the "workhorses" of biocatalysis in organic chemistry. Among most popular transformations are resolutions based on the hydrolysis of racemic esters or other carboxylic acid derivatives for the synthesis of enantiomerically pure acids or alcohols (or, e.g., amines). Under nonaqueous conditions, the same enzymes catalyze enantioselective acylations that can be used to resolve racemic alcohols or amines [2,6,7]. Some of these processes have been industrially applied and run on multithousand ton scales [7–9]. Redox enzymes, though they are less available from commercial sources and require an external cofactor in many cases, are growing in popularity and now account for around one-fourth of biocatalysis examples [6]. It is noteworthy that both reduction and oxidation processes with biocatalysts found broad interest in organic syntheses. In terms of synthetic applicability, however, there are still large differences, for example, in the field of oxidation reactions. Although reactions based on the use of amino acid dehydrogenases (for reductive aminations of α-keto acids), alcohol dehydrogenases (for reduction of ketones), amine oxidases (for amine oxidations to imines), some other oxidases (e.g., D-amino acid oxidase used in industrial antibiotics production), and Baeyer–Villiger monooxygenases have found wide applications in organic synthetic biotransformations and often proceed with high substrate loading, other enzyme classes are still challenging. For example, oxidations with P450-monooxygenases are highly attractive from a synthetic point of view because there are very few alternatives for functionalizing otherwise unactivated hydrocarbons; however, these biocatalysts are often burdened by low stabilities and activities which makes their preparative use for organic transformations difficult.

In summary, a broad range of enzyme types can now be considered as "matured biocatalysts" that are widely applied in organic synthetic transformations on both lab and industrial scales.

3 THE ROLE OF COFACTORS AND PRACTICAL CONCEPTS FOR THEIR *IN SITU* REGENERATION

After having found an enzyme as a "catalyst of choice" one might also have to address aspects of cofactor supply and regeneration. With a few notable exceptions, for example, hydrolases and specific aldolases, most synthetically interesting enzymes depend on cofactors for catalytic function. This does not necessarily mean that for all such enzymes one has to take care of *in situ* cofactor regeneration but for many enzymes this challenge remains a task to be addressed. For example, this is true for many oxidoreductases, particularly for dehydrogenases that utilize nicotinamide cofactors NAD(H) or NADP(H) for their reduction reactions (Scheme 2.2a), and the corresponding oxidized forms accordingly for oxidation reactions (Scheme 2.2b) [10–12].

As the NAD(P)H (Scheme 2.2c) or NAD(P)$^+$-type cofactors represent a severe cost factor (6–7 EUR per gram for NADP$^+$) [13], it is not economically feasible to use them in stoichiometric amounts required for biocatalytic reduction and oxidation reactions, respectively (in contrast to, e.g., typical reducing agents in organic chemistry such as hydrogen, silanes, sodium borohydride, and other hydrides or Hantzsch-type esters). Thus, the appropriate oxidation state of the cofactor must be regenerated *in situ* during the reaction by means of a process that uses a second, sacrificial substrate in the complementary oxidation state [10–12]. In the case of carbonyl reductions, this cosubstrate is the reducing agent required in stoichiometric amounts and thus should be a cheap and readily available chemical. Numerous regeneration methods have been developed in recent years, with some even applied to industrial scales [2]. Examples for practical cofactor regeneration are given in the following. Among them, easy-to-use methods for organic chemists who are interested in applying oxidoreductases to lab-scale applications are also described.

As noted previously, nicotinamide cofactor regeneration is based on coupling, for example, the desired enzymatic reduction process, for example, carbonyl reduction, with a second reaction in which a cosubstrate is oxidized, which reforms the reduced nicotinamide cofactor for reuse in the desired biotransformation. The general concept is shown in Scheme 2.2a, exemplified for asymmetric ketone reduction and reductive amination of α-keto acids as representative processes. These are particularly valuable for organic chemistry because they yield enantiomerically pure alcohols and α-amino acids, respectively. Because the cosubstrate is required in stoichiometric amounts, it should be readily available and inexpensive and its oxidation product should be innocuous. The most

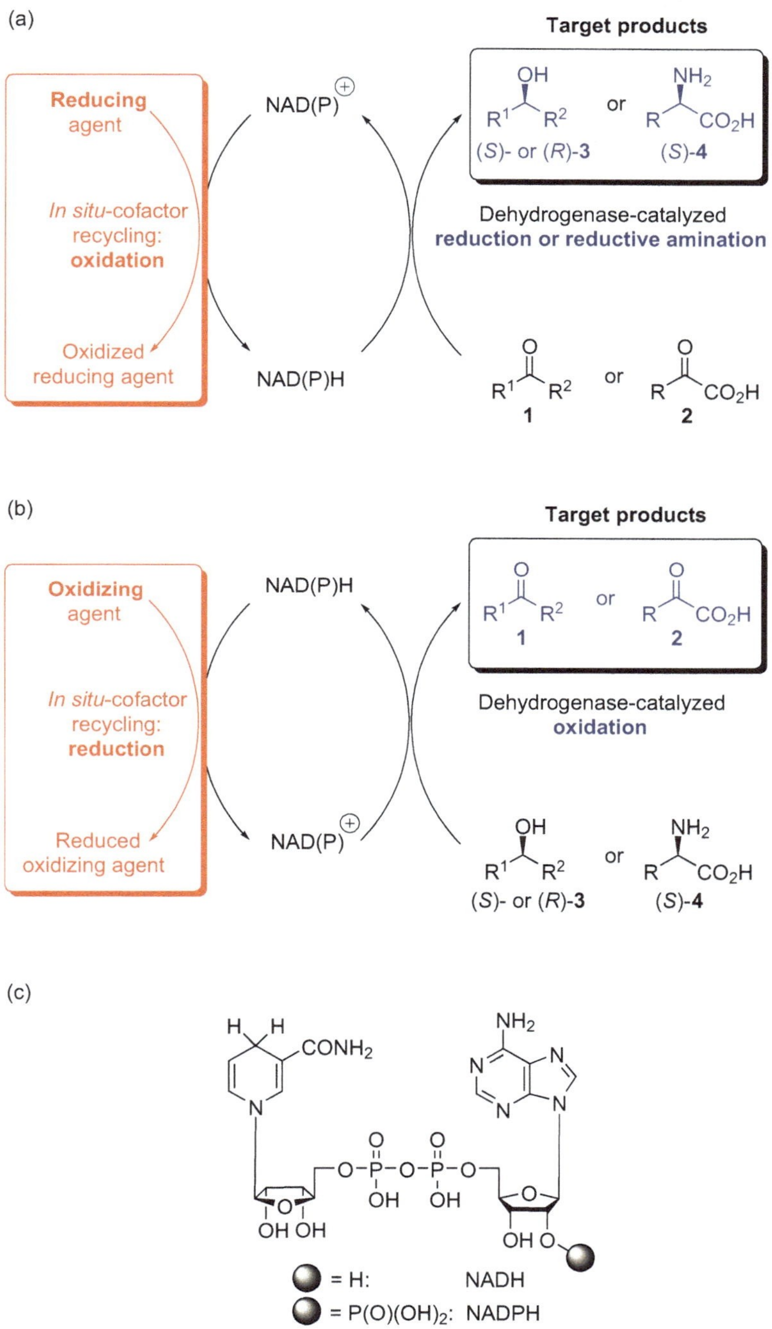

SCHEME 2.2
(a) *In situ* cofactor recycling for enzymatic *reductive* processes. (b) *In situ*-cofactor recycling for enzymatic *oxidative* processes. (c) Cofactor structure (exemplified for reduced cofactor forms NAD(P)H).

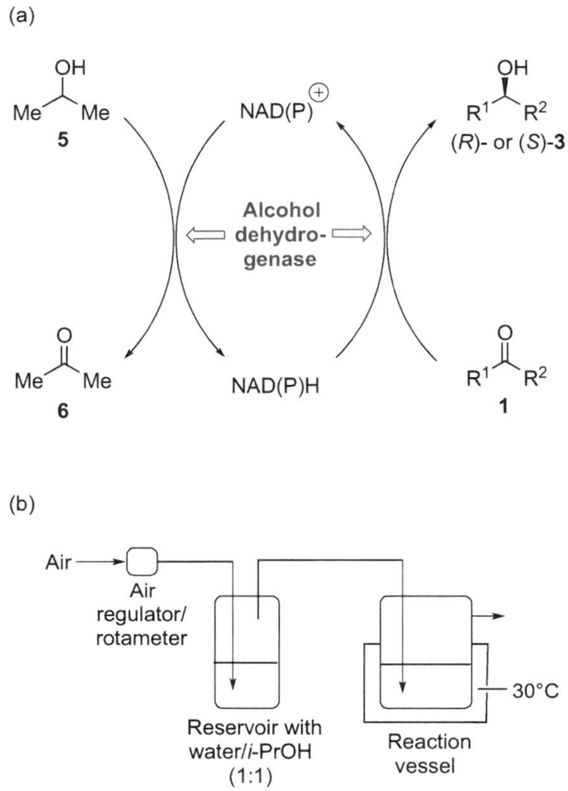

SCHEME 2.3
(a) Concept of the substrate-coupled cofactor recycling with isopropanol. (b) Process and reactor design for an efficient by-product removal in cofactor recycling with isopropanol. *(Figure in part (b) is taken in a slightly modified form from reference [15].)*

popular *in situ* cofactor recycling strategies for NAD(P)H are based on the oxidations of isopropanol, formate or D-glucose.

For organic chemists, isopropanol is an especially attractive hydride source, and reactions catalyzed by alcohol dehydrogenases correspond to the well-known Meerwein–Verley–Ponndorf reaction forming acetone as a by-product (Scheme 2.3a). A general advantage of such an approach is that only one enzyme component, namely the alcohol dehydrogenase that catalyzes both transformations, has to be added. In addition, isopropanol and acetone can help solubilize substrates in the aqueous reaction mixture. A general drawback, however, is thermodynamic of this process since in an equilibrium the formed acetone and desired alcohol product undergo the reverse reaction. Much effort has gone into overcoming this problem so that the equilibrium is shifted strongly to the product side. The most popular approach is to use isopropanol in excess, sometimes in combination with *in situ* removal of the by-product acetone. The latter can be accomplished by conducting the reaction under reduced pressure or by means of membrane technologies [14]. A further, very elegant method that is highly practical also for lab-scale applications by nonspecialists has been developed by the Liese group. A stream of isopropanol-saturated air is bubbled through the reaction mixture to strip off acetone and shift the final equilibrium toward the product side (Scheme 2.3b) [15].

Other practical ways of enzymatic cofactor recycling are based on the use of a second enzyme enabling an irreversible *in situ*-cofactor regeneration and, thus, consequently also an irreversibility of desired reduction or oxidation process [10–12]. The most popular "second enzymes" for this purpose are formate dehydrogenase (FDH) and glucose dehydrogenase (GDH), which are capable to regenerate NADH and NADPH, respectively. Both enzymes are sold by a number of commercial suppliers. The formate dehydrogenase-catalyzed oxidation of formate yields carbon dioxide and coproduces NADH from NAD^+. Glucose dehydrogenase can regenerate both NADH and NADPH by oxidizing D-glucose into D-gluconolactone in the presence of $NAD(P)^+$. Gluconolactone spontaneously hydrolyzes under normal process conditions, which shifts the equilibrium strongly to the direction of product formation and makes it an effectively irreversible route to NAD(P)H. What also makes FDH and GDH attractive (apart from the resulting irreversibility of the processes) is the low price and ready availability of the substrates required in stoichiometric amounts – formic acid and D-glucose. FDH is particularly well suited to coupling with dehydrogenases that catalyze the reductive amination of α-keto acids. The industrial process for L-tert-leucine is based on the cooperative use of a leucine dehydrogenase and formate dehydrogenase, which not only underlines the efficiency of this methodology on large scale but also represents the first *in situ* cofactor regeneration on industrial scale with isolated enzymes (Scheme 2.4) [16].

The *in situ* cofactor regeneration by means of a second, enzyme-catalyzed reaction also plays a key role in oxidations by monooxygenases. In contrast to oxidases (whose cofactor is an active-site prosthetic group that allows redox

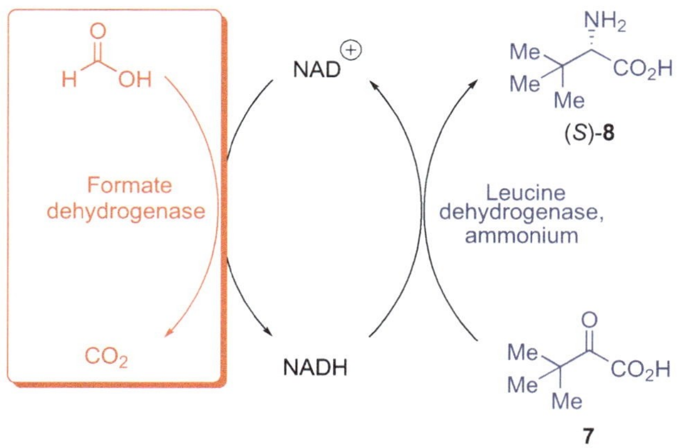

SCHEME 2.4

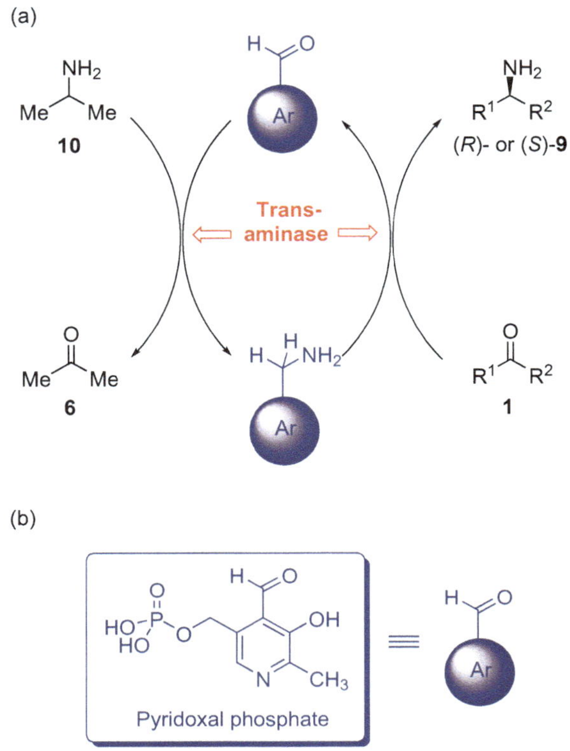

SCHEME 2.5
(a) Concept of the *in situ* cofactor recycling of pyridoxal phosphate. (b) Cofactor structure of pyridoxal phosphate

recycling with no need for an "external" regeneration process), monooxygenases need *in situ*-recycled NAD(P)H for the oxidation process. At first glance, it might be surprising that an oxidation reaction would require the same cofactor as used for reductions, namely the reduced form NAD(P)H. This can be, however, easily explained with the need to reduce one atom of molecular oxygen to water while the other is inserted into a C—H or C—C bond of an organic molecule during a hydroxylation or Baeyer–Villiger oxidation, respectively. The same "second-enzyme" strategies described previously for NAD(P)H can also be coupled to monooxygenases.

Cofactor regeneration is not only a task when using oxidoreductases but also for some representatives from other enzyme classes, for example, transferases and lyases. Transaminases require pyridoxal phosphate and transform a ketone into the chiral amine on the expense of a donor amine that is the cosubstrate needed in stoichiometric amount [17]. A representative example for pyridoxal phosphate regeneration is shown in Scheme 2.5, exemplified for the use of isopropylamine as an amine donor. This reagent is especially useful for two reasons. First, it is a small (and thus atom economical), cheap, and readily available amine.

In addition, the by-product is acetone that can be removed by the strategies described previously, thus shifting the overall equilibrium toward the desired amine product. A very noteworthy example of this strategy was jointly reported by Merck and Codexis in the production of sitagliptin using a transaminase [18,19]. Besides developing a highly efficient transaminase by protein engineering, Codexis and Merck researchers also realized a process running with impressive 100 g/L substrate loading. This process technology, which has made it already to an industrial scale application, will be described in more detail in Section 8 of this chapter.

Several carbon–carbon bond forming reactions depend on a cofactor as well [20]. Enzymes can catalyze a range of these types of reactions which is an important domain of classic organic chemistry. The benzoin condensation and related reactions are good examples of these processes [21]. Indeed, the biocatalytic synthesis of (*R*)-phenylacetcarbinol (an intermediate in the synthesis of ephedrine) probably represents the first industrial synthetic application of enzymes, a process that has been running on an industrial scale since the 1930s [22]. Although this process is based on the use of (wild-type) whole-cells for economic reasons, isolated lyase enzymes are also suitable and numerous syntheses have been developed based on free or immobilized lyases. Supplementing reactions with the cofactor thiamine diphosphate (ThDP, alternative name: thiamine pyrophosphate) turned out to be beneficial. One major difference between ThDP and the nicotinamide cofactors described previously is that the former does not need to be regenerated prior to the next catalytic cycle so no sacrificial cosubstrate is needed. This considerably simplifies synthetic applications of these lyases and a typical reaction based on the use of this cofactor ThDP is shown in Scheme 2.6.

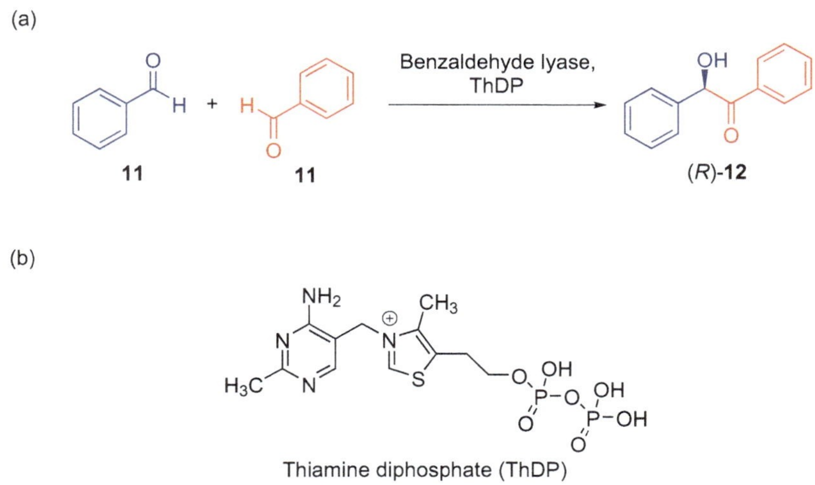

SCHEME 2.6
(a) Representative reaction for a biotransformation with ThDP. (b) Cofactor structure of thiamine diphosphate

However, when the cofactor is bound strongly to the enzyme as a prosthetic group, cofactor addition and additional enzymatic processes for cofactor regeneration are not needed. A representative example for such enzymes is amino acid oxidases that catalyze the oxidation of amino acids to α-keto acids under simultaneous reduction of molecular oxygen to hydrogen peroxide.

4 ENZYME FORMULATION

As in case of chemocatalysts, different types of "formulations" are conceivable for enzymes as well [5]. The type of enzyme formulation chosen is often strongly related to the issue of enzyme purity. There is almost always a tradeoff between purity (which minimizes side reactions) versus the cost and effort required for purification. So long as no extraneous reactions are catalyzed by contaminants, even low-purity enzymes can be tolerated in chemical processes. Thus, the most straightforward and at the same time cheapest "formulation" of an enzyme is a whole cell taken directly from a fermentation and used in the desired biotransformation process. If intact cell membranes interfere with the reaction of interest (usually because of slow transport of substrates and/or products), cells can be permeabilized or disrupted to provide crude extracts. Since cell disruption and isolation of crude extracts are well-established and economically attractive tools, crude extracts belong to the most economical forms (formulation) of an enzyme. At the same time, however, it has to be considered that the whole cells as well as isolated crude extracts contain significantly higher levels of impurities compared to the purified proteins which might interfere with the enzymatic reaction and product isolation. Among the most purified enzymes in biotransformations are diagnostic enzymes; however, these are also usually the most expensive, particularly when compared with crude lysates. Scheme 2.7 provides an overview of the cost versus benefit relationships for enzyme formulations.

High-level recombinant overproduction of enzymes has been an important advance for biocatalysis. Having the target enzyme present at very high levels

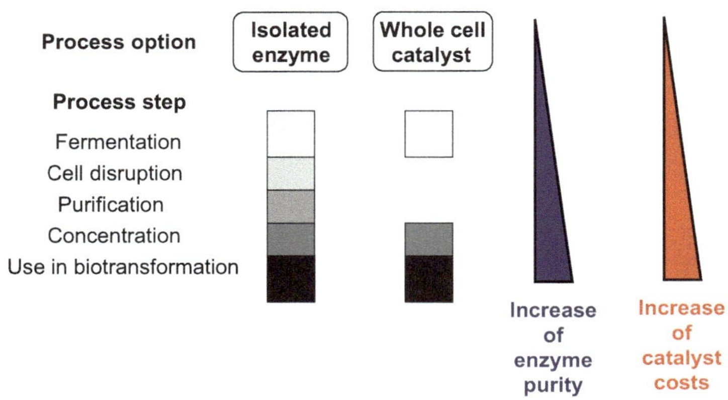

SCHEME 2.7

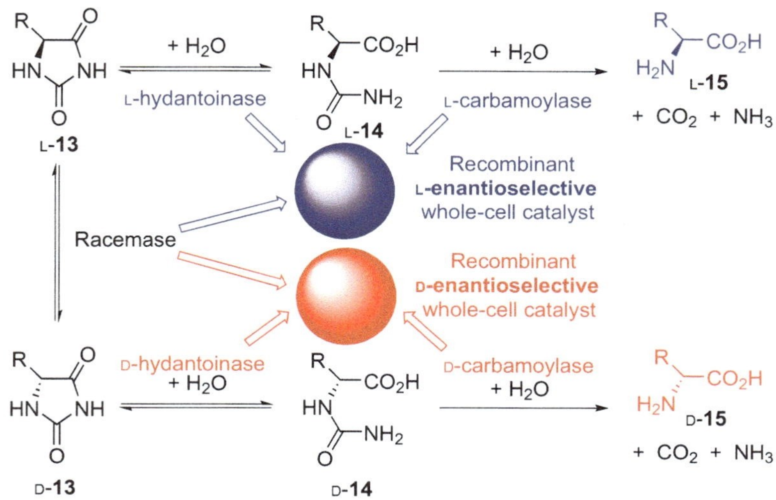

SCHEME 2.8

(typically 10–20% of total protein) means that even crude extracts represent a semipurified form of the biocatalyst. This often makes it possible to employ whole cells or crude extracts thereof, without the typical drawbacks of wild-type whole cells or crude extracts, for example, side reactions and low productivity. The high percentage of the desired enzyme compared to the overall amount of protein also limits the amount of biomass needed for a biotransformation. The latter issue is beneficial for downstream processing as well. Furthermore, in combination with high cell-density fermentations of recombinant cells, the production of an economically attractive catalyst is achieved. Today, recombinant whole cells or crude extracts thereof with an optimized overexpression of the desired enzyme are often the "catalyst of choice".

Recombinant whole cells in particular turned out to be very attractive for biotransformations in which more than one recombinant enzyme is needed such as redox reactions with *in situ* cofactor regeneration or hydrolysis with multiple enzymes. With respect to the latter one, the dynamic kinetic resolution of hydantoins by a whole-cell catalyst that simultaneously overexpresses a racemase, a hydantoinase and a carbamoylase is a popular and industrially relevant example (Scheme 2.8) [23,24]. These cells convert a racemic hydantoin (an easily accessible substrate) to the corresponding enantiomerically pure D- or L-amino acid with both high conversion and enantioselectivity.

Besides recombinant whole cells, crude extracts are also widely applied, and typically most of the enzyme formulations used for biotransformations in organic synthesis are crude extracts or partially purified proteins rather than highly purified enzymes (which are more needed in the field of diagnostic applications).

Besides the direct use of whole cells or isolated enzymes (in the form of crude extracts or purified samples), immobilization of enzymes and the application of such biocatalysts represent a further valuable and attractive alternative option. Since enzyme immobilization has gained broad interest and a wide range of synthetic applications have been reported, an overview of enzyme immobilization and its use is given in the following. This topic will also be discussed separately in detail in this book (Chapter 4).

5 IMMOBILIZATION AND REUSE OF CATALYSTS

A further method often used to make biotransformations more efficient is immobilization (heterogenization) that offers two major advantages. First, after the reaction is complete, immobilized catalysts can typically be collected simply by filtration that facilitates their reuse. One prerequisite, however, is that the catalyst must retain sufficient catalytic activity of the enzyme, thus requiring a high stability of the enzyme under process conditions. Although recycling of the catalyst eliminates its replacement costs for each new batch, these benefits must be weighed against the time, money, and effort required to immobilize the enzyme as well as the loss of enzyme activity that nearly always occurs during the immobilization process. Thus, from this perspective, an immobilization is of particular interest when enzyme costs play a significant role and have to be significantly reduced. The second benefit of enzyme immobilization is that it can tremendously simplify downstream processing. Soluble (free) enzymes can complicate product isolation steps (often by forming emulsions during extraction steps that are difficult to separate), however, biotransformations using an immobilized enzyme can be easily cleared of proteins by a simple filtration step prior to product isolation. Such a facile separation-unit operation step also can contribute to better economics and a more practical overall process.

One of the most prominent examples illustrating the advantages of enzyme immobilization for efficient and scalable biotransformations is the penicillin amidase-catalyzed hydrolytic cleavage of the penicillin G side chain that forms the industrially important 6-aminopenicillanic acid (Scheme 2.9). This key

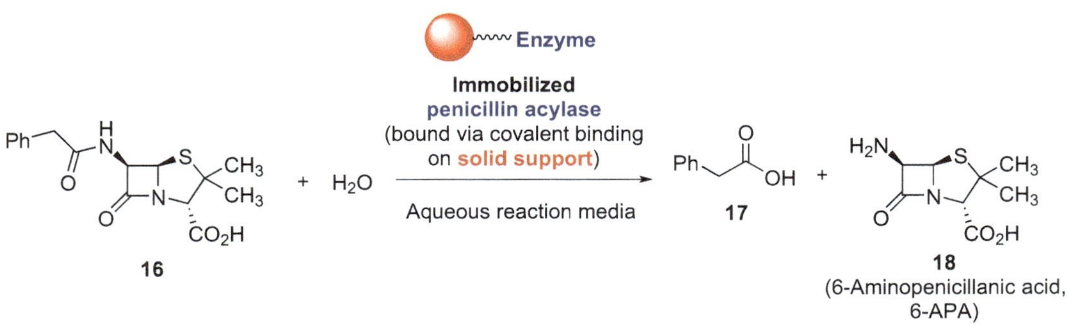

SCHEME 2.9

intermediate for several semisynthetic antibiotics is manufactured on an annual production volume that exceeds 10,000 tons. The enzyme is used in immobilized form, and Eupergit-type oxirane acrylic beads have been especially successful [25,26]. The immobilization concept is based on a ring-opening reaction of the epoxy moieties with amino functionalities of lysine residues at the enzyme, thus leading to a covalent immobilization of the enzyme on the solid support. The macroscopic size of the beads (0.1–0.2 mm diameter), allows enzyme recovery by simple filtration after the reaction. Impressively, the catalyst can be recycled more than 850 times, and activity decreases by only 50% during this period [26]. The overall biotransformation process is also environment friendly and replaced the previous chemical process based on hazardous compounds [27]. This method of immobilization has also been successfully extended to other enzymes [25,26].

A range of other methods and solid supports are known for efficient enzyme immobilization [2] and they are reviewed in detail in Chapter 5 of this book. The following will therefore not focus on a comprehensive description of immobilization methods but rather on a few selected, recently developed methods that address particular practical issues encountered when conducting biotransformations.

As noted previously, enzyme-catalyzed organic synthesis can also be carried out in organic media (see Chapter 3 of this book for additional details). This is particularly relevant for lipase-catalyzed transformations that account for the largest number of enzyme applications in organic chemistry. These lipases are typically used in immobilized forms that are often commercially available, for example, Novozym 435. Although well-suited to laboratory-scale reactions, abrasion of the immobilisate might become a problem on larger scales. An elegant solution was jointly developed by an academic-industrial collaboration by Thum, Ansorge-Schumacher and coworkers [28–30]. Inexpensive silicone building blocks were used to coat immobilized lipase Novozym 435 and yield a very mechanically stable and robust biocatalyst preparation. Even under harsh leaching conditions, very little enzyme desorption was found. The activity remaining after the coating process was also high (92%), as measured by the esterification of lauric acid with 1-propanol. Its potential for industrial applications was evaluated by studying the syntheses of myristyl myristate, poly(ethylene glycol) 400-coconut fatty acid monoester and ethylene oxide and propylene oxide copolymer (EO/PO)-oleic acid diester [31]. Total turnover number (TTN) increased by up to 50-fold compared to the nonsilicone-coated biocatalyst and a representative example is shown in Scheme 2.10. This technique was also successfully used for different types of enzymes, for example, hydrolases other than lipases as well as redox enzymes [30–32].

Another successful example in which an immobilized enzyme was stabilized for biocatalysis in organic media was reported by Merck researchers for a transaminase-catalyzed biotransformation [33]. The transaminase was immobilized on a range of commercially available polymer-based resins. The adsorption support SEPABEAD EXE-120 for binding enzymes through hydrophobic interactions turned out to be most suitable and gave a residual activity of 45%

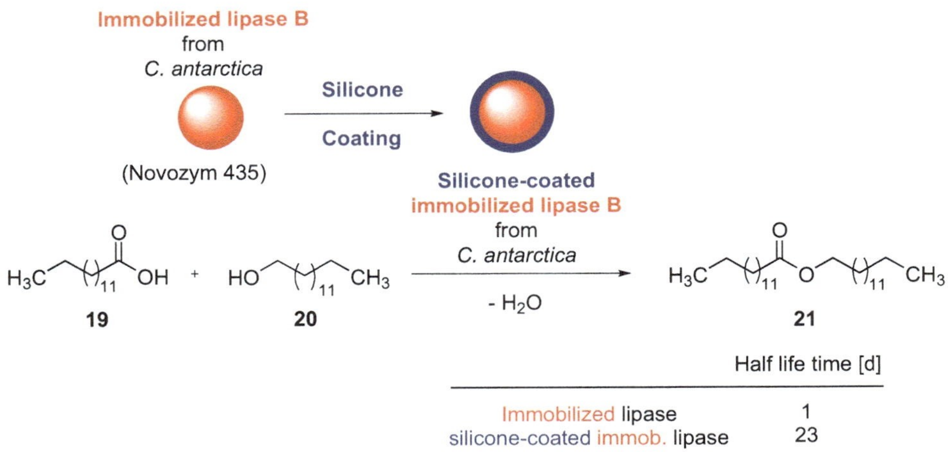

SCHEME 2.10

as well as a high tolerance toward the organic reaction medium. In contrast, the lyophilized, nonimmobilized form of the transaminase was completely denatured and showed no activity under the same conditions. The immobilized transaminase was successfully used to convert the ketone precursor into the corresponding amine, which served as the final key intermediate in the synthesis of the drug sitagliptin. Using an immobilized enzyme in pure organic media avoids the need for extraction processes (compared to an aqueous reaction media), thereby increasing overall process efficiency. A simple recovery of the catalyst is also feasible by means of this methodology.

Enzyme-catalyzed redox processes with *in situ* cofactor recycling are usually carried out in aqueous media (either as pure aqueous media or two-phase systems; for exceptions, see Section 7) because the dissociable nicotinamide cofactors are only soluble in water. An immobilization method that enables enzymatic redox processes with *in situ* recycling of NAD(P)H-type cofactors to be conducted in pure hydrophobic organic solvent was recently reported jointly by the Gröger, Berkessel, and Hummel groups (Scheme 2.11) [34]. The idea of this concept was to integrate the aqueous media into the immobilisate. Toward this end, a superabsorber was treated with aqueous buffer, followed by enzyme and cofactor. The immobilization of an enzyme in a superabsorber was developed earlier by Jeromin and applied to reactions in aqueous and highly water-miscible solvents [35]. As for the enzyme, an alcohol dehydrogenase suitable for ketone reduction was used. When applying superabsorber-entrapped alcohol dehydrogenases in pure organic media, isopropanol (25% v/v) was added and dissolved in the "immobilized" aqueous phase together with the enzyme and cofactor. By means of such an immobilized biocatalytic system, ketone reductions have been carried out using a pure organic liquid phase with a high logP value (in order to prevent leaching of water and isopropanol). For example, ketone (*R*)-**22** was reduced to the desired alcohol (1*R*,3*S*)-**23** by this immobilized catalyst system with high

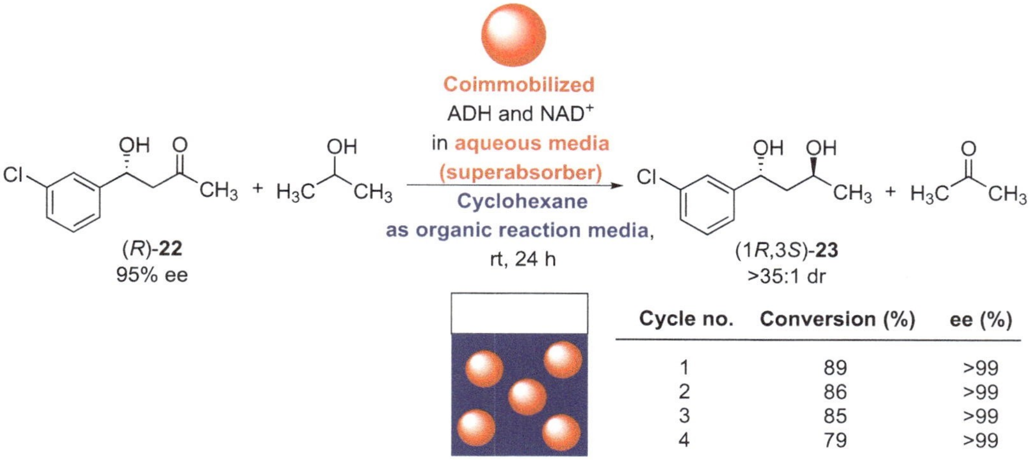

Cycle no.	Conversion (%)	ee (%)
1	89	>99
2	86	>99
3	85	>99
4	79	>99

SCHEME 2.11

conversion and more than 99% ee [34]. In addition, recycling of the enzyme and cofactor immobilisate was demonstrated as well.

6 REAGENT/SUBSTRATE ENGINEERING

Besides choosing a suitable enzyme and enzyme formulation, some biotransformations also allow chemists to "fine tune" the synthesis of a specific target molecule by modifying a reagent or leaving group in the substrate component. Although typically the structure of a substrate is fixed due to the need for specific target molecule, there might be some flexibility in terms of the choice of leaving groups as well as easily removable protecting groups in the substrate. The idea is to modify and optimize aspects of the molecule that influence reaction performance but do not appear in the final product structure.

Biotransformations in which a reagent was optimized for the reaction are discussed first. This strategy is well known, for example, in the asymmetric biocatalytic reduction of ketones by using various reducing agents [36]. Apart from availability and economy, other key issues to consider when optimizing reagents are safety and by-product removal. Eliminating a coproduct during the biotransformation shifts the equilibrium to the desired side, thus enabling higher overall conversion and yield. Two examples are detailed further, one each from the hydrolase and oxidoreductase fields.

Lipase-catalyzed acylation is a "first choice" approach for resolving racemic secondary alcohols [6,37]. Since at least the alcohol moiety of the acylation agent will not be part of the final target molecule (namely the resolved chiral alcohol or the corresponding ester as its O-acylated derivative), there is significant latitude in choosing the acylation reagent (Scheme 2.12). This flexibility is useful since only such acylating agents should be chosen that provide a high driving force that shift the equilibrium in the desired direction. For example,

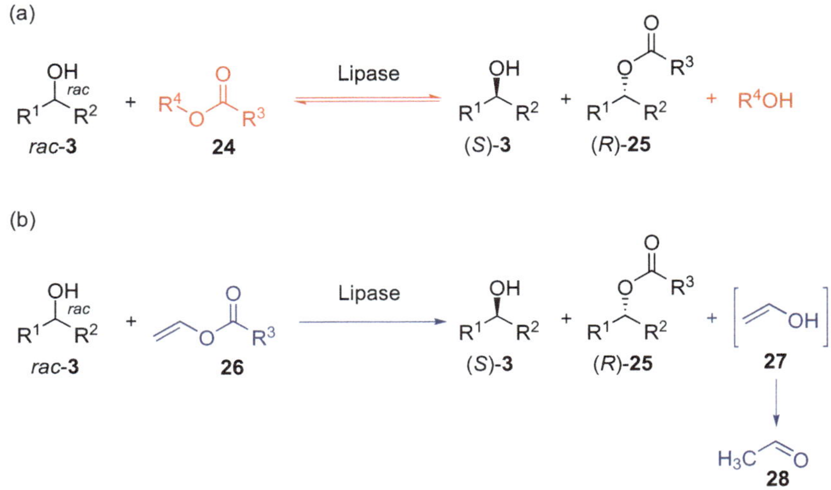

SCHEME 2.12
(a) Concept of lipase-catalyzed acylation using alkyl esters as acylation reagent. (b) Concept of lipase-catalyzed acylation using vinyl esters as acylation reagent.

an alkyl ester such as ethyl acetate as acylating agent would likely lead to an unsatisfactory equilibrium mixture since the resulting desired ester and released ethanol by-product can react in the reverse direction, thus forming substrate and acylation reagent again (Scheme 2.12a). To avoid this reverse reaction, and thereby shift the equilibrium in the desired direction, vinyl acetate instead of alkyl acetate would be a far superior acyl donor since the "vinyl alcohol" by-product tautomerizes to form acetaldehyde. The "removal" of the formed vinyl alcohol from the equilibrium in such a way then shifts the equilibrium toward the desired direction and enables high conversions in this resolution process (Scheme 2.12b). The lipase-catalyzed resolution with vinyl acetate as an acylation agent has widely been applied in organic synthesis and demonstrated in numerous successful examples.

A further related example that has recently garnered tremendous attention when applied to large scale are chiral amine resolutions by lipase-mediated acylations [7–9,38]. For many years, these reactions were plagued by long reaction times when typical acyl donors such as ethyl acetate or other alkyl carboxylic acids esters were employed. BASF researchers made a breakthrough finding that acetic ester derivatives bearing an electronegative heteroatom in the α-position showed greatly increased reaction rates while maintaining excellent enantioselectivities. This provided both the unreacted amine as well as the amide in enantiomerically pure form [7–9]. A representative example is given in Scheme 2.13. Methoxy acetate is a particularly useful acyl donor. The method has been applied on a multithousand ton scale, with the resolution of racemic 1-methoxyisopropyl-2-amine as the major application (2500 tons annually) [7].

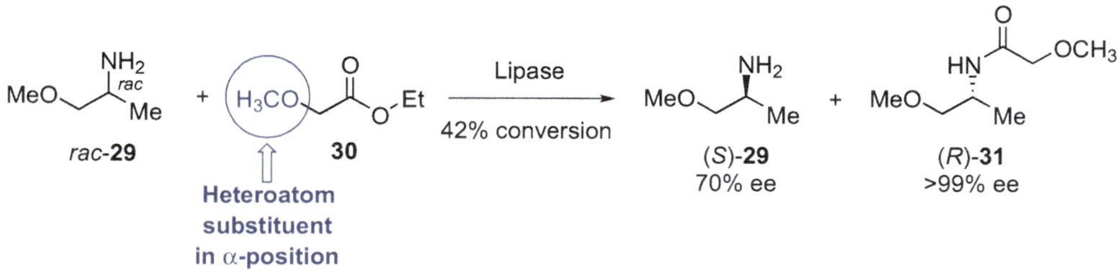

SCHEME 2.13

Enzymatic carbonyl reductions are another type of biotransformation with high flexibility for reagent choice. When isopropanol is used as the stoichiometric hydride donor for *in situ* cofactor regeneration of NAD(P)H, acetone is formed as the corresponding by-product (along with the desired alcohol product). Acetone is generally the most volatile component in the reaction mixture by far, and it can easily be removed by running the reaction under vacuum or by employing a selective membrane. This method has already been described in more detail (Section 3).

An alternative option to changing reagent for process optimization is to modify the substrate. A representative and widely applied example for such a strategy is the resolution of racemic esters *via* hydrolysis [6,37]. In such a hydrolytic ester cleavage an acid and alcohol moiety is formed which in case of a chiral acid enables the choice of an optimized (achiral) alcohol moiety in the substrate. Similarly, when forming a chiral alcohol, a preferred (achiral) acid moiety in the ester substrate can be chosen. A typical example is the synthesis of chiral β-hydroxy nitrile (*R*)-**33** as shown in Scheme 2.14 [39]. In this study, conducted by the Itoh group, different acyl moieties were screened and it was found that enantioselectivity depended critically on the choice of this (achiral) leaving group.

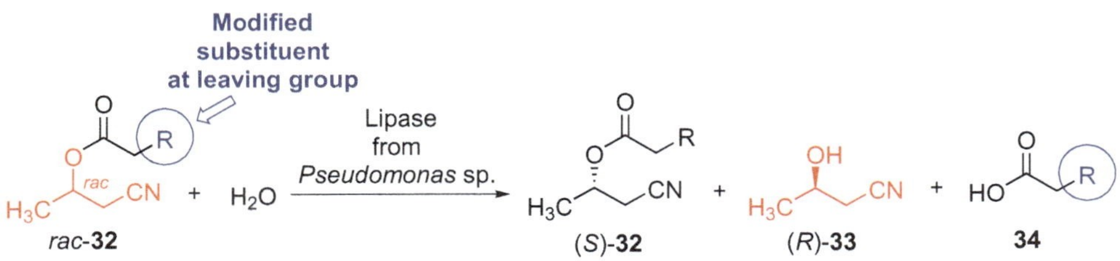

R	E-value
H	7
C_2H_5	2
n-C_3H_7	3
Cl	6
OCH_3	14
SCH_3	29
SBu	13

SCHEME 2.14

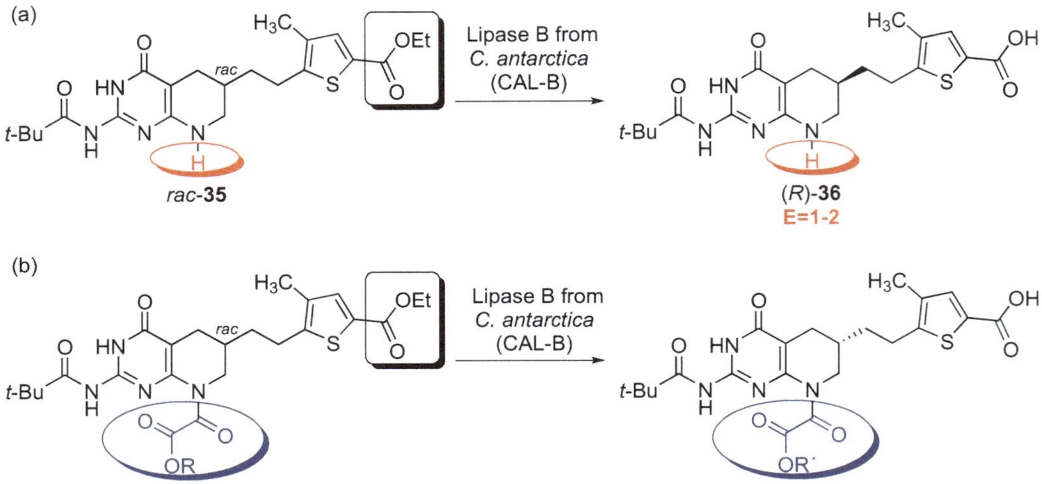

SCHEME 2.15

For example, the highest enantioselectivity (*E* value of 29) was observed for the methylthioacetyl moiety as the leaving group. The more commonly employed acetyl moiety gave a significantly lower E value of 7.

Another way to modify the substrate that facilitates a biotransformation is to reversibly introduce a functional group that influences asymmetric induction by enhancing the differences in the transition states leading to the two product enantiomers. To be practically useful, the groups introduced must also be easily removable after the enantioselective reaction step. A representative example illustrating the preparative value of this synthetic strategy can be found in the preparation of the Pfizer drug pelitrexol [40]. The key chiral intermediate bears a so-called "remote stereogenic center" and its enantioselective synthesis is based on resolution by an enzymatic hydrolysis process starting from a suitable ester. However, none of the hydrolases gave useful enantioselectivities (E-values of 1-2) when the nonacylated racemate *rac*-**35** was used as the substrate (Scheme 2.15a). In contrast, when using the racemic *N*-substituted derivative *rac*-**37**, the hydrolysis proceeded with significantly improved enantioselectivities (E-values of 59). This illustrates the tremendous influence that a "spectator group" can have in enantioselective discrimination in the resolution step (Scheme 2.15b).

7 REACTION MEDIA ENGINEERING

Reaction medium (solvent) engineering is a major tool for optimizing biocatalytic processes. The reaction substrates as well as the target molecules are often highly hydrophobic (and, thus, water immiscible) whereas enzymes typically require water as a solvent (apart from notable exceptions like lipases). Thus, besides purely aqueous solvent (the "classic" reaction medium), the addition of organic solvent(s) has been helpful in many cases.

Two general, complementary strategies are available when organic solvents are applied to biotransformation running in aqueous media. First, water-miscible solvents can be used to increase the solubility of poorly soluble organic compounds. Second, aqueous-organic two-phase solvent mixtures created by mixing a water-immiscible organic solvent component have also been a very useful approach. It should be noted that a range of other solvent strategies have also found wide interest and applications. One of the most useful is a two-phase, emulsion-type system consisting of an aqueous phase with the hydrophobic substrate and/or product providing the "organic" phase. This avoids the need to add an organic solvent as an additional reaction component. A further attractive strategy consists of the use of a pure organic reaction media by means of a water-immiscible organic solvent only. Solvent-free processes, in which the reaction media consists only of the substrate and product, have also been developed.

Success stories can be found for all of these process options. The different concepts for reaction media engineering are briefly discussed and illustrated by case studies as representative examples. The major focus will be on the reaction medium engineering of aqueous media by adding organic solvents as well as solvent-free processes whereas biotransformations carried out in pure organic media will be the specific subject of another chapter of this book (Chapter 3).

7.1 One-Phase Systems Using Water-Miscible Organic Solvents

The use of water-miscible organic solvents is a process option when the organic substrate has little or no aqueous solubility. Adding cosolvents not only increases substrate solubility but can also prevent precipitation in the case of solid substrates. Typical water-miscible cosolvents are short-chain aliphatic alcohols (C1–C4 alcohols), dimethylsulfoxide (DMSO), and dimethylformamide (DMF). Increasing substrate solubility is particularly beneficial with enzymes having a high K_m value for the substrate. In such cases, increased solubility goes hand-in-hand with increased reaction rate. Enzyme deactivation by organic cosolvents, however, might be a major drawback.

An impressive example of a biotransformation process in aqueous media supplemented with a water-miscible cosolvent can be found in the transaminase-catalyzed synthesis of the sitagliptin key intermediate (*S*)-**40** developed jointly by Codexis and Merck researchers (Scheme 2.16) [18,19]. As might be expected, the ketone substrate was only soluble in water at less than 1 g/L. Various cosolvents were evaluated and both methanol and DMSO were suitable candidates with respect to both enzyme performance and increase of ketone solubility. The latter was ultimately chosen, and the final process operates at 50% (v/v) DMSO, which allows a remarkable substrate input of 200 g/L of ketone **39**. This

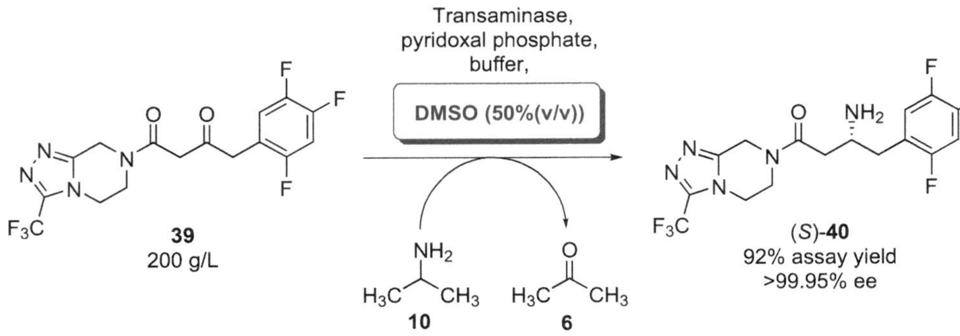

SCHEME 2.16

provides the desired amine (*S*)-**40** in a 92% assay yield and with excellent more than 99.95% ee (Scheme 2.16).

It is noteworthy that the cosolvent DMSO also turned out to be beneficial when using other enzymes, for example, P450-monooxygenases [41]. The use of short-chain alcohols as water-miscible solvents in biotransformations has been reported widely in the field of esterase-catalyzed reactions, for example, in PLE-catalyzed asymmetric syntheses [42].

7.2 Two-Phase Systems Using Water-Immiscible Organic Solvents

The use of a second, organic phase under formation of a two-phase system can be beneficial from different perspectives. For solid substrates that are poorly soluble in water, dissolving in the organic phase can result in a more rapid phase transfer (compared to suspending the solid in an aqueous reaction mixture). Furthermore, the second organic phase can decrease the level of substrate in the aqueous phase (compared to the purely aqueous counterpart). Such a lowered aqueous level (which can be calculated according to the partition coefficient), however, may be advantageous or disadvantageous, depending on the specific case. For example, it is nearly always inadvisable to lower the substrate concentration in the aqueous phase below the K_m value since this means that the enzyme must operate at sub-V_{max} efficiency. On the other hand, lowering the concentration in the aqueous phase to around twice the K_M value can be an advantage to minimize any nonenzymatic "background" reaction occurring simultaneously in the aqueous solution (when the rate of this background reaction depends on the substrate concentration in the aqueous phase).

A further advantage of organic aqueous two-phase reaction media is related to downstream processing and enzyme recovery as well as its (potential) recycling. Since the enzyme is water soluble, in the case of hydrophobic substrates and products, the aqueous phase might be easily separated from the organic phase (in case no emulsion problems occur), thus enabling an elegant way to separate as well as to reuse the enzyme catalyst. As noted previously, however, it is also

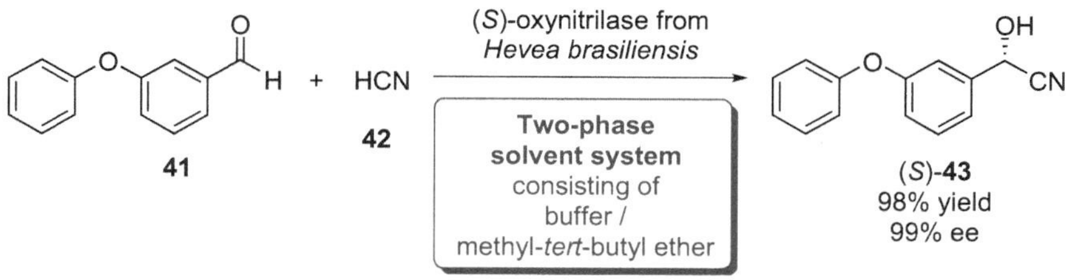

SCHEME 2.17

necessary to use a robust enzyme that is stable toward deactivation by the organic solvent as well as by hydrophobic substrate and product.

The asymmetric hydrocyanation of aldehydes catalyzed by oxynitrilases provides an excellent example of how a two-phase solvent system facilitates biocatalysis [43–45]. The process was developed by the Griengl group in cooperation with DSM and utilizes methyl *tert*-butyl ether as the organic phase along with an aqueous buffer. These conditions provided the desired chiral cyanohydrins with high conversions as well as excellent enantioselectivities by avoiding the nonenzymatic addition reaction to the racemic product in water. An industrial example of this methodology is the production of (*S*)-*meta*-phenoxy benzaldehyde cyanohydrin (Scheme 2.17) that is formed in 98% yield and with an excellent enantioselectivity of 99% ee and a spectacular space–time-yield of 2.1 mol/L/h [46]. In addition, the enzyme can be easily recycled by separating (after simple extraction) and reusing the aqueous phase that has been demonstrated by five reaction cycles.

The key finding of the positive impact of organic solvent on the reaction was made by the Effenberger group using a pure organic reaction medium [47]. It should be noted that the undesired chemical (non-enantioselective) hydrocyanation reaction can also be suppressed in purely aqueous media by maintaining a pH of 3.25, as demonstrated by the Kula group [48].

Organic aqueous two-phase solvent systems are also widely applied for other biotransformations, for example, in hydrolase-catalyzed processes [37,49]. A typical example is the enantioselective hydrolysis of a racemic, water-immiscible ester as a substrate. The reaction yields a water-soluble carboxylic acid (in deprotonated form as a salt). Ionic liquids have also been used in biphasic systems, a topic that is discussed in detail in Chapter 3 in this book [50].

7.3 Two-Phase Systems Using Substrate/Product as Organic Phase (Emulsion/Slurry Systems)

High substrate loading, however, can be also achieved in the absence of an organic solvent even when hydrophobic compounds with low water solubility are used. The resulting emulsions (in case of liquid substrates) or slurry systems (in case of solid substrates) represent suitable reaction media under the prerequisite that mass transfer from the organic phase of the emulsion or slurry into the

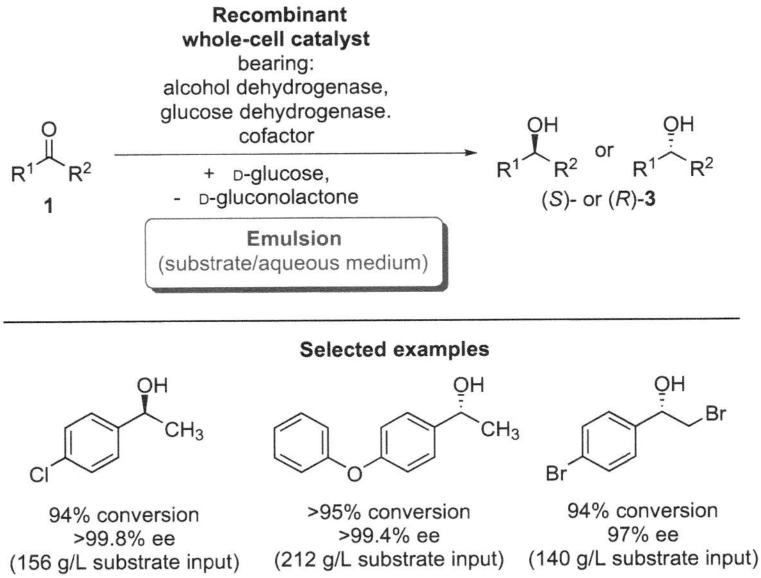

SCHEME 2.18

aqueous phase does not represent a rate-limiting step. Homogeneity of the reaction mixture is best in emulsions based on organic components with low viscosities and such systems can be easily mixed by "standard" stirrer equipment. One advantage of this strategy is that the organic solvent used in any downstream-processing (extraction) steps can be chosen independently of the reaction solvent. In addition, the emulsion or slurry approach maximizes substrate solubility in the aqueous phase since the latter is always constantly saturated. This can be particularly advantageous for enzymes whose K_M values are relatively high for the substrate of interest. Assuming that the K_m value is below 50% of the saturation concentration of substrate, one would still be able to operate at a maximum velocity whereas when adding an organic solvent in such a case leading to a decrease of substrate concentration in the aqueous phase would result in a decreased velocity. The emulsion/slurry strategy has been applied to asymmetric ketone reduction by Degussa researchers along with the Hummel group [51], who found both high conversion and excellent enantioselectivity when operating at very high substrate loading (>100 g/L) (Scheme 2.18).

In cases where mass transfer is rate limiting, one can consider two options. First, a phase-transfer catalyst might be included (thus leading to improved mass transfer from the organic to the aqueous phase) and/or the interface area might be increased. The latter can be accomplished by a miniemulsion system whose organic droplets are in the range of 50–500 nm. This concept was demonstrated successfully for a lipase-catalyzed hydrolysis of racemic amino acid esters operating at the extremely high substrate loading of 827 g/L [52]. Lipases are interface-dependent enzymes and show rate accelerations in the presence of a water/organic phase interface. The much larger interfacial surface area in a miniemulsion might be a (major) reason for the dramatic rate increase observed in this system as compared to a "standard" two-phase system.

7.4 Fully Solvent-Free Syntheses

The ultimate solvent-reduction strategy in biocatalysis is to eliminate it completely and rely solely on the substrate and product as the (liquid) phase. This is particularly attractive for low-price, high-volume chemicals where solvent usage (and subsequent removal within the downstream-processing unit operations) would be a significant cost contributor. A completely solvent free process would also be "Green" and environmentally attractive. One example of a solvent-free biotransformation can be found in the lipase-catalyzed formation of fatty esters (see also Section 5), used by the cosmetic industry as emollient esters [30]. An immobilized lipase catalyzes the ester formation under mild conditions, in contrast to the alternative, "chemical" method that requires high temperature conditions. This process has been further improved by the Liese group in collaboration with Evonik Industries. A new reactor concept based on a bubble column prevents mechanical erosion of the immobilized biocatalyst that was a problem when the reaction mixture was mechanically stirred [53]. When using *Candida antarctica* lipase B in this reactor, myristyl myristate and polyglycerol-3 laurate, **46**, were formed with excellent space–time-yields of 6731 and 3042 $gL^{-1}d^{-1}$, respectively (Scheme 2.19).

8 REACTION ENGINEERING: BATCH *VERSUS* CONTINUOUSLY RUNNING PROCESSES

A final major choice in biotransformation reaction development is the decision between a batch or a continuously running process [2]. The choice depends not only on the specific needs of the process but also on the reaction scale. Small-scale biotransformations carried out in organic chemistry labs are most conveniently conducted in batch mode since this simplifies reaction setup and makes minimal demands on control technologies. On larger scales, batch processes are commonly employed in the synthesis of fine chemicals and pharmaceuticals due to the (relatively) small volumes and the general preference for production in multipurpose plants. On the other hand, manufacturing a specific product over an extended time period on industrial scales in dedicated plants makes a continuously running process more attractive. The aminoacylase-catalyzed enantioselective synthesis of L-amino acids provides a good case study for these choices, since the same reaction has been conducted in different fashions (batch versus continuous process), depending on the application and scale.

The aminoacylase-catalyzed resolution of *N*-acetyl amino acids yields enantiomerically pure amino acids by enantioselective hydrolysis. This has been a "workhorse reaction" in enantioselective amino acid chemistry and has been applied on industrial scales [2,54]. Moreover, because both L- and D-aminoacylases are available, both amino acid enantiomers are accessible by means of this methodology. Due to its broad substrate scope (that has been demonstrated by several

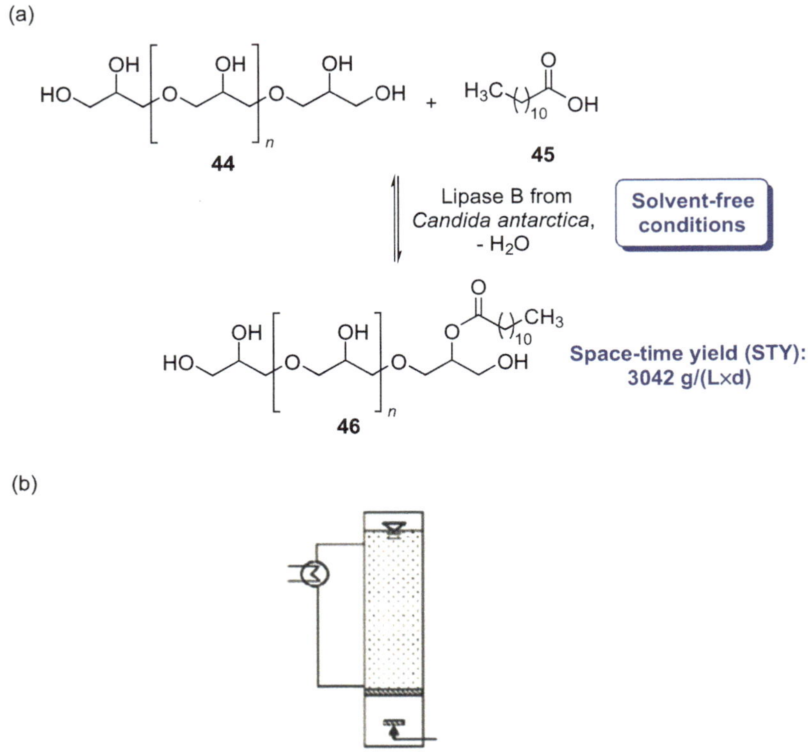

SCHEME 2.19
(a) Solvent-free lipase-catalyzed synthesis of polyglycerol-3 laurate. (b) Reactor concept of the bubble column reactor. *(Figure in part (b) printed from reference [53].)*

groups including the Whitesides group [55]) as well as the high efficiency of such enzymes, this method gained industrial attention even at an early stage. Boehringer Ingelheim Pharmaceuticals developed a pilot production route to the aliphatic L-amino acid L-**48**, required as an intermediate for the synthesis of a protease inhibitor, using this enzyme in a batch mode (Scheme 2.20) [56]. In the presence of the aminoacylase, the resolution of racemic substrate *rac*-**47** that was used as a slurry in water, proceeds under formation of the desired L-amino

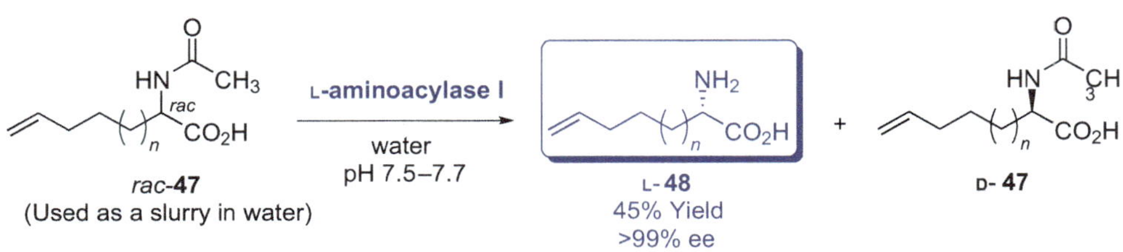

SCHEME 2.20

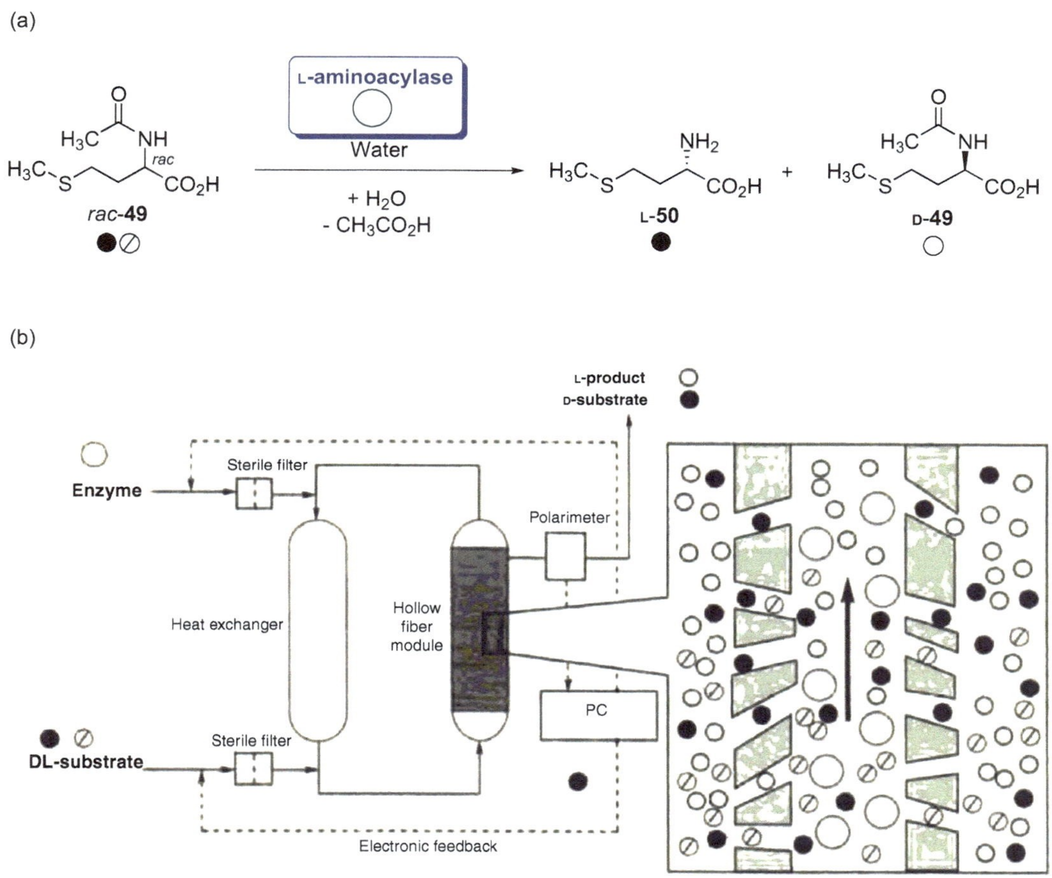

SCHEME 2.21
(a) Synthesis of L-methionine *via* L-aminoacylase-catalyzed resolution. (b) Process concept of the enzyme membrane reactor for L-amino acid synthesis. *(Figure in part B printed from Ref. [54].)*

acid L-**48** in more than or equal to 45% yield and with an excellent enantioselectivity of more than 99% ee.

In contrast, Degussa AG developed a continuous process for manufacturing L-methionine on a multihundred ton scale utilizing aminoacylase in a membrane reactor [54,57,58]. This type of reactor enables simple recovery and separation of the enzyme from the reaction mixture. In this way, the strategy combines advantages of heterogeneous catalysis (namely easy recovery and reuse of the catalyst) with homogeneous catalysis (namely use of dissolved catalyst and avoidance of mass transfer limitations) (Scheme 2.21).

In general, continuously running process are of particular interest when it comes to regular and permanent applications at a large scale, which ideally are conducted in a sophisticated production plant for a specific product. In turn, a batch process is often the preferred option when a product should be prepared in a multipurpose plant within a short production time and with irregular production cycles. Thus, not only the concept and efficiency of the biotransformation itself but also the application influences the decision whether a process is preferably carried out in a batch or continuous mode.

9 SUMMARY

Numerous synthetic tools are available for organic chemists wishing to design efficient reactions for the synthesis of valuable organic compounds using biocatalysts. Besides identifying a suitable enzyme, one must also consider whether a suitable enzyme formulation is available (in the form of whole cells or enzymes, whether free or immobilized). Additional desirable aspects from an organic synthesis perspective are high product concentrations, conversions, selectivities (enantio-, diastereo-, regio-, and chemoselectivity) and space–time yields. Downstream processing plays a crucial role in moving from an enzyme identified in a screening effort to a practical and economically attractive process. Such process criteria have already been successfully addressed already in many cases, thus leading to the development of a broad range of both, lab, and industrial-scale processes.

References

[1] U.T. Bornscheuer, G.W. Huisman, R.J. Kazlauskas, S. Lutz, J.C. Moore, K. Robins, Nature 485 (2012) 185–194.

[2] K. Drauz, H. Gröger, O. May (Eds.), Enzyme Catalysis in Organic Synthesis, vol. 1–3, Wiley-VCH, Weinheim, 2012.

[3] J. Pleiss, in: K. Drauz, H. Gröger, O. May (Eds.), Enzyme Catalysis in Organic Synthesis, vol. 1, Wiley-VCH, Weinheim, 2012, pp. 89–117 Chapter 4.

[4] M.T. Reetz, in: K. Drauz, H. Gröger, O. May (Eds.), Enzyme Catalysis in Organic Synthesis, vol. 1, Wiley-VCH, Weinheim, 2012, pp. 119–190 Chapter 5.

[5] Y. Hirose, in: K. Drauz, H. Gröger, O. May (Eds.), Enzyme Catalysis in Organic Synthesis, vol. 1, Wiley-VCH, Weinheim, 2012, pp. 191–215 Chapter 6.

[6] K. Faber, Biotransformations in Organic Chemistry, fifth ed., Springer, Berlin, (2004) Chapter 2.1, p. 29–176.

[7] M. Breuer, K. Ditrich, T. Habicher, B. Hauer, M. Keßeler, R. Stürmer, T. Zelinski, Angew. Chem.116 (2004) 806–843; Angew. Chem. Int. Ed. 43 (2004) 788–824.

[8] F. Balkenhohl, K. Ditrich, B. Hauer, W. Ladner, J. Prakt. Chem. 339 (1997) 381–384.

[9] K. Ditrich, Synthesis (2008) 2283–2287.

[10] A. Weckbecker, H. Gröger, W. Hummel, Adv. Biochem. Engin./Biotechnol. 120 (2010) 195–242.

[11] H. Wu, C. Tian, X. Song, C. Liu, D. Yang, Z. Jiang, Green Chem. 15 (2013) 1773–1789.

[12] W. Hummel, H. Gröger, J. Biotechnol. 191 (2014) 22–31.

[13] W. Zhang, K. O'Connor, D.I.C. Wang, Z. Li, Appl. Environm. Microb. 75 (2009) 687–694.

[14] K. Goldberg, K. Edegger, W. Kroutil, A. Liese, Biotech. Bioeng. 95 (2006) 192–198.

[15] T. Stillger, M. Bönitz, M.V. Filho, A. Liese, Chem. Ing. Tech. 74 (2002) 1035–1039.

[16] A.S. Bommarius, M. Schwarm, K. Drauz, J. Mol. Catal. B: Enzym. 5 (1998) 1–11.

[17] M. Höhne, U.T. Bornscheuer, in: K. Drauz, H. Gröger, O. May (Eds.), Enzyme Catalysis in Organic Synthesis, vol. 2, Wiley-VCH, Weinheim, 2012, pp. 779–820 Chapter 19.

[18] C.K. Savile, J.M. Janey, E.C. Mundorff, J.C. Moore, S. Tam, W.R. Jarvis, J.C. Colbeck, A. Krebber, F.J. Fleitz, J. Brands, P.N. Devine, G.W. Huisman, G.J. Hughes, Science 329 (2010) 305–309.

[19] A. A. Desai, Angew. Chem. 123 (2011) 2018–2020; Angew. Chem. Int. Ed. 50 (2011) 1974–1976.

[20] J. Sukumaran, U. Hanefeld, Chem. Soc. Rev. 34 (2005) 530–542.

[21] M. Pohl, C. Dresen, M. Beigi, M. Müller, in: K. Drauz, H. Gröger, O. May (Eds.), Enzyme Catalysis in Organic Synthesis, vol. 2, Wiley-VCH, Weinheim, 2012, pp. 919–945 Chapter 22.

[22] B. Schäfer, Naturstoffe der chemischen Industrie, Elsevier, München, (2007) p. 389.

[23] O. May, P.T. Nguyen, F.H. Arnold, Nat. Biotechnol. 18 (2000) 317–320.

[24] O. May, S. Verseck, A. Bommarius, K. Drauz, Org. Process Res. Dev. 6 (2002) 452–457.

[25] T. Boller, C. Meier, S. Menzler, Org. Process Res. Dev. 6 (2002) 509–519.

[26] E. Katchalski-Katzir, D.M. Kraemer, J. Mol. Catal. B Enzym. 10 (2000) 157–176.

[27] K. Buchholz, V. Kasche, U.T. Bornscheuer, Biocatalysts and Enzyme Technology, Wiley-VCH, Weinheim, (2005).

[28] L.O. Wiemann, R. Nieguth, M. Eckstein, M. Naumann, O. Thum, M.B. Ansorge-Schumacher, ChemCatChem 1 (2009) 455–462.

[29] L.O. Wiemann, P. Weisshaupt, R. Nieguth, O. Thum, M.B. Ansorge-Schumacher, Org. Process Res. Dev. 13 (2009) 617–620.

[30] M.B. Ansorge-Schumacher, O. Thum, Chem. Soc. Rev. 42 (2013) 6475–6490.

[31] R. Nieguth, M. Eckstein, L.O. Wiemann, O. Thum, M.B. Ansorge-Schumacher, Adv. Synth. Catal. 353 (2011) 2522–2528.

[32] A. Scholz, M. Eckstein, M.B. Ansorge-Schumacher, ChemCatChem 5 (2013) 815–821.

[33] M.D. Truppo, H. Strotman, G. Hughes, ChemCatChem 4 (2012) 1071–1074.

[34] M. Heidlindemann, G. Rulli, A. Berkessel, W. Hummel, H. Gröger, ACS Catal. 4 (2014) 1099–1103.

[35] G.E. Jeromin, Biotechnol. Lett. 31 (2009) 1717–1721.

[36] H. Gröger, W. Hummel, R. Metzner, in: E.M. Carreira, H. Yamamoto (Eds.), Synthetic Methods/Comprehensive Chirality, vol 7, Elsevier, Amsterdam, 2012, pp. 181–215 Chapter 7.10.

[37] M. Paravidino, P. Böhm, H. Gröger, U. Hanefeld, in: K. Drauz, H. Gröger, O. May (Eds.), Enzyme Catalysis in Organic Synthesis, vol. 1, Wiley-VCH, Weinheim, 2012, pp. 251–362 Chapter 8.

[38] S. Buchholz, H. Gröger, in: R.N. Patel (Ed.), Biocatalysis in the Pharmaceutical and Biotechnology Industries, CRC Press, New York, 2006, pp. 829–847 Chapter 34.

[39] T. Itoh, K. Kuroda, M. Tomosada, Y. Takagi, J. Org. Chem. 56 (1991) 797–804.

[40] S. Hu, S. Kelly, S. Lee, J. Tao, E. Flahive, Org. Lett. 8 (2006) 1653–1655.

[41] J. Kuper, T.S. Wong, D. Roccatano, M. Wilmanns, U. Schwaneberg, J. Am. Chem. Soc. 129 (2007) 5786–5788.

[42] P. Dominguez de Maria, C.A. Garcia-Burgos, G. Bargeman, R.W. van Gemert, Synlett (2007) 1439–1452.

[43] T. Purkarthofer, W. Skranc, C. Schuster, H. Griengl, Appl. Microbiol. Biotechnol. 76 (2007) 309–320.

[44] H. Griengl, A. Hickel, D.V. Johnson, C. Kratky, M. Schmidt, H. Schwab, Chem. Commun. (1997) 1933–1940.

[45] T. Purkarthofer, W. Skranc, H. Weber, H. Griengl, M. Wubbolts, G. Scholz, P. Pochlauer, Tetrahedron 60 (2004) 735–739.

[46] A. Liese, K. Seelbach, C. Wandrey, Industrial Biotransformations, Wiley-VCH, Weinheim, (2006) p. 311.

[47] F. Effenberger, T. Ziegler, S. Förster, Angew. Chem. 99 (1987) 491–492; Angew. Chem. Int. Ed. Engl. 26 (1987) 458–460.

[48] U. Niedermeyer, M.-R. Kula, Angew. Chem. 102 (1990) 423–425; Angew. Chem. Int. Ed. Engl. 29 (1990) 386–387.

[49] A. Liese, K. Seelbach, C. Wandrey, Industrial Biotransformations, Wiley-VCH, Weinheim, (2006).

[50] U. Kragl, M. Eckstein, N. Kaftzik, Curr. Opin. Biotechnol. 13 (2002) 565–571.

[51] H. Gröger, F. Chamouleau, N. Orologas, C. Rollmann, K. Drauz, W. Hummel, A. Weckbecker, O. May, Angew. Chem. 118 (2006) 5806–5809; Angew. Chem. Int. Ed. 45 (2006) 5677–5681.

[52] H. Gröger, O. May, H. Hüsken, S. Georgeon, K. Drauz, K. Landfester, Angew. Chem. 118 (2006) 1676–1679; Angew. Chem. Int. Ed. 45 (2006) 1645–1648.

[53] L. Hilterhaus, O. Thum, A. Liese, Org. Process Res. Dev. 12 (2008) 618–625.

[54] H. Gröger, K. Drauz, Methods for the Enantioselective Biocatalytic Production of L-Amino Acids on an Industrial Scale, in: E. Schmidt, H.U. Blaser (Eds.), Asymmetric Catalysis on Industrial Scale, Wiley-VCH, Weinheim, 2004, pp. 131.

[55] H.K. Chenault, J. Dahmer, G.M. Whitesides, J. Am. Chem. Soc. 111 (1989) 6354–6364.

[56] X.-j. Wang, L. Zhang, L.L. Smith-Keenan, I.N. Houpis, V. Farina, Org. Process Res. Dev. 11 (2007) 60–63.

[57] C. Wandrey, E. Flaschel, Adv. Biochem. Eng. 12 (1979) 147–218.

[58] C. Wandrey, R. Wichmann, W. Leuchtenberger, M.R. Kula, US 4,304,858, 1981.

CHAPTER 3

Biocatalysis in Organic Solvents, Supercritical Fluids and Ionic Liquids

Chen Cao, Tomoko Matsuda
Department of Bioengineering, Graduate School of Bioscience and Biotechnology, Tokyo Institute of Technology, Nagatsuta-cho, Midori-ku, Yokohama, Japan

1 INTRODUCTION

Enzymes catalyze reactions in aqueous solvents in natural environments in living organisms. However, the aqueous solvents are not always best when they are used as a catalyst for organic synthesis [1–3]. The initial success of using enzymes in nonaqueous solvents or in mixtures of aqueous solvent and water miscible organic solvents includes the improvement in the rate by dissolving hydrophobic substrates and the discovery of enzymatic reactions in organic solvents, which are not feasible in aqueous solvents. For example, nonaqueous solvents have been widely used to shift the equilibrium positions of the reactions as shown in Figure 3.1. Lipase-catalyzed esterification and transesterification can be conducted effectively only in nonaqueous solvents by eliminating water from the system to drive the equilibrium to the desired synthetic direction (Figure 3.1). Interestingly, a very small concentration of water is usually necessary to maintain the enzyme activity. Alcohol dehydrogenase catalyzed reductions of carbonyl group in aqueous buffer can be conducted in the presence of a large excess of 2-propanol that can act as both a solvent and a hydride donor to recycle the nicotinamide cofactor using the same enzyme. Finally, decarboxylases can operate in the reverse (carboxylation) direction when CO_2 is employed as a cosolvent. The interest in nonaqueous solvents developed with these discoveries and also partly because enzymes have been replacing traditional chemical catalysts that have been conducted in organic solvents. Now, biocatalysis in nonaqueous solvents has been widely investigated mainly using not only hydrolases such as lipases but also oxidoreductase, etc.

2 KEY ATTRIBUTES OF REACTIONS IN NONAQUEOUS SOLVENTS

Wide varieties of reaction systems using nonaqueous solvents have been studied and developed due to the advantages shown in this section. However, there are also some disadvantages that have been solved partly with recent extensive studies.

Organic Synthesis Using Biocatalysis. http://dx.doi.org/10.1016/B978-0-12-411518-7.00003-2

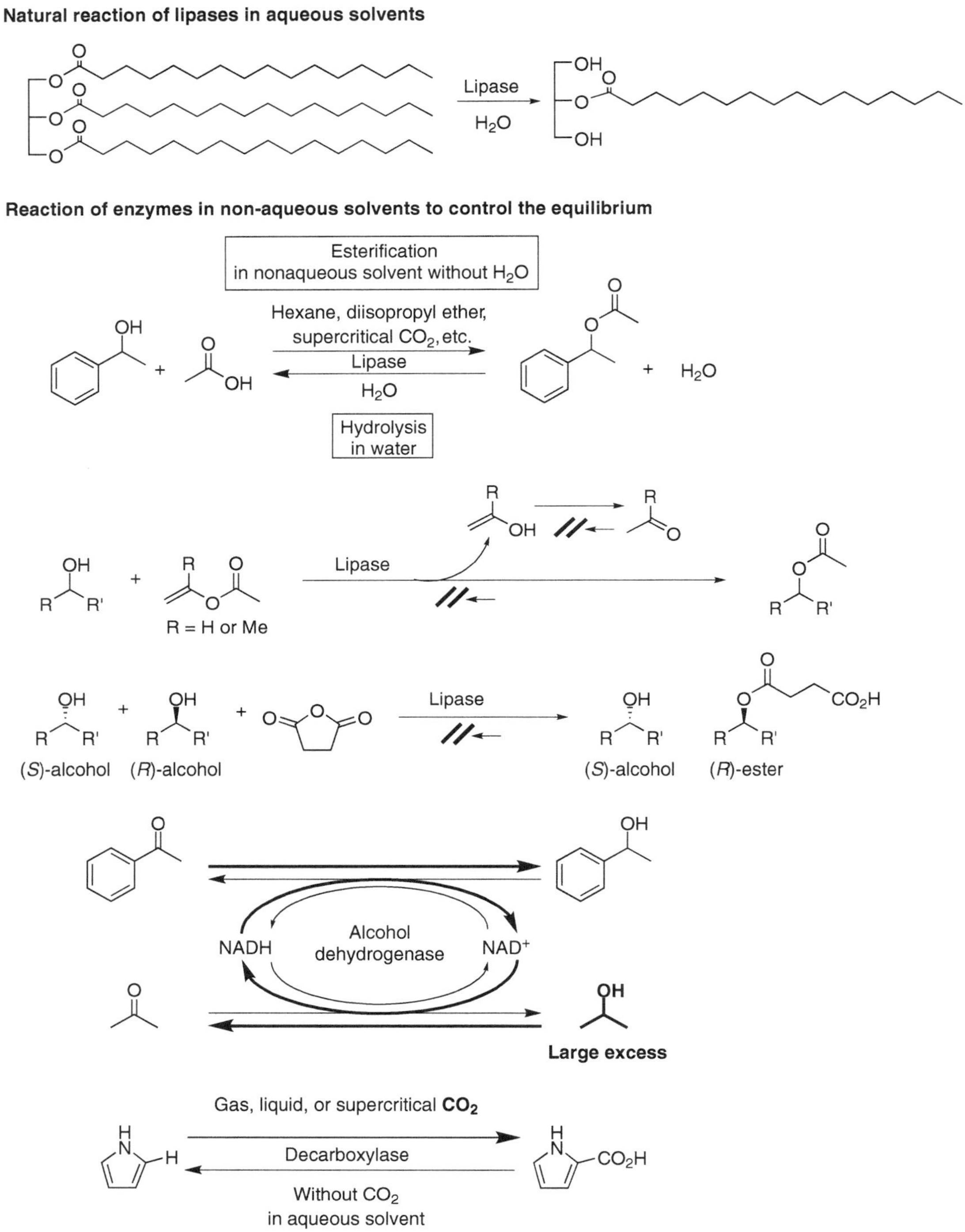

FIGURE 3.1
Nonaqueous solvents to shift the equilibrium position of the reaction.

Advantages

- Catalysis of a wide variety of chemical reactions that are not feasible in aqueous media, for example, unfavourable thermodynamic equilibrium in water as shown in Figure 3.1.
- Easy work-up procedure to separate biocatalyst and organic solvent containing products followed by the evaporation of the organic solvent. (Organic solvent is easy to be evaporated in reduced pressure, and supercritical CO_2 can be removed very easily with depressurization, whereas separation of water and product is not easy.)
- Increased solubility of hydrophobic substrates in hydrophobic nonaqueous solvents.
- Suppression of water-dependent side reactions.
- Improvement and control of chemo-, regio-, and stereo-selectivities.
- Recovery and reusability of enzyme even without immobilization.
- Alteration of the partition of substrates/products: aids separations, improve yields, elimination of substrate, and product inhibition of biocatalysts.
- Enhancement of thermostability of biocatalyst in nearly anhydrous organic solvent system.
- Elimination of microbial contamination.

Disadvantages

- Inactivation of enzymes.
- Labour and cost-intensive preparation of immobilized biocatalysts.
- Mass-transfer limitations in the case of biphasic systems or viscous solvents.
- Water activity control needed for processes involving condensation reactions.

3 KINDS OF NONAQUEOUS SOLVENTS USED FOR BIOCATALYSIS

The choice of solvents for enzymatic reaction has been widened from organic solvents to various types of solvents such as supercritical fluids, ionic liquids, etc. The enzymatic reaction in organic solvent has been reported already in 1970s, the first biocatalysis in ionic liquids [4] was in 2000, and the first biocatalysis in supercritical fluids [5] was in 1985. Currently four kinds of liquid or fluid solvents, aqueous, organic solvents, ionic liquids, and supercritical fluids, are available for biocatalysis as shown in Figure 3.2. Moreover, biphasic or triphasic solvent systems consisting of two or more kinds of the solvents are also often employed for biocatalysis. Solid phase of immobilized enzymes and/or hydrophobic polymer to adsorb substrate and product may also exist. The performance of a biocatalyst depends significantly on the solvent system. The best medium should provide optimum reaction rates and simplify work-up procedure to make the process both economical and environment friendly.

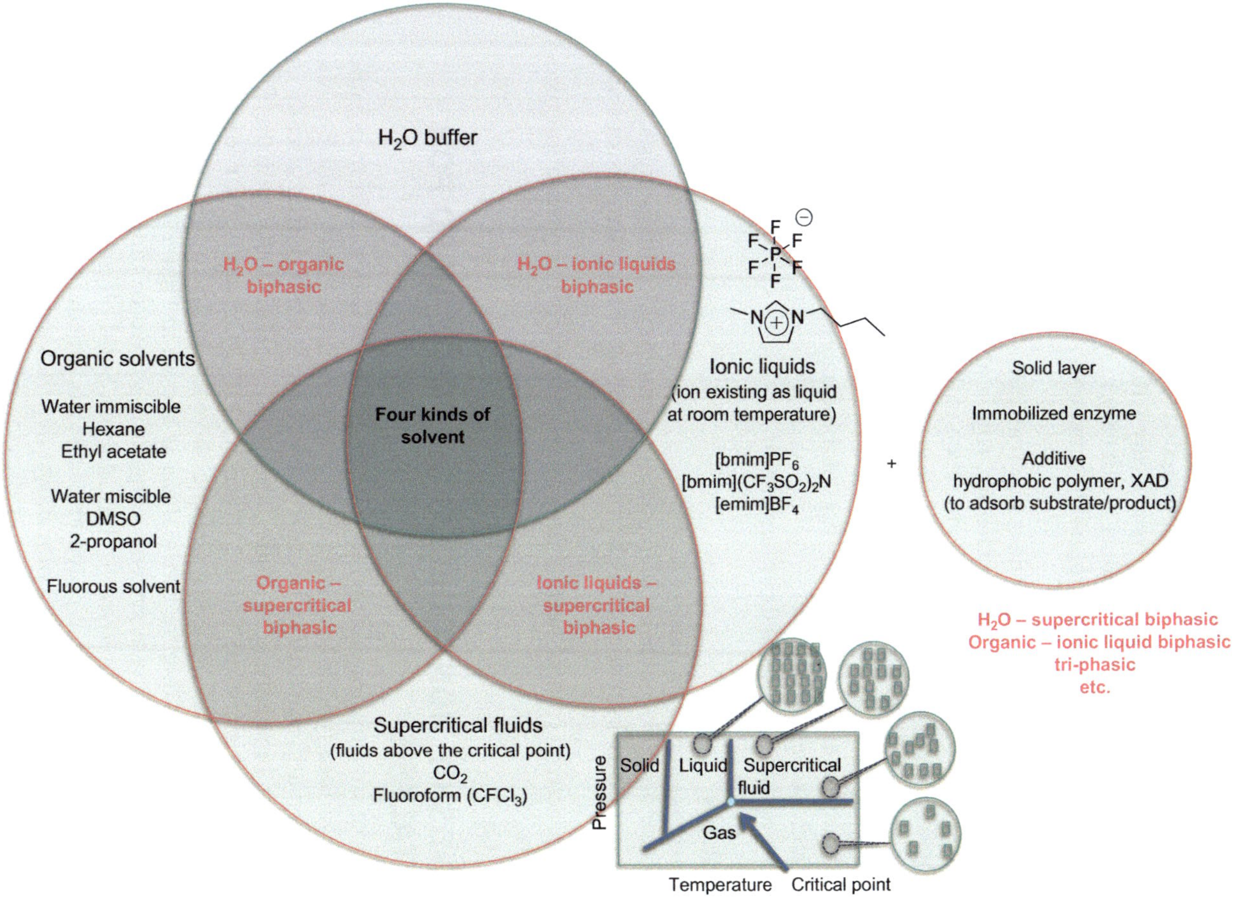

FIGURE 3.2
Four kinds of liquid or fluid solvents and solid layer for biocatalysis.

The properties and characteristics of nonaqueous solvents are as follows. Organic solvents are most common for the nonaqueous solvents for biocatalysis. For example, water immiscible solvents such as hexane, ethyl acetate, and isopropyl ether, and water miscible solvents such as DMSO and 2-propanol have been used widely. The concentration of the organic solvents has to be tuned to optimize the reaction to dissolve substrate and not to inhibit enzyme.

One of the interesting organic solvents system recently developed is fluorous solvents that dissolve highly fluorinated materials. Its phase separations with conventional organic solvent depend on the temperature, so that separation of the product can be efficiently conducted. For example, in fluorous biphasic system, at low temperature, the fluorous and organic phases are immiscible. On warming above the consolute temperature, the two phases become miscible and a homogeneous phase forms allowing a reaction to occur. On cooling, the phases separate and the catalyst is retained in the fluorous phase, where it is easily separated from the product and recycled [6].

Supercritical (sc) fluid is a fluid above its critical point with gas-like low viscosity and high diffusivity to accelerate diffusion-limited reaction and liquid-like solubilizing power. Most of the gas at ambient conditions has suitable critical points for biocatalysis, whereas the critical points for liquids such as H_2O and alcohols are too high for biocatalysis. The supercritical fluid when used as a solvent makes the separation of the product from solvent very easy because the fluid becomes gas and evaporates just by depressurization.

For example, the critical point of CO_2 is 31°C and 7.4 MPa (73 atm), and that of CHF_3 is 26°C and 4.9 MPa (48 atm). CO_2, existing abundantly, is an environment-friendly solvent. The polarity of supercritical CO_2 is lower than hexane, so that hydrophobic substrate dissolves well in supercritical CO_2, while cosolvent may be necessary to dissolve hydrophilic compounds. CHF_3 is an interesting solvent because of its great tunability of properties; small changes in pressure or temperature lead to significant changes in density-dependent solvent properties such as the relative permittivity, the solubility parameter, and the partition coefficient. For example, the polarity of supercritical CHF_3 changes largely by changing the temperature and pressure.

Ionic liquid consists of ion pair and is a liquid at the ambient conditions. It is a nonvolatile solvent, so that it can be used forever theoretically. The use of ionic liquids as solvents in biocatalysis has significant potential due to some of their excellent properties, such as nonflammability, low toxicity, and high solubility for both organic and inorganic materials.

Hydrophobic polymer (XAD, etc.) can be an additive for the reaction in aqueous buffer, but the role can be considered to be same as that for solvent. As shown in Figure 3.3, it promotes solubilization of hydrophobic substrates by adsorbing them reversibly onto the surface of the polymer. The substrate, after desorption from the surface of the polymer, is converted enzymatically to product that is then adsorbed on the surface of the polymer. In so doing, it reduces the substrate and product concentrations in the aqueous layer that often prevents substrate and product inhibition and/or enzyme inactivation.

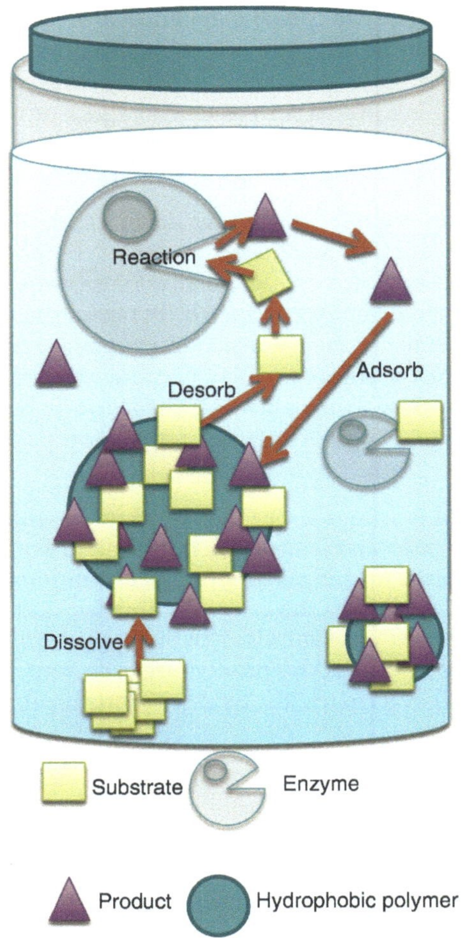

FIGURE 3.3
Role of hydrophobic polymer as an additive to adsorb hydrophobic compounds.

4 SOLUBILITY OF ENZYMES, SUBSTRATES, AND PRODUCTS IN NONAQUEOUS SOLVENTS

When many layers coexist in a reaction system, the partition coefficients of substrates and products between the different solvents must be carefully considered. Hydrophobic substrates and products partition preferentially into organic solvents, supercritical CO_2 and adsorb strongly onto hydrophobic polymers. By contrast, enzymes nearly always remain in an aqueous environment. It is also important to note that the partition coefficients of the substrate and product often differ. For example, as shown in Figure 3.4, 1-phenylethanol is more soluble in the aqueous layer than acetophenone, although it should be noted both species have much higher solubilities in the organic versus the aqueous layer.

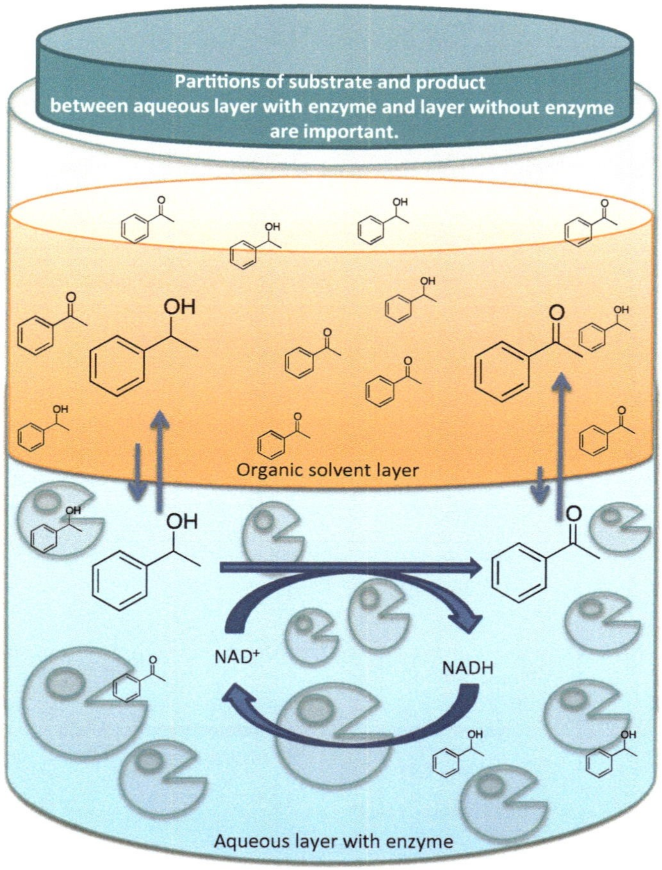

FIGURE 3.4

Partition of substrate and product between organic and aqueous layers [1d]. *Experimental procedure:* In a typical experiment in water, racemic 1-phenylethanol (0.08 mmol) was added to *G. candidum* NBRC 4597 wet cell (0.5 g) in water (3.0 mL), and shaken for 24 h at 30°C. The alcohol was oxidized to give a mixture of the alcohol (88%) and the corresponding ketone, acetophenone, in 12% yield. In a typical experiment in hexane, racemic 1-phenylethanol (0.08 mmol) and cyclohexanone (0.2 mmol) was added to the immobilized cell (wet cell (0.5 g) in water (3.0 mL) was immobilized on the water-adsorbent polymer, BL-100, (0.5 g)) in hexane (6.0 mL), and shaken for 24 h at 30°C. The biocatalyst was filtered and washed with ether and the filtrates were combined. The alcohol was oxidized to give a mixture of the alcohol (50%, 99% ee (*R*)) and the corresponding ketone (50%) determined using a chiral GC-column. The reaction in the organic solvent without adding cyclohexanone gave a mixture of the alcohol (52%, 86% ee) and the corresponding ketone in 48% yield [1d].

This can be a problem if it leads to insufficient substrate concentrations in the aqueous layer for efficient bioconversion rates, whereas this can give benefit if the substrate and product inhibition to the enzyme is problematic. For example, in the oxidation of 1-phenylethanol by alcohol dehydrogenase from *Geotrichum candidum*, adding hexane to the aqueous buffer increases the partition of the

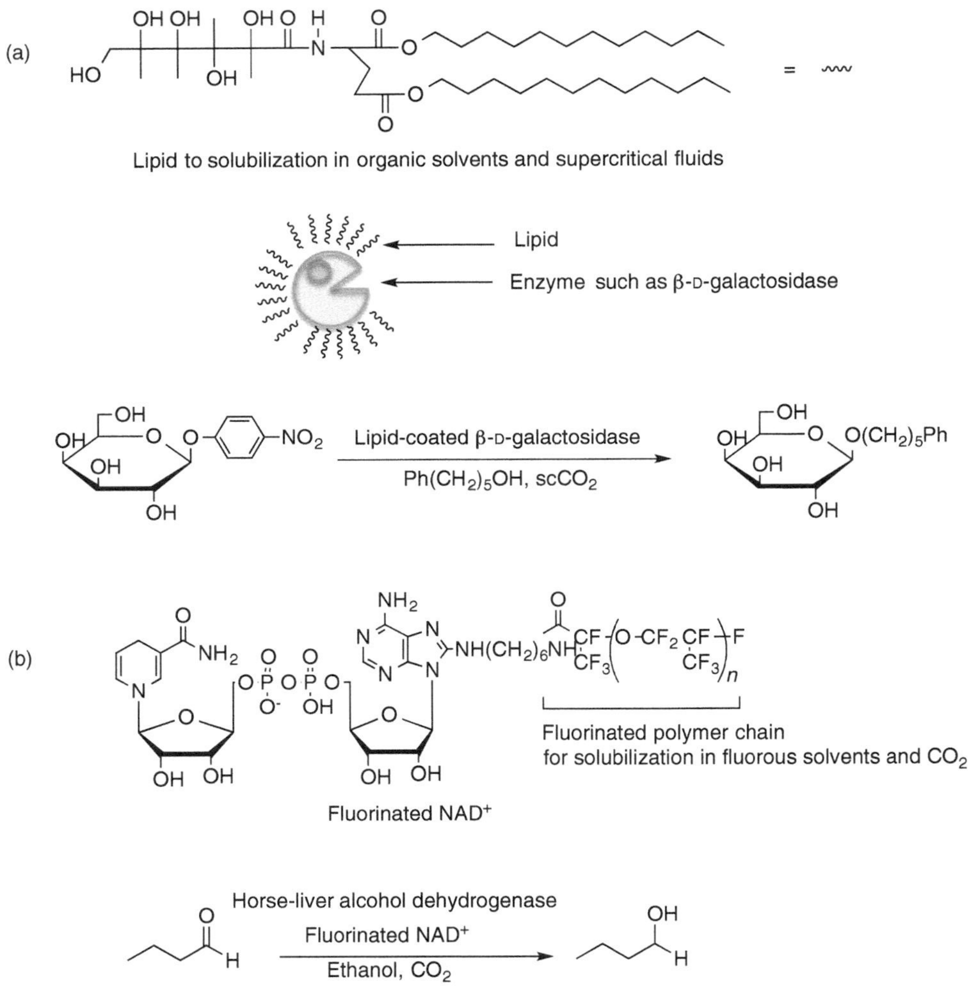

FIGURE 3.5
Solubilization of (a) enzymes [7] and (b) coenzyme [8] in nonaqueous solvents.

substrate 1-phenylethanol into the aqueous phase (containing the enzyme) and also increases the removal of the product acetophenone from the aqueous phase due to its higher partition to the organic phase (hexane) and thereby making the reaction more favorable [1d].

As noted previously, enzymes and cofactors do not normally dissolve in nonaqueous solvents. This can be altered by coating or covalently modifying the surfaces of these species with materials (often polymers) that are miscible with nonaqueous solvents. For example, lipid-coated enzymes can be dissolved in organic solvents and supercritical fluids (Figure 3.5a) [7]. Fluorinated coenzyme can be dissolved in fluorous solvents and supercritical CO_2 (Figure 3.5b) [8].

5 STABILITY OF ENZYMES IN NONAQUEOUS SOLVENTS

Enzyme stability under nonaqueous conditions depends on (1) the nature of the enzymes, (2) whether they are free or immobilized, (3) the presence or absence of additives such as sugars, (4) on the water content, and (5) type of nonaqueous solvent. There are no general rules that can be used by chemists and enzymologists to judge the likely stabilities of enzymes they might use since the prediction of the stability from the amino acid sequence of the enzyme has not been possible yet. However, there are accumulations of experimental data in the literature, so that there is an increasing number of ready-to-use and stable biocatalysts in nonaqueous solvent, and they are commercially available as a reagent for organic chemists.

5.1 The Nature of the Enzymes

Enzymes that are used in a natural environment with high concentrations of hydrophobic organic molecules are often tolerant toward hydrophobic organic solvents. For example, lipases that catalyze hydrolysis of triacylglycerol in nature, work in an environment with a high concentration of triacylglycerols, so they have high stability toward organic compounds or solvents with properties similar to triacylglycerol. Immobilized lipases generally have excellent stabilities in organic solvents; for example, an immobilized *Candida antarctica* lipase B, Novozym 435, and an immobilized *Burkholderia cepacia* (formally *Pseudomonas cepacia*) lipase, Amano PS, are among the most frequently used biocatalyst in nonaqueous solvents for organic synthesis. These lipases had been used widely for organic synthesis in nonaqueous solvents because they have excellent enantioselectivities beside the excellent stabilities. Other immobilized lipases have also been used for organic synthesis, especially for chiral synthesis.

An understanding at the molecular level of how enzymes can preserve their structure and activity in nonnatural environment is very important [9a]. Molecular dynamics (MD) computer simulations were used to study molecular aspects of the stability of the *C. antarctica* lipase B in supercritical CO_2. It was found that water and supercritical CO_2 cover the protein surface quite heterogeneously (Figure 3.6). The more hydrophobic region of the active site is mainly solvated by CO_2, whereas the region across the active site is mainly solvated by water. In the first solvation layer, the hydrophilic residues at the surface of the protein are able to pin down patches of water, whereas CO_2 solvates preferentially hydrophobic surface residues. In the outer solvation shells, water molecules tend to cluster predominantly on top of the larger water patches of the first solvation layer. This exposes the substrate-binding region of the enzyme to CO_2, possibly facilitating diffusion of nonpolar substrates into the catalytic funnel, resulting in the enhanced accessibility of hydrophobic substrates to the active site, while preserving the functional structure of the enzyme.

The experimental data to describe the structures and structural changes that occur when enzymes are placed in supercritical CO_2 unfortunately cannot be taken

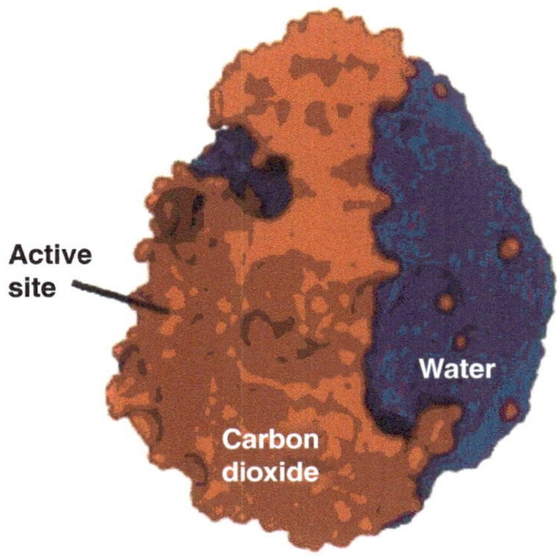

FIGURE 3.6
Molecular surface of the solvent around *C. antarctica* lipase B showing a heterogeneous distribution of H_2O and CO_2. *(Reprinted with permission from American Chemical Society [9a].)*

because it is under high pressure. However, it is shown that the rate of the reaction by the immobilized *C. antarctica* lipase B in supercritical CO_2 for kinetic resolution of *rac*-1-phenylethanol derivatives is improved in supercritical CO_2 [9b]. Here, three different preparations of the lipase, immobilized enzyme on resin (Novozym 435), and cross-linked enzyme aggregate (CLEA 103 and CLEA 301) were used. Section 11 shows large-scale continuous kinetic resolution of alcohols using scCO_2-flow reactor using Novozym 435.

Another class of enzyme with excellent stability in nonaqueous solvent is from thermophiles [10]. High levels of thermal stability are positively correlated with stability in the presence of denaturing agents, including detergents and organic solvents [10a]. Such enzymes have excellent thermostability with rigid structure that also gave excellent stabilities in nonaqueous solvents. Therefore, many enzymes from thermophiles have been used for biocatalysis [10a–d].

The close relationship between the thermostability and organic-solvent tolerance were also proved with enzyme engineering such as rational and random mutations. First, an example of the rational mutations to improve both the thermostability and organic-solvent tolerance is shown. A protein engineering approach based on analyses of structural and sequence data was taken. As a result, the modulation of one specific loop segment near the active site of *Zymomonas mobilis* ene reductase showed an increased tolerance to organic solvents along with an enhanced thermostability [10e]. Further, an example of the random mutations to improve both the thermostability and organic-solvent tolerance is shown. With random mutations, the enzyme is randomly mutated to create thousands of mutant libraries, and then, the mutants with profitable properties are selected with high-throughput screening methods. Such laboratory evolution of the enzyme is often called "directed evolution." With such random approach, mutants of the lipase from *Bacillus subtilis*, previously engineered for enhanced thermostability using directed evolution, was found to show significantly increased tolerance to organic solvents also [10f].

5.2 Free Enzyme versus Immobilized Enzyme

Immobilization can improve enzyme stability under nonaqueous conditions [11]. Enzymes can be immobilized covalently or physically on/in various supports including organic or inorganic polymers. For example, one of the most important immobilized enzyme forms, cross-linked enzyme aggregates (CLEAs), can be prepared by physical aggregation of the enzyme followed by the reaction of surface NH_2 groups on enzyme with a bifunctional chemical cross-linker, such as glutaraldehyde [11a]. Various kinds of CLEAs have been prepared and used in aqueous as well as nonaqueous solvents with the first examples of CLEAs being derived from penicillin G amidase, an industrially important enzyme used in the synthesis of semisynthetic penicillin and cephalosporin antibiotics. Although the free enzyme is known to have a limited thermal stability and a very low tolerance toward organic solvents, CLEAs of the enzyme catalyzed the synthesis of ampicillin in a broad range of organic solvents [11d]. Another example shows that pyruvate decarboxylase stability in a carbon dioxide/aqueous biphasic system was improved by immobilization on ion-exchange polymer (Figure 3.7) [11e]. Other examples, including a sol–gel-encapsulated alcohol dehydrogenase-catalyzed asymmetric reduction and continuous sustainable processes for enzymatic synthesis of biodiesel in hydrophobic ionic liquids/supercritical carbon dioxide biphasic systems, are shown later. Physical immobilization of cells containing the enzyme on the surface of water-absorbing polymer with hydrogen bonding is shown in Figure 3.4. In that case, the immobilized cell can be only used in nonaqueous solvents. Further details on immobilization of enzymes and their applications are discussed separately in Chapter 4.

5.3 The Presence or Absence of Additives such as Sugars

Enzyme stabilization can be achieved also by additives such as sugars. Enzymes are most stable in the cell where they are produced, and isolated and purified enzymes are generally less stable. Though stability can be increased by low temperature and freezing, the stability of frozen isolated and purified enzymes are still not great. To enhance stability under aqueous conditions, polyols and sugars have been used by the enzymologists. Additives such as sugars can be also used for the organic synthesis to stabilize the enzyme in nonaqueous solvents. As shown in Figure 3.7, pyruvate decarboxylase stability in a carbon dioxide/aqueous biphasic system was improved by adding additive such as glycerol and trehalose as well as by immobilization.

5.4 Water Content

The water content is one of the most important factors effecting enzyme stability. Enzymes always need essential water molecules, so complete dehydration

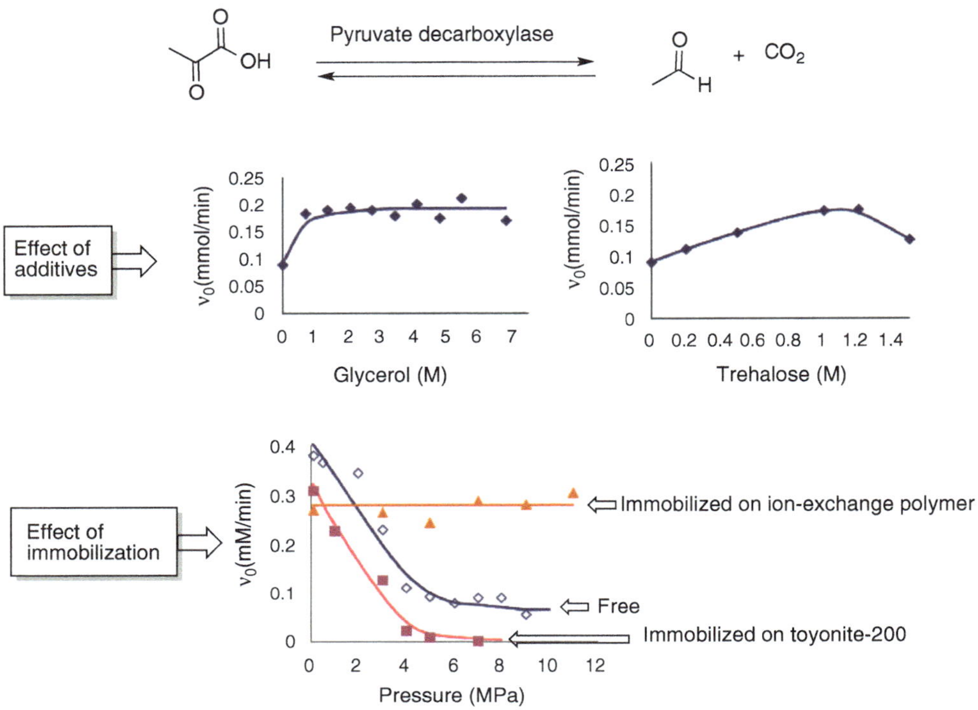

FIGURE 3.7
Stabilization of pyruvate decarboxylase by additive and immobilization in a pressurized CO_2/H_2O biphasic system. High-pressure treatment of the enzyme and analysis of residual activity were conducted [11b].

always deactivates enzymes. The water content most suitable for enzyme stabilization depends largely on the kind of enzyme. For example, the immobilized lipases are more stable under very limited water content than in aqueous solvents. A lipase can function from 100 °C to −80 °C in an organic solvent [12], whereas lipase inactivate soon in boiling water at 100 °C. When a lipase is used for esterification, water produced by the reaction can also cause the reverse reaction (hydrolysis) as shown in Figure 3.1. Therefore, a vinyl ester, acid anhydride, or so forth is often used to avoid water production and thereby shift equilibrium to the desired products (Figure 3.1). Vinyl acetate and solvents such as hexane are usually dehydrated using molecular sieves; however, the handling of the theses compounds is usually done under air, since complete dehydration is not suitable.

Although lipases do not need bulk water in the reaction system, generally, other enzymes need more water. For example, alcohol dehydrogenase can work in hexane but it is immobilized with a lot of water, so it is a two-layer reaction as shown in Figure 3.4.

5.5 Type of Nonaqueous Solvent

The type of nonaqueous solvent largely affects the stability of the biocatalyst. Generally, when the water on the surface of the enzyme is completely removed, this alters hydrogen-bond patterns, changing the protein structure and deactivating the enzyme. On the other hand, if the hydrogen bonding by water in a natural environment is replaced by hydrogen bonding by other solvents, the enzyme can keep its active structure. Therefore, the hydrophobicity of the solvents, or the ability to not destroy the hydrogen bonding, is very important. In the case of lipases, the reaction proceeded smoothly in hydrophobic solvents such as hexane, isopropyl ether, and supercritical and liquid CO_2.

6 SELECTION OF NONAQUEOUS SOLVENTS FOR A DESIRED REACTION

To select the most suitable solvent for a desired reaction, screening of various solvents is usually necessary, since the prediction of the best solvents by the simulation is still difficult. The screening has to be conducted to optimize the reaction conditions with respect to (1) the type and scale of the reaction, (2) the stability of the biocatalyst, (3) the hydrophobicity of the solvents, (4) the solubilities of the substrate and product, (5) the recovery of the product and enzyme, (6) environmental and safety concerns, and (7) the cost of the biocatalyst, substrate, and product. For example, to conduct a dehydration reaction using hydrolytic enzyme, a water-free nonaqueous solvent must be selected instead of water. On the other hand, the level of the dehydration should not be too much, as noted previously, to take away the water from enzyme necessary for its activity. Supercritical CO_2 is better to be used in relatively large scale using a flow reactor than using in the very small-scale batch reaction because product recovery in the small-scale reaction in batch CO_2 reactor needs extraction with organic solvents.

The reaction rate and selectivity also largely depend on the solvents. Detailed studies using lipases have been conducted to investigate the effect of solvents on the reaction. Basically, the enhanced rate using hydrophobic solvents can be explained as follows. Lipase has a lid at the entrance of the substrate-binding site, and the lid is open when a hydrophobic triacyl glycerol comes close to the enzyme in its natural environment. Therefore, the lid is always open in a nonpolar solvent like hexane that is also hydrophobic like triacyl glycerol.

Examples of different activities or enantioselectivities in kinetic resolutions by lipase observed by changing the solvents are shown in Figure 3.8.

6.1 Hydrophobicity and Cyclic versus Acyclic Organic Solvent

Enantioselectivity (*E*-value) of the acetylation of a nitro alcohol by lipase (AK) was affected by the hydrophobicity parameter of solvents (log *P*) where *P* is the partition coefficient of solvent between 1-octanol and water [*P* = (concentration in

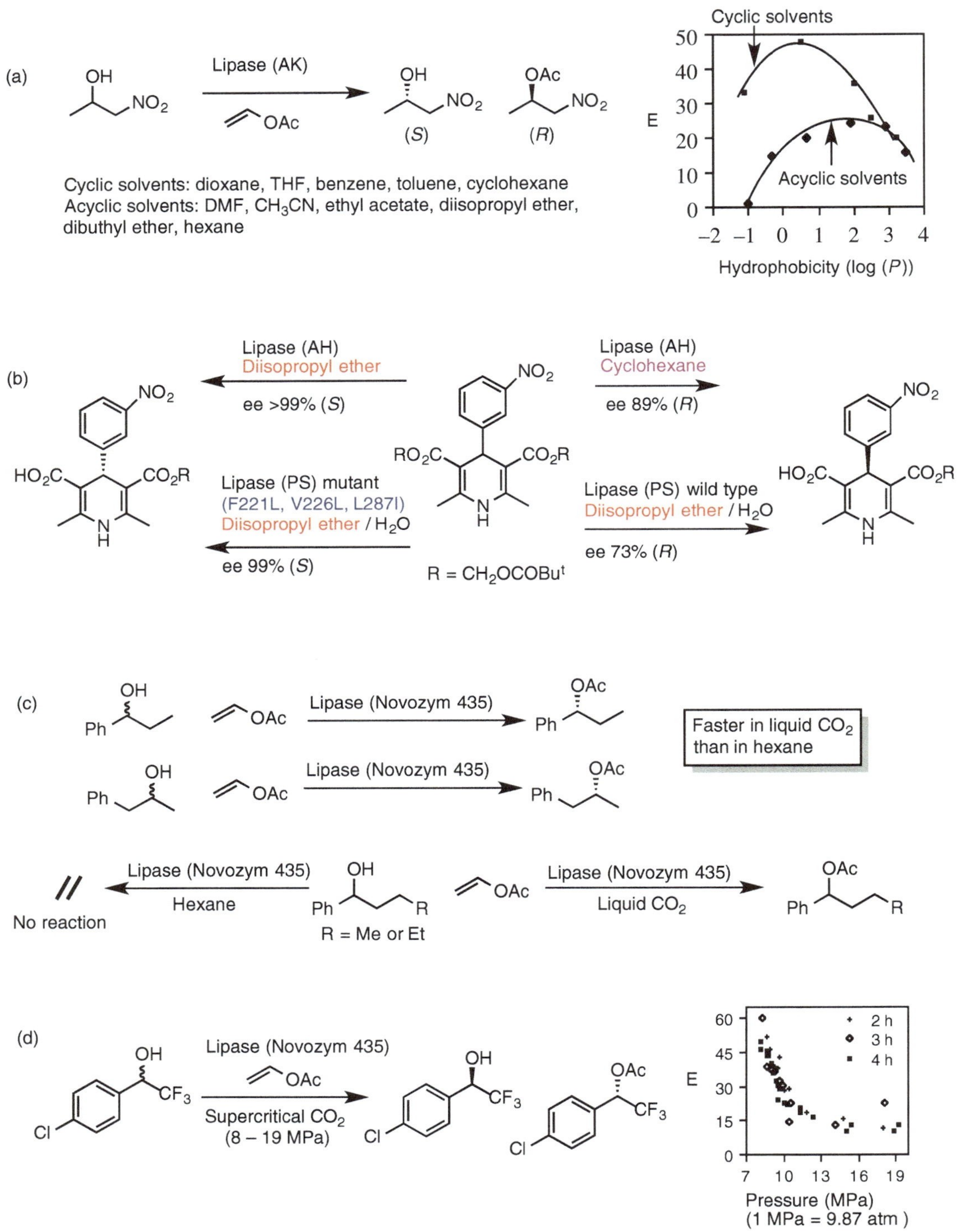

FIGURE 3.8
Selection of nonaqueous solvents for lipase catalyzed reactions.

1−octanol)/(concentration in water)]. Though the trends of E versus $\log P$ are similar, cyclic and acyclic solvents show different absolute values (Figure 3.8a) [13].

6.2 Effect of Solvent on the Enantioselectivity – Diisopropyl Ether versus Cyclohexane for the Inversion in Enantioselectivities

As shown in Figure 3.8b, enantioselectivities of the hydrolysis of 1,4-dihydropyridine, cardiovascular drug, by lipase (AH) in diisopropyl ether (ee >99% (*S*)) was completely opposite to that in cyclohexane (ee 89% (*R*)) [14]. However, when different kind of lipase is used, different enantioselectivity was observed. The wild-type lipase PS in diisopropyl ether gave (*R*)-product, whereas a mutant of lipase (PS) in which three amino acids (F221L, V226L, L287I) were replaced by site-specific mutagenesis showed an inversion of enantioselectivity in the hydrolysis of 1,4-dihydropyridine (Figure 3.8b). The positions of two of the mutations, V226L, L287I, were at the vicinity of reaction center (catalytic triads) of the lipase, but F221L located at the end of β-sheet remote from the reaction center. The double mutant (V226L, L287I) showed no change in enantioselectivity [15].

6.3 Hexane versus Liquid CO_2

Reaction rate and yield of the enantioselective acetylation of racemic phenylpropanols by lipase (Novozym 435) in liquid CO_2 was much higher than in hexane despite the fact that liquid CO_2 and hexane are both hydrophobic solvents. Moreover, the reaction of 1-phenylbutanol and 1-phenylpentanol proceeded in liquid CO_2 but not in hexane (Figure 3.8c) [16].

6.4 Density of Supercritical CO_2 Dependency

Enantioselectivity of the esterification of *rac*-1-(*p*-chlorophenyl)-2,2,2-trifluoroethanol in supercritical CO_2 was largely affected by pressure, which changes the density of the solvent (Figure 3.8d) [17].

7 EXAMPLES OF REACTIONS IN ORGANIC SOLVENTS

A wide range of organic solvents have been applied to biocatalysis, with ethanol, propanol, hexane, acetone, diisopropyl ether being very common. Examples of reactions in organic solvents that illustrate the key points are given further.

An interesting new approach for making carboxylic acid amides is shown in Figure 3.9. Monsan and coworkers [18] used lipases from *C. antarctica* and *Rhizomucor miehei* as catalysts in the direct amidation reaction between free fatty acids and a secondary sugar amine (*N*-methyl-glucamine) in hexane. Such a reaction would be impossible in water since amide hydrolysis would predominate for thermodynamic reasons. Because free *N*-methyl-glucamine was insoluble in hexane, it was added as an ionic complex with the free fatty acid. The enzymatic

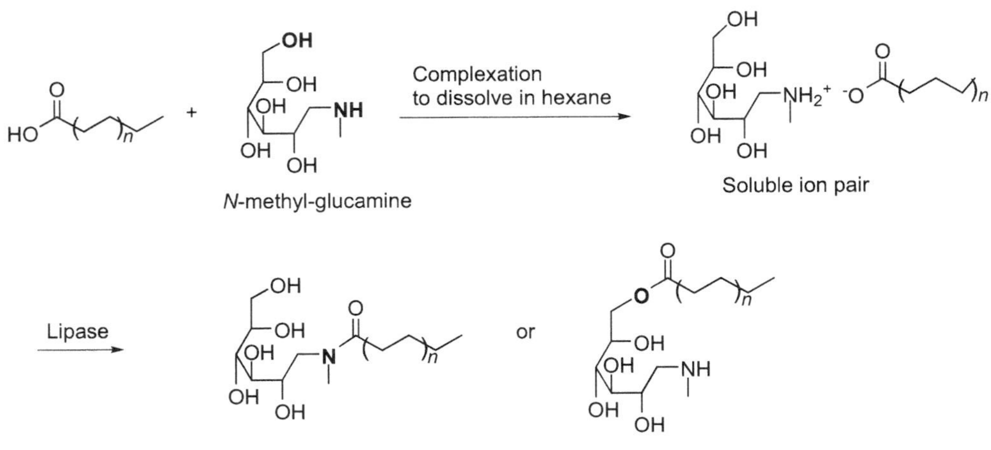

FIGURE 3.9
Lipase-catalyzed synthesis of carboxylic acid amides in hexane [18].

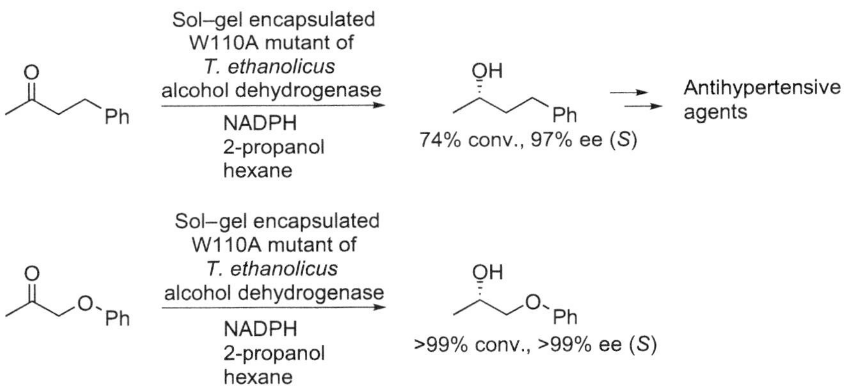

FIGURE 3.10
Encapsulation of W110A mutant of *Thermoanerobacter ethanolicus* alcohol dehydrogenase in sol–gel for asymmetric reduction of ketones [20].

reaction proceeded in good yield using the ion pair as the substrate. This example also illustrates the chemoselectivity of lipases since both O- and *N*-acylation can occur with the sugar amine substrate. The esterification:amidation ratio was controlled from 100:0 to 0:100 by changing the acid:sugar ratio [19].

Figure 3.10 shows the use of sol–gel-encapsulated alcohol dehydrogenase (W110A mutant of *Thermoanaerobacter ethanolicus* alcohol dehydrogenase, *Te*ADH) for asymmetric reduction of 4-phenyl-2-butanone to (*S*)-4-phenyl-2-butanol, a precursor for both bufeniode and labetalol (antihypertensive agents) [20]. Immobilization made it possible to use the biocatalyst in hexane containing 2-propanol and increase the substrate concentrations to 140 mM (vs. 35 mM in an aqueous buffer system). It also allowed the biocatalyst to be reused

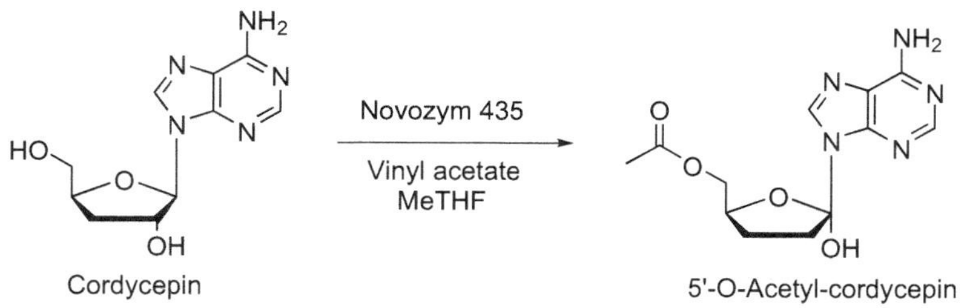

FIGURE 3.11
Regioselective acylation of cordycepin with vinyl acetate catalyzed by Novozym 435 in 2-methyltetrahydrofuran (MeTHF) [21].

three times. 1-Phenoxy-2-propanone was also reduced with a very high yield and enantioselectivity to the corresponding (*S*)-alcohol [20].

As shown in Figure 3.11, an immobilized *C. antarctica* lipase B (Novozym 435) was used to synthesize cordycepin acetate in 2-methyltetrahydrofuran (96.2% isolated yield) [21]. The biocatalyst displayed excellent regioselectivity and high operational stability during the transformation. The 5′-substituted cordycepin derivative was the sole detectable product from the acylation reaction. Novozym 435 could be recycled for the synthesis of cordycepin derivative on a 25-g scale and 63% of its original activity was maintained after seven batches.

Homogeneous biocatalysis in both fluorous biphasic and supercritical CO_2 systems has been demonstrated [6]. By forming protein–surfactant complexes by hydrophobic ion pairing with a highly fluorinated anionic surfactant, cytochrome *c* can be solubilized in perfluoromethylcyclohexane (PMFC) and in $scCO_2$. The secondary structure of the proteins within these ion-paired complexes has been shown to remain intact, and particle size analysis indicated that small aggregates of protein molecules surrounded by surfactant molecules are formed. The presence of the KDP (perfluoropolyether carboxylate surfactants) ion paired with α-chymotrypsin appears to enhance its catalytic activity as compared to that of the native enzyme in a fluorous biphasic system. The facile recycling of the α-chymotrypsin–KDP complex in a fluorous biphasic system has been demonstrated with retention of enzyme activity over four reaction cycles.

Dynamic kinetic resolution (DKR) of a racemic alcohol with lipase and metal catalysts is often conducted in organic solvents. Since the maximum yield of kinetic resolution of racemic alcohol by lipase is only 50% with 100% ee, metal catalysts to racemize the unreactive enantiomer substrate is necessary to achieve 100% yield and 100% ee [22]. For example, ruthenium catalysts have been widely used for this process as shown in Figure 3.12a [22b]. Vanadium has also been used, and to improve both catalytic activity and compatibility of the oxovanadium catalysts with the lipases, a novel oxovanadium catalyst (V-MPS) immobilized inside mesoporous silica (MPS) with pores of approximately 3 nm in diameter was prepared. With this immobilization preparation, a complete division of the racemization site and the enzymatic reaction site was achieved.

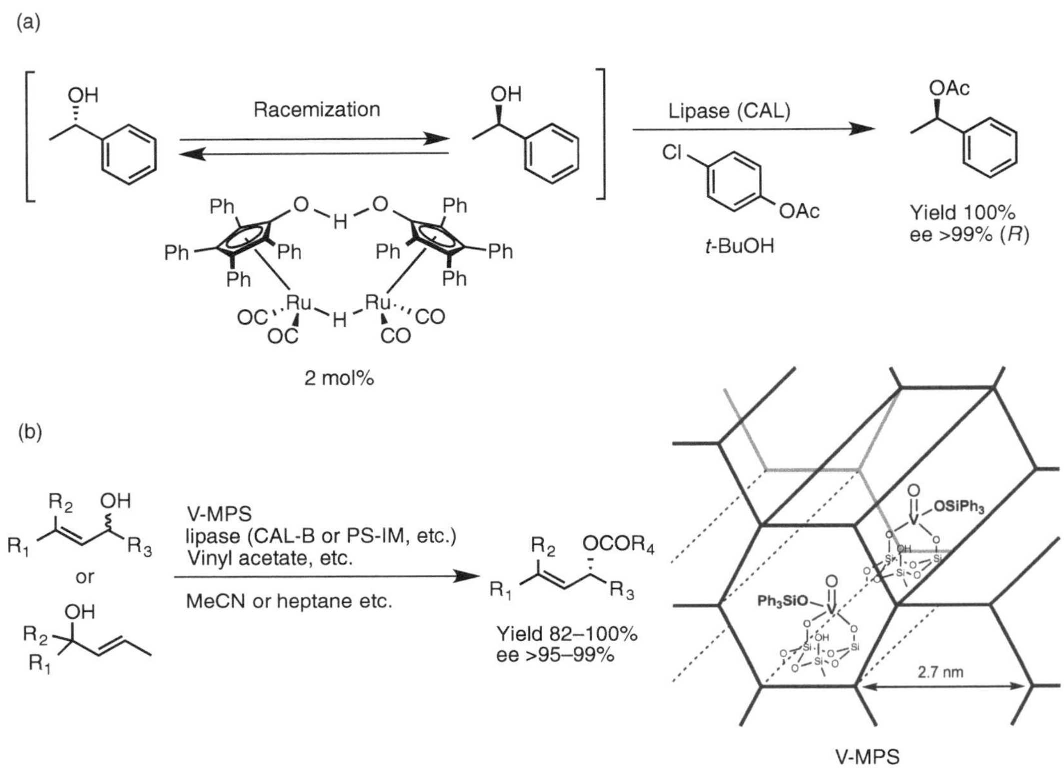

FIGURE 3.12
Dynamic kinetic resolution by biocatalyst in organic solvent coupled with (a) ruthenium or (b) oxovanadium catalyst. V-MPS structure in (b) is reprinted with permission from WILEY-VCH Verlag GmbH & Co. KGaA, Weinheim [22d].

The combined lipase–V-MPS catalyst achieved DKR of a wide range of racemic alcohols with excellent yield and enantioselectivity and is reusable up to six times [22d] (Figure 3.12b).

8 EXAMPLES OF REACTIONS IN AQUEOUS – ORGANIC BIPHASIC SYSTEMS

In general, microbial conversions must be carried out in aqueous media since cells rarely tolerate nonaqueous solvents. Unfortunately, bioconversions of hydrophobic substrates in aqueous systems often yield poor results due to the poor substrate solubility under the reaction conditions. The use of organic–aqueous biphasic systems can be an effective way to overcome these problems. Some examples are given further.

Hydroxynitrilase-catalyzed additions of HCN or the carbanion of nitromethane to an aldehyde (Figure 3.13) proceeded successfully in an aqueous–organic biphasic

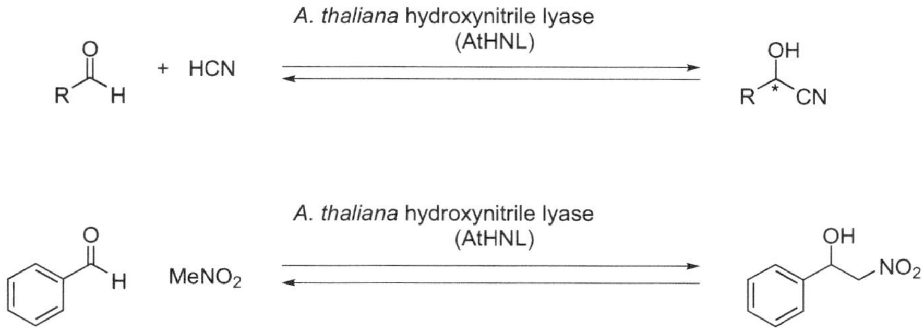

FIGURE 3.13
Enantioselective *A. thaliana* hydroxynitrile lyase (*At*HNL)-catalyzed (a) synthesis of cyanohydrins and (b) Henry reaction of benzaldehyde and $MeNO_2$ [23].

system [23]. An *R*-selective hydroxynitrile lyase (HNL) from the noncyanogenic plant *Arabidopsis thaliana* (*At*HNL) catalyzed the nitro aldol (Henry) reaction of nitromethane with aromatic aldehydes to yield (*R*)-β-nitro alcohols, the first example of this reaction type by an *R*-HNL. *n*-Butyl acetate (AcOBu) in a 1:1 mixture with the aqueous phase gave the highest enantioselectivity (Figure 3.14).

A diketone was reduced by a ketoreductase using an aqueous–organic biphasic system to yield a diol with high de and ee (both greater than 99.5%) (Figure 3.12) [24]. Formate dehydrogenase (FDH) was used to regenerate NADH. Best results were obtained with an aqueous-hexane biphasic system, which allowed a substrate concentration of 105 g/L and provided a building block for the chiral side chain of statin drugs [24].

Hollow-fiber membrane reactors consisting of an aqueous layer, a sponge layer containing the enzyme and toluene layers was developed for the synthesis of a key intermediate of a useful coronary vasodilator, Diltiazem [25]. (−)-3-Phenylglycidic acid ester, (−)-MPGM, was synthesized through lipase-catalyzed optical resolutions (Figure 3.15). (+/-)-MPGM was hydrolysed by lipase to the corresponding (2*S*, 3*R*)-acid, leaving the desired unreacted ester (−)-(2*R*, 3*S*)-MPGM. The undesired acid was removed efficiently by the spontaneous decarboxylation followed by

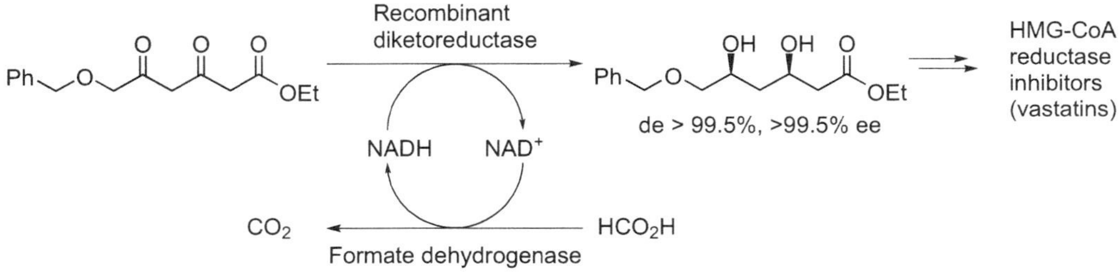

FIGURE 3.14
Biocatalytic synthesis of statin side chain [24].

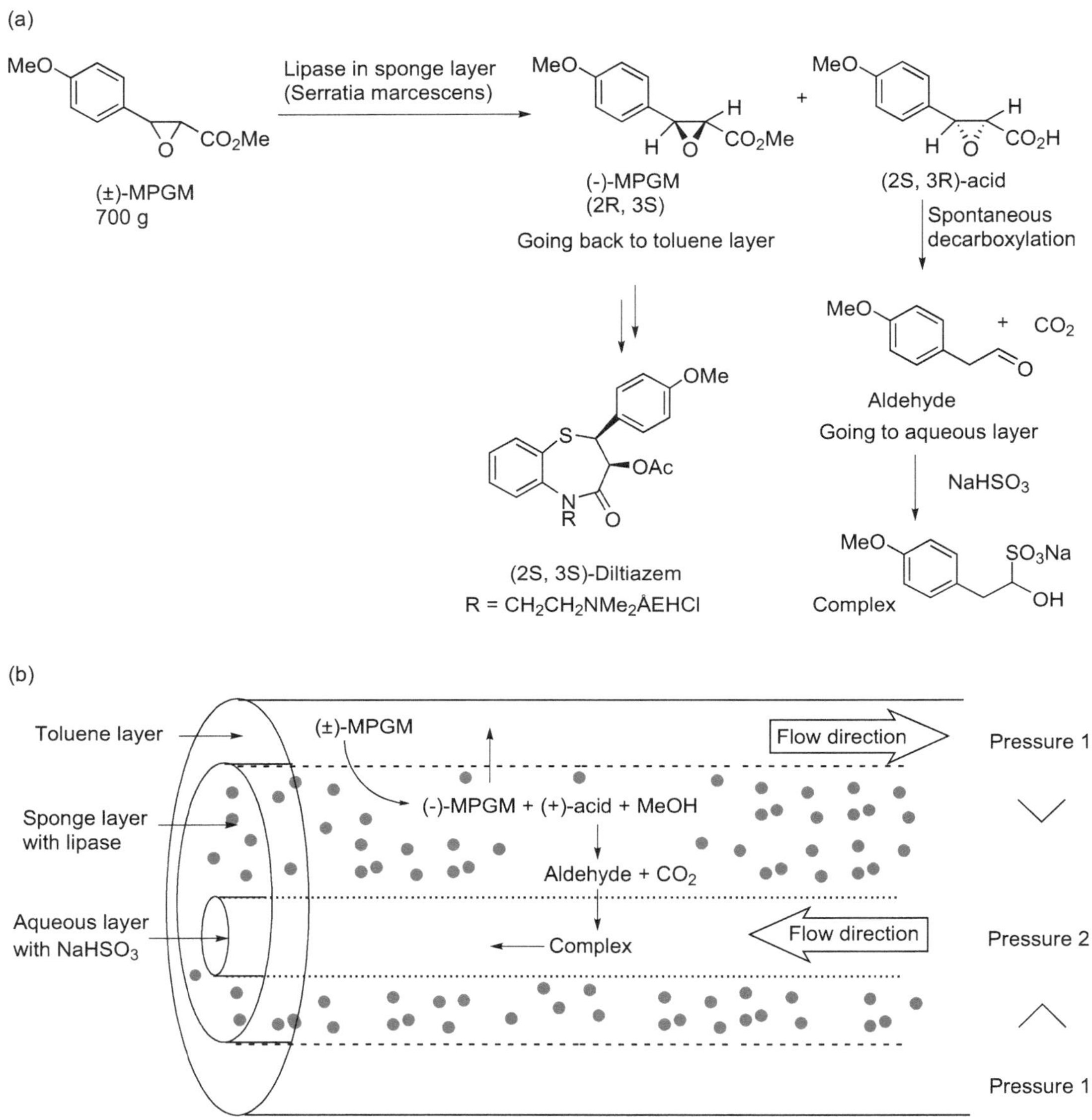

FIGURE 3.15
Three-layered hollow-fiber membrane reactor consisting of aqueous, enzyme-sponge, and organic layers for optical resolution of 3-phenylglycidic acid methyl ester (MPGM) by lipase and separation of the optically active product, (–)-MPGM, from unnecessary acid [25].

the formation of water-soluble complex with $NaHSO_3$, while recovering the necessary (−)-(2*R*, 3*S*)-MPGM going back to toluene layer. The pressure of the toluene layer was kept slightly higher than that of the aqueous layer, so that the substrate moved to the sponge layer but the water did not move to the toluene layer, preventing the decomposition of MPGM.

9 EXAMPLES OF REACTIONS IN SUPERCRITICAL FLUIDS

Supercritical fluids are promising solvents with high functionalities. They have been used as solvents for organic synthesis as well as for extraction and chromatography by taking advantage of their gas-like low viscosities and high diffusivities, and their liquid-like solubilizing power. Some examples of biocatalysis in supercritical CO_2 are shown further.

Candida rugosa lipase (CRL) catalyzed the regioselective acylation of methyl 6-O-trityl-D-glucopyranoside in $scCO_2$, using vinyl acetate as the acyl donor (Figure 3.16) [26]. The effect of pressure on yield and regioselectivity was examined and appropriate conditions were established to obtain the 3-O-acetyl 6-O-trityl β-D-glucopyranoside regioisomer with 100% selectivity.

The immobilized form of *C. antarctica* lipase B (Novozym 435) was used for the large-scale synthesis of several optically active compounds using a $scCO_2$-flow reactor [27]. This strategy is particularly useful for the large-scale production of optically active alcohols. In one example (Figure 3.17a), the reactants and $scCO_2$ were passed through the reactor packed with the immobilized enzyme to carry out the reaction and the product was recovered simply by removal of CO_2 after depressurizing the effluent. This system was used to kinetically resolve 1-phenylethanol in a completely organic solvent-free process and with a significant improvement in productivity. The space–time yield was improved by over 400-fold compared to the corresponding batch reaction using $scCO_2$. The biocatalyst maintained its performance in terms of the reactivity and selectivity during 3 days of operation under supercritical conditions, and resulted in a quantitative transformation of *rac*-1-phenylethanol (221 g) to (*S*)-phenylethanol with 99% ee and the corresponding (*R*)-acetate with 99% ee using 1.73 g of the immobilized enzyme (Figure 3.17b). Furthermore, it was found that aliphatic alcohols, 2-undecanol, and tetralin-1-ol were also resolved with Novozym 435 in $scCO_2$ to give a mixture of their respective optically active (*R*)-esters and unreacted (*S*)-alcohols (Figure 3.14c and d).

As noted previously, the use of supercritical CO_2 allows decarboxylases to catalyze the reverse, carboligation reaction because the solvent is also the reactant and thus present at a very high concentration. In one example, *Thauera aromatica* phenylphosphate carboxylase was used to carboxylate phenylphosphate in the *para*-position without any detectable *ortho*-product in supercritical CO_2 without any added H_2O (Figure 3.18) [28].

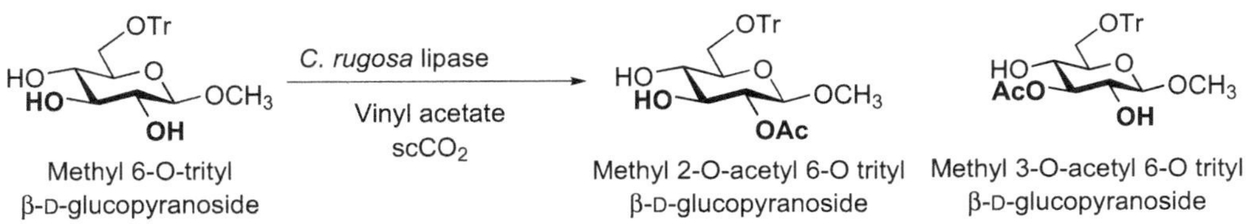

FIGURE 3.16
Lipase-catalyzed regioselective acetylation of sugar [26].

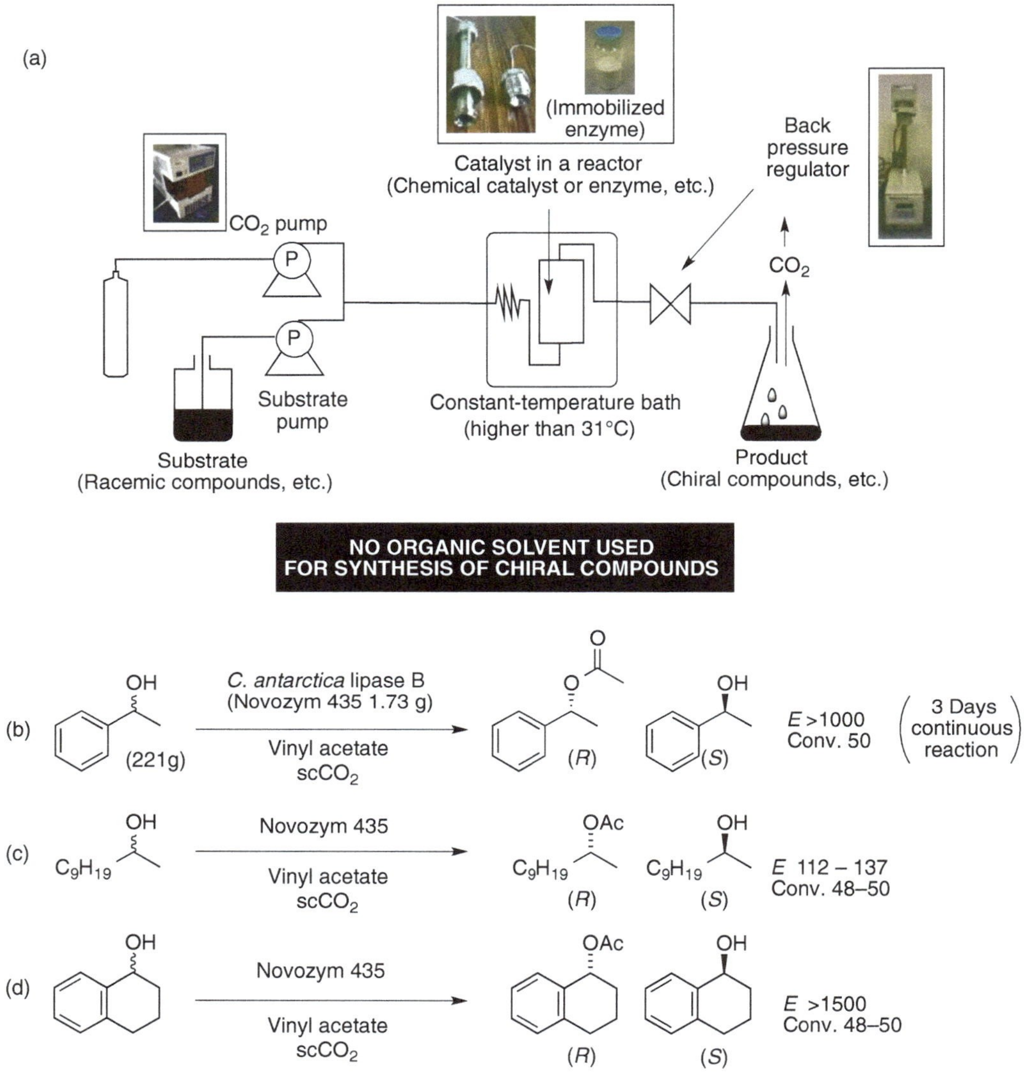

FIGURE 3.17

Continuous lipase-catalyzed kinetic resolution of alcohols using $scCO_2$-flow reactor. (a) Reaction apparatus; (b) kinetic resolution of *rac* 1-phenylethanol; (c) kinetic resolution of *rac*-2-undecanol; (d) kinetic resolution of *rac*-Tetralin-1-ol [27].

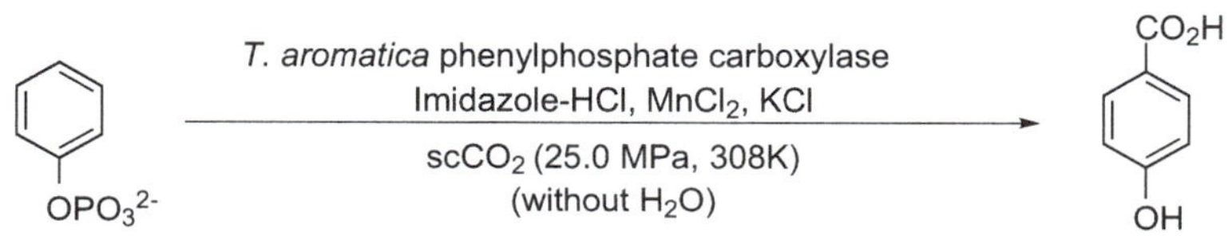

FIGURE 3.18

Carboxylation in supercritical CO_2 without H_2O by *T. aromatica* phenylphosphate carboxylase [28].

10 EXAMPLES OF REACTIONS IN AQUEOUS – SUPERCRITICAL FLUIDS BIPHASIC SYSTEMS

Biocatalytic reaction in aqueous-supercritical fluids biphasic systems have also been investigated and some examples are shown further.

Employing supercritical CO_2 under high pressure improved the yield of pyrrole-2-carboxylate from pyrrole using a decarboxylase from *Bacillus megaterium* PYR 2910 by 12-fold (Figure 3.19) [29].

Asymmetric reduction of ketones by crude alcohol dehydrogenase from *G. candidum* was carried out in aqueous-supercritical fluids biphasic systems (Figure 3.20) [30]. It was found that the lowered pH caused by $scCO_2$ irreversibly inactivated the enzyme. In order to overcome this undesired side effect of $scCO_2$, basic salts were added to the reaction mixture. Sodium bicarbonate was found to be the best among several tested. For the reduction of acetophenone, adding sodium bicarbonate improved the yield from 4% to 25%. The asymmetric reduction of *o*-fluoroacetophenone with the immobilized

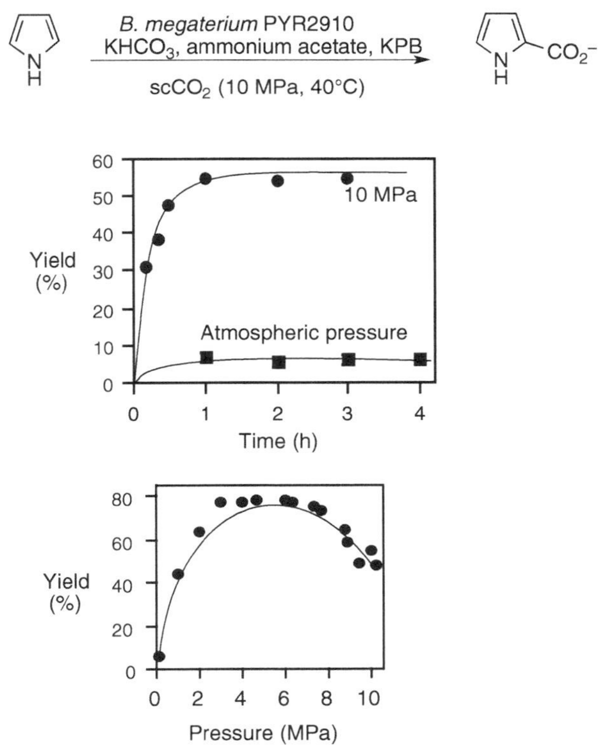

FIGURE 3.19
Carboxlation of pyrole by decarboxylase from *B. megaterium* PYR 2910 in aqueous – supercritical fluids biphasic systems [29].

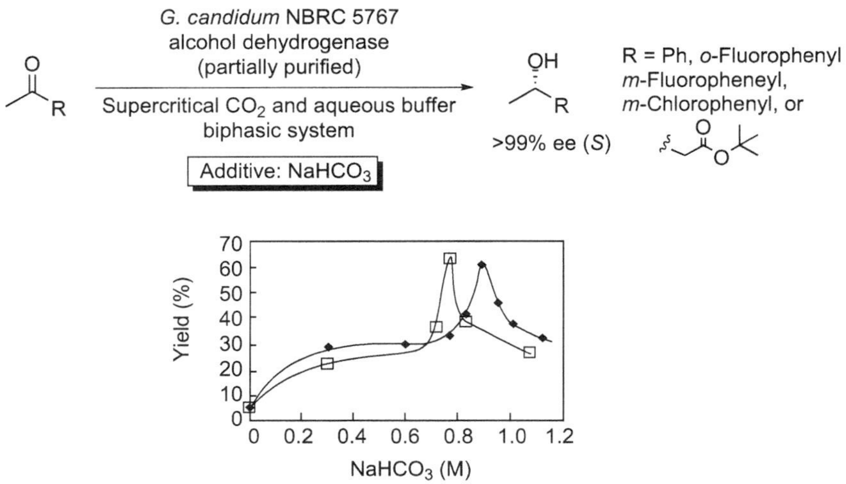

FIGURE 3.20
Effect of $NaHCO_3$ on the reduction of *o*-fluoroacetophenone in the $scCO_2/H_2O$ biphasic system. Open squares: 8 MPa, 35°C; closed squares: 10 MPa, 35°C [30].

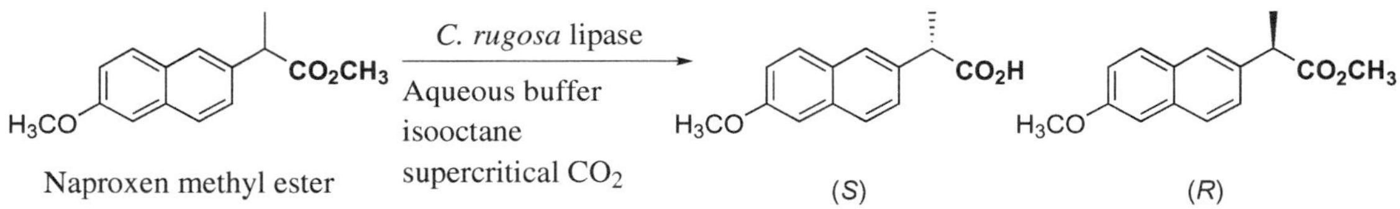

FIGURE 3.21
Hydrolysis of racemic naproxen methyl ester by *C. rugosa* lipase [31].

enzyme in a $scCO_2$/water biphasic system showed the best results, providing 75% yield of the (*S*)-alcohol in more than 99% ee.

The enantioselective hydrolysis of racemic naproxen methyl ester by *C. rugosa* lipase (CRL) was studied in an aqueous buffer solution/isooctane reaction system in the presence of supercritical CO_2 (Figure 3.21) [31]. The conversion reached 41.3% at 120 bar at 37°C after 2 h of reaction and provided enantiomeric purities of product and substrate of 97.9 and 68.8%, respectively. This corresponded to an enantiomeric ratio (*E* value) of 193. The lipase remained active for at least 12 h at 37°C and 120 bar in the supercritical CO_2 medium.

11 EXAMPLES OF REACTIONS IN IONIC LIQUIDS

Ionic liquids (ILs) have many advantages for enzymatic reactions including high conversion rates, high enantioselectivities, better enzyme stabilities, along with simplified recoverability and recyclability. To date, a wide range of approaches have been studied to further improve the performance of enzymes in ionic liquids. ILs are composed entirely of ions that are liquids at ambient or subambient

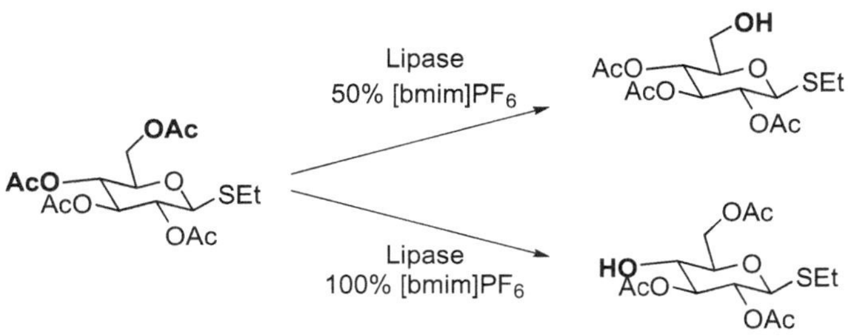

FIGURE 3.22
***C. cylindracea* lipase catalysed hydrolysis [32].**

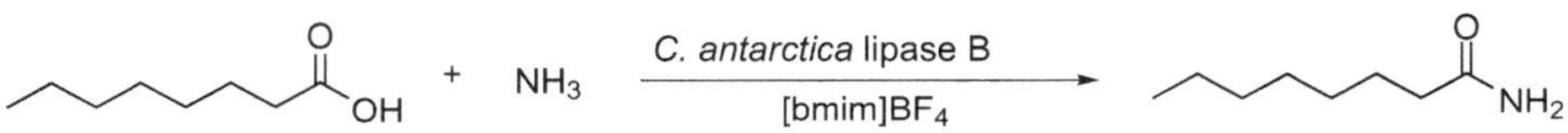

FIGURE 3.23
***C. antarctica* lipase catalyzed ammoniolysis of ethyl octanoate [4a].**

temperature. They have been extensively explored as potential alternatives to toxic, hazardous, flammable, and highly volatile organic solvents. A wide number of enzymes have been subjected to ILs to test their catalytic activity including lipases, alcohol dehydrogenases, proteases, and oxidoreductases. Several respresentative examples are summarized further.

An easy preparation of mono-deprotected thioglucopyranosides via a *Candica cylindracea* lipase-catalyzed selective hydrolysis of a commercially available peracetylated precursor is described in Figure 3.22 [32]. Ethyl 2,3,4-tri-O-acetyl-1-thio-β-D-glucopyranoside and ethyl 2,3,6-tri-O-acetyl-1-thio-β-D-glucopyranoside were obtained in 100% and 54% isolated yields, respectively.

C. antarctica lipase catalyzes alcoholysis, ammonolysis (shown in Figure 3.23) [4a], and hydrolysis reactions in the ionic liquids 1-butyl-3-methylimidazolium tetrafluoroborate or hexafluorophosphate. Reaction rates were generally same as or better than those observed in organic media. For example, the direct reaction of octanoic acid and ammonia was catalyzed by Novozym 435 by bubbling ammonia through a suspension in an ionic liquid. A quantitative conversion was obtained after 4 days.

12 EXAMPLES OF REACTIONS IN AQUEOUS – IONIC LIQUIDS BIPHASIC SYSTEMS

As is the case for organic solvents, several workers have investigated two-phase systems in which water-immiscible ionic liquids are mixed with aqueous buffers. Such strategies offer all of the advantages of organic/aqueous biphasic mixtures, but without the drawbacks of handling volatile and flammable solvents. Some examples are shown in this section.

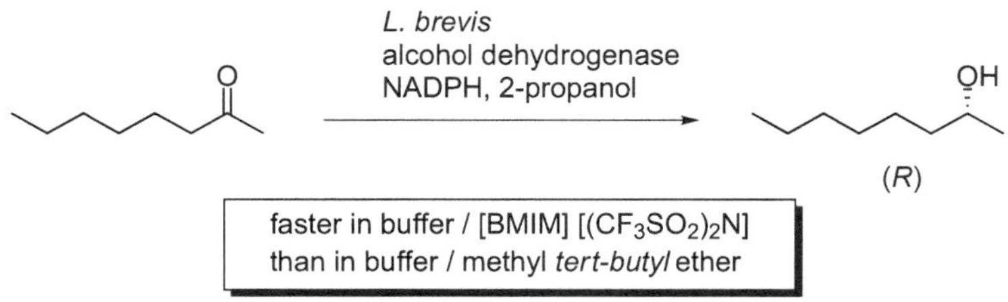

FIGURE 3.24
Reduction of 2-octanone catalyzed by alcohol dehydrogenase from *L. brevis* [33].

The highly enantioselective reduction of 2-octanone catalyzed by *Lactobacillus brevis* alcohol dehydrogenase is faster in a biphasic system containing buffer and the ionic liquid [BMIM][(CF(3)SO(2))(2)N] compared to the reduction in a biphasic system containing buffer and methyl *tert*-butyl ether due to favorable partition coefficients (Figure 3.24) [33].

Biphasic systems that include ionic liquids can also be applied to whole-cell biocatalysis. The ability of these solvents to act as a substrate reservoir and *in situ* extracting agent was demonstrated by an efficient asymmetric ketone reduction. 4-Chloroacetophenone was reduced to the key pharmaceutical intermediated (*R*)-1-(4-chlorophenyl) ethanol using *Lactobacillus kefir* cells in ILs. The indigenous cellular cofactor regeneration system remained active, which allowed high product concentrations without cofactor supplementation [34].

The selective hydrolysis of glycyrrhizin (GL) to glycyrrhetic acid 3-O-mono-β-D-glucuronide (GAMG) catalyzed by recombinant β-D-glucuronidase from *Escherichia coli* BL21 (PGUS-E) was successfully performed in an IL/buffer biphasic system (Figure 3.25) [35]. An obvious enhancement in the initial reaction rate, substrate conversion, GAMG yield, and chemical bond selectivity (*S*cb) was observed using 40% (v/v) [BMIM]PF_6/buffer as the reaction medium when compared to the acetate buffer medium. Under the optimized conditions (pH 6.0, 50°C, 6 mM substrate concentration and shaking speed 200 rpm), the initial reaction rate, the GAMG yield, and the *S*cb reached 3.15 mM/h, 74.36% and 98.12%, respectively. The recyclability of [BMIM]PF_6 was also studied and it was found to be reusable for five batches with high recovery percentage (≥92%). Furthermore, the desired product and by-product were easily separated since they were distributed in different phases. Finally, a higher V_{max} (3.14 vs. 2.24 mM/h), a lower apparent K_M (1.21 vs. 1.80 mM) and activation energy (E_a, 25.97 vs. 32.60 kJ/mol) were achieved in the [BMIM]PF_6/buffer biphasic system as compared to the monophasic buffer system.

13 EXAMPLES OF REACTIONS IN IONIC LIQUIDS – SUPERCRITICAL FLUIDS BIPHASIC SYSTEMS

When biocatalysis is conducted in ionic liquid/supercritical fluids biphasic systems, the major function of ILs is to protect the enzyme, whereas the supercritical fluid dissolves the substrate and extracts the product. Both enzyme stability and

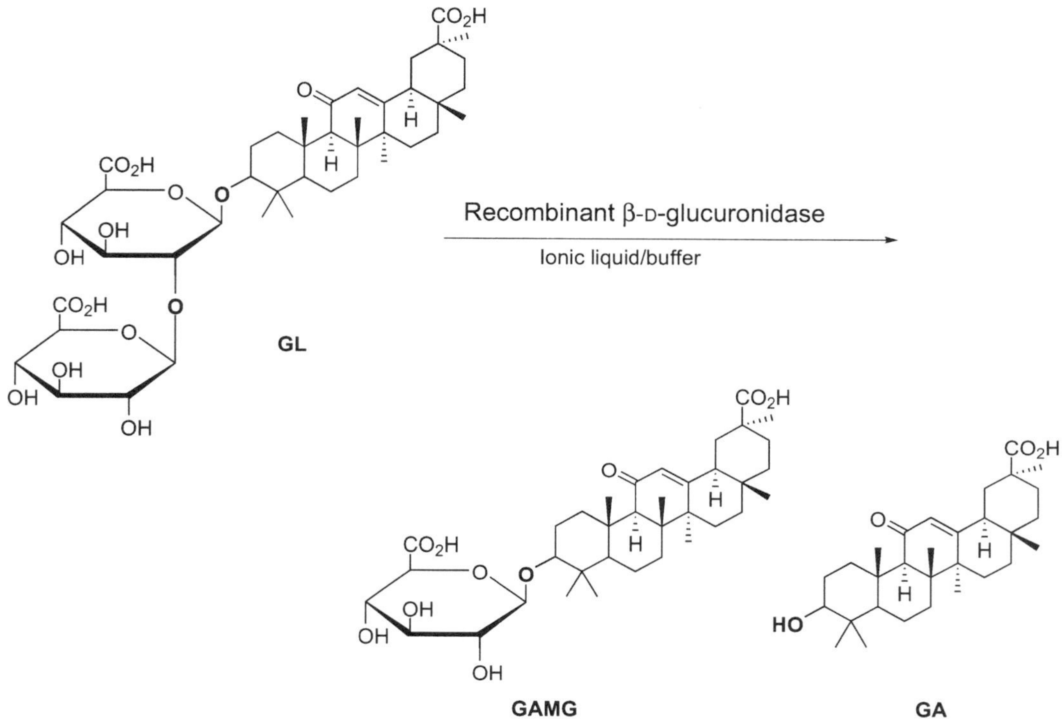

FIGURE 3.25
Hydrolysis of glycyrrhizin in ionic liquid/buffer biphasic system [35].

reaction efficiency have been improved by employing this strategy and some examples are shown further.

Methyl oleate (biodiesel) was synthesized by methanolysis of triolein using immobilized *C. antarctica* lipase B (Novozym 435) (Figure 3.26) [36]. The operational stability of the immobilized lipase was improved by coating with ILs such as 1-methyl-3-octadecylimidazolium hexafluorophosphate. A continuous enzymatic reactor for biodiesel synthesis in supercritical carbon dioxide at 60 °C and 180 bar was evaluated. This provided up to 82% biodiesel yield after 12 cycles at 4 h per cycle.

The synthesis of butyl propionate in a recirculating bioreactor was carried out in an IL/supercritical carbon dioxide biphasic system at 50 °C and 80 bar using immobilized *C. antarctica* lipase B (Figure 3.27) [37]. Selectivity increased (reaching more than 99.5%) when the ionic liquid/supercritical carbon dioxide biphasic systems were used rather than supercritical carbon dioxide alone.

14 CONCLUSIONS

The types and numbers of solvents suitable for carrying out enzymatic reactions have been widened in recent years and examples of types of enzymatic reactions performed in such nonaqueous solvents have been expanding. Further discovery

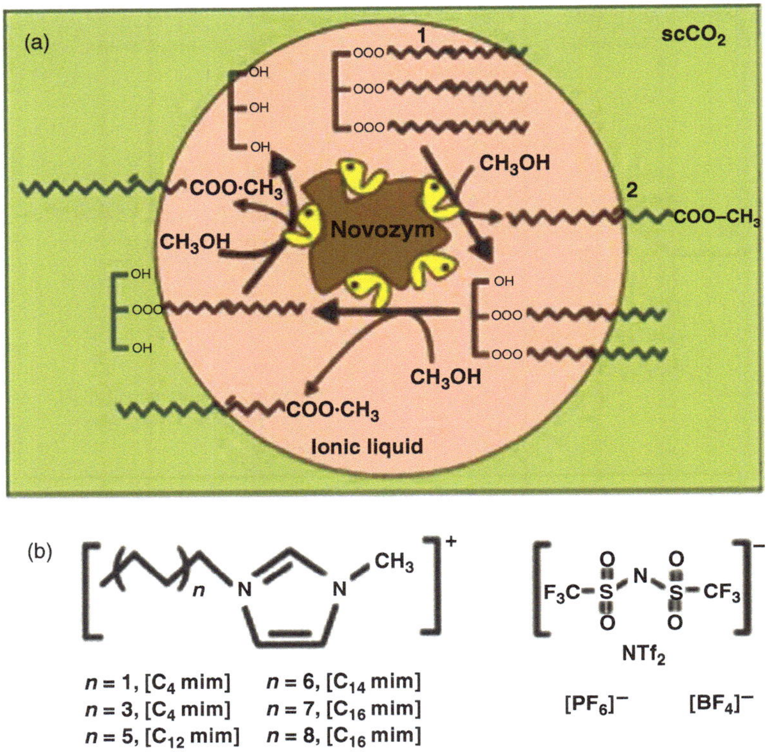

FIGURE 3.26
(a) Immobilized lipase-catalyzed synthesis of biodiesel (**2**, methyl oleate) by methanolysis of triolein (**1**) in IL/$scCO_2$ biphasic systems. (b) Cations and anions of the ILs. *Reprinted with permission from Elsevier [36].*

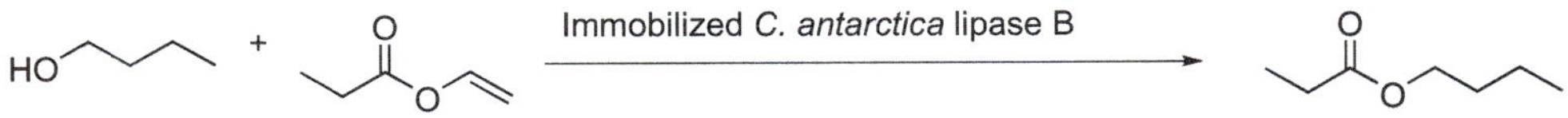

FIGURE 3.27
Synthesis of butyl propionate from vinyl propionate and 1-butanol catalyzed by CAL-B [37].

of robust systems of different enzymes and various nonaqueous environments will enhance the application of enzymatic reactions for organic synthesis.

References

[1] (a) N. Doukyu, H. Ogino, Organic solvent-tolerant enzymes, Biochem. Engineer. J. 48 (2010) 270–282. (b) A.M. Klibanov, Improving enzymes by using them in organic solvents, Nature 409 (2001) 241–246. (c) A.L. Serdakowski, J.S. Dordick, Enzyme activation for organic solvents made easy, Trends Biotech. 26 (2008) 48–54. (d) T. Matsuda, R. Yamanaka, K. Nakamura, Recent progress in biocatalysis for asymmetric oxidation and reduction, Tetrahedron Asymm. 20 (2009) 513–557.

[2] (a) T. Itoh, Biotransformation in ionic liquid, in: T. Matsuda (Ed.), Future Direction in Biocatalysis, Elsevier, Amsterdam, 2007, pp. 1–20. (b) M. Moniruzzaman, N. Kamiya, M. Goto, Activation and stabilization of enzymes in ionic liquids, Org. Biomol. Chem. 8 (2007) 2887–2899.

[3] (a) T. Matsuda, Recent progress in biocatalysis using supercritical carbon dioxide, J. Biosci. Bioeng. 115 (2013) 233–241. (b) T. Matsuda, T. Harada, K. Nakamura, T. Ikariya, Asymmetric synthesis using hydrolytic enzymes in supercritical carbon dioxide, Tetrahedron Asymm. 16 (2005) 909–915. (c) H.R. Hobbs, N.R. Thomas, Biocatalysis in supercritical fluids, in fluorous solvents, and under solvent-free conditions, Chem. Rev. 107 (2007) 2786–2820. (d) J.M. DeSimone, W. Tumas, Green chemistry using liquid and supercritical carbon dioxide, Oxford University Press, New York, (2003).

[4] (a) R.M. Lau, F. van Rantwijk, K.R. Seddon, R.A. Sheldon, Lipase-catalyzed reactions in ionic liquids, Org. Lett. 2 (2000) 4189–4191. (b) T. Itoh, E. Akasaki, K. Kudo, S. Shirakami, Lipase-catalyzed enantioselective acylation in the ionic liquid solvent system: reaction of enzyme anchored to the solvent, Chem. Lett. 3 (2001) 262–263. (c) S.H. Schöfer, N. Kaftzik, P. Wasserscheid, U. Kragl, Enzyme catalysis in ionic liquids: lipase catalysed kinetic resolution of 1-phenylethanol with improved enantioselectivity, Chem. Commun. (2001) 425–426.

[5] (a) T.W. Randolph, H.W. Blanch, J.M. Prausnitz, C.R. Wilke, Enzymatic catalysis in a supercritical fluid, Biotechnol. Lett. 7 (1985) 325–328. (b) D.A. Hammond, M. Karel, A.M. Klibanov, V.J. Krukonis, Enzymatic reactions in supercritical gases, Appl. Biochem. Biotechnol. 11 (1985) 393–400. (c) K.Y. Nakamura, M. Chi, Y. Yamada, T. Yano, Lipase activity and stability in supercritical carbon dioxide, Chem. Eng. Commun. 45 (1986) 207–212.

[6] H.R. Hobbs, H.M. Kirke, M. Poliakoff, N.R. Thomas, Homogeneous biocatalysis in both fluorous biphasic and supercritical carbon dioxide systems, Angew. Chem. Int. Ed. 46 (2007) 7860–7863.

[7] (a) T. Mori, Y. Okahata, Effective biocatalytic transgalactosylation in a supercritical fluid using a lipid-coated enzyme, Chem. Commun. (1998) 2215–2216. (b) T. Mori, M. Li, A. Kobayshi, Y. Okahata, Reversible control of enzymatic transglycosylations in supercritical fluoroform using a lipid-coated beta-D-galactosidase, J. Am. Chem. Soc. 124 (2002) 1188–1189. (c) T. Mori, M. Funasaki, A. Kobayshi, Y. Okahata, Reversible activity changes of lipid-coated lipase for enantioselective esterification in supercritical fluoroform, Chem. Commun. (2001) 1832–1833.

[8] J.L. Panza, A.J. Russell, E.J. Beckman, Synthesis of fluorinated NAD as a soluble coenzyme for enzymatic chemistry in fluorous solvents and carbon dioxide, Tetrahedron 58 (2002) 4091–4104.

[9] (a) R.L. Silveira, J. Martínez, M.S. Skaf, L. Martínez, Enzyme microheterogeneous hydration and stabilization in supercritical carbon dioxide, J. Phys. Chem. B 116 (2012) 5671–5678. (b) T. Matsuda, K. Tsuji, T. Kamitanaka, T. Harada, K. Nakamura, T. Ikariya, Rate enhancement of lipase-catalyzed reaction in supercritical carbon dioxide, Chem. Lett. 34 (2005) 1102–1103.

[10] (a) D.A. Cowan, Thermophilic proteins: stability and function in aqueous and organic solvents, Biochem. Physiol. 118A (1997) 429–438. (b) A.D. Frock, R.M. Kelly, Extreme thermophiles: moving beyond single-enzyme biocatalysis, Curr. Opin Chem. Eng. 1 (4) (2012) 363–372 12. (c) P. Falcicchio, S. Wolterink-Van Loo, M.C. Franssen, J. van der Oost, DHAP-dependent aldolases from (hyper)thermophiles: biochemistry and applications, Extremophiles. 18 (2014) 1–13. (d) W. Haoran, W. Zhi, Z. Hong, C. Ge, Y. Hong, W. Lei, Enzyme catalytic promiscuity: asymmetric aldol addition reaction catalyzed by a novel thermophilic esterase in organic solvent, Green Chem. Lett. Rev. 7 (2014) 145–149. (e) S. Reich, N. Kress, B.M. Nestl, B. Hauer, Variations in the stability of NCR ene reductase by rational enzyme loop modulation, J. Struct. Biol. 185 (2014) 228–233. (f) M.T. Reetz, P. Soni, L. Ferna'ndez, Y. Gumulya, J.D. Carballeira, Increasing the stability of an enzyme toward hostile organic solvents by directed evolution based on iterative saturation mutagenesis using the B-FIT method, Chem. Commun. 46 (2010) 8657–8658.

[11] (a) R.A. Sheldon, Enzyme immobilization: the quest for optimum performance, Adv. Synth. Catal. 349 (2007) 1289–1307. (b) P.V. Iyer, L. Ananthanarayan, Enzyme stability and stabilization – aqueous and non-aqueous environment, Process Biochem. 43 (2008) 1019–1032. (c) E. Magner, Immobilisation of enzymes on mesoporous silicate materials, Chem. Soc.

Rev. 42 (2013) 6213–6222. (d) L. Cao, F.v. Rantwijk, R.A. Sheldon, Cross-linked enzyme aggregates: a simple and effective method for the immobilization of penicillin acylase, Org. Lett. 2 (2000) 1361–1364. (e) T. Matsuda, K. Nakayama, T. Abe, M. Mukouyama, Stabilization of pyruvate decarboxylase under pressurized carbon dioxide and water biphasic system, Biocatal. Biotrans. 28 (2010) 167–171. (f) T. Matsuda, R. Marukado, M. Mukouyama, T. Harada, K. Nakamura, Asymmetric reduction of ketones by *Geotrichum candidum*: immobilization and application to reactions using supercritical carbon dioxide, Tetrahedron Asymm. 19 (2008) 2272–2275.

[12] T. Sakai, Temperature control of the enantioselectivity in the lipase-catalyzed resolutions, in: T. Matsuda (Ed.), Future Direction in Biocatalysis, Elsevier, Amsterdam, 2007, pp. 21–50.

[13] K. Nakamura, Y. Takebe, T. Kitayama, A. Ohno, Effect of solvent structure on enantioselectivity of lipase-catalyzed transesterification, Tetrahedron Lett. 32 (1991) 4941–4944.

[14] Y. Hirose, K. Kariya, I. Sasaki, Y. Kurono, H. Ebiike, K. Achiwa, Drastic solvent effect on lipase-catalyzed enantioselective hydrolysis of prochiral 1,4-dihydropyridines, Tetrahedron Lett. 33 (1992) 7157–7160.

[15] Y. Hirose, K. Kariya, Y. Nakanishi, Y. Kurono, K. Achiwa, Inversion of enantioselectivity in hydrolysis of 1,4- dihydropyridines by point mutation of lipase PS, Tetrahedron Lett. 36 (1995) 1063–1066.

[16] H.N. Hoang, T. Matsuda, Liquid carbon dioxide as an effective solvent for immobilized *Candida antarctica* lipase B catalyzed transesterification, Tetrahedron Lett. 56 (2015) 639–641.

[17] (a) T. Matsuda, R. Kanamaru, K. Watanabe, T. Kamitanaka, T. Harada, K. Nakamura, Control of enantioselectivity of lipase catalyzed esterification in supercritical carbon dioxide by tuning the pressure and temperature, Tetrahedron Asymm. 14 (2003) 2087–2091. (b) T. Matsuda, R. Kanamaru, K. Watanabe, T. Harada, K. Nakamura, Control on enantioselectivity with pressure for lipase catalyzed esterification in supercritical carbon dioxide, Tetrahedron Lett. 42 (2001) 8319–8321.

[18] T. Maugard, M. Remaud-Simeon, D. Petre, P. Monsan, Lipase catalyzed chemoselective *N*-acylation of amino-sugar derivatives in hydrophobic solvent: acid-amine ion-pair effects, Tetrahedron 53 (1997) 7587–7594.

[19] Y.L. Khmelnitsky, J.O. Rich, Biocatalysis in nonaqueous solvents, Curr. Opin. Chem. Biol. 3 (1999) 47–53.

[20] M.M. Musa, K.I. Ziegelmann-Fjeld, C. Vieille, J.G. Zeikus, R.S. Phillips, Xerogel-encapsulated W110A secondary alcohol dehydrogenase from *Thermoanaerobacter ethanolicus* performs asymmetric reduction of hydrophobic ketones in organic solvents, Angew. Chem. Int. Ed. 46 (2007) 3091–3094.

[21] Z.-G. Chen, D.-N. Zhang, L. Cao, Y.-B. Han, Highly efficient and regioselective acylation of pharmacologically interesting cordycepin catalyzed by lipase in the eco-friendly solvent 2-methyltetrahydrofuran, Bioresource Technol. 133 (2013) 82–86.

[22] (a) M.-J. Kim, Y. Ahn, J. Park, Dynamic kinetic resolutions and asymmetric transformations by enzymes coupled with metal catalysis, Curr. Opin. Biotechnol. 13 (2002) 578–587. (b) B.A. Persson, A.L.E. Larsson, M.L. Ray, J.-E. Bäckvall, Ruthenium- and enzyme-catalyzed dynamic kinetic resolution of secondary alcohols, J. Am. Chem. Soc. 121 (1999) 1645–1650. (c) S. Akai, Dynamic kinetic resolution of racemic allylic alcohols via hydrolase-metal combo catalysis: an effective method for the synthesis of optically active compounds, Chem. Lett. 43 (2014) 746–754. (d) M. Egi, K. Sugiyama, M. Saneto, R. Hanada, K. Kato, S. Akai, A mesoporous-silica-immobilized oxovanadium cocatalyst for the lipase-catalyzed dynamic kinetic resolution of racemic alcohols, Angew. Chem. Int. Ed. 52 (2013) 3654–3658.

[23] K. Fuhshuku, Y. Asano, Synthesis of (*R*)-β-nitro alcohols catalyzed by *R*-selective hydroxynitrile lyase from *Arabidopsis thaliana* in the aqueous–organic biphasic system, J. Biotechnol. 153 (2011) 153–159.

[24] X. Wu, C. Chen, N. Liu, Y. Chen, Preparation of ethyl 3*R*,5*S*-6-(benzyloxy)-3,5-dihydroxy-hexanoate by recombinant diketoreductase in a biphasic system, Bioresource Technol. 102 (2011) 3649–3652.

[25] H. Matsumae, M. Furui, T. Shibatani, T. Tosa, Production of optically active 3-phenylglycidic acid ester by the lipase from Serratia marcescens on a hollow-fiber membrane reactor, J. Ferment. Bioeng. 78 (1994) 59–63.

[26] C. Palocci, M. Falconi, L. Chronopoulou, E. Cernia, Lipase-catalyzed regioselective acylation of tritylglycosides in supercritical carbon dioxide, J. Supercrit. Fluids 45 (2008) 88–93.

[27] T. Matsuda, K. Watanabe, T. Harada, K. Nakamura, Y. Arita, Y. Misumi, S. Ichikawa, T. Ikariya, High-efficiency and minimum-waste continuous kinetic resolution of racemic alcohols by using lipase in supercritical carbon dioxide, Chem. Commun. (2004) 2286–2287.

[28] A. Dibenedetto, R.L. Noce, C. Pastore, M. Aresta, C. Fragale, First *in vitro* use of the phenylphosphate carboxylase enzyme in supercritical CO_2 for the selective carboxylation of phenol to 4-hydroxybenzoic acid, Environ. Chem. Lett. 3 (2006) 145–148.

[29] T. Matsuda, Y. Ohashi, T. Harada, R. Yanagihara, T. Nagasawa, K. Nakamura, Conversion of pyrrole to pyrrole-2-carboxylate by cells of *B. megaterium* in supercritical CO_2, Chem. Commun. (2001) 2194–2195.

[30] T. Harada, Y. Kubota, T. Kamitanaka, K. Nakamura, T. Matsuda, A novel method for enzymatic asymmetric reduction of ketones in a supercritical carbon dioxide/water biphasic system, Tetrahedron Lett. 50 (2009) 4934–4936.

[31] U. Salgın, S. Salgın, S. Takaç, Hydrolysis of racemic naproxen methyl ester by *Candida rugosa* lipase, J. Supercrit. Fluids 43 (2007) 310–316.

[32] C. Gervaise, R. Daniellou, C. Nugier-Chauvin, V. Ferrières, Influencing the regioselectivity of lipase-catalyzed hydrolysis with [bmim]PF6, Tetrahedron Lett. 50 (2009) 2083–2085.

[33] M. Eckstein, M.V. Filho, A. Liese, U. Kragl, Use of an ionic liquid in a two-phase system to improve an alcohol dehydrogenase catalysed reduction, Chem. Commun. (2005) 1084–1085.

[34] H. Pfruender, M. Midjojo, U. Kragle, D. Weuster-Botz, Efficient whole-cell biotransformation in a biphasic ionic liquid/water system, Angew. Chem. Int. Ed. 43 (2004) 4529–4531.

[35] D.-M. He, I. Kaleem, S.-Y. Qin, D.-Z. Dai, G.-Y. Liu, C. Li, Biosynthesis of glycyrrhetic acid 3-O-mono-β-D-glucuronide catalyzed by β-D-glucuronidase with enhanced bond selectivity in an ionic liquid/buffer biphasic system, Process Biochem. 45 (2010) 1916–1922.

[36] P. Lozano, J.M. Bernal, M. Vaultier, Towards continuous sustainable processes for enzymatic synthesis of biodiesel in hydrophobic ionic liquids/supercritical carbon dioxide biphasic systems, Fuel 90 (2011) 3461–3467.

[37] F.J. Hernández, A.P.d.l. Ríos, D. Gómez, M. Rubio, G. Víllora, A new recirculating enzymatic membrane reactor for ester synthesis in ionic liquid/supercritical carbon dioxide biphasic systems, Appl. Catal. B Environ. 67 (2006) 121–126.

CHAPTER 4

Enzyme Immobilization for Organic Synthesis

Melissa L.E. Gutarra*, Leandro S.M. Miranda†, Rodrigo O.M.A. de Souza†
*Escola de Química, Federal University of Rio de Janeiro, Pólo Xerém, Estrada de Xerém, Xerém, Duque de Caxias, Rio de Janeiro, Brazil
†Biocatalysts and Organic Synthesis Lab, Organic Chemistry Department, Chemistry Institute, Federal University of Rio de Janeiro, Rio de Janeiro, Brazil

1 Introduction

During the past years, the use of enzymes for organic chemistry has become more common in research laboratories [1,2] and is now in place even in undergraduate labs [3,4]. Some enzymatic reactions have also served as the basis of green and sustainable manufacturing processes [5] in the pharmaceutical, flavor/fragrance and fine chemicals industries. Unfortunately, of the many biocatalytic processes reported in the research literature, only a relative few have been applied on production scales so far. One reason for the lack of impact on larger scales is the often low stabilities of enzymes along with difficulties in their recovery and reuse [6]. Immobilization offers an elegant solution to these problems. Since the 1916 report by Nelson and Griffin [7], chemists have sought efficient methods for enzyme immobilization [8,9], which offers several potential improvements:

- Easier handling;
- Minimized protein contamination in the final product;
- Efficient enzyme recovery and recycling;
- Enhanced enzyme stability;
- Low allergenicity;
- Higher productivities and consequently lower costs;

It should also be noted that immobilizing enzymes also has some disadvantages for industrial applications:

- Cost of the carrier and fixing agents;
- Loss of enzyme activity during the immobilization process;
- Slower mass transfer through the solid support, both during the immobilization process and during the reaction of interest.

Organic Synthesis Using Biocatalysis. http://dx.doi.org/10.1016/B978-0-12-411518-7.00004-4

Until recently, these drawbacks limited the broader application of immobilized enzymes on industrial scales. Nowadays, large-scale production processes (10^3–10^7 tons per year) using immobilized enzymes are practiced with a small number of enzymes such as lipase [10–12], lactase [13], penicillin G acylase [14], aspartase [15], and glucose isomerase [16], among others. DiCosimo and coworkers have recently published an interesting review article on the industrial use of immobilized enzymes [5].

The growing interest of organic and inorganic chemists in developing new immobilized enzymes has significantly improved the toolbox of immobilization techniques. These efforts have also yielded a better understanding of the final biocatalyst preparation using methods originally devised to study heterogeneous catalysts [17–23]. A number of analytical techniques have recently been used including fluorescence and infrared spectroscopies, X-ray diffraction, Brunauer–Emmett–Teller (BET) and thermogravimetric analyses (TG), differential scanning calorimetry (DSC), transmission electron microscopy (TEM), and X-ray photoelectron spectroscopy (XPS). All have provided valuable insights into the structure of immobilized biocatalysts.

2 Immobilization techniques

Protocols for enzyme immobilization include entrapment, cross-linking, and adsorption or linkage to porous or nanosupports. It should also be noted that enzyme immobilization can be used as a step in enzyme isolation and purification in many cases.

2.1 Enzyme Entrapment

Entrapment describes methods in which the enzyme is confined inside the inner cavities of a solid organic or inorganic matrix. Enzyme entrapment is one of the simplest immobilization strategies and is least aggressive with regard to changing enzyme structure since the protein/matrix interactions are weak and the conditions during the preparation are usually mild [24–26]. Entrapment promotes enzyme insolubilization and thereby enables easy recovery of the biocatalyst from the reaction media. Depending on the gel material and its modification, the matrix can also provide an advantageous microenvironment [27] that enhances enzyme stability and/or other useful properties. Specifically, the electrostatic, hydrophilic *versus* hydrophobic character of the matrix interacting with the surroundings can make the concentrations of substrates, products, and/or hydrogen ions in the micro environment different from those of the bulk macroenvironment [28]. Microenvironment effects are also observed in enzymes anchored in membranes or cell walls. This occurs naturally in cells and has been extensively explored to produce, by genetic modification, several anchored enzymes for application as biocatalysts such as lipases displayed on *Pichia pastoris* or *Escherichia coli* cell surfaces [29] so that whole cells can be used in biodiesel production.

When using immobilization onto supports, one potential problem is that the matrix structure can slow diffusion rates, an effect that depends critically on pore size and microenvironment. It is therefore essential to consider both the chemical and physical properties of the matrix since these can alter thermal stability, kinetic behaviors, and the optimal pH and temperature of the enzyme for a given reaction [30].

Common organic materials used for enzyme entrapment are gelatin, alginate, polyacrylamide, chitosan, xanthan, and agarose with alginate being the most frequently used for enzyme entrapment [24,31]. All of these supports enjoy the advantages of being both biocompatible and abundant. On the other hand, they also have low mechanical stabilities and their large pores may allow enzyme to leach out from the matrix [32]. Entrapment within these organic matrices can be improved covering their surfaces with a polymer, by cross-linking or by modifying their structures by other techniques. For example, calcium alginate was covered by silicate after entrapment of a thermoalkalophilic esterase, which improved immobilization yield and efficiency, reuse and stability [33]. The stability of laccase for dye decolorization was also improved by preparing interpenetrating polymer networks or semi-interpenetrating polymer networks of two different polymers such as polyacrylamide and alginate [32]. A hybrid gel composed of both alginate and starch immobilized peroxidase both by entrapment as well as by covalent linkage to make a more stable biocatalyst [24]. Cellulose biopolymer composite hydrogel beads were produced for lipase immobilization. Cellulose was chosen as main polymer because of its thermal and mechanical properties as well as its abundance and biocompatibility. During synthesis, cellulose and the copolymers,for example, agarose, chitosan, carrageenan, and agar, were solubilized by using ionic liquids. The best yield of immobilization, 52.6%, was achieved with cellulose–agar and the best residual activity after reuse (95.9%) was achieved with cellulose–agarose. These results, together with morphology analysis of the matrix, indicated that better properties could be reached by combining two or more polymers [34]. It is also possible to improve entrapment immobilization by physical methods. For example, using ultrasound during dextranase immobilization into calcium alginate resulted in better properties and a higher recyclability [35].

Inorganic materials, particularly those based on silica, are also very useful for enzyme entrapment. Silica is biocompatible and has excellent mechanical properties. Some other important parameters such as surface area and pore size can be modulated [36,37]. The sol–gel process involves hydrolyzing silane precursors (tetra-alkoxysilanes and alkoxysilane) followed by their polycondensation to form ceramic materials [38]. Conventional sol–gel matrices have tight pores of approximately 1.5 nm diameter [28,39]. These prevent diffusion of the enzyme out of the matrix and can also stabilize enzyme structures, thereby improving thermal and chemical stabilities and increasing storage abilities [40]. Lim et al. immobilized alcalase from *Bacillus licheniformis*

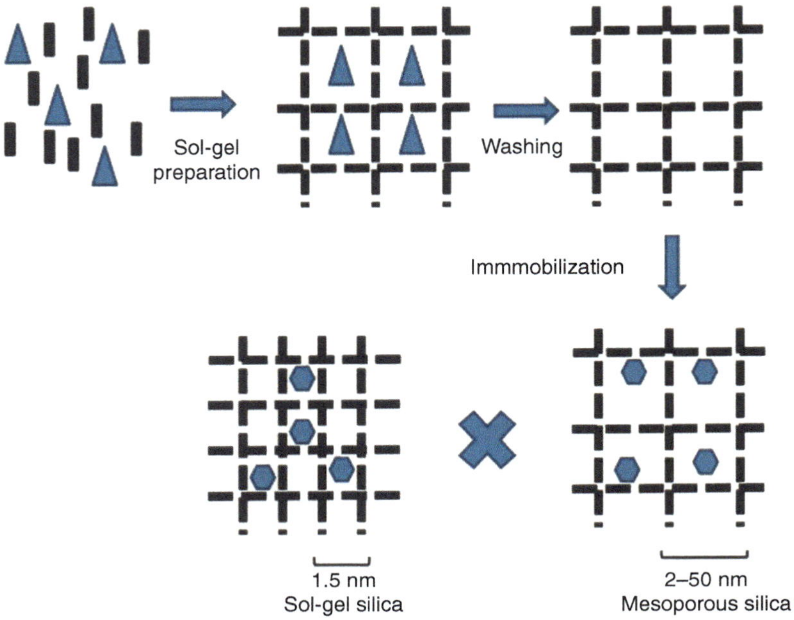

FIGURE 4.1
Synthesis of sol–gel mesoporous silica for enzyme immobilization and comparison with conventional sol–gel silica. Triangles, surfactants; hexagon, enzymes.

in a sol–gel matrix and found 100% retention of activity after 14 reaction cycles, high stability at 70 °C, and also after storage for 50 days in ambient temperature [41]. A silica sol–gel/β-galactosidase preparation possessed higher pH and temperature stabilities and greater retained activity during storage. These properties were very useful in formulating new pharmaceutical preparations for the treatment of lactose intolerance [36]. In some applications, mesoporous sol–gel matrices were prepared [42]. An additive (usually a surfactant) is used to increase the pore size to produce polymers with larger internal diameters (2–50 nm) that allow better substrate and product diffusion rates (Figure 4.1) [22]. In one example, cellobiase was immobilized in mesoporous sol–gel silica so that it could be used for multiple reaction cycles in biomass hydrolysis. In this case, fructose, a nonsurfactant molecule, was used as the template. After sol–gel synthesis, this additive was easily removed from the preparation [42].

Entrapment has also been used extensively to immobilize enzymes for biosensor applications, for example, glucose oxidase into a polyacrymamide microgel [43] and L-aminoacid oxidase into carbon nanotubes aqueous sol–gel composite [44].

Example 1

Enzyme immobilization by sol–gel methodology

Sol–gel methodology for enzyme entrapment is based on the formation of a tetraethoxyorthosilicate (TEOS) matrix. For the matrix formation, 40 μL polyethylene glycol (PEG-6000) (20% in phosphate buffer (50 mM, pH 7.0)), 200 μL TEOS, and 10 μL NaF (0.2 M) in a final volume of 350 μL are mixed and incubated for 3 min at 4°C. Afterward, 100 μL of enzyme solution (0.5–3.5 units in the same buffer) was added and the mixture is stirred on ice for 15 min until a homogeneous emulsion is formed. This mixture was made using the following proportion: PEG:TEOS:NaF:Enzyme/0.4:2:0.1:1. Then, the entrapment matrix was washed three times with 1 mL of HEPES/NaOH buffer (1 mM, pH 8.2), containing 20 mM Na_2SO_4 and 1 mM EDTA and then centrifuged at 13,000 rpm for 1 min. The enzyme entrapped by sol–gel was stored at 4°C [45].

2.2 Cross-Linked Enzyme Aggregates (CLEAs)

Cross-linked enzyme aggregates (CLEAs) are a type of enzyme immobilization that does not require exogenous solid supports (Figure 4.2). Compared with the earlier cross-linked enzyme crystals (CLECs) methodology, CLEAs are far simpler to prepare since no prior enzyme crystallization is required [46]. CLEAs are formed by adding both protein precipitating and cross-linking reagents to produce stabilized aggregates. The cross-linker prevents the aggregate from redissolving after the precipitant is removed. The precipitation step utilizes conventional techniques such as salting-out, solvent, or polymer precipitation. Ammonium sulfate is used most commonly and glutaraldehyde is the usual cross-linker [47]. The quantities of precipitating and cross-linking agents along with the enzyme amount impact both the recovered catalytic activity and the properties of the resulting biocatalysts. These generally need to be optimized for each enzyme [48]. CLEAs show several advantages over other types of protein immobilization strategies: (1) CLEAs have very high activity per gram of biocatalyst as almost all the mass is protein, rather than inert carrier. (2) CLEA production protocols are very simple and inexpensive. In particular, there are no costs associated with a solid support. (3) Since no support is used, diffusion problems and mechanical stresses are lower. (4) It is not necessary to use purified enzyme extracts. (5) CLEAs are easy to recover for reuse [46]. For these reasons, CLEAs have been used extensively for immobilizing many different enzymes including lipases [48–50], amylases [51,52], tyrosinase [53], epoxide hydrolase [54], penicillin acylase [55], peroxidase [56], laccases [57,58] as well as multimeric enzymes [59]. In almost every case, the CLEA preparations showed improved storage properties and enhanced thermal and solvent stabilities. Enzymes in CLEA form have shown higher activity, specificity and enatioselectivity than the corresponding free enzymes [50].

CLEAs do have some limitations, however. The low number of lysine residues on some enzyme surfaces may hinder with glutaraldehyde cross-linking, which

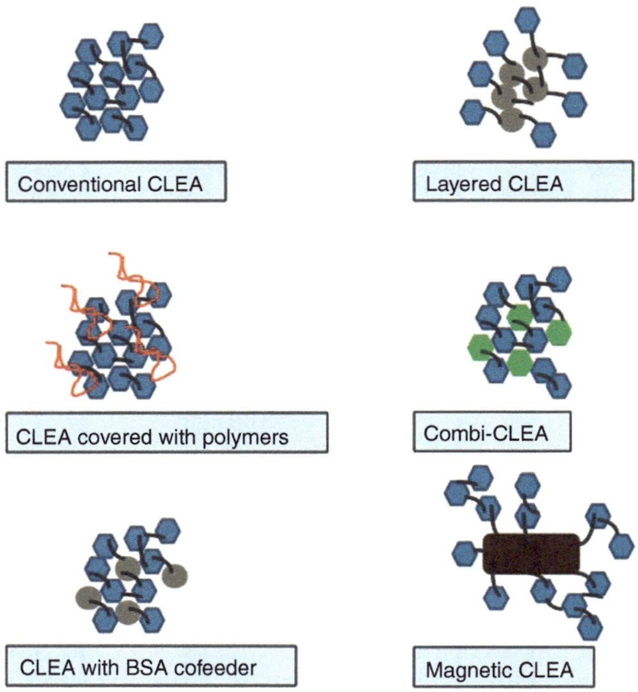

FIGURE 4.2
Cross-linked enzyme aggregates (CLEAs). Blue hexagon, enzyme; green hexagon, enzyme; gray sphere, BSA; black line, glutaraldehyde linkage; red line, polymers; rectangle, magnetic nanoparticle.

can result in enzyme leakage during reactions. In other cases, the covalent linkage between glutaraldehyde and a surface lysine residue, particularly near the active site, can interfere with the enzyme's catalytic ability. CLEAs can also have lower activity when the substrate cannot access the enzyme active sites in the interior of the aggregate [26]. To improve CLEAs stability and activity, the coprecipitation of the enzyme with polymers or adjuvant protein or additives were proposed.

2.2.1 CLEAS WITH ADJUVANT

Coprecipitation of enzymes and a polymer rich in amino groups such as polyethyleneimine or pentaethylenehexamine were proposed as one solution to the problems outlined above [56,60,61]. Lipases from *Alcaligenes* sp. and *Candida antarctica* fraction B were coprecipitated with polyethyleneimine and polyethyleneimine sulfate-dextran, which changed the enzyme microenvironment and also physically stabilized the biocatalyst. The greater stability of coprecipitated lipase CLEA was apparent from the absence of free lipase after samples were boiled in detergent (sodium dodecylsulfate) and analyzed by gel electrophoresis. By contrast, conventional CLEAs released free enzyme into solution under the same conditions. Even better, coprecipitating enzyme and polymer also improved the lipase catalytic activity and enantioselectivity [61]. CLEAs can also be covered by a polymer to avoid enzyme leakage and enhance recyclability (Figure 4.2).

CLEAs have also been prepared with additives such as *n*-heptane or oleic acid. The hypothesis was that these reagents would activate the immobilized lipase by locking it into an active conformation by the presence of a hydrocarbon/water interface and by bioimprinting with fatty acids, respectively [48]. Bovine serum albumin (BSA) has also been used as an adjuvant to enhance precipitation, particularly when the target enzyme concentration was low (Figure 4.2). BSA is an inexpensive protein with a large number of surface lysine side chains that facilitate incorporation into CLEA networks. It also helps avoid excessive cross-linking of the target enzyme, particularly near its active site, when high glutaraldehyde concentrations are needed. In the case of *Thermomyces lanuginosa* lipase, the addition of BSA increased the recovered activity [48]. Ovalbumin has also been used as a coprecipitated protein in CLEA formation for peroxidase immobilization where it enhanced the recovered activity [56].

2.2.2 LAYERED CLEAS

To improve both the recovered activity and substrate access to the enzyme, a modification of the conventional CLEAs preparation was developed. Layered CLEAs are formed by precipitating enzymes over a previously precipitated and cross-linked core of BSA or similar inexpensive protein (Figure 4.2). Lipase was immobilized in layered CLEAs with one and three layers showing higher recovered activity with one-layered CLEA (75%) compared to nonlayered CLEA (8%) and to three-layered CLEA (25%) [48].

2.2.3 COMBI-CLEAS

Such CLEAs can also be produced that contain two or more different enzymes in the same aggregates. These so-called combi-CLEAs allow multiple reactions (either cascade or noncascade) to be carried out in the same reaction space (Figure 4.2). This preparation strategy also has some advantages in promoting high specific-enzyme activity and catalytic productivity since reaction rates can be improved by close spatial proximity between the different enzymes and substrates.

Using combi-CLEAs allows multiple process steps in a single operation, thereby minimizing solvent usage, waste generation, and costs. For example, a combi-CLEA combining glucoamylase and pullulanase were used in the saccharification of starch with high efficiency and recyclability (8 cycles) [51]. This was improved even further by synthesizing a combi-CLEA that contained three enzymes (α-amylase, pullulanase, and glucoamylase), which performed starch liquefaction and saccharification in a single step. The one-pot process with a three-enzyme combi-CLEA had higher conversion (100%) as compared to a mixture of separate CLEAs (60%) and the free enzyme mixture (40%) [52].

2.2.4 *MAGNETIC CLEAS*

Recycling of conventional CLEAs may result in increased aggregate size due to clumping during separation from reaction mixture by centrifugation or filtration. Larger CLEA particle sizes may present difficulties in mass transfer even when the substrate is not a polymer. Using magnetic CLEAs allows the particles to be recovered without centrifugation or filtration (Figure 4.2). A magnetic CLEA of α-amylase was prepared by adding amino-functionalized magnetite nanoparticles. Surface amino groups on the nanoparticles also improved amylase immobilization since this protein has a relatively small number of surface lysine residues. These magnetic CLEAs could be recovered with 100% efficiency by a magnetic field and no loss of activity was observed after 6 cycles. A conventional CLEA showed only 45% recovered activity and only 25% of the initial catalytic activity remained after 6 cycles (mostly due to α-amylase leaching from the CLEAs) [62]. Magnetic nanoparticles cross-linked with amyloglucosidase were prepared and showed enhanced thermal stability (57 °C for 24 h with 82% of recovered activity) and reusability (31 cycles with 90% of recovered activity) [63].

Example 2

Enzyme immobilization by CLEA methodology

The formation of CLEAs was divided in two steps. The first was responsible for the precipitation of the proteins of the crude extract and the second for coagulation. Addition of ammonium sulfate to enzyme solution containing 5–25 mg of protein/mL was done to reach 60–90% of salt saturation and the mixture was magnetically stirred at 4°C for 10 min to 2 h before a solution of glutaraldehyde (final concentration of 2.4–40 mM) was added. The mixture was subjected to magnetic stirring at 4°C for 2–24 h. Then the enzyme aggregates were separated by centrifugation (14,000 rpm, 10 min), washed with buffer and stored at 4°C [53,57].

2.3 Enzyme Immobilization on Porous Supports

Adsorbing or linking enzymes to the surfaces of porous materials is one of the most common methodologies for enzyme immobilization. The interactions can be reversible, for example, by hydrophobic and ionic interactions or irreversible if covalent bonds are formed.

2.3.1 *TYPES OF INTERACTIONS*

The interaction between an enzyme and a material surface can either increase or decrease its catalytic activity, and the actual outcome depends on the nature of the enzyme, the nature of the material surface, and the microenvironment produced by their interaction. Changes in enzyme conformation caused by immobilization may alter its interactions with different substrates and its stability with respect to solvent, temperature, and pH. It is therefore highly valuable to understand how an enzyme unfolds in order to choose an adequate support for immobilization and to develop strategies of enzyme or surface modification to drive a successful immobilization and stabilization of the biocatalyst [64].

Supports with widely different degrees of hydrophilicity/hydrophobicity have been used for enzyme immobilization. This ratio is an important variable since the support can cause conformational changes in enzymes. Hydrophilic surfaces offer the possibility of hydrogen bonding between the support and surface amino-acid side chains on the enzyme. Hydrophobic supports may interact with nonpolar amino-acid side chains; because these are normally sequestered within the protein interior, their interaction with a solid support almost always requires an enzyme conformational change. In most cases, these changes are detrimental to catalytic activity, although the converse is often true for lipases. These hydrolases normally function at an aqueous/organic interface and their active sites are often covered by a loop of protein (usually referred to as the "lid") that opens best when in contact with an aqueous/organic interface. Interactions with a hydrophobic support can keep the lipase in the open conformation that is most efficient for catalysis (Figure 4.3). In such cases, the immobilized enzyme has an even higher activity compared to the free enzyme, a phenomenon known as hyperactivation [65,66]. Due to these structural features, lipase immobilization on hydrophobic supports at low ionic strength occurs preferentially compared to other types of enzymes, which allows both purification and immobilization to occur in a single step [65].

Acidic and basic amino acids from a protein impart a net charge that varies with external pH. Ionic adsorption on a negatively charged support occurs by complementary positive surface charges on the protein and vice versa (Figure 4.3). Depending on the distribution of charged moieties on the enzyme surface,

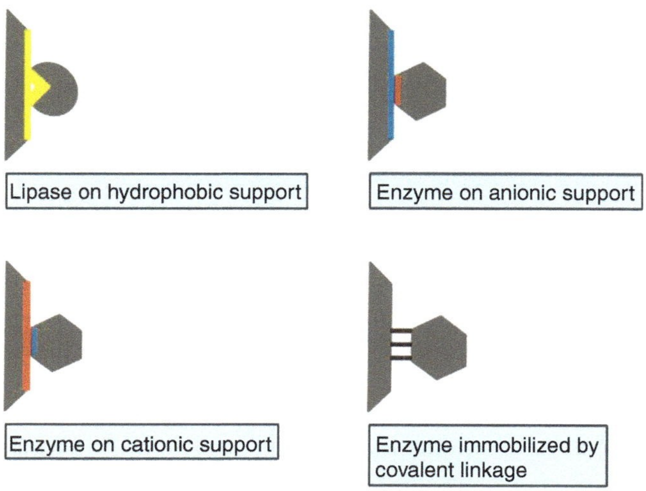

FIGURE 4.3
Lipase immobilization on hydrophobic, ionic and covalent supports. Yellow, hydrophobic area; red, positive charged areas; blue,negative charged areas; black lines, covalent linkage.

ionic immobilization can also affect enzyme orientation on the solid support. In addition, electrostatic attraction or repulsion may distort the enzyme structure, resulting in changes in activity and other properties. To improve ionic immobilization efficiency or promote better-defined (or different) orientations on a charged support, the charge distribution on the enzyme surface can be altered by genetic engineering or by chemical modification. The most common chemical modifications are the introduction of amines by reaction of ethylendiamine on carboxylic acids to produce a positively charged protein or the introduction of acids by reaction of succinic anhydride on amines to produce a negative charge [64].

Covalent bonding between an enzyme and a solid support is often associated with enzyme stabilization and high reusability. In this strategy, the enzyme is permanently fixed in a particular conformation because immobilization almost always involves multiple covalent bonds (multipoint linkage) on the surface of the enzyme (Figure 4.3). Covalent attachment requires that either the enzyme or the support (or both) be activated. The reaction of aldehydes (glyoxyl-agarose beads) on the support with amino groups on the enzyme is the most commonly used attachment methodology. This involves Schiff's base formation between amino moieties of surface lysine side chains with aldehydes from the support, usually at or near pH 10 (to maintain a reasonable concentration of deprotonated lysine side chains available for the reaction). If the conjugation is carried out at pH 7, only the amino terminus is reactive that yields only one enzyme/support bond (unless the enzyme is multimeric). Once formed, the hydrolytically unstable Schiff's bases are normally reduced by sodium borohydride to eliminate the possibility of enzyme release. This process also converts residual aldehyde groups on the support into much less-reactive hydroxyl groups [29,30,66,67]. Glutaraldehyde, a dialdehyde, has also been used to form Schiff's base bridges between amino groups on both the enzyme surface and those on a solid support. Subsequent reduction by sodium borohydride is also commonly employed in this approach as well [68]. Although less reactive than aldehydes, surface epoxy groups have also been used for multipoint immobilization as they react with amino groups in proteins under very mild experimental conditions (pH 7.5) forming a very stable secondary amine [69].

To achieve high covalent-immobilization efficiency, several important factors must be optimized including protein load, pH, ionic strength, temperature, mass: volume ratio, and agitation rate [70,71]. Moreover, it is sometimes necessary to add protective molecules such as carbohydrates and polyethyleneglycol to avoid enzyme denaturation and/or inactivation during conjugations.

2.3.2 SUPPORT CHARACTERISTICS

Porous supports are commonly used for enzyme immobilization. One key advantage they enjoy over nonporous materials is a much higher loading capacity. In addition, by locating enzymes within a porous structure, they are prevented from interactions that might lead to aggregation, autolysis, or hydrolysis by proteases that are often present in crude enzyme mixtures. A porous structure also helps prevent inactivation by immiscible organic solvents if they cannot penetrate inside the porous material. Immobilization inside a porous structure can also create a microenvironment that stabilizes the enzyme under operational conditions [66]. On the other hand, when compared with the free enzyme, immobilizing an enzyme on a porous support may introduce problems related to hindered mass transfer and diffusion that reduces reaction rates. Mass transfer and diffusion problems depend mostly on the size of pores, protein loading, substrate solubility, and the nature of the support (interface between protein, material and bulk substrate solution) [72]. Porous supports such as cross-linked agarose contain much bigger pores (up to 200 nm depending on agarose concentration in the hydrogel), which would not cause problems to protein diffusivity inside the porous structures. On the other hand, other supports such as acrylic, for example, Sepabeads, have much smaller pores sizes (10–20 nm).

It is important that the support material be stable under operational conditions with respect to mechanical forces, solvent, temperature, pH, etc. A wide variety of different supports have been used for enzyme immobilization including natural polymers (cellulose, agarose, chitosan), synthetic polymers (acrylic resins, styrene–divinylbenzene, polypropylene), and inorganic materials (silica, porous glass, Celite). Each offers advantages and disadvantages, and the optimal support for a given enzyme and reaction must be chosen individually.

2.3.3 MATRIX SURFACE FUNCTIONALIZATION OR ENZYME MODIFICATION

Many matrices used for enzyme immobilization require prior surface modification to optimize enzyme interactions. This can involve introducing groups to increase hydrophobicity, to add or alter surface charge or to add reactive groups for covalent linkage. Some synthetic polymers such as styrene-divinylbenzene and polypropylene are naturally hydrophobic and can be used directly for enzyme immobilization [73]. Modifications of the surface and/or enzyme can also be performed to improve immobilization, stabilize the biocatalyst, promote enzyme orientation on the support, or better control the process. Although all support modifications involve chemical reactions, enzymes can either be modified chemically, by modifying amino-acid side chains with polymers, small molecules, peptides, or by genetic engineering by changing specific amino acids of the protein [74].

2.3.4 GREEN MATERIALS

Natural, inexpensive, and easily available alternative materials have been used for the immobilization of enzymes as alternatives to traditional matrices. Such materials facilitate subsequent utilization at large scales and reduce biocatalyst production cost. Some industrial by-products of natural origin such as chitin and chitosan as well as lignocellulosic agroindustrial residues have been used for enzyme immobilization. Chitosan is a material with interesting chemical and biological properties for enzyme immobilization. It has high biodegradability, biocompatibility, and renewability that allow applications in many fields. This polymer also has reactive amino and hydroxyl groups that can be used for chemical modifications. Chitin and chitosan can be obtained at a relatively low cost from shells of crabs and shrimps that are wastes of the seafood industry. Chitin can be obtained by deproteinization and decalcification of the raw material. Chitosan is produced by partial or complete *N*-deacetylation of chitin and is available with a variety of different amino group contents. Chitosan has been used to immobilize several enzymes such as cellulases, α-amylase, β-amylase, catalase, α-chymotrypisin, alcohol dehydrogenase, and alkaline protease [75]. Corn stalk and other lignocelulosic agroindustrial residue can be used as enzyme carrier. They can be used for adsorption or can be modified to allow covalent immobilization [76]. Green coconut fiber has been used for laccase immobilization by physical adsorption [77].

One limitation of these natural enzyme supports is their often-heterogeneous compositions, which can result in different and unknown immobilization behaviors, making reproducibility an issue.

Example 3

Enzyme immobilization on porous support by hydrophobic interaction methodology

The immobilization procedure is divided in two steps: support washing and immobilization. 1 g of support was suspended in 10–40 mL of pure solvent and gently stirred for 20 min to 1 h and then filtered by vacuum and the procedure is repeated with solutions with decreased concentration of solvent (50%, 25%, and 0%). After that, 1 g of support is suspended in 10–40 mL enzyme solution (0.5–35 mg of protein/g of support) in 10mM sodium phosphate buffer at pH 7 and 20°C for 1–24 h. After that the support is washed by vacuum filtration with water and stored at 4°C [65].

For immobilization by hydrophobic interaction of *C. antarctica* B lipase on porous styrene–divinylbenzene beads supports, 1 g support was suspended in 40 mL of an enzyme solution containing 0.5 mg protein/mL in 10mM sodium phosphate at pH 7 and 20°C. The mixture was incubated for 2 h reaching almost 100% of immobilization. After that, the support was washed by vacuum filtration with water and stored at 4°C [73].

2.4 Nanostructured Supports

Enzymes immobilized on nanostructures have many applications in biocatalysis, biosensors and for biomedical purposes. Nanostructures for enzyme immobilization can be particles with spherical shapes, fibers, or tubes. Although most nanostructures used for enzyme immobilization are nonporous materials, their very high surface-to-volume ratios still allows them to incorporate significant quantities of enzyme per unit mass of carrier. Moreover, because all enzymes are exposed on the surface, nanostructures offer several advantages over porous macrobeads such as minimum diffusion limitation, less mass transfer problems as well as higher mechanical strength [78]. One problem with nanosupported enzymes is that nearly all protein molecules are solvent exposed, making them susceptible to environmental factors such as temperature and pH that can decrease activity and stability since the enzyme is not protected by a porous microenvironment. One solution is to cover the immobilized enzymes with polymers [64]. Other disadvantages of nanosupports that their production costs can be relatively high and they can be difficult to recover as a consequence of their small sizes [79].

In some cases, nanosupports innately possess functional groups suitable for enzyme immobilization; in others, their surfaces must first be modified. Enzymes can be immobilized on nanostructured supports either by adsorption or by covalent linkage. For covalent linkage, epoxy groups or amines (combined with glutaraldehyde) are most common. Enzymes immobilized on nonporous nanoparticles are attached on the surface of such particles. In this case, they interact more easily with external interfaces, enzyme molecules immobilized on other particles, and molecules from the bulk of substrate.

Magnetic particles made of iron oxide are the most popular nanostructure for enzyme immobilization in biocatalysis since they can be recovered easily by proximity to a magnetic field and reused for multiple cycles. Iron oxide nanoparticles have been used to immobilize a wide variety of enzymes including lipases, cellulases, pectinase, esterase, glucose oxidase, catalase, and others [80]. One disadvantage is that unmodified iron oxide nanoparticles tend to agglomerate, which changes their magnetic properties and diminishes their homogeneity. This problem has been addressed by encapsulating magnetic nanoparticles with different biomaterials such as cellulose and gum acacia or by copolymerization of iron and biomaterials such as chitosan [18,31,81]. Moreover, in biocatalysis, this strategy allows a controlled localization of the immobilized enzyme along the height of the reactor during reaction to locate the biocatalyst near substrate and to favor conversion of product as described by Dussan et al. [82] on enzymatic production of biodiesel.

Nonmagnetic nanoparticles made from gold, silica, chitosan, zirconia, and other materials have also been widely used for enzyme immobilization [79]. Gold nanoparticles are nontoxic and biocompatible, allowing them to be used for medical applications such as drug delivery. Chitosan is used for enzyme immobilization most commonly in the form of membranes, fibers, and particles. This material is also biocompatible and has reactive groups for enzyme immobilization as described in Section 3.4. Nanochitosan has great physical and chemical properties such as high surface area, porosity, strength, conductivity, and increased mechanical properties. Silica is largely used to produce enzyme carriers as silica nanoparticles, porous silica nanotubes along with the mesoporous silica preparations described previously. Silica is also applied for enzyme entrapment as described in Section 1. Zirconia nanoparticles have attractive properties such as high chemical, thermal, and pH stability and they can be modified to allow covalent enzyme linkage. A few studies on enzyme immobilization on zirconia supports have been reported such as the immobilization of lipase by Chen et al. [83]

In addition to nanoparticles, nanotubes have also been used to immobilize several enzymes. Because they are relatively simple to prepare, carbon nanotubes have been used most frequently. These consist of graphite sheets organized into a cylindrical shape with diameters up to 100 nm and lengths ranging up to micrometers. They offer advantages such as great mechanical and thermal stabilities as well as biocompatibility. Immobilization can be performed by adsorption or covalent attachment [84]. Nanofibers have interesting properties when compared with other nanostructures such as reduced-mass-transfer problems due to their thinness and easy recovery [77]. Cellulose nanofibers have been used most often for enzyme immobilization.

Example 4

Enzyme immobilization on magnetic nanoparticles by covalent linkage

Enzymes can be immobilized on different types of magnetic nanoparticles conjugated with organic materials as chitosan or gum acacia. The magnetic nanoparticles (250–500 mg) are dispersed in 10 mL of 2.5% glutaraldehyde solution in ultrasonic bath and then the suspension is rinsed twice with phosphate buffer (pH 7.0). After that 10 mL of the enzyme solution (0.5–5 mg of enzyme) is added to nanoparticles and the suspension is incubated at room temperature for 1 h with gentle stirring. The biocatalyst is separated by magnetic field and washed with buffer [79].

Magnetic nanoparticles can be acquired as commercial supports or can be prepared. Commercial magnetic nanoparticles usually are available in the range of 10–500 nm with many different functionalization. Noncommercial magnetic nanoparticles can be produced by magnetotatic bacteria or by synthetic routes. The most common synthetic route is the precipitation or coprecipitation of iron to produce magnetite. For gum acacia, magnetic nanoparticles a coprecipitation strategy was used. To produce nanoparticles, 13.51 g of ferric chloride hexahydrate and 6.95 g of ferrous sulphate heptahydrate were added to a 250 mL flask and dissolved in 70 mL of water. Then 10 g of gum acacia in aqueous ammonium hydroxide solution was slowly added to the solution over 30 min. The mixture was stirred vigorously for 1 h. The precipitate obtained was separated using magnetic decantation. The precipitate was washed with water six times. Then nanoparticles was dried in vacuum oven at 60°C for 12 h and stored in vacuum desiccators [79].

3 Specific examples of immobilized biocatalysts

Table 4.1 presents representative examples of immobilized enzymes and their applications to synthetic transformations. Examples of commercially available immobilized enzymes were omitted in order to emphasize recently developed immobilization technologies. In addition, the reactions catalyzed by immobilized whole cells were also omitted since these lie outside the scope of this chapter. With few exceptions, the examples date from 2010 at the earliest since several recent reviews have covered the prior literature [1,2,5,6,17,19–21, 23,32–38, 85–91].

Table 4.1 Transformations with Immobilized Enzymes

	Enzyme	Support	References
1) Lipases			
Jatropha Oil + MeOH ⟶ Biodiesel	*Burkholderia cepacia* lipase	3-Aminopropyl triethoxysilane Attapulgite	[92]
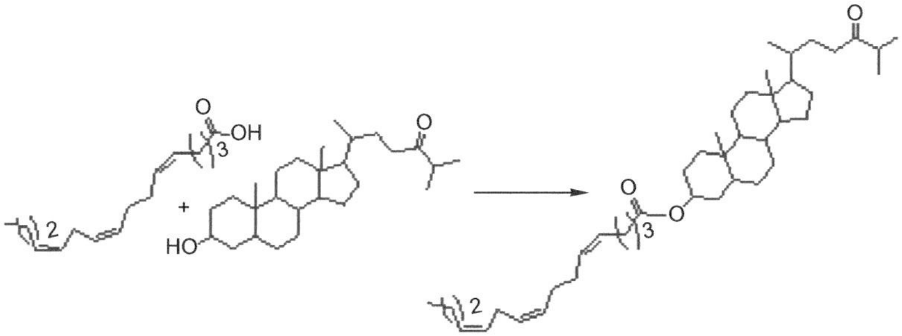	*Candida rugosa* lípase	Lewatit VP OC 1600	[93]
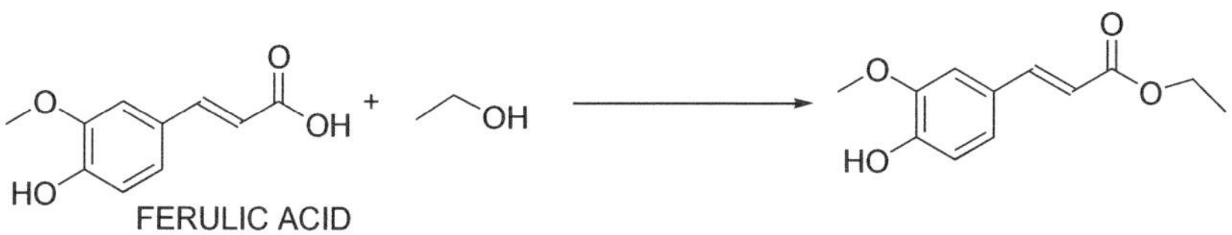	*Rhizopus oryzae* lipase	Magnetic Fe_3O_4-chitosan beads	[94]
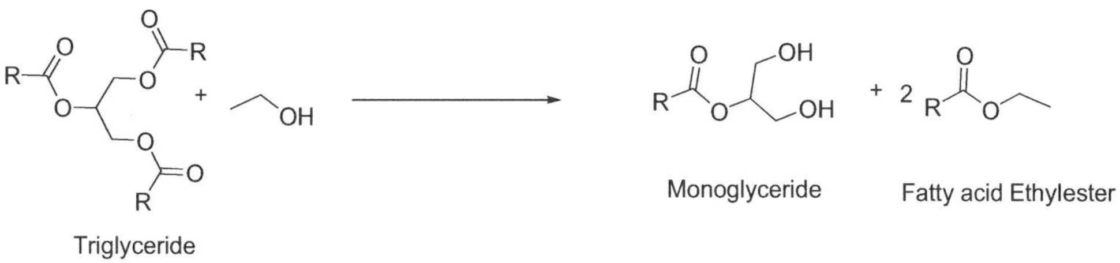	Pig pancreatic lipase	$AlPO_4$	[95]

(Continued)

Table 4.1 Transformations with Immobilized Enzymes *(cont.)*

	Enzyme	Support	References
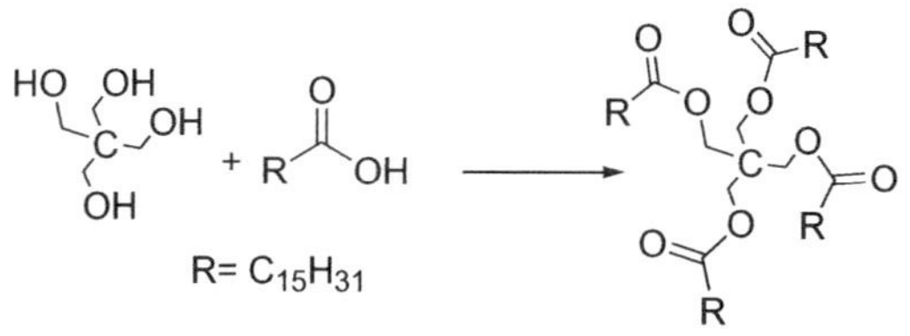	*Candida* sp.	Polyurethane foam	[96]
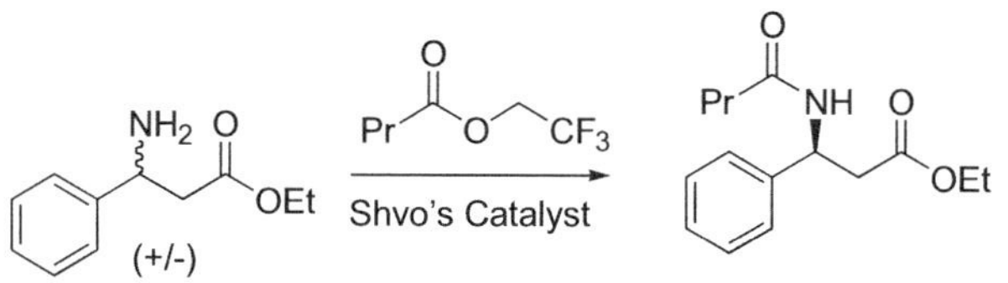	*C. antarctica* lipase A	Functionalized mesocellular foam	[97]
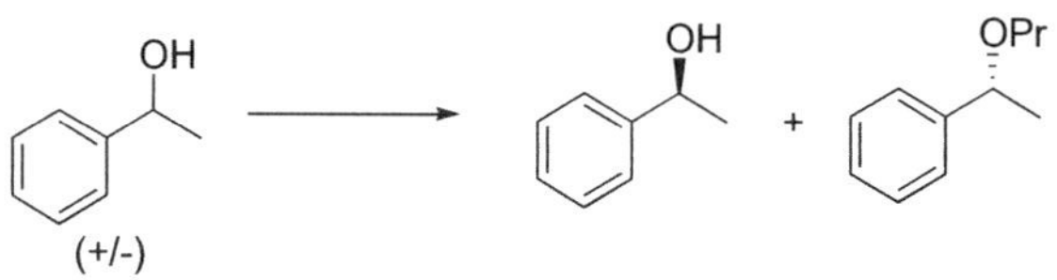	Lipase AK Amano 20 From *Pseudomonas fluorescens*	Sol–gel matrix	[98]
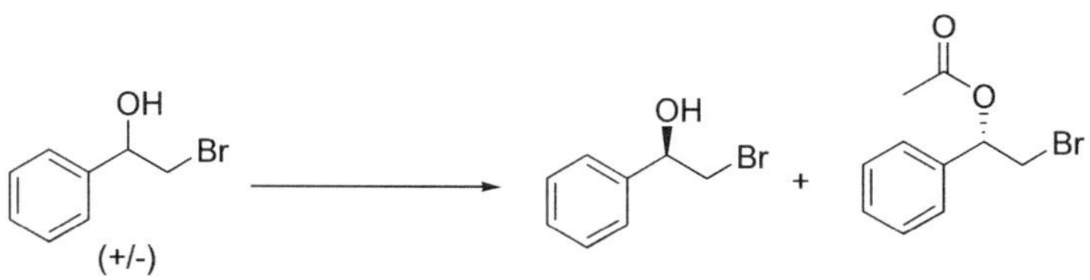	*B. cepacia* lipase	Superpara-magnetic nanoparticles	[99]
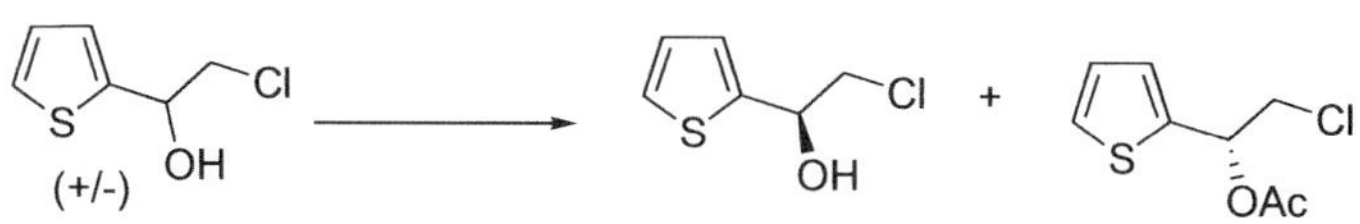	*Alcaligenes* sp	Magnetic Fe_3O_4 nanoparticles	[100]

	Enzyme	Support	References
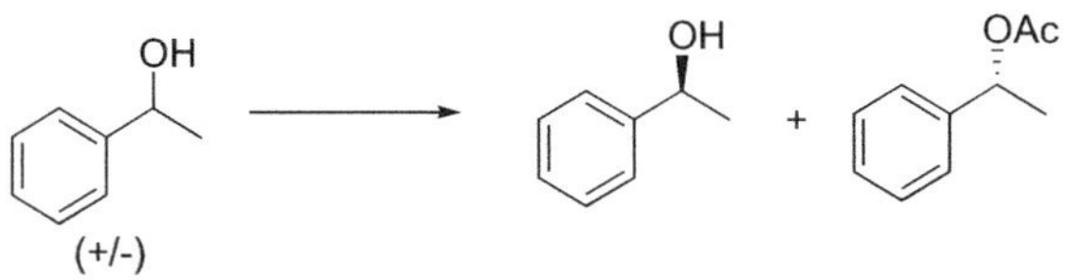	*C. antarctica* lipase B	Alginate hydrogel beads	[101]
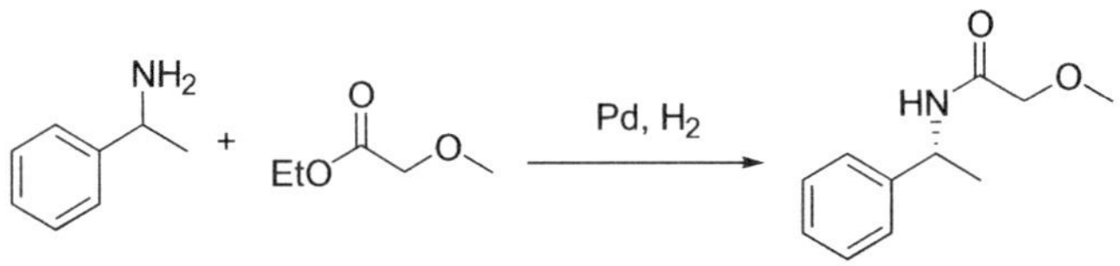	*C. antarctica* lipase B	*Aminopropyl-functionalized* mesostructured cellular foam (*MCF*) silica	[102]
2) Acylase			
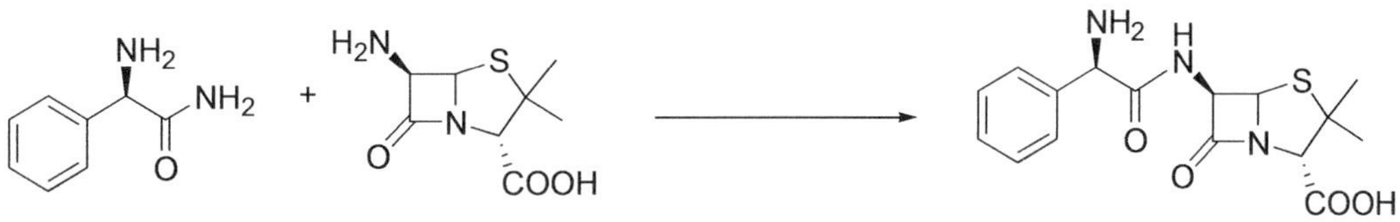	*Pencillin acylase* from *Alcaligenes faecalis*	Cross-linked enzyme aggregate	[103]
3) Epoxide Hydrolase			
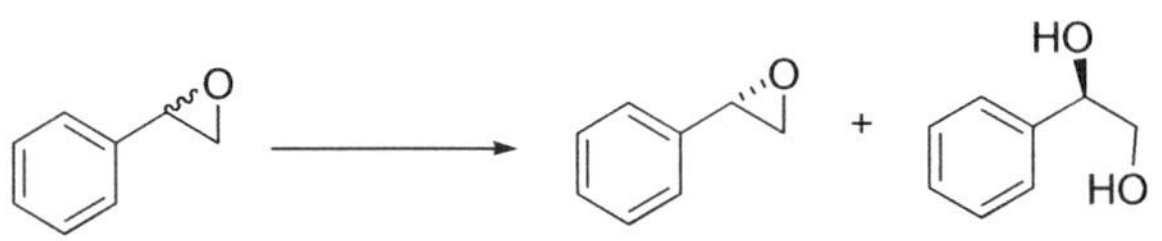	Epoxide hydrolase from *Aspergillus niger*	Eupergit C 250 L	[104]
	Fish microsomal epoxide hydrolase (mEH)	NiO nanoparticles	[105]

(Continued)

Table 4.1 Transformations with Immobilized Enzymes *(cont.)*

	Enzyme	Support	References
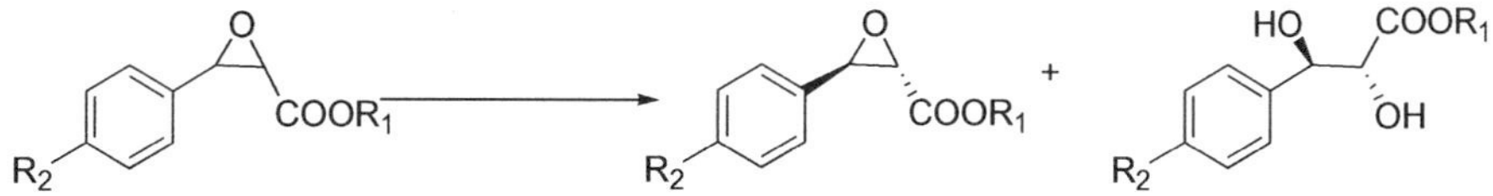 R_1= Et, R_2= H R_1= Me, R_2= OMe	Epoxide hydrolase *Phaseolus radiatus*	Gelatin gel	[106]
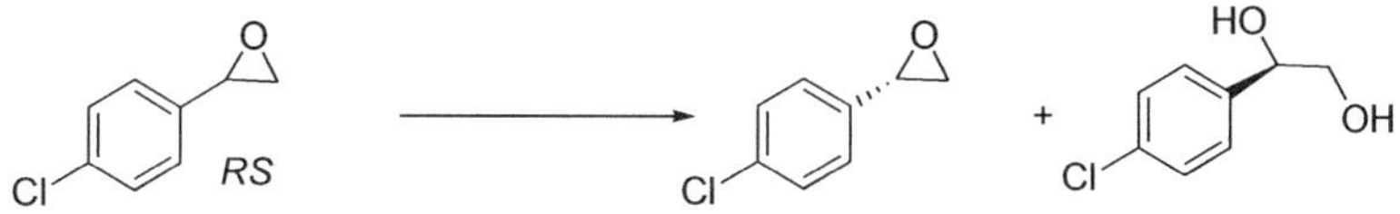	Epoxide hydrolase recombinant *A. niger*	Accurel EP	[107]
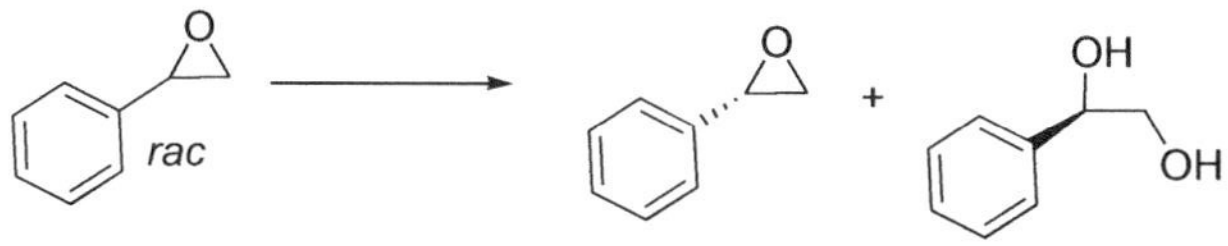	Epoxide hydrolase *Mugil cephalus*	Mesoporous silica with bottle-neck mesopores	[108]
	Epoxide hydrolase from *A. niger*	Eupergit C 250 L	[109]
4) Nitrilase			
	Nitrilase from *A. niger* K10	Butyl sepharose	[110]

Reaction	Enzyme	Support	References
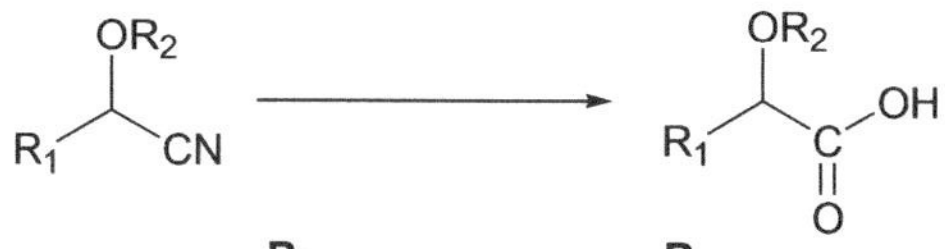	*A. faecalis* ATCC8750	Not disclosed	[111]
5) *Isomerase*			
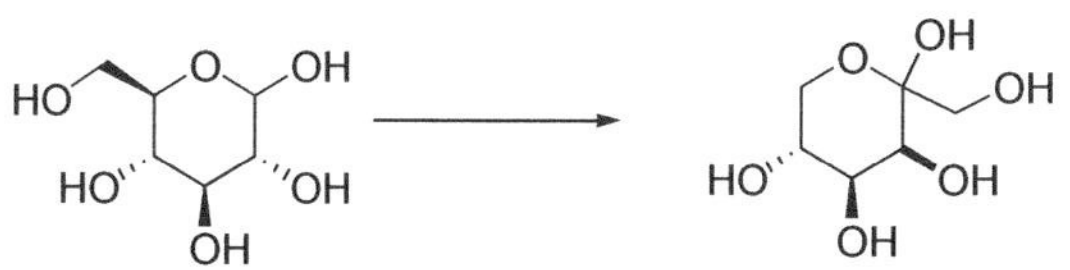	Glucose isomerase produced from *Streptomyces murinus*	Silica	[112]
	Glucose isomerase (GI) from *Streptomyces rubiginosus*	GAMM support	[39]
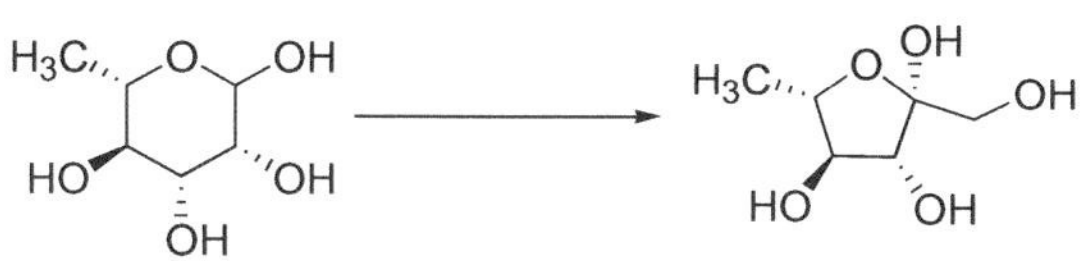	L-Rhamnose isomerase from *Dictyoglomus turgidum*	Duolite A568 beads	[40,113]

R_1	R_2
Ph	OH
PH	OAc
$(CH_2)_2SMe$	OH
$(CH_2)_2SMe$	OAc
$(CH_2)_2SMe$	MOM

(Continued)

Table 4.1 Transformations with Immobilized Enzymes *(cont.)*

	Enzyme	Support	References
6) Transaminase			
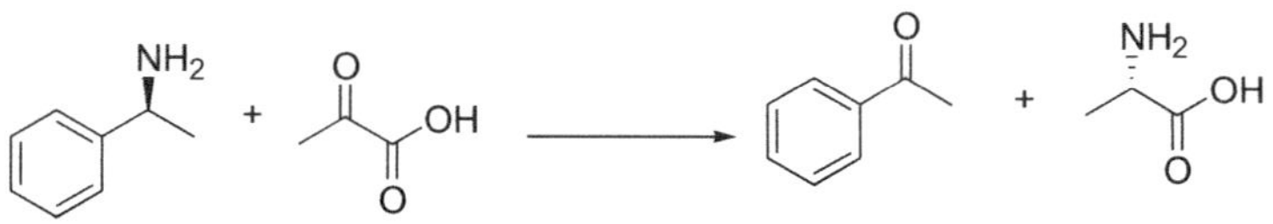	ϖ-TA from *Burkholderia cenocepacia*	Catechol-chitosan iron oxide nanoparticles	[114]
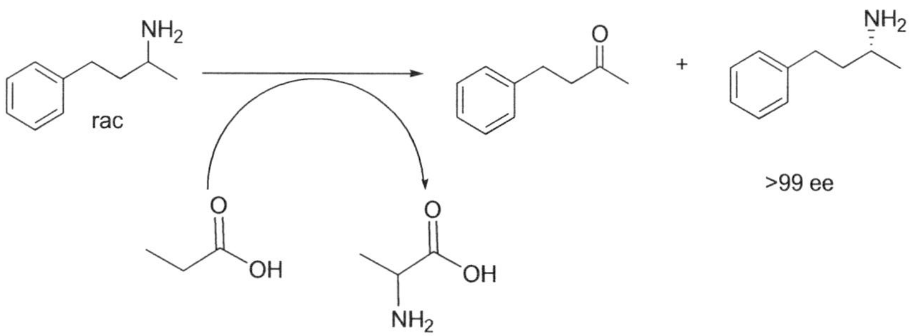	ϖ-transaminase from *Vibrio fluvialis*	Sol–gel matrix	[115]
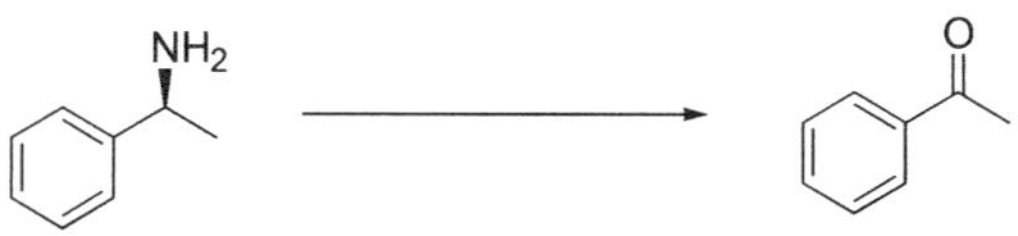	ϖ-transaminase from *V. fluvialis*	Chitosan	[116]
	Phenylalanine dehydrogenase from *Rhodoccocus* sp.	Eupergit CM	[117]
	L-aspartate aminotransferase from porcine heart	Eupergit®C and MANA agarose	[118]

Reaction	Enzyme	Support	References
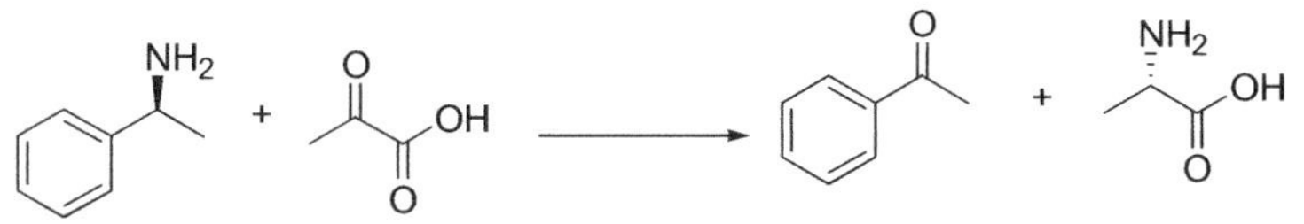	(R)-selective amine transaminases from *Gibberella zeae* (GibZea) and from *Neosartorya fischeri* (NeoFis)	MANA-agarose	[119]
7) Aldolase			
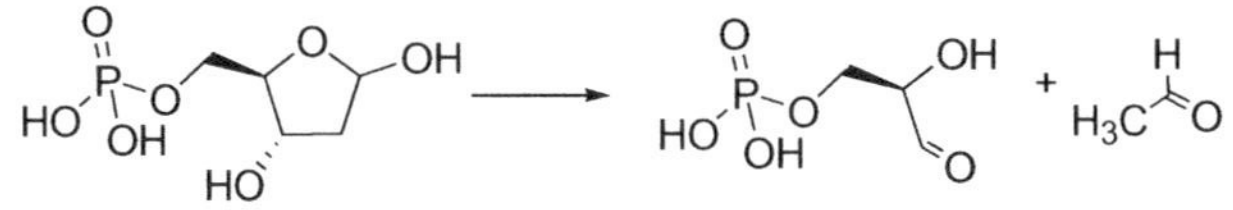	2-Deoxy-D-ribose-5-phosphate aldolase	Mesocellular siliceous foams (MCFs)	[120]
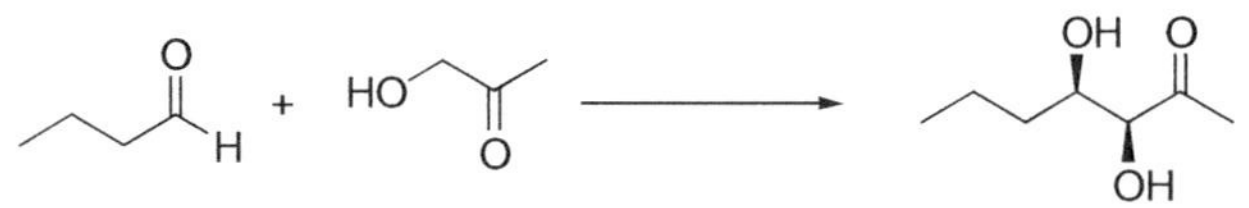	Fructose-biphosphate aldolase	Layered double Hydroxide (FSA@LDH)	[121]
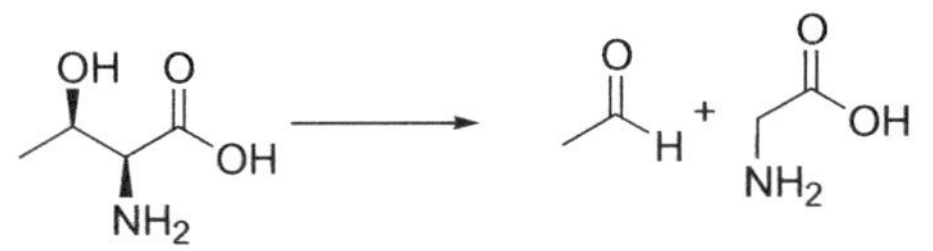	L-Threonine aldolase	Eupergit CM	[122]
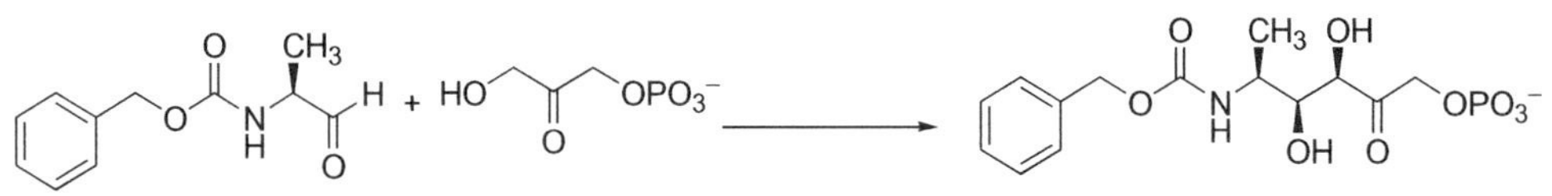	Rhamnulose-1-phosphate aldolase	Iminodiacetic acid-functionalized agarose charged with cobalt (Co-IDA)	[123]

(Continued)

Table 4.1 Transformations with Immobilized Enzymes *(cont.)*

	Enzyme	Support	References
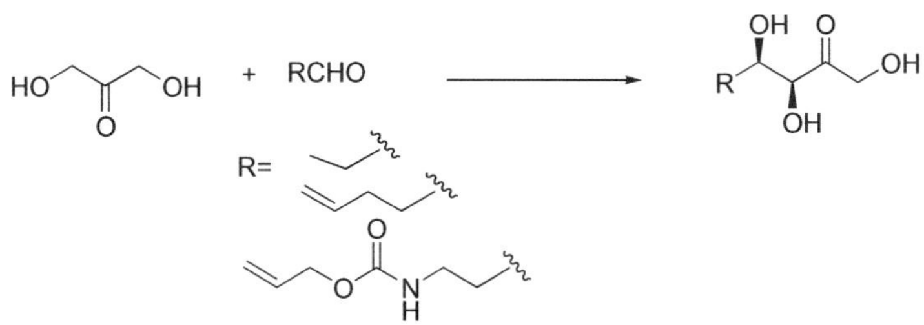	Rhamnulose-1-phosphate aldolase	Sepabeads EC-EP and Relizyme EP-403	[124]
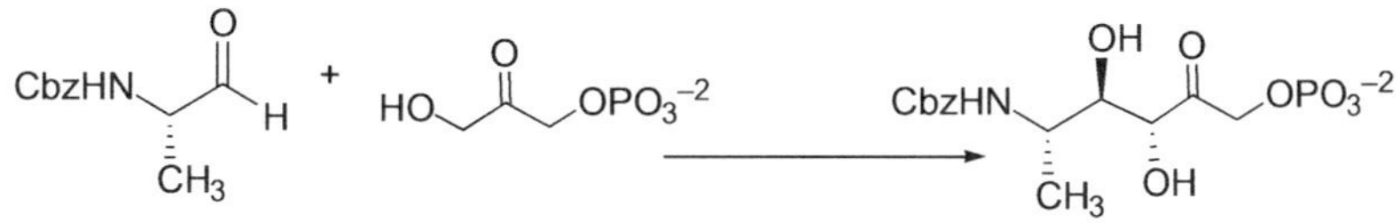	Fuculose-1-phosphate aldolase	Glyoxal agarose gels	[125]
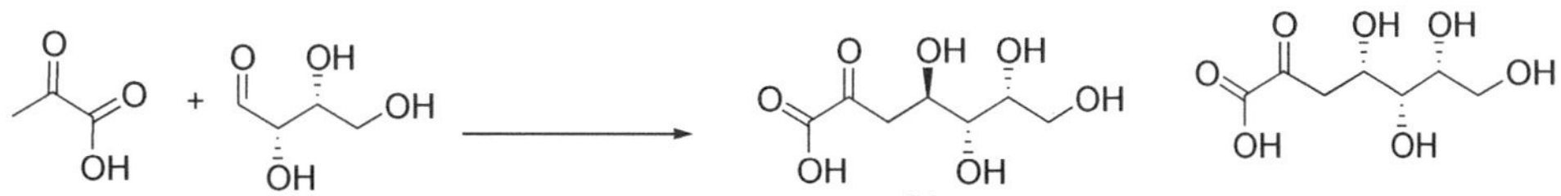 90 10	2-Keto-3-deoxygluconate aldolase	His-bind resin	[126]
8) Lyases			
(9Z,11E,15Z)-(13S)-hydroperoxyoctadeca-9,11,15-trienoate → *(9Z,15Z)-(13S)*-12,13-epoxyoctadeca-9,11,15-trienoate + H_2O	Amaranthus tricolor hydroperoxide lyase	1,6-hexamethylenediamine attached chitosan–carrageenan	[127]

	Enzyme	Support	References
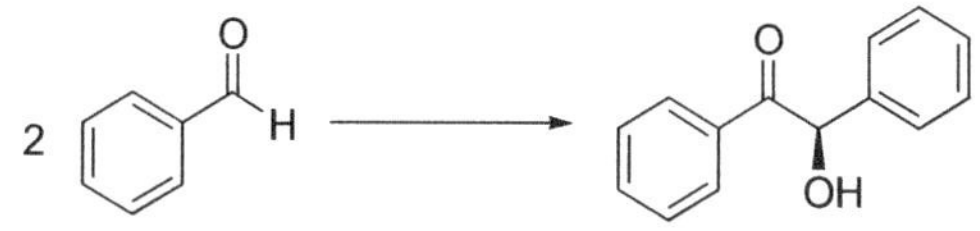	Pseudomonas fluorescens Biova benzaldehyde lyase	Magnetically responsive epoxy-Chelate	[128]
	Hexahistidine-tagged benzaldehyde lyase from *E. coli* SG13009/BAL_{HIS}	Nickel(II)–nitrilotriacetic acid derivatized carrier	[129]
	Benzaldehyde lyase from *P. fluorescens*	Entrapped in polyvinyl alcohol	[130]
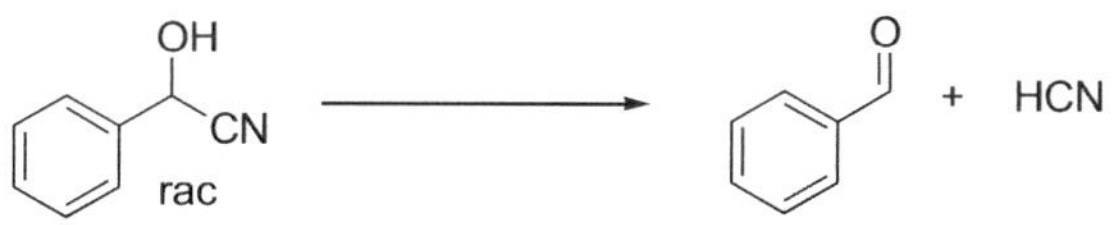	Seeds of *Prunus pseudoarmeniaca* Hydroxynitrile lyase	Eupergit C and eupergit C 250	[131]
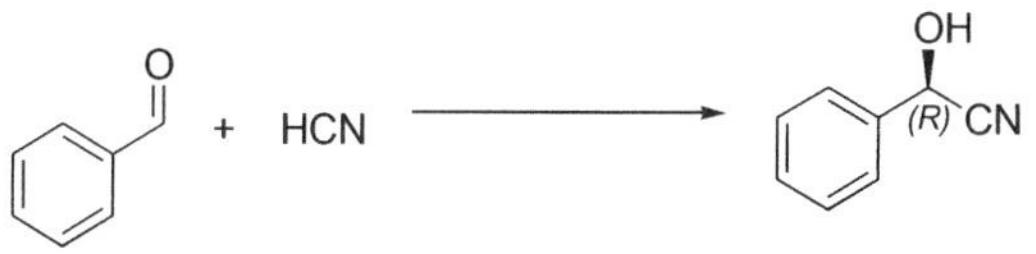	Hydroxynitrile lyase	Cross-linked and poly(vinyl alcohol)-entrapped	[132]
	Hydroxynitrile lyase	Sol–gel matrix	[41,42, 133,134]
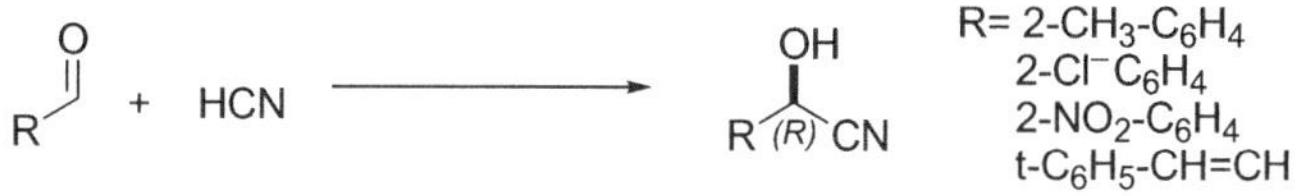	Hydroxynitrile lyase from *Prunus amygdalus*	Cross-linking using glutaraldehyde	[135]

(Continued)

Table 4.1 Transformations with Immobilized Enzymes *(cont.)*

	Enzyme	Support	References
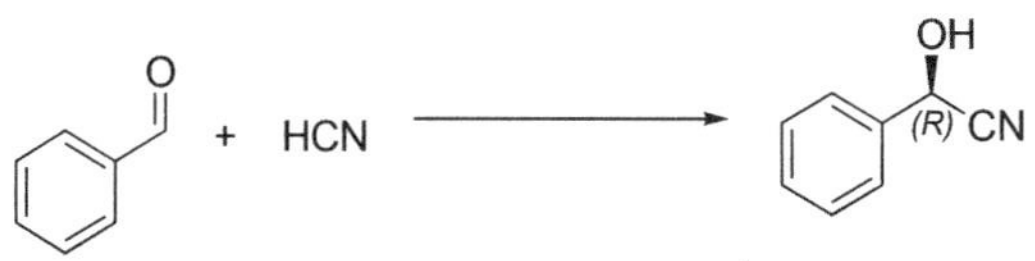	Hydroxynitrile lyase from *Manihot esculenta*		[136]
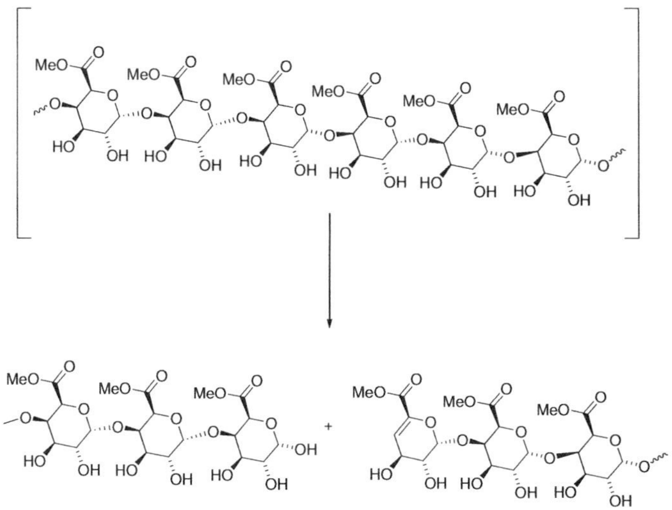	Pectin lyase *Penicillium italicum*	Nylon 6	[137]
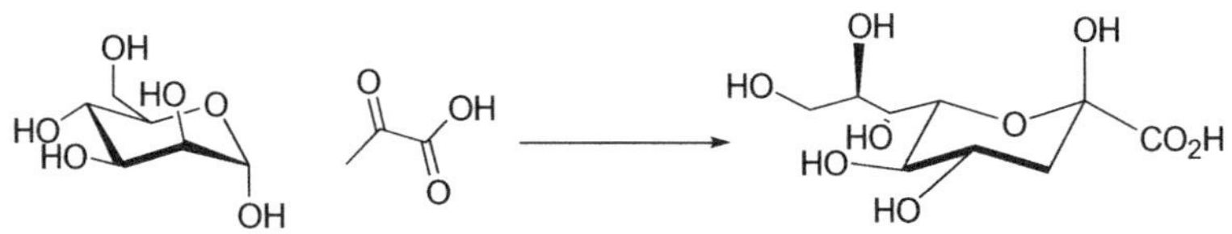	Acylneuraminate pyruvate lyase	Agarose	[138]

Reaction	Enzyme	Support	References
9) Naringinase			
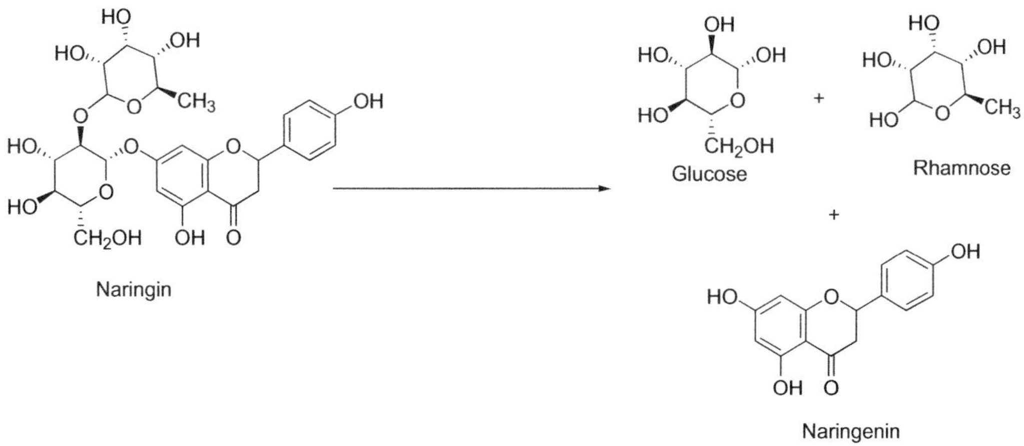	Naringinase from *A. niger*	Hollow fiber reactor	[139]
	Naringinase from *Penicillium decumbens*	Alginate	[140,141]
	Naringinase from *P. decumbens*	K-carrageenan	[142]
		PVA	[143]
10) Oxidases			
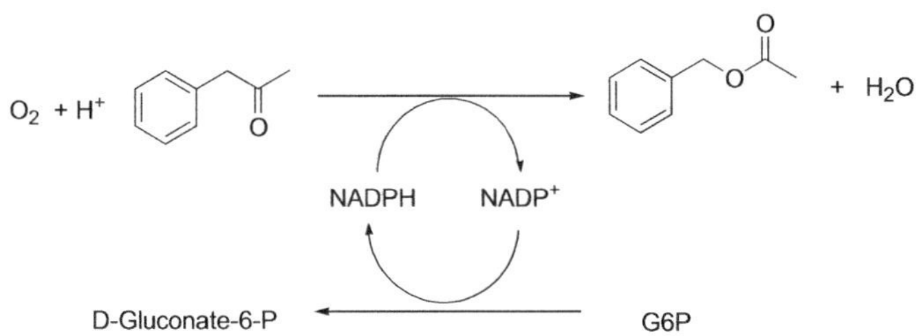	Phenylacetone monooxygenase from *Hermobifida fusca*	Polyphosphazene	[144]
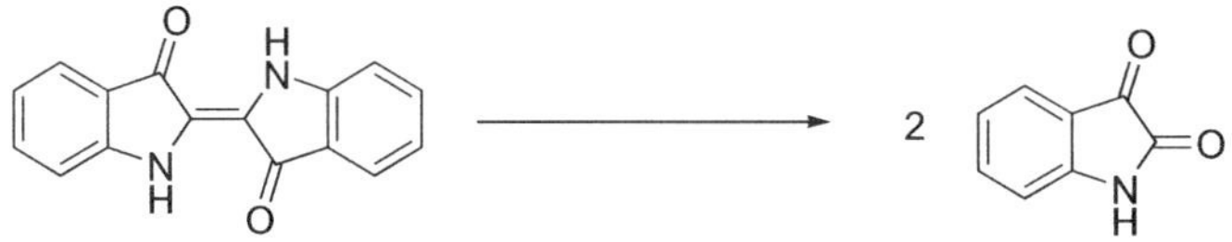	Cerrena unicolor Laccase	Silica gel	[145]

(Continued)

Table 4.1 Transformations with Immobilized Enzymes *(cont.)*

	Enzyme	Support	References
11) Other			
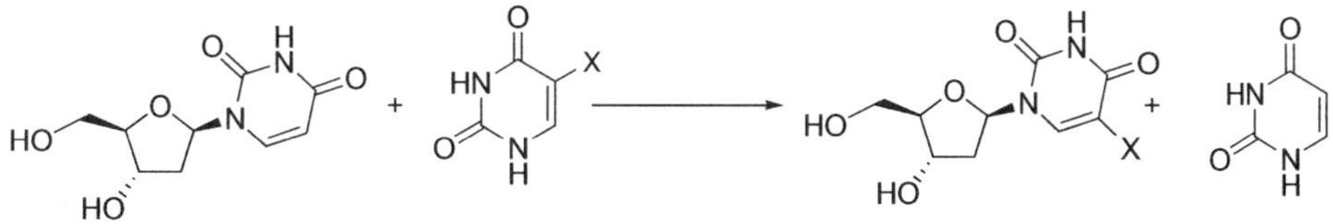 B. subtilis / E. coli X= CH_3: 67 % / 67% F: 68 % / 62% CF_3: 57 % / 51% Br: 57% / 56% I: 57% / 56%	Thymidine phosphorylase *Bacillus subtilis*	Sepabeads EC-EP derivatized with PEI	[146]

References

[1] R.C. Rodrigues, C. Ortiz, A. Berenguer-Murcia, R. Torres, R. Fernandez-Lafuente, Modifying enzyme activity and selectivity by immobilization, Chem. Soc. Rev. 42 (2013) 6290–6307.

[2] R.A. Sheldon, S. van Pelt, Enzyme immobilisation in biocatalysis: why, what and how, Chem. Soc. Rev. 42 (2013) 6223–6235.

[3] H.D. Colon, D.R. Walt, Immobilization of enzymes in polymer supports, J. Chem. Educ. 63 (1986) 368.

[4] S. Kulkarni, V. Tran, M.K.-M. Ho, C. Phan, E. Chin, Z. Wemmer, M. Sommerhalter, Biocatalysis with Sol–Gel Encapsulated Acid Phosphatase, J. Chem. Educ. 87 (2010) 958–960.

[5] R. DiCosimo, J. McAuliffe, A.J. Poulose, G. Bohlmann, Industrial use of immobilized enzymes, Chem. Soc. Rev. 42 (2013) 6437–6474.

[6] A.S. Bommarius, M.F. Paye, Stabilizing biocatalysts, Chem. Soc. Rev. 42 (2013) 6534–6565.

[7] J.M. Nelson, E.G. Griffin, Adsorption of Invertase, J. Am. Chem. Soc. 38 (1916) 1109–1115.

[8] J.B. Sumner, Denaturation of Urease Without Inactivation, Science 108 (1948) 410–413.

[9] N. Grubhofer, L. Schleith, Modifizierte Ionenaustauscher als spezifische Adsorbentien, Naturwissenschaften 40 (1953) 508.

[10] R.C. Rodrigues, R. Fernandez-Lafuente, Lipase from Rhizomucor miehei as an industrial biocatalyst in chemical process, J. Mol. Catal B: Enzym. 64 (2010) 1–22.

[11] K.-E. Jaeger, M.T. Reetz, Microbial lipases form versatile tools for biotechnology, Trends Biotechnol. 16 (1988) 396–403.

[12] D. Cowan, H.C. Holm, Lipid Technol. 18 (2006) 101–104.

[13] P.S. Panesar, R. Panesar, R.S. Singh, J.F. Kennedy, H. Kumar, Microbial production, immobilization and applications of β-D-galactosidase, J. Chem. Technol. Biotechnol. 81 (2006) 530–543.

[14] A.I. Kallenberg, F. van Rantwijk, R.A. Sheldon, Immobilization of Penicillin G Acylase: The Key to Optimum Performance, Adv. Synth. Catal. 347 (2005) 905–926.

[15] I. Chibata, Industrial Applications Of Immobilized Biocatalysts And Biomaterials, Adv. Mol. Cell Biol. 15a (1996) 151–160.

[16] P.B. Poulsen, European and American trends in the industrial application of immobilized biocatalysts, Enzym. Microbial Technol. 3 (1981) 271–273.

[17] Z. Zhou, M. Hartmann, Progress in enzyme immobilization in ordered mesoporous materials and related applications, Chem. Soc. Rev. 42 (2013) 3894–3912.

[18] K. Pospiskova, I. Safarik, Low-cost, easy-to-prepare magnetic chitosan microparticles for enzymes immobilization, Carbohydr. Polym. 96 (2013) 545–548.

[19] M. Hartmann, X. Kostrov, Immobilization of enzymes on porous silicas - benefits and challenges, Chem. Soc. Rev. 42 (2013) 6277–6289.

[20] E. Magner, Immobilisation of enzymes on mesoporous silicate materials, Chem. Soc. Rev. 42 (2013) 6213–6222.

[21] O. Barbosa, R. Torres, C. Ortiz, A. Berenguer-Murcia, R.C. Rodrigues, R. Fernandez-Lafuente, Heterofunctional supports in enzyme immobilization: from traditional immobilization protocols to opportunities in tuning enzyme properties, Biomacromolecules 14 (2013) 2433–2462.

[22] N. Carlsson, H. Gustafsson, C. Thörn, L. Olsson, K. Holmberg, B. Åkerman, Enzymes immobilized in mesoporous silica: a physical–chemical perspective, Adv. Colloid Interface Sci. (2013).

[23] D.I. Fried, F.J. Brieler, M. Froeba, Designing inorganic porous materials for enzyme adsorption and applications in biocatalysis, Chemcatchem 5 (2013) 862–884.

[24] M. Matto, Q. Husain, Entrapment of porous and stable Con A-peroxidase complex into hybrid calcium alginate-pectin gel, Chem. Technol. Biotechnol. 81 (2006) 1316–1323.

[25] R.A. Sheldon, Enzyme immobilization: the quest for optimum performance, Adv. Synth. Catal. 349 (2006) 1289–1307.

[26] A. Illanes, A. Cauerhff, L. Wilson, G.R. Castro, Recent trends in biocatalysis engineering, Biores. Technol. 115 (2012) 48–57.

[27] K. Kawakami, S. Yoshida, Thermal stabilization of lipase by sol-gel entrapment in organically modified silicates formed on Keiselguhr, J. Ferm. Bioeng. 82 (1996) 239–245.

[28] Y. Wei, J. Xu, Q. Feng, H. Dong, M. Lin, Encapsulation of enzymes in mesoporous host materials via the nonsurfactant-templated sol-gel process, Mater. Lett. 44 (2000) 6–11.

[29] C. Mateo, J.M. Palomo, G. Fernandez-Lorente, J.M. Guisan, R. Fernandez-Lafuente, Improvement of enzyme activity, stability and selectivity via immobilization techniques, Enzym. Microbial Technol. 40 (2007) 1451–1463.

[30] C. Mateo, J.M. Palomo, M. Fuentes, L. Betancor, V. Grazu, F. López-Gallego, B.C.C. Pessela, A. Hidalgo, G. Fernandez-Lorente, R. Fernández-Lafuente, J.M. Guisan, Glyoxyl agarose: a fully inert and hydrophilic support for immobilization and high stabilization of proteins, Enzym. Microbial Technol. 39 (2006) 274–280.

[31] V. Swarnalatha, R.A. Esther, R. Dhamodharan, Immobilization of α-amylase on gum acacia stabilized magnetitenanoparticles, an easily recoverable and reusable support, J. Mol. Catal B: Enzym. 96 (2013) 6–13.

[32] S. Cantone, V. Ferrario, L. Corici, C. Ebert, D. Fattor, P. Spizzo, L. Gardossi, Efficient immobilisation of industrial biocatalysts: criteria and constraints for the selection of organic polymeric carriers and immobilisation methods, Chem. Soc. Rev. 42 (2013) 6262–6276.

[33] M.C.R. Franssen, P. Steunenberg, E.L. Scott, H. Zuilhof, J.P.M. Sanders, Immobilised enzymes in biorenewables production, Chem. Soc. Rev. 42 (2013) 6491–6533.

[34] U. Hanefeld, Immobilisation of hydroxynitrile lyases, Chem. Soc. Rev. 42 (2013) 6308–6321.

[35] A. Liese, L. Hilterhaus, Evaluation of immobilized enzymes for industrial applications, Chem. Soc. Rev. 42 (2013) 6236–6249.

[36] F. Secundo, Conformational changes of enzymes upon immobilisation, Chem. Soc. Rev. 42 (2013) 6250–6261.

[37] S. Talekar, A. Joshi, G. Joshi, P. Kamat, R. Haripurkar, S. Kambale, Parameters in preparation and characterization of cross linked enzyme aggregates (CLEAs), Rsc Adv. 3 (2013) 12485–12511.

[38] D.-H. Zhang, L.-X. Yuwen, L.-J. Peng, Parameters affecting the performance of immobilized enzyme, J. Chem. (2013).

[39] H. Yu, Y. Guo, D. Wu, W. Zhan, G. Lu, Immobilization of glucose isomerase onto GAMM support for isomerization of glucose to fructose, J. Mol. Catal B: Enzym. 72 (2011) 73–76.

[40] Y.S. Kim, K.C. Shin, Y.R. Lim, D.K. Oh, Characterization of a recombinant l-rhamnose isomerase from Dictyoglomus turgidum and its application for l-rhamnulose production, Biothecnology Lett. 35 (2013) 259–264.

[41] L. Veum, U. Hanefeld, P. Alain, The first encapsulation of hydroxynitrile lyase from Hevea brasiliensis in a sol–gel matrix, Tetrahedron 60 (2004) 10419–10425.

[42] F.L. Cabirol, U. Hanefeld, Immobilized Hydroxynitrile Lyases for Enantioselective Synthesis of Cyanohydrins: Sol-Gels and Cross-Linked Enzyme Aggregates, Adv. Synth. Catal. (2006) 1645–1655.

CHAPTER 5

Hydrolysis and Formation of Carboxylic Acid and Alcohol Derivatives

Romas Kazlauskas
Department of Biochemistry, Molecular Biology & Biophysics and The Biotechnology Institute, University of Minnesota, Saint Paul, MN, USA

1 INTRODUCTION – HYDROLASES COMMONLY USED IN ORGANIC SYNTHESIS

Hydrolases catalyze cleavage of substrates by the addition of water. This chapter focuses on hydrolases that cleave carboxylic acid derivatives such as esters and amides: lipases, esterases, and proteases. By carrying out reactions in organic solvents without water, hydrolases can also catalyze acyl transfer reactions to make esters and amides.

Lipases favor water-insoluble substrates (triglycerides are their natural substrates), whereas esterases and proteases favor water-soluble substrates, Table 5.1. Tributyrin or *p*-nitrophenyl acetate are typical esterase substrates, whereas peptides are typical protease substrates. Lipases contain a hydrophobic active site covered by a hydrophobic lid. During catalysis, this lid opens and binds the lipase to a hydrophobic interface where an organic-soluble substrate can partition into the active site and react. This lid opening and binding explains the higher reaction rates of lipase-catalyzed reactions at an organic-water interface than in aqueous solution and is called interfacial activation. Lipase-catalyzed hydrolyses are best run as emulsions or two-phase mixtures of water and organic solvent. In contrast, esterases and proteases dissolve in the aqueous phase, so their substrates must also dissolve in the aqueous phase. The poor solubility of typical organic substrates in water can slow esterase- and protease-catalyzed hydrolyses.

Lipases and esterases have deeply buried active sites that bind substrates in a folded conformation, whereas proteases have shallow, active sites that bind substrates in an extended conformation. In both cases, large substrates extend into the solvent, but lipases and esterases bury more of the substrate in the active site. For example, during resolution of esters of secondary alcohols, lipases bind both substituents in hydrophobic pockets within the active site, but proteases

Organic Synthesis Using Biocatalysis. http://dx.doi.org/10.1016/B978-0-12-411518-7.00005-6

Table 5.1 General Properties of Lipases, Esterases, and Proteases

	Lipase	Esterase	Protease
Natural substrates	Triglycerides	Various esters	Proteins
Unnatural substrate types	Water-insoluble esters	Water-soluble esters	Water-soluble esters and amides
	Deep pocket binds substrate in folded conformation	Deep pocket binds substrate in folded conformation	Shallow pocket binds substrate in extended conformation
Catalytic machinery	Ser-His-Asp	Ser-His-Asp	Ser-His-Asp, but in a mirror image orientation
Enantioselectivity	High toward alcohol/amine moiety, usually favor (*R*)-secondary alcohol	High toward alcohol/ amine moiety, usually favor (*R*)-secondary alcohol	High toward acyl moiety; moderate toward alcohol/ amine moiety; usually favor (*S*)-secondary alcohol, solvent influences enantioselectivity
Best applications	Resolution of secondary alcohols, primary alcohols, hydrophobic-water-insoluble substrates.		Amides, sterically hindered esters, polar substrate, synthesis of peptides

bind only one substituent, leaving the other substituent in contact with the solvent. The shallow active site allows proteases to accept more sterically hindered substrates than lipases and to accept polar substrates since one substituent can remain in water [1].

Lipases and esterases typically show selectivity toward the alcohol or amine part of a carboxylic acid derivative. Favored acyl groups are simple straight chains like acetate or butyrate. In contrast, proteases show higher specificity for the acyl part of a carboxylic acid derivative. Proteases contain a specificity pocket for the acyl group. For example, subtilisins and chymotrypsin favor ester and amides of phenylalanine esters. Another difference between the enzyme classes is that lipases and esterases catalyze hydrolysis of only esters, whereas proteases catalyze hydrolysis of both amides and esters. Several good books on hydrolases in organic synthesis are available [2–4].

The application of enzymes to organic synthesis is a small market, so most commercial enzymes are intended for other applications. For example, laundry detergents contain subtilisin to aid removal of food and blood stains. Commercial enzymes are sold by their properties that are not based on structure. Many commercial enzymes are mixtures of partly purified proteins and inert carriers, Table 5.2. Their composition is often proprietary, making it difficult to connect properties to specific structures.

Crude preparations often work better for biocatalysis than purified enzymes. First, the crude commercial enzymes are less expensive than pure enzymes.

Table 5.2 Five Commercial Hydrolases Commonly Used in Organic Synthesis

Name	Commercial Source	Comments
Lipase B from *Candida antarctica* (CAL-B)	Sigma-Aldrich, Novozymes, other sources	Recombinant, immobilized and thermostable Novozym 435 Synonyms: *Pseudozyma antarctica*
Pseudomonas cepacia lipase (PCL)	Sigma-Aldrich, Amano Enzyme Inc. (lipase PS SD) http://www.amano-enzyme.co.jp/aeu/index.html	Nonrecombinant Synonyms: *Burkholderia cepacia* lipase
Candida rugosa lipase (CRL)	Sigma-Aldrich, Meito Sangyo (lipase MY) http://www.meito-sangyo.co.jp/kaseihin/index_e.html	Nonrecombinantprotein powder diluted with lactose Synonyms: *Candida cylindracea* lipase
Pig liver esterase (PLE)	Sigma-Aldrich, various sources	Commercial samples are mixtures of isozymes; recombinant pure forms also available
Subtilisins: subtilisin BL *(Bacillus lentus)*, subtilisin BPN' (*B. amyloliquefaciens*), subtilisin Carlsberg (*B. licheniformis*)	Sigma-Aldrich, Dupont (Genencor), Novozymes	Recombinant, many variants optimized for laundry and other cleaning applications

During manufacture, the microbes usually secrete hydrolases into the medium. Purification usually involves removing the cells and concentrating the growth media. Contaminating proteins that do not react with the substrate or product are innocuous. Second, the crude preparations contain sugars and other inert carriers that increase the surface area and stabilize the lipases, especially for reactions in organic solvents.

The names of enzymes can be confusing. Unlike the names of chemical structures that never change, the names of microorganisms do change as microbiologist learn more about them. For this reason the names of hydrolases sometimes change while the enzyme itself stays the same. For example, the yeast *Candida antarctica* has been renamed *Pseudozyma antarctica*. Commercial enzyme manufacturers may also change the name of the same enzyme to reflect different application or different additives in the preparation.

The most commonly used hydrolase for biocatalysis is lipase B from *C. antarctica* (CAL-B). The commercial material is usually Novozym 435, which is protein immobilized noncovalently on an acrylic resin. This immobilization is suitable for use in organic solvents, but in water, the lipase desorbs from the support. CAL-B is a recombinant protein from the yeast *C. antarctica* produced in a strain of the fungus *Aspergillus oryzae* [5]. CAL-B shows little or no interfacial activation and hydrolyzes long chain triglycerides only slowly. For this reason, it may be better classified as an esterase. It shows high activity and enantioselectivity toward a wide range of alcohols, especially secondary alcohols. Its enantioselectivity is usually low toward carboxylic acids. It retains activity in organic solvents even at 70 °C, but is less stable in water.

Lipase from *Candida rugosa* (CRL) is a nonrecombinant protein that is typically sold as a powder with 2–11 wt% protein [6], the rest is sugars and inert carriers. Gel electrophoresis of commercial samples of CRL show a major band at an apparent molecular weight of 60.1 kDa and a minor one at 58.4 kDa. The major band corresponds to the Lip 1 isozyme, whereas the minor band is a mixture of isozymes Lip 2–Lip 5. Commercial samples of CRL also contain a small amount of contaminating protease [7]. The isozymes differ in their reactivity and enantioselectivity toward secondary alcohols, but all favor the same enantiomer [8].

Lipase from *Pseudomonas cepacia* (PCL) from Amano (lipase P or PS-30; the current product is PS SD) is a nonrecombinant protein. It contains 1–25 wt% protein as well as diatomaceous earth, dextran, and $CaCl_2$. LPL-80 and LPL-200S are diagnostic grades that contain glycine. LPL-200S contains no detectable amounts of any other proteins. Four groups have cloned and expressed very similar lipases from different *Pseudomonas* strains [9,10].

Commercial pig liver esterase (PLE) is a mixture of isozymes, whose composition may vary with the pig's diet. PLE is a trimer of varying combinations of α, β, and γ subunits. The isozymes differ in their substrates specificities and enantioselectivities. The γ-isoenzyme of PLE has been cloned and expressed in yeast. It shows higher, and sometimes reversed, enantioselectivity as compared to isozyme mixtures [11].

Subtilisins are a family of bacterial serine proteases secreted by various *Bacillus* species. Subtilisins are endopeptidases with a broad specificity and contain a structural calcium ion. Subtlisins favor a large uncharged residue as the acyl group (P_1 position), for example, phenylalanine esters and amides. Subtilisins are used in laundry detergents to remove protein-based stains and therefore are produced on a multiton scale and are inexpensive. The most important commercial subtilisins are subtilisin BL (subtilisin from *Bacillus lentus*), subtilisin Carlsberg (subtilisin from *Bacillus licheniformis*) and subtilisin BPN′ (subtilisin from *Bacillus amyloliquefaciens*). In spite of differences in amino acid sequence (84 of 275 amino acids differ between subtilisins Carlsberg and BPN′) [12], the structures and substrate specificities are very similar. Subtilisins are alkaline serine proteases, that is, proteases that show maximum activity at alkaline pH. The reaction rate increases in the pH range 6–9 and then remains constant.

2 CHOOSING A HYDROLASE – SUBSTRATE SPECIFICITY AND ENANTIOSELECTIVITY

The synthetic usefulness of hydrolases rests in their ability to accept unnatural substrates that are useful to organic synthesis. Lipases and esterases accept esters of a wide range of primary and secondary alcohols. Primary alcohols react faster.

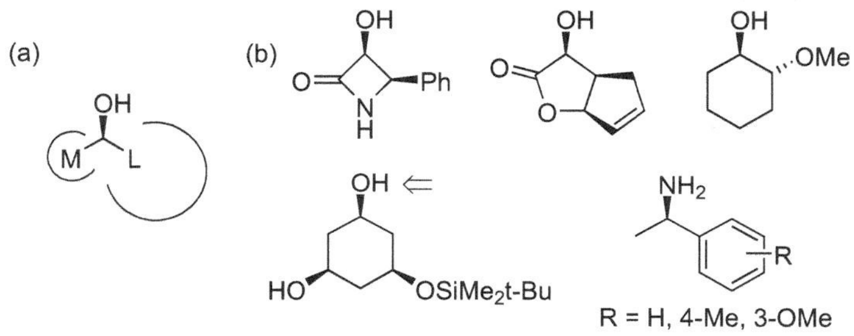

FIGURE 5.1

Enantiopreference of lipases and esterase toward secondary alcohols and the corresponding primary amines. (a) A general model showing the shape of the fast-reacting enantiomer. L represent a large substituent, while M represents a medium-sized substituent. This enantiomer reacts faster in both acylation and hydrolysis reactions. (b) Examples of well-resolved alcohols and amines show M substituents of a no more than a few carbons and significantly larger L substituents. The alcohol marked with the arrow reacts in the *meso*-diol. Solvent has little effect on the enantioselectivity of lipases and esterases because the substrate is buried.

In most cases, tertiary alcohols do not react. A few lipases and esterases have a larger active site and react slowly with esters of tertiary alcohols [13]. Esters of straight chain carboxylic acids (acetate, butyrate) are the best substrates. Esters with substituents at the α-carbon (relative to the carboxylic acid) and esters of aryl acids typically do not react.

An empirical rule based on relative sizes of substituents can predict the enantiopreference of most lipases and esterases toward secondary alcohols and the corresponding primary amines [14–16] (Figure 5.1). When the Cahn–Ingold–Prelog priority of *L* is higher than that of *M*, the fast-reacting enantiomer has the (*R*)-configuration. Besides predicting enantiopreference, this rule also suggests that increasing the relative size difference of the substituents should increases the enantioselectivity.

Such rules are less successful and more limited for other substrates. For example, an empirical rule for primary alcohols works only for one lipase – *Pseudomonas cepacia* lipase – and only for primary alcohols without an oxygen at the stereocenter [17,18].

Subtilisins are enantioselective toward secondary alcohols (HOCHRR′) and the isosteric primary amines (H_2NCHRR'), but subtilisin has a shallow binding site and leaves part of the substrate in the solvent (Figure 5.2). For this reason, the enantioselectivity of subtilisin is typically lower than that of lipases and the solvent strongly affects the enantiopreference of subtilisin. A general model shows that one substituent (R_{S1}') binds in the hydrophobic S_1'-pocket, which is large enough to fit a *para*-substituted phenyl group, while the other substituent (R_{SOLV}) remains in the solvent [1].

In organic solvents, both the S_1'-pocket and the solvent have similar polarity, so steric effects determine the enantioselectivity. The general model simplifies

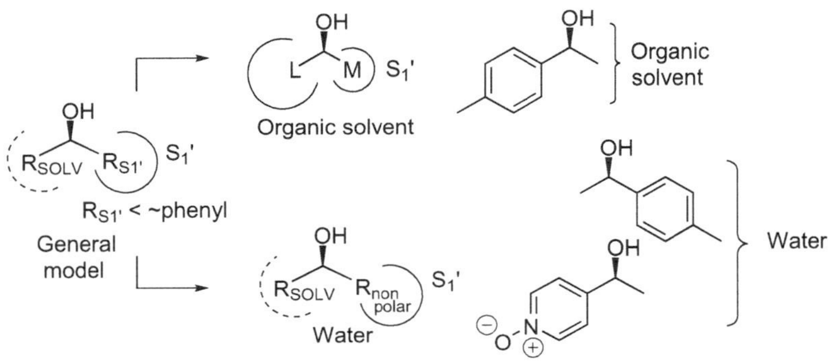

FIGURE 5.2
Enantiopreference of subtilisins toward secondary alcohols and primary amines varies with solvent because part of the substrate remains in water. A general model (left) shows that one substituent ($R_{S1'}$) binds in the hydrophobic S_1'-pocket, which is large enough to fit a *para*-substituted phenyl group, while the other substituent (R_{SOLV}) remains in the solvent. In organic solvents, both the pocket and solvent are non polar, so the size of the substituents determines the enantiopreference. In water, polar substituents favor the solvent, while non polar substituents favor the pocket. This difference explains the reversals in enantiopreference water and organic solvents for 1-(4'-methylphenyl) ethanol and the different behavior of 1-(4'-methylphenyl)ethanol and the isosteric *N*-oxide in water.

to a rule based on size where *L* is the large substituent and *M* is the medium substituent. The larger substituent avoids the S_1' pocket to minimize steric hindrance [19,20]. The enantiopreference of subtilisin in organic solvent is opposite to that for lipases, due to differences in protein structure [20]. Like lipases, subtilisin contains a catalytic triad, an oxyanion hole and follows an acyl enzyme mechanism. However, the three-dimensional arrangement of the catalytic triad in subtilisin is the mirror image of that in lipases. Folding the protein to assemble the catalytic machinery creates a pocket with restricted size, the "*M*" or stereoselectivity pocket for lipases and the S_1' pocket for subtilisin. The mirror image relationship places the catalytic histidine on opposite sides of this pocket in lipases and subtilisin. This difference in structure accounts for the opposite enantiopreference.

In water, the polarity of the substituents determined the enantiopreference [1]. Placing a nonpolar substituent in water is unfavorable. Reactions in water involving methyl and nonpolar aryl substituents will favor the nonpolar aryl substituent in the S_1' pocket, opposite to that predicted based on size alone. Thus, subtilisin favors the (*S*)-enantiomer of 1-(*p*-tolyl)ethanol in organic solvents (*p*-tolyl is larger than methyl), but the (*R*)-enantiomer in water (methyl is more polar than *p*-tolyl). On the other hand, with a polar aryl group such as (4-pyridine *N*-oxide), the (*S*)-enantiomer is favored both in water, where solvation of the pyridine *N*-oxide is favorable, and in organic solvent, where placing the pyridine *N*-oxide in the solvent avoids steric interactions in the S_1' pocket.

3 ENANTIOPURE COMPOUNDS: RESOLUTIONS, DESYMMETRIZATIONS, AND DYNAMIC KINETIC RESOLUTIONS

The most common application of hydrolase-catalyzed reactions is the preparation of enantiopure compounds. The three possible routes are kinetic resolutions, desymmetrizations and dynamic kinetic resolutions (Figure 5.3 [21–23]). The choice of substrate determines the possible routes. If the substrate is a racemate, then the choice is either a kinetic resolution or a dynamic kinetic resolution. If the substrate is a *meso* or prochiral compound, then the route is a desymmetrization. Since racemates are more numerous than *meso* or prochiral compounds, the most common routes are resolutions.

3.1 Kinetic Resolution

A kinetic resolution is a separation so the maximum yield of each enantiomer is 50%. A kinetic resolution starts with a racemic substrate. As the enantioselective

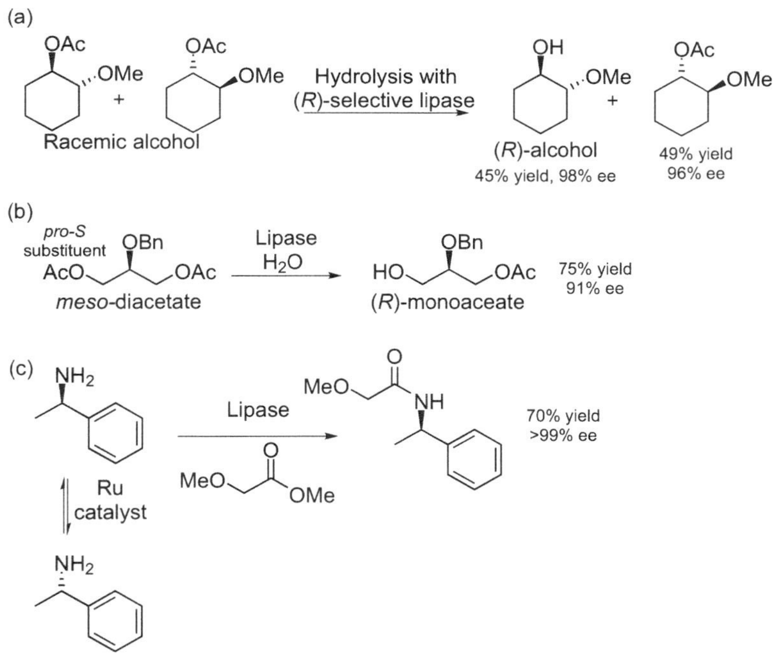

FIGURE 5.3

Three routes to enantiopure compounds using hydrolase-catalyzed reactions. (a) Kinetic resolution starts with racemic substrate and converts one enantiomer into product. This separation yields one enantiomer as the product alcohol and one as the starting acetate, both with a maximum yield of 50%. (b) Desymmetrization of a prochiral compound transforms one of prochiral groups to yield a chiral product with a maximum yield of 100%. (c) A dynamic kinetic resolution combines rapid racemization of racemic starting material with a hydrolase catalyzed acylation of one enantiomer. The maximum yield is 100%.

reaction proceeds, one enantiomer in converted to product, so the enantiomeric purity of the unreacted starting material starts to increase. If the hydrolase is highly enantioselective, then at 50% conversion one enantiomer exists as product, the other as remaining starting material and the reaction slows or even stops. To compare different kinetic resolutions, one compares the enantioselectivity, *E*. A nonselective reaction has an *E* of 1, whereas resolutions with *E* values above 20 are useful for synthesis. It is not appropriate to compare enantiomeric purities because the enantiomeric purity of the product and starting material varies as the reaction proceeds. To measure enantioselectivity, one measures two of these three variables: enantiomeric purity of the starting material (ee_s), enantiomeric purity of the product (ee_p), and extent of conversion (c) and uses one of the following three equations [24]. Often enantiomeric purities are more accurately measured than conversion; in these cases, the third equation is more accurate.

$$E=\frac{\ln[1-c(1+ee_P)]}{\ln[1-c(1+ee_P)]};\quad E=\frac{\ln[(1-c)(1-ee_S)]}{\ln[(1-c)(1+ee_S)]};\quad E=\frac{\ln\left[\dfrac{1-ee_S}{1+(ee_S/ee_P)}\right]}{\ln\left[\dfrac{1+ee_S}{1+(ee_S/ee_P)}\right]} \tag{5.1}$$

3.2 Desymmetrization and Dynamic Kinetic Resoluion

Hydrolase-catalyzed desymmetrizations start with *meso* compounds or prochiral compounds and yield chiral products in up to 100% yield. This high yield is a big advantage in most cases, but demands a symmetrical substrate, so it does not fit most synthetic problems. In a desymmetrization the enantiomeric purity of the product remains constant as the reaction proceeds and is given by ee = $(E - 1)/(E + 1)$, where *E* is the enantioselectivity. For example, an enantioselectivity of 50 yields product with 96% ee. Rearrangement of this equation gives $E = (1 + ee)/(1 - ee)$, useful to calculate the enantioselectivity from the enantiomeric purity of the product.

A dynamic kinetic resolution extends the high yield advantage of desymmetrizations to racemic substrates. A dynamic kinetic resolution is a kinetic resolution combined with rapid *in situ* racemization of the substrate. The requirements for a dynamic kinetic resolution are: (1) the substrate must racemize at least as fast as the subsequent enzymatic reaction, (2) the product must not racemize, and (3) as in any asymmetric synthesis, the enzymic reaction must be highly stereoselective. The equations relating product enantiomeric purity and enantioselectivity are the same as those for desymmetrizations.

The key step in dynamic kinetic resolutions is the *in situ* racemization. A number of reactions can racemize organic substrates, but most conditions are too harsh to allow a simultaneous enzyme-catalyzed reaction. In the past, the difficulty of racemizing normal alcohols and carboxylate esters restricted these dynamic kinetic resolutions to special cases. However, recent discovery of organometallic catalysts that can racemize a wide range of secondary alcohols have extended the range of these reactions [23].

4 CHOOSING REACTION CONDITIONS – WATER OR ORGANIC SOLVENT

Although hydrolysis of esters and amides is the natural reaction of hydrolases, placing them in organic solvents allows formation of esters and amides. These reactions are usually transesterification reactions where an acyl group from an acyl donor is added to an alcohol or amine. The alternative is a condensation reaction between a carboxylic acid and alcohol or amine. Condensations are slower, reversible, and only used when the cost of the acyl donor is important.

Hydrolases, especially lipases, which normally act at a water–organic interface, remain active in many organic solvents. Nonpolar solvents like hexane, toluene or *t*-butyl methyl ether are best. Hydrolase activity in polar solvents like methanol or DMSO is poor, likely because they hydrogen bond to the protein main chain and disrupt the folded conformation. Hydrolases do not dissolve in organic solvents, but are used as lyophilized enzyme powders suspended in the organic solvent or as immobilized enzymes.

4.1 Acyl Donors

The ideal acyl donor would be inexpensive, acylate quickly and irreversibly in the presence of hydrolase, be completely unreactive in the absence of hydrolase and, for kinetic resolutions, yield a product that is easily separated from the starting material. No acyl donor fulfills all of these criteria. The irreversibility is important for enantioselective reactions because reversibility lowers the enantiomeric purity of the product [24].

A useful class of activated acyl donors are enol esters, such as vinyl acetate or isopropenyl acetate (Figure 5.4a). The product alcohol tautomerizes to a carbonyl compound, thereby driving the reaction toward product and eliminating possible inhibition by the released alcohol. Another activated acyl donor is succinic anhydride that yields a monosuccinate ester that can be easily separated from the starting alcohol by extraction with a basic aqueous phase.

For transformations of inexpensive chemicals, cost is most important, so researchers use acids and simple esters, for example, methyl or glyceryl. Acylations with these donors are often slow and reversible with an equilibrium constant near one. To drive reactions to completion, researchers removed the water or alcohol by evaporation or azeotropic distillation. In other cases, crystallization of the product drives the reaction.

4.2 Hydrolysis Versus Acylation to Yield Desired Enantiomer as Ester or Alcohol

One reason to choose acylation in an organic solvent over hydrolysis in water is to get the desired enantiomer in either the alcohol or ester form. A kinetic resolution of an alcohol yields one enantiomer as the alcohol and the other as the ester. Although the same enantiomer is favored in both hydrolysis and acylation, in hydrolysis the fast-reacting enantiomer is the product alcohol, whereas

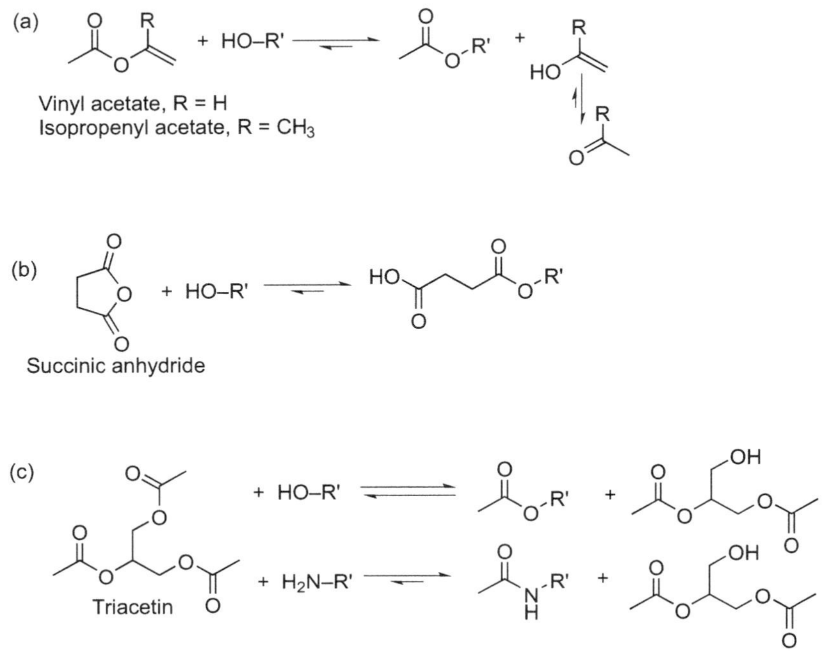

FIGURE 5.4
Acyl donors for hydrolase-catalyzed acylations. Reactive acyl donors like (a) vinyl acetate and (b) succinic anhydride react effectively irreversibly with alcohols. The tautomerization of the vinyl alcohol to the keto form makes the acylation favorable. These acyl donors are not suitable for amines because they react spontaneously. Simple esters like (c) triacetin are less expensive acyl donors. All three acetyl groups from triacetin can react. For alcohols, these simple ester acyl donors are reversible so are not suitable when kinetically controlled selectivity is important. Simple esters are good, effectively irreversible, acyl donors for amines.

in acylation it is the product ester. One of these is likely more convenient for subsequent use.

For example, Glaxo uses (1*S*,2*S*)-*trans*-2-methoxycyclohexanol, a secondary alcohol, for the synthesis of a tricylic β-lactam antibiotic [25]. The secondary alcohol rule predicts that this is the slow-reacting enantiomer for lipase-catalyzed reactions. If Glaxo chose to resolve this compound by lipase-catalyzed hydrolysis of the acetate, they would first need to make the acetate, then the desired enantiomer would be the unreacted acetate that would require another step to convert the acetate to the alcohol. (In principle, a hydrolysis with an (*S*)-enantiomer selective subtilisin is an alternative, but the enantioselectivity of subtilisins is often not high enough for many applications.) Instead, Glaxo chose to resolve this compound by acetylation of the alcohol with vinyl acetate (Figure 5.5). This approach yields the desired alcohol directly, thereby avoiding the added deacetylation required by the hydrolysis approach.

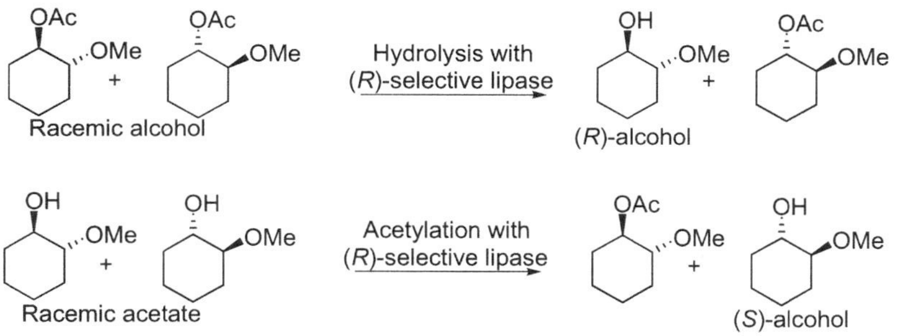

FIGURE 5.5
Lipase favors the (*R*)-enantiomer in both hydrolysis of the racemic acetate and in acetylation of racemic alcohol; the product alcohol is the (*R*)-enantiomer in hydrolysis and the (*S*)-enantiomer in acetylation.

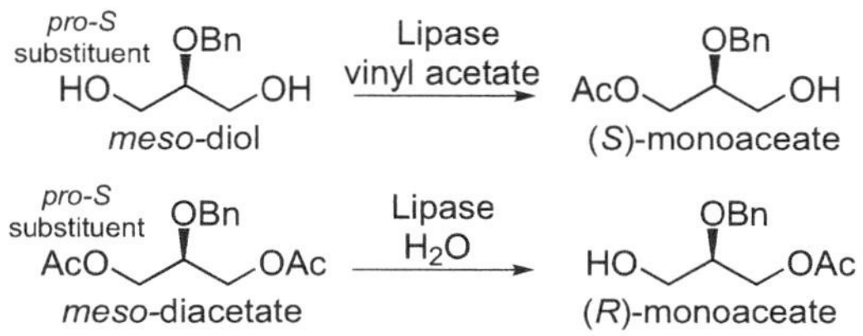

FIGURE 5.6
Lipases favor the *pro-S* substituent in both cases; acetylation of 2-benzyl glycerol yields the (*S*)-monoacetate, while hydrolysis of the diacetate yields the (*R*)-monoacetate.

In a similar manner, desymmetrization of a *meso* compound yields opposite enantiomers depending on whether the reaction is an acylation of the diol or a hydrolysis of the corresponding diacetate (Figure 5.6).

4.3 Other Reasons to Use Organic Solvents

Another reason to use organic solvents is to increase the selectivity of the hydrolase-catalyzed reaction. Changing the solvent changes the solvation of the enzyme–substrate complex. If solvation differs significantly for different substrates, then the enzyme selectivity can change in different solvents. The altered selectivity of subtilisin in water and organic solvents discussed previously is one such example.

A third reason to use organic solvents is that other features of the reaction require it to be an acylation and not a hydrolysis. For example, some dynamic kinetic resolutions require an alcohol substrate because organometallic complexes racemize secondary alcohols, but not esters. The substrate must be an alcohol for the dynamic kinetic resolution to proceed. In another example, lipase- and esterase-catalyzed hydrolysis of amides to amines is too slow for practical use. However, the lipase-catalyzed acylation of amines in organic solvents is fast. BASF manufactures chiral amines by a lipase-catalyzed acylation [26].

Other reasons to avoid water are to prevent decomposition of water-sensitive compounds such as organometallics or to simplify the workup of hydrophilic compounds that are difficult to recover from water. Finally, some organic chemists favor reactions in organic solvents simply because they are less familiar with reactions in water.

5 EXAMPLE REACTIONS

5.1 Enzyme-Catalyzed Hydrolysis of Ester or Amide

The simplest hydrolase-catalyzed reaction is ester or amide hydrolysis in water or biphasic mixtures of water and an organic solvent. Although proteases and esterases require a soluble substrate, it is enough that some fraction of the substrate dissolve in the aqueous phase. For lipases, a second phase is desirable because it activates most lipases by 10–100-fold, probably due to lid opening as discussed previously. Liquid substrate can also act as the organic phase.

A general procedure is as follows. Add 100 mg of liquid ester or amide such as acetate or butyrate (or 1 mL of a solution of solid ester or amide in a water-immiscible solvent such as toluene or ethyl ether) to 5 mL of 50 mM sodium phosphate buffer at pH 7. Monitor the progress of the reaction and stop it when it reaches ~30% conversion for a resolution of a racemate (see further note) or ~80% for an asymmetric synthesis. Thin layer chromatography is accurate enough for preliminary screening, but gas chromatography or HPLC is needed to optimize reactions by stopping at a precise conversion. If available, a pHstat can be used, which is a pH meter with an automatic burette that can add base to maintain the pH at a constant value. Hydrolysis of an ester at pH 7 causes the pH to drop due to release of acid. A pHstat is not suitable to monitor hydrolysis of an amide because amide hydrolysis yields both an acid and an amine. The acid releases a proton, but the amine picks it up, so there is no change in pH. This procedure can be used to screen a dozen or so enzymes. For the enzymes that catalyze hydrolysis, work up the reaction. For a kinetic resolution, measure two of these three quantities: the fraction of starting material converted to product, the enantiomeric purity of the remaining starting material and the enantiomeric purity of the product. Use one of the equations for kinetic resolutions above to calculate the enantiomeric ratio, *E*. If you measure all three quantities, you can check by using two of the equations. For a desymmetrization, measure the enantiomeric purity of the product. Calculate *E* using the simpler equation for desymmetrizations.

Note: In many cases the reaction slows and stops near 50% conversion for a kinetic resolution. If the remaining starting material is the desired enantiomer, then extending the conversion to more than 50% will increase its enantiomeric purity at the expense of a lower yield. If the product is the desired enantiomer, then stopping the hydrolysis at less than 50% will give higher enantiomeric purity. Choose the best trade-off between enantiomeric purity and yield for a given enantioselectivity.

Three examples of hydrolyses are given further.

5.1.1 KINETIC RESOLUTION OF 1,1'-BINAPHTHOL BY CHOLESTEROL ESTERASE CATALYZED HYDROLYSIS OF THE PENTANOATE DIESTER[27]

In this two-step resolution, the enzyme catalyzes the hydrolysis of the diester to monoester, then the monoester to the diol. Each step contributes to the overall enantioselectivity. The reaction mixture is biphasic: buffered water containing the crude enzyme and an ether solution of the diester. The crude enzyme is bovine pancreas acetone powder that contains many enzymes, but the substrate reacts only with the cholesterol esterase so the other enzymes do not interfere, other than to make the work-up messy. The enzyme cholesterol esterase requires a bile salt, taurocholate, for full activity. This bile salt helps emulsify the two phases. Cholesterol esterase seems to behave more like a lipase than an esterase in this example since it works with the substrate in the organic phase. The dipentanoate ester is chosen to simplify the separation of starting diester and product diol (Figure 5.7).

In a 12 L, round-bottomed flask equipped with an overhead stirrer, stir a two-phase mixture of an ether solution of binaphthol dipentanoate (4 L, 0.71 mol, prepared from 203 g of racemic binaphthol) and an aqueous solution of sodium taurocholate (60 g, crude grade in 4 L of 0.1 M phosphate buffer, pH 7.5) to form an opaque emulsion. Add bovine pancreas acetone powder (150 g), stopper the flask to minimize evaporation of ether and continue stirring slowly. The pH of the emulsion is measured occasionally and readjusted to 7.2 ± 0.3

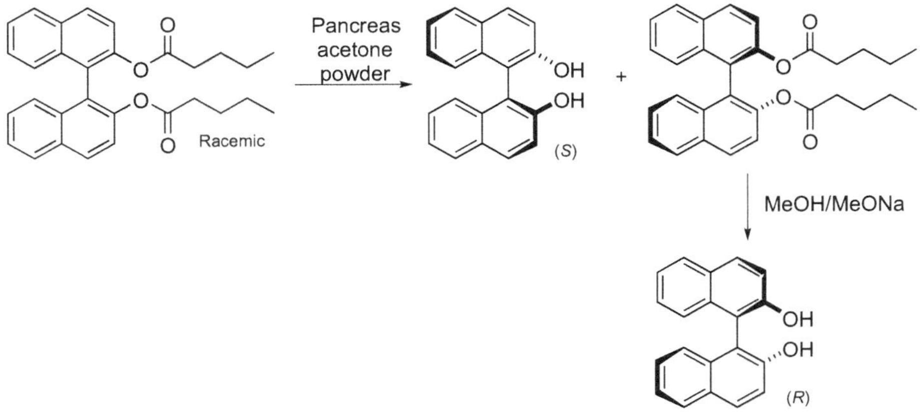

FIGURE 5.7
Kinetic resolution of 1,1'-bi-2-naphthol with a crude extract containing cholesterol esterase.

by adding aqueous 1 M sodium hydroxide. Approximately 250 mL of base is consumed during the first 3 h; an additional 400 mL is consumed over the next 20 h. Although the consumption of base virtually ceases after 24 h, stirring is continued for a total of 3 days. To break the emulsion 400 mL of ethanol is added and the mixture is transferred to a separatory funnel and allowed to settle for 4 h. Three layers form: at the top – a clear yellow ether phase, at the bottom – a brown aqueous phase and in between – an opaque emulsion layer. The brown aqueous layer is discarded. The emulsion layer (~1 L) is transferred to a flask and broken up by addition of magnesium sulfate (200 g added in portions). Heat is evolved, the ether boils and two layers form. This ether layer is combined with the first ether layer, dried over 50 g of magnesium sulfate, filtered, and concentrated by rotary evaporation to ~300 mL of an orange oil. Toluene (500 mL) is added and the solution cooled to 4 °C overnight. Fine white crystals (70–71 g) are collected by filtration and washed twice with 20-mL portions of cold toluene. The filtrate is concentrated to ~600 mL by rotary evaporation and cooled once again to 4 °C. An additional 10 g of crystals form and are collected by filtration and washed twice with 20-mL portions of toluene. Recrystallization of the combined crystals from 375–400 mL of toluene yields 65–68 g (64–67%) of white crystals of (*S*)-1,1′-bi-2-naphthol, mp 211–213.5 °C; >99% diol, >99% ee.

Toluene in the aforementioned filtrate is removed by rotary evaporation, and the residue is recrystallized from methanol (500 mL) overnight at 4 °C. Yellow crystals (125–141 g) form and are collected by filtration and washed twice with 20 mL of hexane. Recrystallization from 500 mL of methanol yields pure (*R*)-binaphthol dipentanoate, mp 63–65 °C, 89–102 g, >99% ee (*R*). This recrystallization step is critical to increase the enantiomeric purity of the dipentanoate. Recrystallization of the product diol does not increase the enantiomeric purity. If desired, an additional 25 g of dipentanoate can be isolated from the filtrate by column chromatography on 1 kg of silica gel eluted with methylene chloride followed by crystallization. The crystalline dipentanoate (89–102 g, 0.20 mol) is dissolved in 1 L of methanol containing sodium methoxide (6.6 g, 0.12 mol). After 4 h at room temperature, analysis of the solution by thin layer chromatography shows only traces of the mono- and diester. The solution is neutralized to pH < 7 (test paper) with ~10 mL of conc. hydrochloric acid. The solution is diluted with 1 L of 0.1 M phosphate buffer (pH 7), transferred to a 4-L separatory funnel and extracted with a mixture of 1 L of ethyl ether and 500 mL of toluene. The organic layer is washed with a 1-L portion of water, dried over magnesium sulfate, concentrated to 300 mL and cooled to 4 °C. White crystals (48–64 g) of (*R*)-1,1′-bi-2-naphthol separate and are collected by filtration and washed twice with 20-mL portions of cold toluene, 99% chemical purity, >99% enantiomeric purity. Chemical purity is determined by HPLC on a reverse-phase column eluted with a water-acetonitrile gradient (50–100% over 10 min) or by thin layer chromatography on silica gel eluted with 1:4 ethyl acetate/cyclohexane: 1,1′-bi-2-naphthol, R_f 0.39; monopentanoate, R_f 0.56; dipentanoate, R_f 0.71. Enantiomeric purity of binaphthol is determined using chiral stationary phase HPLC: poly(triphenylmethyl) methacrylate on silica gel (Chiralpak OT, Daicel Corp.) eluted with methanol.

5.1.2 DESYMMETRIZATION OF MESO cis-3,5-*DIACETOXYCYCLOPENTENE TO (1R,4S)-(+)-4-HYDROXY-2-CYCLOPENTENYL ACETATE CATALYZED BY LIPASE B FROM* C. ANTARCTICA

This next procedure is another hydrolysis, but differs in several ways from the first example. First, this is an enantioselective hydrolysis of a secondary alcohol ester, one of the most common and useful applications of hydrolases (Figure 5.8). Second, the reaction is not a kinetic resolution, but an asymmetric synthesis or desymmetrization, which yields up to 100% of the desired enantiomer (96% isolated yield in this example). The product (1*R*,4*S*)-(+)-4-hydroxy-2-cyclopentenyl acetate is a precursor for enantiopure 4-hydroxy-2-cyclopentenones. This compound was historically a key intermediate for synthesis of prostaglandins, but its application has been extended to a wide range of complex natural products [28].

A procedure in *Organic Syntheses* used acetylcholine esterase from electric eel as the enantioselective catalyst [29], but the much less expensive and more common enzyme, *C. antarctica* lipase B can also be used for this reaction [30]. Desymmetrization of 15 g of diacetate within ~10 h requires 2 g of Novozym 435 (~US$ 40) or 18 mg of acetylcholine esterase from electric eel (~US$ 500). The procedure given below uses Novozym 435, which is *C. antarctica* lipase B immobilized on acrylic resin. Note that this immobilization is by ionic absorption, which is effective in organic solvents, but not in water, which this procedure uses. The lipase washes off the resin in water and cannot be recovered. Lipase from *P. cepacia* (Amano PS-30) also catalyzes this desymmetrization [31], but the yield was only ~60% because the hydrolysis stopped for unknown reasons.

If the other enantiomer is needed, one can use an enzyme-catalyzed acetylation of the diol in organic solvent catalyzed by porcine pancreatic lipase in 72% isolated yield [30] (Figure 5.8).

Novozym 435 lipase (2.0 g, Sigma-Aldrich) was added to a suspension of *meso* diacetate (15.0 g, 81.5 mmol) in phosphate buffer (pH 8, 400 mL, 1 M). (Note: Hydrolysis liberates acetic acid. The high concentration of buffer maintains the pH near 8.) The mixture was stirred at room temperature and monitored by thin layer chromatography (silica gel eluted with ethyl ether, stained with *p*-anisaldehyde; R_f (monoacetate) = 0.44, R_f (diacetate) = 0.72). After 18 h, the diacetate was completely converted to monoacetate, so the reaction was filtered to

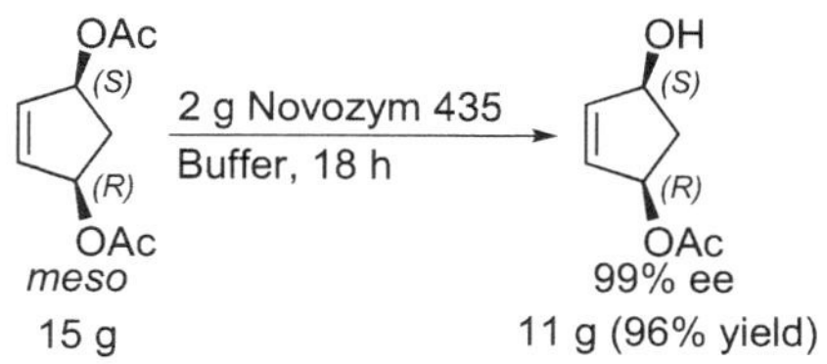

FIGURE 5.8
Enantioselective hydrolysis catalyzed by lipase B from *C. antarctica* desymmetrized a *meso* diacetate to the chiral monoacetate.

remove the resin beads, washed with water (100 mL) and EtOAc (2 × 200 mL). The aqueous layer was extracted with EtOAc (3 × 200 mL) and the combined organic fractions were dried (Na_2SO_4), filtered and concentrated under reduced pressure to afford the title monoacetate as white crystals (11.1 g, 78.1 mmol, 96, 99% ee). The enantiomeric excess was measured by gas chromatography (methylsilicone column) of the corresponding Mosher ester [29].

5.1.3 *KINETIC RESOLUTION OF P-TOLYLSULFINAMIDE BY PROTEASE-CATALYZED HYDROLYSIS OF THE N-3-(3-PYRIDINE)PROPIONYL AMIDE*

This example uses a protease, chymotrypsin, to catalyze hydrolysis of an amide (Figure 5.9). Chymotrypsin is used because it is commercially available, but subtilisin BPN′ [32] or subtilisin E are preferred proteases because they are more enantioselective for this reaction and much less expensive. Commercially available subtilisins, for example, Alcalase and Savinase from Novozymes, are widely used as detergent proteases and may be suitable, but have not been tested for this reaction. In our experiments, we prepared our own subtilisin [1], but this may be inconvenient for some laboratories.

The acyl group of the substrate is designed to increase reaction rate and simplify separation of product and starting materials. The 3-(3-pyridine)propionyl group mimics phenylalanine that is a good substrate for subtilisins and chymotrypsin. This mimicry increases binding of the substrate to the protease. This acyl group also increases the substrate solubility in water, which is important for proteases and esterases, but not for lipases. Both of these features increase the rates of protease-catalyzed reactions. In addition, the 3-(3-pyridine)propionyl group eliminates chromatography since mild acid extraction separates the remaining starting material and product. This example also resolves a sulfur stereocenter.

This example also uses change in pH to monitor the progress of the hydrolysis. Typical hydrolysis of an amide does not change the solution pH because the liberated acid donates its proton to the liberated amine resulting in no release

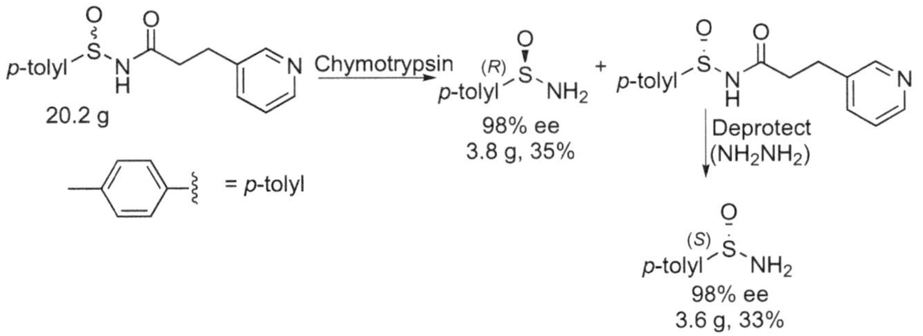

FIGURE 5.9

The 3-(3-pyridyl)propionyl acyl group is a convenient acyl group for protease catalyzed hydrolysis. This resolution of a chiral sulfur compound uses chymotrypsin, but subtilisin can also be used.

or consumption of protons. In this case, the amine released is not basic and does not protonate, so hydrolysis of amide released one proton. The reaction progress can be monitored by tracking the amount of 1 M NaOH needed to maintain the pH constant at 7.2. A total of 35 mL will be required; hydrolysis is approximately first order – faster at the beginning and slower as it reaches 50% conversion. If available, a pHstat (pH meter with automatic burette) can automatically maintain the pH and track the amount of base added (Figure 5.9).

α-Chymotrypsin from bovine pancreas (12 g, Sigma-Aldrich, C4129) was added to a solution of BES (*N,N-bis*-[2-hydroxyethyl]-2-aminoethane sulfonic acid) buffer (3.15 L, 1 mM, pH 7.2) and 100 mM KCl (needed to prevent formation of inactive dimers of chymotrypsin) and stirred for 15 min to ensure complete dissolution. *N*-3-(3-pyridine)propionyl-*p*-toluene sulfinamide [33] (20.2 g, 70 mmol) dissolved in dimethylformamide (350 mL) was added drop-wise to the enzyme solution over 1–2 h. The rate of hydrolysis was monitored by pH stat that maintained the pH at 7.2 by automatic titration with 1 N NaOH. At ~50% conversion (4 d), the reaction was stopped by extraction of remaining starting material and product with CH_2Cl_2 (3 × 500 mL). The combined organic layers were washed with dH_2O (3 × 500 mL), sat. NaCl (1 × 500 mL), dried over $MgSO_4$, and concentrated *in vacuo*. The crude mixture was dissolved in EtOAc (250 mL) and unreacted starting material was extracted with ice-cold 0.1 N HCl (2 × 100 mL). The combined aqueous layers were then back-extracted with EtOAc (50 mL). The combined EtOAc layers were washed with sat. $NaHCO_3$ (100 mL), sat. NaCl (100 mL), dried over $MgSO_4$, and concentrated *in vacuo* to give (*R*)-*p*-toluenesulfonamide as a white solid (4.48 g, 41% yield) with 87% ee. The combined aqueous layers were neutralized with solid $NaHCO_3$ and extracted with CH_2Cl_2 (2 × 200 mL). The combined CH_2Cl_2 layers were washed with sat. $NaHCO_3$ (100 mL), sat. NaCl (100 mL) and dried over $MgSO_4$. The solution was concentrated *in vacuo* to give (*S*)-*N*-3-(3-pyridine)propionyl-*p*-toluene sulfinamide that was subsequently treated with hydrazine hydrate (35 mL). After stirring for 3 h, the solution was diluted with CH_2Cl_2 (100 mL) and washed with 1 N HCl (50 mL), sat. $NaHCO_3$ (50 mL), sat. NaCl (50 mL), and concentrated *in vacuo* to gave (*S*)-*p*-toluenesulfonamide (4.29 g, 40% yield) with 92% ee. Recrystallization of the lower ee materials from hexanes/ethyl acetate gave (*R*)-toluenesulfonamide (3.81 g, 35% yield) with 98% ee and (*S*)-toluenesulfonamide (3.58 g, 33% yield) with 98% ee. The enantiomers were separated using HPLC (Chiralcel OD-H column, 85:15 hexanes/EtOH, 0.75 mL/min, 254 nm; (*R*)-enantiomer t_R = 20.0 min; (*S*)-enantiomer, t_R = 22.5 min).

5.2 Reactions in Organic Solvents: Acyl Transfer Reactions

A general procedure for enantioselective acylation of an alcohol with enol acetate or succinic anhydride is as follows. Dissolve 1 g of substrate alcohol in solvent (35 mL of hexane, toluene, or tetrahydrofuran) containing vinyl or isopropenyl acetate (3 mL) and triethylamine (0.5 mL) (see following note). Add 0.5 g of hydrolase powder or immobilized hydrolase; the enzyme powder will not dissolve. Stir the suspension with a magnetic stirrer and monitor the progress of the reaction by thin layer chromatography or by HPLC. When the reaction reaches the desired conversion, filter the mixture to remove hydrolase and stop the reaction. The hydrolase may be reused for another reaction. Measure enantiomeric purity of starting material and product and conversion by HPLC or gas chromatography. Concentrate the reaction mixture by rotary evaporation. Separate the starting alcohol from the product acetate by chromatography on silica gel.

Note: Dry reaction conditions maximize the rate of the hydrolase-catalyzed reaction and minimize the competing hydrolysis of vinyl or isopropenyl acetate, which generates acetic acid. Serine hydrolases are active only above pH 6, where the side chain imidazole of the active site histidine is in the uncharged form. Triethylamine prevents acidification of the reaction mixture due to hydrolysis and may have other effects [34–36].

5.2.1 KINETIC RESOLUTION OF A SECONDARY ALCOHOL (4-TRIMETHYLSILYL-3-BUTYN-2-OL) BY LIPASE-CATALYZED ACETYLATION WITH VINYL ACETATE

This typical acetylation example uses vinyl acetate as the acyl donor and yields the acetate and unreacted alcohol [37,38]. The oven-dried glassware, nitrogen atmosphere, and molecular sieves minimize the amount of water in the reaction mixture. Distillation of the vinyl acetate removes any polymer and also any acetic acid from hydrolysis during storage. A similar procedure in *Organic Syntheses* resolves other similar secondary alcohols [39] (Figure 5.10).

Racemic 4-trimethylsilyl-3-butyn-2-ol (10.0 g, 69.9 mmol) and pentane (250 mL) are combined in an oven dried, 1-L, round-bottomed flask, containing a large magnetic stir bar and sealed with a rubber septum with nitrogen inlet. To this stirred solution is added Amano Lipase AK (from *Pseudomonas fluorescens*) (2.00 g), freshly distilled vinyl acetate (50 mL), and pulverized, activated 4-Å molecular sieves (1 g). The mixture is stirred at room temperature for 3 days when the reaction reached 50% conversion as determined by ^{1}H-NMR or gas chromatography (Carbowax, 110 °C, 1 °C ramp/min; acetate 5.75; alcohol 8.22 min). The reaction was stopped by filtering the mixture through a medium porosity sintered glass funnel, washed with pentane, and concentrated via rotary evaporation affording 11.5 g of a nearly 1:1 mixture of alcohol and acetate. Separation by silica gel column chromatography eluted with 4–10% acetone/

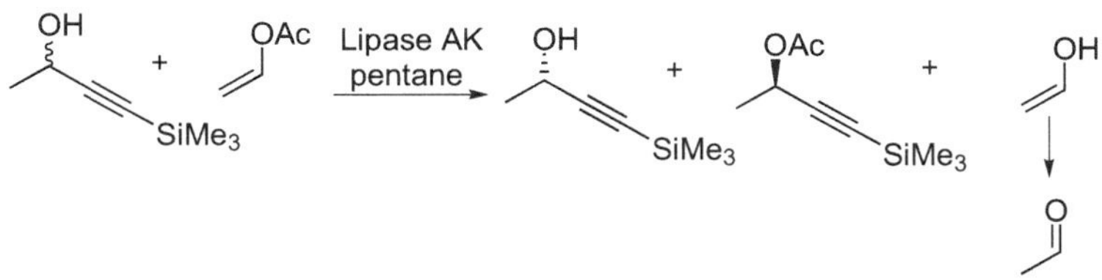

FIGURE 5.10
Kinetic resolution of 4-trimethylsilyl-3-butyn-2-ol by lipase AK-catalyzed acetylation with vinyl acetate.

hexane yielded the *S*-alcohol, (*S*)-4-trimethylsilyl-3-butyn-2-ol and *R*-acetate, (*R*)-4-trimethylsilyl-3-butyn-2-yl acetate. Enantiomeric purity of acetate: GC on β-Dex (80°C, 0.2°C ramp/min): (*S*) 31.7; (*R*) 32.4 min. To measure the enantiomeric purity of the alcohol, it is converted to the acetate by treatment with acetic anhydride and dimethylaminopyridine in methylene chloride.

Instead of column chromatography, this mixture of acetate and alcohol can be reacted with succinic anhydride that converts the alcohol to the succinate [38]. The resulting mixture of acetate and succinate can be separated by bulb-to-bulb distillation.

5.2.2 KINETIC RESOLUTION OF A SECONDARY ALCOHOL (1-INDANOL) BY LIPASE-CATALYZED SUCCINYLATION WITH SUCCINIC ANHYDRIDE

Kinetic resolution by acylation of an alcohol with succinic anhydride yields the unreacted alcohol and the hemisuccinate of the fast-reacting enantiomer. These two products are easily separated by extraction with aqueous base [40]. The procedure given further [41] uses immobilized CAL-B that can be recovered and reused. The work-up details include procedures from Gutman et al. [42]. The small-scale procedure given further still includes chromatography on silica gel for the final purification, but on a larger scale, this can be omitted and the alcohol purified by distillation [42]. Although the acylation generates acid and the reaction mixture does not contain a base, the enzyme remains active. The dry conditions and organic solvent may prevent ionization of the carboxylic acid formed (Figure 5.11).

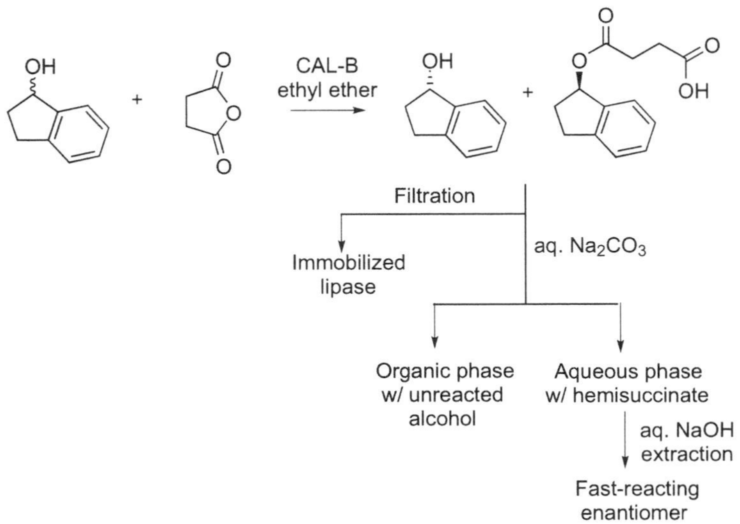

FIGURE 5.11
Kinetic resolution 1-indanol by acylation with succinic anhydride yields products that are easily separated by liquid-liquid extraction.

To a flame-dried round bottom flask containing racemic 1-indanol (2 mmol, 270 mg) dissolved in diethyl ether (10 mL) was added succinic anhydride (2 mmol, 200 mg; a twofold excess since only one enantiomer reacts), followed by immobilized *C. antarctica* lipase B (Novozym 435, 300 mg; can be reduced to 6 mg). The suspension was stirred at room temperature for 22–24 h to reach ~50% conversion (44% in this example) as determined by HPLC of the mixture on a Chiralcel OD-H capillary column. The reaction was stopped by filtration to remove the immobilized enzyme that was washed with ethyl ether, dried at room temperature in air and saved for reuse up to seven times. The reaction rate decreases with subsequent uses, but the enantioselectivity remains high. The combined filtrates were extracted with aqueous sodium carbonate (1 M, 3 × 5 mL) and water (1 × 5 mL) to remove the succinate. The remaining ethyl ether solution containing unreacted 1-indanol was concentrated by rotary evaporation and purified by chromatography on silica gel eluted with hexane/ethyl acetate 80:20 – 48% isolated yield of (*S*)-1-indanol, 79% ee. Enantiomeric purity was measured after acetylation of the alcohol (Ac_2O, DMAP, NEt_3) by HPLC on a Chiralcel OD-H capillary column. To recover the fast-reacting enantiomer, the combined aqueous extract (~20 mL) was extracted with ethyl ether (2 × 5 mL). Sodium hydroxide (2 g, 50 mmol) was added to the aqueous phase and stirred for 6 h at room temperature. The solution was extracted with dichloromethane (3 × 5 mL), the combined extracts washed with water (5 mL), dried over Na_2SO_4, concentrated by rotary evaporation and purified by silica gel chromatography: 25% isolated yield (*R*)-1-indanol, >99% ee.

References

[1] C.K. Savile, R.J. Kazlauskas, How substrate solvation contributes to the enantioselectivity of subtilisin toward secondary alcohols, J. Am. Chem. Soc. 127 (2005) 12228–12229.

[2] K. Faber, Biotransformations in Organic Chemistry, sixth ed., Springer, Berlin, (2011).

[3] C.-H. Wong, G.M. Whitesides, Enzymes in Synthetic Organic Chemistry, Pergamon Press, Oxford, (1994).

[4] U.T. Bornscheuer, R.J. Kazlauskas, Hydrolases in Organic Synthesis, second ed., Wiley, Weinheim, (2004).

[5] (a) I. Hoegh, S. Patkar, T. Halkier, M.T. Hansen, Two lipases from *Candida antarctica* – cloning and expression in *Aspergillus oryzae*, Can. J. Botan. 73 (1995) S869–S875. (b) O. Kirk, M.W. Christensen, Lipases from *Candida antarctica*: unique biocatalysts from a unique origin, Org. Process. Res. Dev. 6 (2002) 446–451.

[6] H.K. Weber, H. Stecher, K. Faber, Some properties of commercially available crude lipase preparations, in: S.M. Roberts (Ed.), Preparative Biotransformations, Wiley, New York, 1995.

[7] J.J. Lalonde, C. Govardhan, N. Khalaf, A.G. Martinez, K. Visuri, A.L. Margolin, Cross-linked crystals of *Candida rugosa* lipase: highly efficient catalysts for the resolution of chiral esters, J. Am. Chem. Soc. 117 (1995) 6845–6852.

[8] K. Lundell, T. Raijola, L.T. Kanerva, Enantioselectivity of *Pseudomonas cepacia* and *Candida rugosa* lipases for the resolution of secondary alcohols: The effect of *Candida rugosa* isoenzymes, Enzyme Microb. Technol. 22 (1998) 86–93.

[9] E.J. Gilbert, *Pseudomonas* lipases: Biochemical properties and molecular cloning, Enzyme Microb. Technol. 15 (1993) 634–645.

[10] A. Svendsen, K. Borch, M. Barfoed, T.B. Nielsen, E. Gormsen, S.A. Patkar, Bio- chemical properties of cloned lipases from the *Pseudomonas* family, Biochim. Biophys. Acta 1259 (1995) 9–17.

[11] A. Hummel, E. Brüsehaber, D. Böttcher, H. Trauthwein, K. Doderer, U.T. Bornscheuer, Isoenzymes of pig-liver esterase reveal striking differences in enantioselectivities, Angew Chem. Int. Ed. Engl. 46 (2007) 8492–8494.

[12] R.J. Siezen, J.A.M. Leunissen, Subtilases: the superfamily of subtilisin-like serine proteases, Protein Sci. 6 (1997) 501–523.

[13] E. Henke, J. Pleiss, U.T. Bornscheuer, Activity of lipases and esterases towards tertiary alcohols: Insights into structure-function relationships, Angew Chem. Int. Ed. 41 (2002) 3211–3213.

[14] R.J. Kazlauskas, A.N.E. Weissfloch, A. Rappaport, L. Cuccia, A rule to predict which enantiomer of a secondary alcohol reacts faster in reactions catalyzed by cholesterol esterase, lipase from *Pseudomonas cepacia*, and lipase from *Candida rugosa*, J. Org. Chem. 56 (1991) 2656–2665.

[15] M. Cygler, P. Grochulski, R.J. Kazlauskas, J.D. Schrag, F. Bouthillier, B. Rubin, A.N. Serreqi, A.K. Gupta, A structural basis for the chiral preferences of lipases, J. Am. Chem. Soc. 116 (1994) 3180–3186.

[16] Q. Jing, R.J. Kazlauskas, Determination of absolute configuration of secondary alcohols using lipase-catalyzed kinetic resolutions, Chirality 20 (2008) 724–735.

[17] A.N.E. Weissfloch, R.J. Kazlauskas, Enantiopreference of lipase from *Pseudomonas cepacia* toward primary alcohols, J. Org. Chem. 60 (1995) 6959–6969.

[18] W.V. Tuomi, R.J. Kazlauskas, Molecular basis for enantioselectivity of lipase from *Pseudomonas cepacia* toward primary alcohols. Modeling, kinetics, and chemical modification of Tyr29 to increase or decrease enantioselectivity, J. Org. Chem. 64 (1999) 2638–2647.

[19] P.A. Fitzpatrick, A.M. Klibanov, How can the solvent affect enzyme enantioselectivity?, J. Am. Chem. Soc. 113 (1991) 3166–3171.

[20] R.J. Kazlauskas, A.N.E. Weissfloch, A structure-based rationalization of the enantiopreference of subtilisin toward secondary alcohols and isosteric primary amines, J. Mol. Catal. B Enzymatic 3 (1997) 65–72.

[21] K. Laumen, D. Breitgoff, R. Seemayer, M.P. Schneider, Enantiomerically pure cyclohexanols and cyclohexane-1,2-diol derivatives; chiral auxiliaries and substitutes for (–)-8-phenylmenthol. A facile enzymatic route, J. Chem. Soc. Chem. Commun. (1989) 148–150.

[22] D. Breitgoff, K. Laumen, M.P. Schneider, Enzymatic differentiation of the enantiotopic hydroxymethyl groups of glycerol; synthesis of chiral building blocks, J. Chem. Soc. Chem. Commun. (1986) 1523–1524.

[23] (a) L. Thalén, C. Hoben, J.-E. Bäckvall, Synthesis of (*R*)-2-methoxy-*N*-(1-phenylethyl) acetamide via dynamic kinetic resolution, Org. Synth. 89 (2012) 255–266. (b) J.-E. Bäckvall, B. Martín-Matute, Dynamic kinetic resolution catalyzed by enzymes and metals, Curr. Opin. Chem. Biol. 11 (2007) 226–232.

[24] C.S. Chen, C.J. Sih, General aspects and optimization of enantioselective biocatalysis in organic solvents – the use of lipases, Angew Chem. In. Ed. 28 (1989) 695–707.

[25] P. Stead, H. Marley, M. Mahmoudian, G. Webb, D. Noble, Y.T. Ip, E. Piga, T. Rossi, S. Roberts, M.J. Dawson, Efficient procedures for the large-scale preparation of (1*S*,2*S*)-trans-2-methoxycyclohexanol, a key chiral intermediate in the synthesis of tricyclic β-lactam antibiotics, Tetrahedron Asymmetry 7 (1996) 2247–2250.

[26] F. Balkenhohl, K. Ditrich, B. Hauer, W. Ladner, Optically active amines via lipase-catalyzed methoxyacetylation, J. Prakt. Chem. 339 (1997) 381–384.

[27] R.J. Kazlauskas, (*S*)-(-)-and (*R*)-(+)-1,1′-bi-2-naphthol, Org. Synth. 70 (1991) 60–67.

[28] S.P. Roche, D.J. Aitken, Chemistry of 4-hydroxy-2-cyclopentenone derivatives, Eur. J. Org. Chem. 28 (2010) 5339–5358.

[29] D.R. Deardorff, C.Q. Windham, C.L. Craney, Enantioselective hydrolysis of *cis*-3,5-diacetoxycyclopentene: (1*R*,4*S*)-(+)-4-hydroxy-2-cyclopentenyl acetate – (4-cyclopentene-1,3-diol, monoacetate, (1*R*-*cis*)-), Org. Synth. 73 (1996) 25–35.

[30] L. Tietze, C. Stadler, N. Böhnke, G. Brasche, A. Grube, Synthesis of enantiomerically pure cyclopentene building blocks, Synlett. (2007) 485–487.

[31] P.M. Khan, R. Wu, K.S. Bisht, Pd(0) catalyzed intramolecular alkylation: stereoselective synthesis of furan and isoxazoline-2-oxide analogs, Tetrahedron 63 (2007) 1116–1126.

[32] S.-J. Cho, S.-H. Oh, R.D. Pridmore, M.A. Juillerat, C.-H. Lee, Purification and characterization of proteases from *Bacillus amyloliquefaciens* isolated from traditional soybean fermentation starter, Agric. Food. Chem. 51 (2003) 7664–7670.

[33] C.K. Savile, R.J. Kazlauskas, The 3-(3-pyridine)propionyl anchor group for protease-catalyzed resolutions: *p*-toluenesulfinamide and sterically hindered secondary alcohols, Adv. Synth. Catal. 348 (2006) 1183–1192.

[34] F. Theil, H. Sonnenschein, T. Kreher, Lipase-catalysed resolution of 3,3′-bi-indolizines: The first preparative access to enantiomerically pure samples, Tetrahedron Asymmetry 7 (1996) 3365–3370.

[35] M.-C. Parker, S.A. Brown, L. Robertson, N.J. Turner, Enhancement of *Candida antarctica* lipase B enantioselectivity and activity in organic solvents, Chem. Commun. (1998) 2247–2248.

[36] M.S. Rasalkar, M.K. Potdar, M.M. Salunkhe, *Pseudomonas cepacia* lipase-catalysed resolution of racemic alcohols in ionic liquid using succinic anhydride: role of triethylamine in enhancement of catalytic activity, J. Mol. Catal. B Enzym. 27 (2004) 267–270.

[37] K. Burgess, L.D. Jennings, Enantioselective esterifications of unsaturated alcohols mediated by a lipase prepared from *Pseudomonas* sp, J. Am. Chem. Soc. 113 (1991) 6129–6139.

[38] J.A. Marshall, H. Chobanian, Lipase-catalyzed resolution of 4-trimethylsilyl-3-butyn-2-ol and conversion of the (*R*)-enantiomer to (*R*)-3-butyn-2-yl mesylate and (*P*)-1-tributylstannyl-1, 2-butadiene, Org. Synth. 82 (2005) 45–54.

[39] R.T. Beresis, J.S. Solomon, M.G. Yang, N.F. Jain, J.S. Panek, Synthesis of chiral (*E*)-crotylsilanes: [3*R*- and 3*S*-]-(4*E*)-methyl 3-(dimethylphenylsilyl)-4-hexenoate, Org. Synth. 75 (1998) 78–83.

[40] Y. Terao, K. Tsuji, M. Murata, K. Achiwa, T. Nishio, N. Watanabe, K. Seto, Facile process for enzymic resolution of racemic alcohols, Chem. Pharm. Bull. 37 (1989) 1653–1655.

[41] N. Bouzemi, H. Debbeche, L. Aribi-Zouioueche, J.-C. Fiaud, On the use of succinic anhydride as acylating agent for practical resolution of aryl–alkyl alcohols through lipase-catalyzed acylation, Tetrahedron Lett. 45 (2004) 627–630.

[42] A.L. Gutman, D. Brenner, A. Boltanski, Convenient practical resolution of racemic alkyl-aryl alcohols via enzymatic acylation with succinic anhydride in organic solvents, Tetrahedron Asymmetry 4 (1993) 839–844.

CHAPTER 6

Modern Biocatalytic Ketone Reduction

Thomas S. Moody*, J. David Rozzell†
*Department of Biocatalysis and Isotope Chemistry,
Almac, Craigavon, Northern Ireland, UK
†Provivi, Santa Monica, CA, USA

1 INTRODUCTION

Stereoselective reduction of ketones to form chiral alcohols is one of the most useful reactions in organic synthesis [1]. Chiral alcohol structures are frequently embodied in pharmaceuticals and other high-value compounds, and reduction of the corresponding ketone is often the most practical route available. As a result, methods for stereoselective ketone reduction that are broadly useful and applicable to a wide range of compounds have been widely sought, and significant effort has been made to the development of methods that produce alcohols with high enantiomeric purities [1].

Carbonyl reductases (CREDs, also called ketoreductases or KREDs, alcohol dehydrogenases or ADHs) offer important advantages as catalysts for stereoselective ketone reduction. Recent enzyme discovery efforts have dramatically expanded the number of CREDs available for use [2,3], and cost-effective methods for recycling of the nicotinamide cofactors have been developed [3]. Basic reactions using CREDs typically require no special conditions (20–45 °C, atmospheric pressure), they are compatible with aqueous and organic solvent systems, and they avoid expensive toxic metal–ligand catalysts. High yields and optical purities (>99%) are routinely achieved from CRED-catalyzed reactions, and high titers (>100 g/L) are common, especially after the reaction has been optimized. A further advantage of CREDs is that enzymes capable of catalyzing the synthesis of either the *R*- or *S*-alcohol from the starting ketone are usually available, providing ready access to either stereoisomer of a chiral alcohol [2]. The cost of a biocatalytic chiral alcohol synthesis also compares favorably to that for other chiral catalysts, and a CRED-catalyzed reaction have increasingly been found to be the most practical and cost effective method.

Identification of the best enzyme for a desired application has also been simplified. With the wide availability of "screening sets" of CRED enzymes, finding a

Organic Synthesis Using Biocatalysis. http://dx.doi.org/10.1016/B978-0-12-411518-7.00006-8

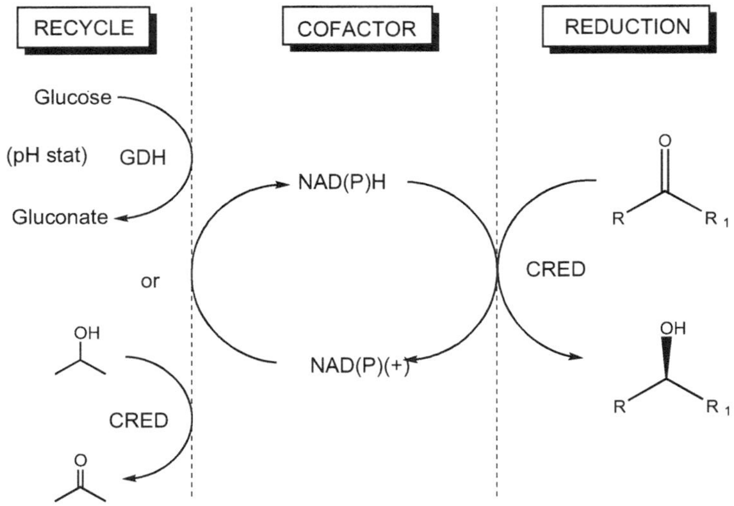

FIGURE 6.1
CRED catalyzed bioreduction of prochiral ketones with cofactor recycling.

suitable enzyme to use as a starting point in developing a CRED-catalyzed route to a chiral alcohol can be accomplished in a few days by any chemist who has a chiral separation method for the desired product. As a result, enzymatic ketone reductions are now an important addition to the synthetic chemist's tool box, expanding and often improving upon the established stereoselective chemical-reduction chemistry.

2 MECHANISM OF ENZYMATIC KETONE REDUCTION

Enzyme-catalyzed ketone reduction proceeds by a well-established mechanism that involves hydride transfer from a nicotinamide cofactor. The cofactor NAD(P)H can either be regenerated by the same CRED or by a second enzyme such as glucose dehydrogenase (GDH) as shown in Figure 6.1 [4].

The cofactor can be either NADH or NADPH that differ structurally only in the level of phosphorylation. Both forms of the nicotinamide cofactor function identically. Hydride is transferred from the reduced form of the cofactor to the carbonyl group whereas the substrate is bound in the active site. The reaction is typically stereoselective, although not always perfectly so, as binding of the ketone substrate to the enzyme is generally lower in energy when binding one face of the ketone toward the cofactor relative to the other. These differences in binding energy translate, in turn, to stereoselectivity in hydride transfer as the chiral alcohol is formed. The newly produced chiral alcohol is released from the enzyme and the cofactor, now in the oxidized NAD(P)+ form, is regenerated, for example, *via* the acceptance of hydride from a different reduced substrate (other regenerations methods are shown in Figure 6.2). This completes the catalytic cycle, and the CRED with reduced cofactor is now ready for reduction of another

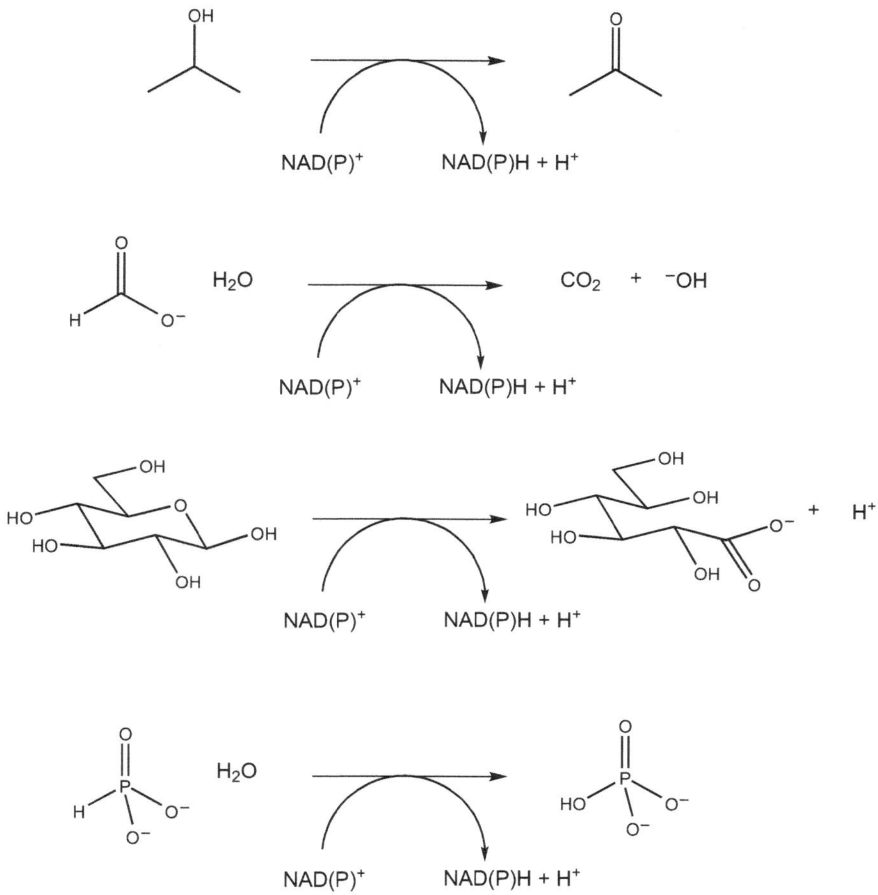

FIGURE 6.2
NAD(P)H cofactor recycle systems for CRED mediated ketone reductions.

ketone molecule. Ideally, the same CRED that catalyzes the reduction of the ketone precursor can also catalyze oxidation of *iso*-propanol or a similar inexpensive alcohol to regenerate the nicotimamide cofactor. For enzymatic ketone reduction to be practical from a commercial standpoint, cofactor regeneration must be low in cost, and 1000 or more catalytic cycles are typically achieved.

Efficient cofactor recycling can also be achieved *via* the participation of a second enzyme. In this process configuration, the oxidized cofactor is released from the CRED and is reduced by the action of a second enzyme. Examples of this type of cofactor regeneration include glucose dehydrogenase (GDH), with reduction equivalent provided by the oxidation of glucose to gluconolactone; formate dehydrogenase (FDH), with reducing equivalents provided by the oxidation of formate to carbon dioxide; with a second alcohol dehydrogenase with a separate inexpensive alcohol as supplier of reducing equivalents; phosphite dehydrogenase (PDH), with reducing equivalents provided by the oxidation of phosphite to phosphate as shown in Figure 6.2 [5].

2.1 Driving Ketone Reductions to Completion

An inexpensive source of reducing equivalents is a prerequisite for a cost-effective commercial-scale process that uses a CRED. Typically this is achieved by using an inexpensive alcohol, for example, *iso*-propanol, as the source of hydride. The commercial example of montelukast, described later, used this approach to achieve excellent final yields, together with an insoluble product that precipitated from the reaction mixture.

Others have driven the bioreduction process to completion by employing an *in situ* product removal (ISPR) technique to overcome the inherent equilibrium problem associated with the coupled-substrate approach to biocatalytic carbonyl reduction. This technique allowed the valuable chiral alcohol, (*S*)-2-bromo-2-cyclohexen-1-ol, to be isolated in 88% yield and 99.8% ee without the need for further purification, validating the general applicability of this experimental setup shown in Figure 6.3 [6].

However, since the overall reaction is effectively a transfer hydrogenation in which reducing equivalents from a donor alcohol (isopropanol) are used to reduce the desired ketone, driving the reaction to completion can be problematic. Typically, a large excess of isopropanol is used that allows conversions of 90–95% to be achieved if high concentrations of isopropanol are compatible with the required overall reaction conditions.

An alternative method was described by Hollman and coworkers [7]. By choosing 1,4-butanediol as the source of reducing equivalents with a CRED that would catalyze its oxidation, a further oxidation of the hemiacetal to the lactone provided an additional molar surplus of hydride equivalents with an essentially irreversible conversion to the lactone, enabling the reaction to be

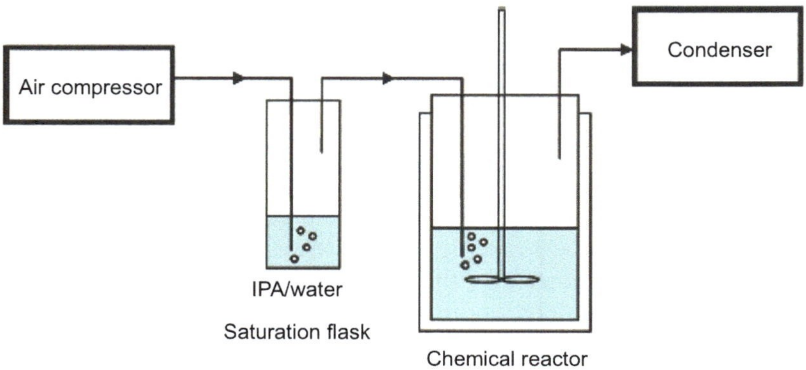

FIGURE 6.3
ISPR technique to drive CRED bioreduction conversion to completion.

driven to completion with a lower excess of the alcohol. Data from the HLADH-catalyzed reduction of 2-phenyl-1-propanal showed that 99% conversion could be achieved with a molar ratio of only 0.5 equivalents of 1,4-butanediol to 2-phenyl-1-propanal. In contrast, to drive the reaction to 99% using ethanol as the hydride donor, a molar ratio of 1000:1 was required.

3 RECENT BIOREDUCTION EXAMPLES

This chapter is aimed at those researchers that are exploring the potential opportunities of enzymes for ketone reductions. The following sections will whet the readers' appetite for the possibilities of ketone reduction using enzymes and should inspire those who want to push further into this field of opportunities.

3.1 Commercially Scaled Reductions

The use of enzymatic ketone reduction has progressed far beyond the level of small-scale laboratory reactions or curiosity. The wide-spread availability of CRED enzymes from a number of sources has resulted in many chemists using them from small-scale supply of chiral alcohol right through to commercial multiton production. An excellent guide for sourcing suppliers of CREDs is detailed in the *"Biocatalysis Enzyme Guide"* [8]. The ability to evolve enzymes for fitness under industrial process conditions has led to the commercialization of a number of CRED-catalyzed processes to produce chiral alcohols at the multiton scale.

3.1.1 *LIPITOR*

Once the largest-selling drug in the world, Lipitor **1** (generic name atorvastatin) saw peak sales exceeding $13 billion per year before going generic in 2013. Atorvastatin has two chiral centers and is sold as the single (*R*,*R*)-diastereoisomer.

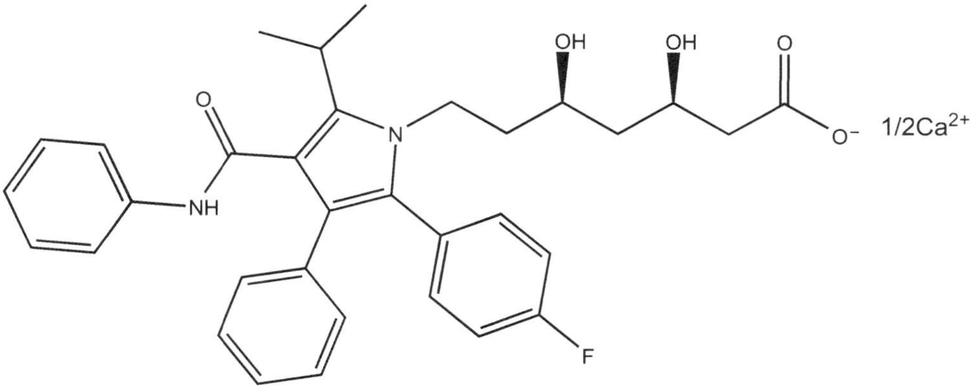

1

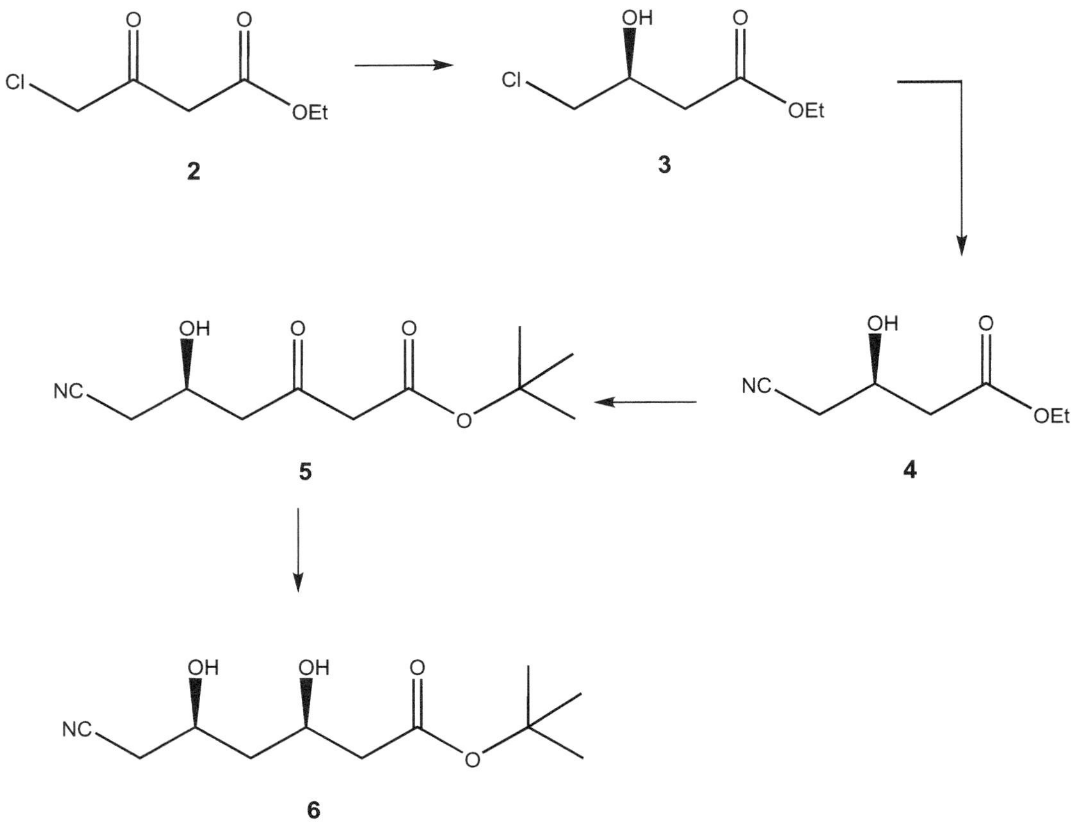

SCHEME 6.1
Bioreduction route to diol intermediate 6.

With two chiral centers, both of which can be installed by stereoselective ketone reduction, atorvastatin is ideally suited to the application of CREDs for its synthesis, a fact not lost on the pharmaceutical industry. Seven different biocatalytic processes have been developed by various groups to produce key intermediates for this large volume drug [9].

The most-commonly used method involves the use of one or two different CREDs to catalyze the reduction of the ketones to produce the hydroxynitrile intermediate **4** and the more advanced TBIN intermediate **6** as shown in Scheme 6.1 [10].

This process has been commercialized to produce more than 100 tons of atorvastatin annually, leading to a more sustainable process that produces dramatically less waste and a final product that has been shown to be of higher purity than the chemically produced alternative. Table 6.1 shows the comparison between the biocatalytic reduction (to produce TBIN) and the chemical process that preceded it.

Table 6.1 Comparison of Chemical Process to CRED-Catalyzed Production of Atorvastatin Intermediate 6

Parameter	Enzymatic Process	Chemical Process
Crude substrate load	300 g/L	100 g/L
Substrate/catalyst (wt/wt)	85:1	1:1
Substrate/NADP (wt/wt)	>1200	NA
Conversion	99.3%	Not provided
Diastereomeric excess	>99.9%	~94%
Work-up	Easy	Difficult
Solvent use	3.2 L/kg	27.5 L/kg

Adapted from Ref. [10].

3.1.2 MONTELUKAST

Montelukast 7 is an orally active selective leukotriene receptor antagonist used for the treatment of asthma. With an annual volume of approximately 20 metric tons per year and a synthetic route that requires both large amount of solvents and a toxic, difficult-to-handle chiral reducing agent, montelukast represented an attractive target for biocatalysis as a way to implement a greener, more efficient process.

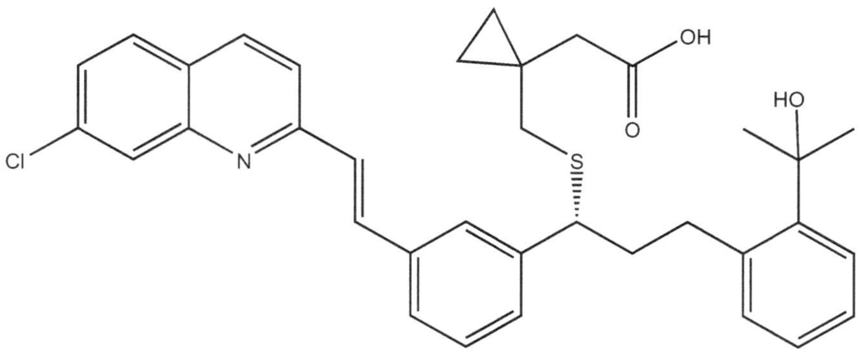

7

The bioreduction step for the production of montelukast key intermediate is shown in Scheme 6.2. The key step is the stereoselective reduction of ketone 8 to the *S*-configured alcohol 9. The alcohol subsequently undergoes an S_N2 displacement with a thiol to give the *R*-configured final product.

The original chemical process required in particular, the toxic and difficult to handle chiral reducing agent (−)-ß-chlorodiisopinocampheylborane [(−)-DIP-chloride]. Replacing (−)-DIP-chloride eliminated a major source of waste and

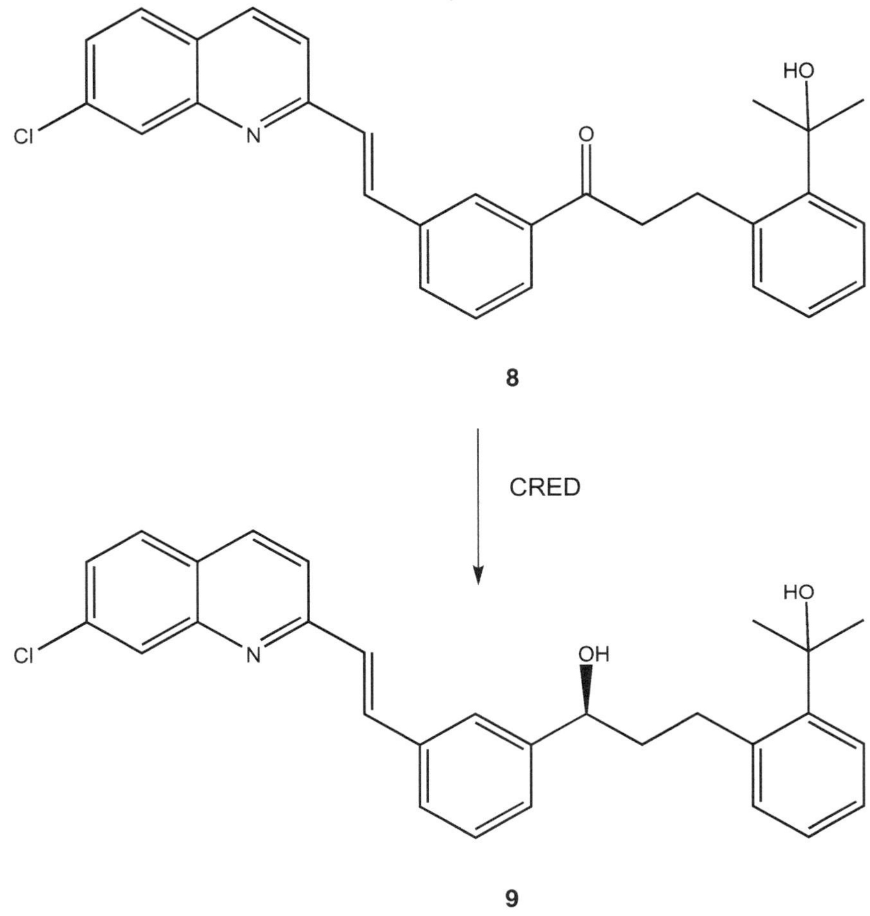

SCHEME 6.2
Bioreduction step required prior to activation and thiol displacement to Montelukast.

cost in the process. Directed evolution methods were used to create a CRED that was capable of reducing the ketone precursor in the presence of an aqueous–organic solvent mixture that was required due to the insolubility of the starting material. Table 6.2 shows the improvement provided by the CRED-catalyzed route. The benefits include dramatically reduced waste, lower solvent consumption, higher volumetric productivity, higher yield, and a product of greater chemical and enantiomeric purity.

3.1.3 (S)-BIS-3,5-TRIFLUOROMETHYLPHENYLETHANOL

Both enantiomers of 3,5-*bis*-trifluoromethylphenylethanol **11** are used as key intermediates in the stereoselective synthesis of NK-1 receptor antagonists. The (*R*)-isomer is a key building block in the synthesis of the Merck drug aprepitant

Table 6.2 Comparison of CRED-Catalyzed Step to Chemical Step in Montelukast Synthesis

Parameter	Biocatalytic Process	(–)-DIP-Cl Process
Ketone 8 concentration	100 g/L	100 g/L
Chiral agent/ketone (kg/kg)	0.03–0.05 (KRED)	1.25 (DIP-Cl, 1.8 eq)
NADP/ketone (%wt/wt)	0.1	Not applicable
Reaction time	24 h	6 h
Temperature	45°C	–25°C
Conversion	99.3%	Not provided
Product isolation	Direct filtration	Extraction with high dilution
Enantiomeric excess	>99.9% (enantiomer not detected)	~99.2% (after recryst.)
Solvent/MLK-III (L/kg)	6	30–50
Solvents used	Isopropanol, water, toluene	Dichloromethane, THF

Adapted from Ref. [10].

(Emend) whereas (*S*)-isomer is incorporated into a number of antagonists under clinical evaluation. The synthesis of (*S*)-*bis*-3,5-trifluoromethylphenylethanol was described by Pollard and coworkers using the CRED from *Rhodococcus erythropolis* with cofactor recycling using formate dehydrogenase and glucose dehydrogenase as shown in Scheme 6.3 [11].

Operating at concentrations of 100 g/L or more, 96% conversion of the ketone starting material was achieved, resulting in *S*-alcohol product **11** of 99% ee. Space–time yields of 260 g/L/day were attained in a stirred-tank reactor.

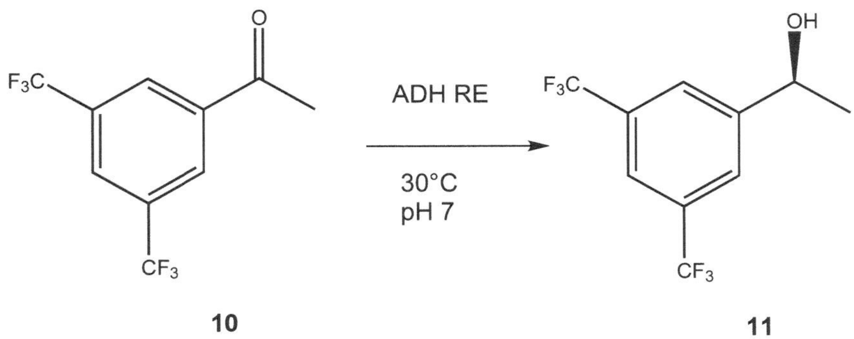

SCHEME 6.3
Bioreduction using ADH RE enzyme from *R. erythropolis*.

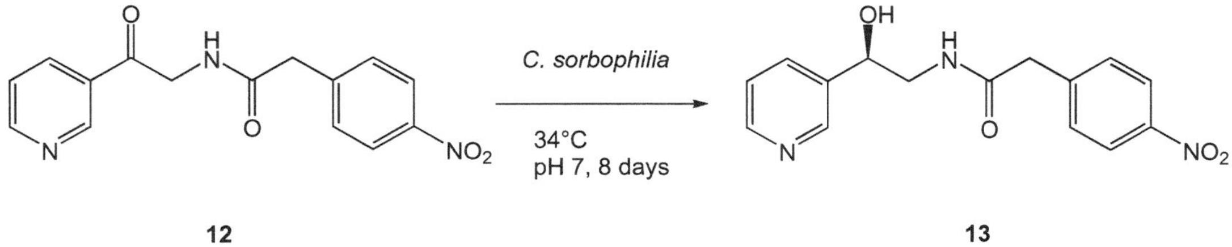

SCHEME 6.4
Whole cell bioreduction using *C. sorbophila*.

3.2 Microbial Bioreductions

Prior to the widespread availability of recombiant carbonyl reductases enzymes, the use of microbial reductions using either actively growing or dormant cells was commonplace. Bakers' yeast, in particular, was a readily available source of stereoselective carbonyl reductases enzymes. Even with the widespread knowledge of the power of recombinant CRED biocatalysts, the literature is still rife with wild-type whole-cell microbial reductions. The reductions presented have advanced well beyond the early Bakers' yeast reduction and have an application even today. When the whole-cell fermentation is developed and finely tuned, high titers of product alcohol are possible and Scheme 6.4 shows an example of a keto-amide **12** bioreduction performing at 100 g/L with more than 98% ee with multi-kg isolation [12]. The bioprocess was performed over 8 days at pH 7 using the yeast *Candida sorbophila*.

The bacterium *Bacillus pumilus* has been used with a large number of different ketones **14**, including both β- and α-keto esters as shown in Schemes 6.5 and 6.6, respectively. The β-keto esters were reduced to the corresponding chiral alcohols with conversions more than 79%, and ee > 90% [13].

The same microbe also gave very high activities for α-ketoesters **17** with conversions more than 82% and ee greater than 93.9%. The isolated product ethyl-(*R*)-2-hydroxy-4-phenylbutyrate is used as a building block for ACE inhibitors including Enalapril and Cilapril. The reaction was performed at 30 mM of ethyl-(*R*)-2-keto-4-phenylbutyrate charge using 12 g of dry cell weight (gdcw)/L of *B. pumilus* cells. The process yielded 95% conversion and 97.1% ee product after 19 h. It was also demonstrated that once the cells were washed, they could be successfully reused, and 80% of activity remained after 6 recycles.

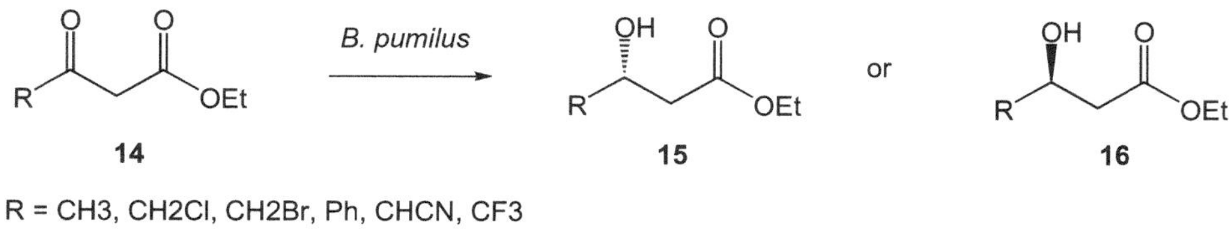

SCHEME 6.5
***B. pumilus* mediated bioreduction of β-keto esters.**

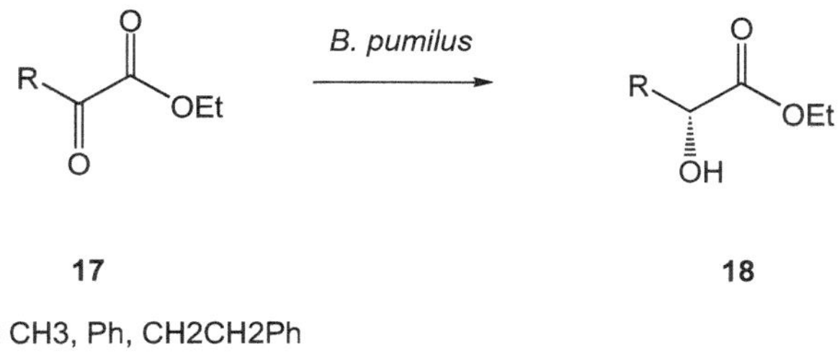

SCHEME 6.6
***B. pumilus* mediated bioreduction of α-keto esters.**

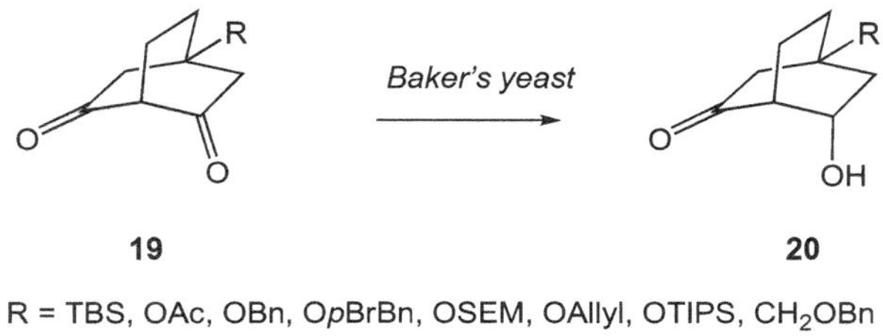

SCHEME 6.7
Monoreduction using Bakers' yeast.

It was shown that the selective bioreduction of diketones **19** to yield mono-keto-alcohols **20** could be achieved using Bakers' yeast as shown in Scheme 6.7 [14].

The enantioselectivity was influenced by the nature of the *R* group, and no reduction was achieved when *R* = OTIPS. On the other hand, when *R* = OTBS, the product was obtained in 87% conversion and 46% ee. The highest ee was observed when *R* = O-Allyl (80, 82% ee) and lowest ee observed when *R* = OAc (73, 10% ee).

Again with Bakers' yeast it was shown that diols **22** and **23** could be produced as shown in Scheme 6.8 [15]. The diketone **21** was dissolved in DMF and when added to the Bakers' yeast media, high enantiomeric excesses were obtained.

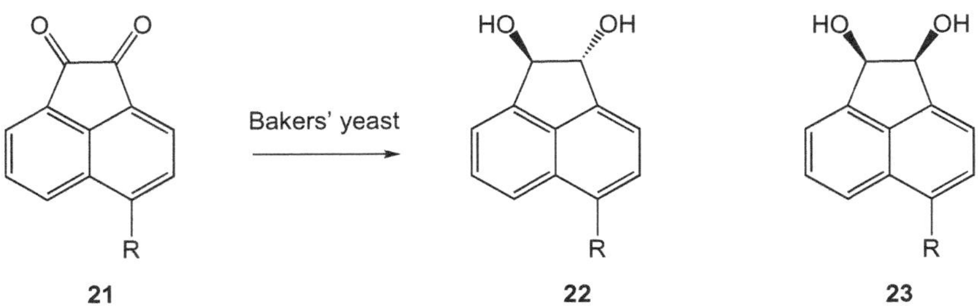

SCHEME 6.8
Transdiol production using Bakers' yeast.

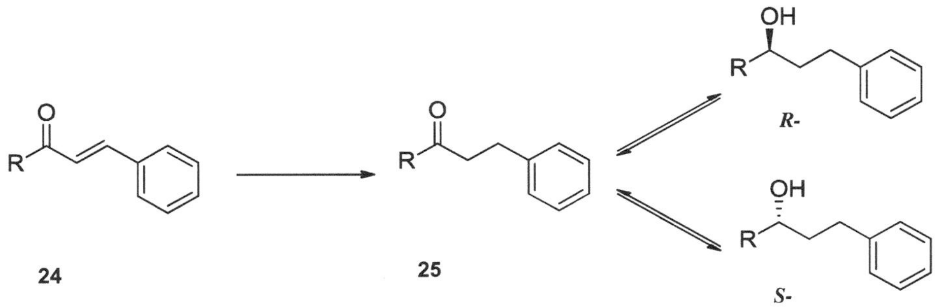

24

25

26

SCHEME 6.9
Yeast- and fungi-mediated reversible bioreductions.

It was also demonstrated that when DMF was used as a cosolvent, enantioselectivity for the trans-diol increased from 87% to 91% ee but its yield decreased from 85% to 42%. DMSO on the other hand, gave 97% ee, and the yields were maintained at 74%. The electronic properties of the substituent at the 5-position had no influence on the enantioselectivity. *Cis*-diols were obtained in low ee; this was attributed to electronic properties of the substituted group at the 5-position: when the group was electron donating, it inhibited the activity of reductases toward the carbonyl group.

Researchers investigated both yeast and fungi bioreductions of enone substrates **24** to yield the corresponding saturated chiral alcohols **26** as shown in Scheme 6.9 [16].

The double bond was saturated within 1 h of exposure to the yeast cells and was observed to take longer with selected fungi. The subsequent ketone reduction was found to be reversible. An interesting change was observed in alcohol enantiopurity while the reaction was being monitored. The microbe initially reduced the ketone in a nonenantioselective manner, but the *R*-enantiomer was subsequently oxidized. This resulted in the ee of the *S* isomer increasing. For example, in the bioreduction using *Rhodotorula glutinis,* after 3 h the *R*-alcohol was formed with 47% ee and 31% substrate conversion. With the same system after 6 days, the *S*-isomer was obtained with 99% ee and 91% conversion.

It was shown with *Candida parapsilosis* that cells induced with glycerol produced (*R*)-ethyl-4-chloro-3-hydroxybutanote (*R*)-**3** with more than 99% ee, and cells induced with sucrose or glucose produced (*S*)-ethyl-4-chloro-3-hydroxybutanote (S)-**3** (>99% ee) as shown in Scheme 6.10 [17]. It was also noted that activity was higher on glucose grown cells and cells grown on glucose or sucrose preferred NADH as a cofactor. Cells grown on glycerol can use either cofactor.

Homann and coworkers screened 300 microbes for their ability to generate chiral secondary alcohols by enantioselective reduction of a series of alkyl aryl

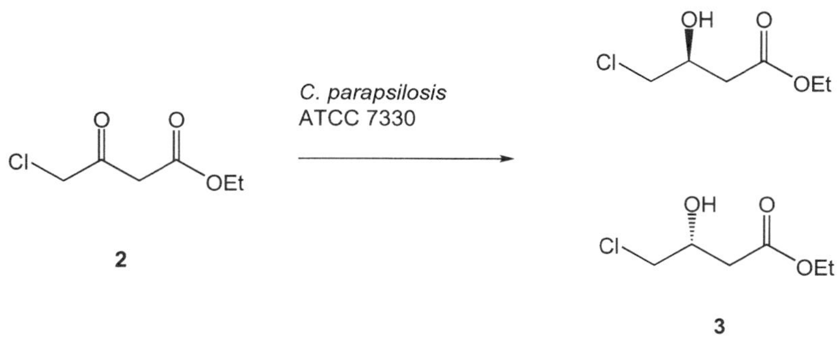

SCHEME 6.10
Tuneable bioreduction of *C. parapsilosis* using different inducers.

ketones. Microbial cultures demonstrating the ability to reduce model ketones were arrayed in multiwell plates and used to rapidly identify specific organisms capable of producing chiral alcohols. The plate format afforded a 5–10-fold reduction in the average screen time compared to a conventional flask-based screening protocol.

Sixty cultures were shown to selectively reduce various ketones providing both the *R* and *S* enantiomers of the corresponding alcohols in 92–99% ee with yields up to 95% [18].

One major and potential drawback of using whole-cell wild-type microbes is that they contain multiple enzymes that can do a range of different biotransformations and can result in multiple side products. Therefore, whole cell biotransformations are usually contaminated with impurities.

It is in addressing the need for high selectivity that the recombinant enzyme comes of age to yield a selective biotransformation based on a single, expressed recombinant enzyme. Furthermore, recombinant enzymes can be evolved to incorporate improvements in selectivity, stability, and reaction rate to meet the demanding needs of commercial processes. The chapter will now focus on recombinant systems, including engineered recombinant CREDs.

3.3 Aromatic Ketone Reductions

Acetophenone-type compounds dominate in the literature and are a successful class of compounds for bioreductions using CRED enzymes. Their simple availability, minimal steric hindrance around the ketone, and excellent solubility in a range of solvents make them useful substrates for further study.

It has been shown through molecular modeling that the enantio-preference of a Prelog CRED from *Thermoanaerobacter ethanolicus* can be inverted by a simple mutation of a single amino acid that lines the small pocket (I86A) for

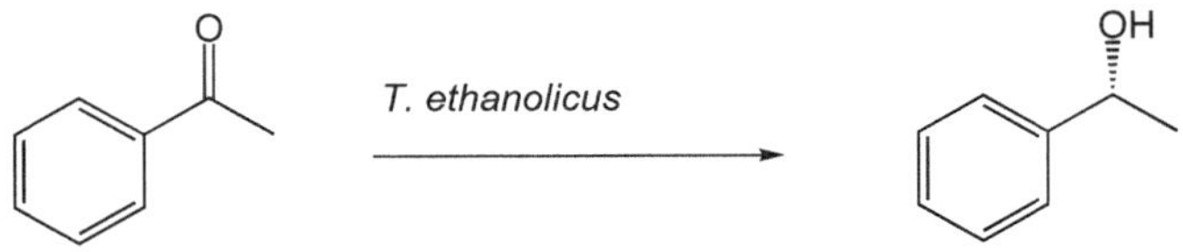

27 28

SCHEME 6.11
Bioreduction of acetophenone using *T. ethanolicus*.

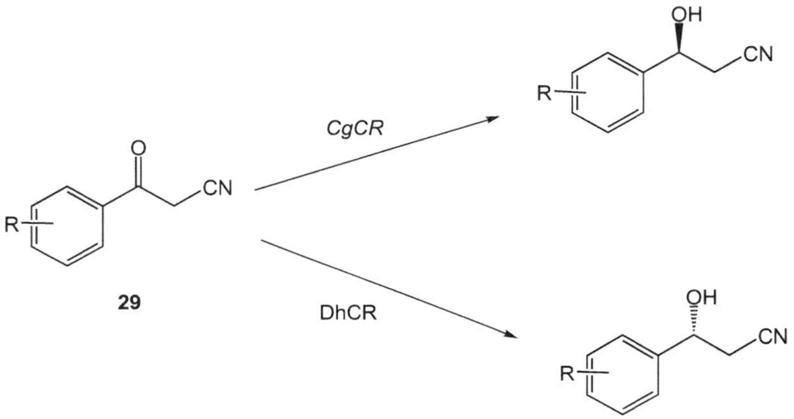

29 30

SCHEME 6.12
CRED reductions yielding β-cyanohydrins.

the bioconversion of acetophenone **27** to chiral phenylethanol **28** as shown in Scheme 6.11. This mutant having a large substrate-binding pocket not only shows anti-Prelog stereospecificity but also accepts more sterically demanding substrates than those accepted by wild-type enzyme [19].

Work performed by Xu et al. demonstrated that through genome mining, CREDs could be identified and actively expressed to yield enzymes that yielded both *R* and *S* selectivity as shown in Scheme 6.12 [20].

When the GDH cofactor recycle system was used with toluene cosolvent, 145 g/L of substrate was tolerable with a space–time yield of 51.6 g/L/day.

When the enzyme cofactor preference was changed from NADPH to NADH through site directed mutagenesis, the enantioselectivity was also changed for the conversion shown in Scheme 6.13. The double-point mutation S67D/H68D of the anti-Prelog CRED from *C. parapsilosis* successfully converted the coenzyme specificity of the short-chain reductase from NADPH to NADH as well as the product enantioselectivity from (*S*)- to (*R*)-configuration [21].

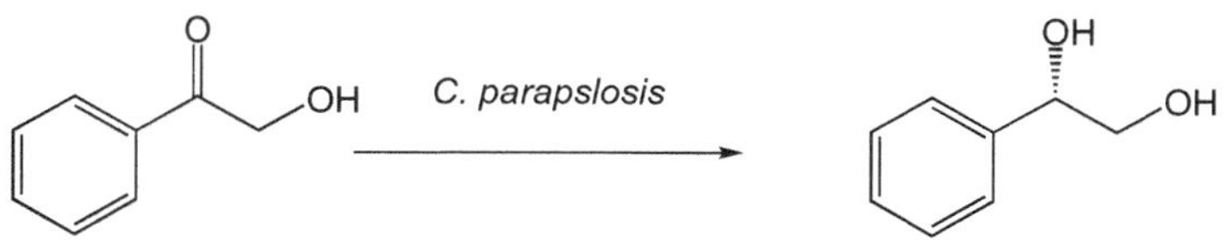

31 32

SCHEME 6.13
CRED reductions of hydroxy ketones.

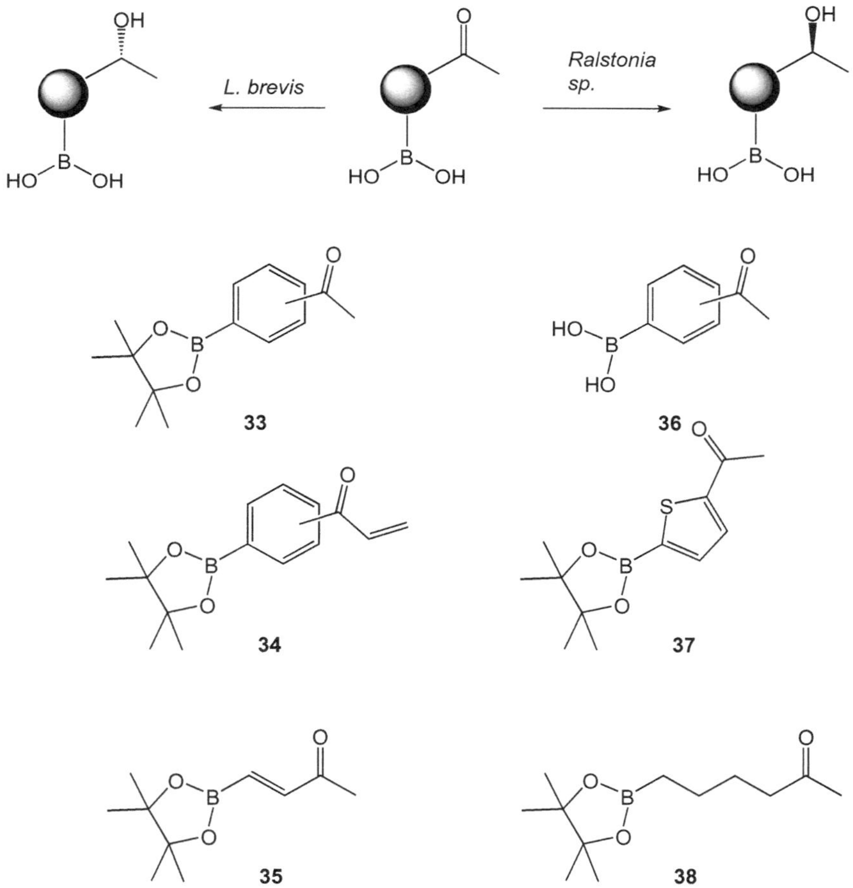

SCHEME 6.14
Bioreduction of Boronic acid and ester containing ketones.

Ou et al. reported the first study of bioreduction with boron containing compounds (Scheme 6.14). The CRED from *Ralstonia* sp. had the highest conversions while maintaining high ee (>95%). It was shown that acetophenone boronic acids and esters were accepted by the enzyme. The C=C double bond of α,β-unsaturated ketones (**34**) bearing a terminal alkene was not reduced [22].

Computational enzyme–substrate docking studies carried out on a CRED from *Sporobolomyces salmonicolor* showed that residues Met 242 and Gln 245 were in close proximity to the *para*-substituent of acetophenones in the substrate binding site. Site-saturation mutagenesis of Met 242 or Gln 245 (replacing these residues with all 19 other possible amino acids) and double mutation of both Met 242 and Gln 245 were carried out to enhance the enzyme's enantioselectivity toward the reduction of *para*-substituted acetophenones. The enantiopreference was inverted from (*R*)- to (*S*)-configuration for three Gln 245 mutants that were obtained. Preference for the formation of (*S*)-enantiomeric alcohols was shown for four Met 242 mutant enzymes when compared to the wild-type enzyme. Gln 245 and Met 242/Gln 245 double variations inverted the enantiopreference and affected the enantiomeric purity of the product alcohols. Residues 242 and 245 also exerted an effect on the catalytic activity of this CRED [23].

C. parapsilosis [(S)-carbonyl reductase II] reduced acetophenone to chiral phenylethanol with a low yield of 3.2%. Site-directed mutagenesis was used to design two mutants Ala 220 Asp and Glu 228 Ser. The two mutations resulted in a significant enantioselectivity shift toward (*R*)-phenylethanol in the reduction of acetophenone. The variant Glu 228 Ser produced (*R*)-phenylethanol with an ee of 99%, with 80.2% yield. This mutant had a 4.6-fold decrease in the K_M value, and a 5-fold and 21-fold increase in the k_{cat} and k_{cat}/K_M values compared to the wild-type organism [24].

Indian researchers investigated the chemoenzymatic route to rivastigmine **41**. Rivastigmine is pharmaceutically active compound that inhibits acetylcholinase enzymes. The key step in the synthesis was the bioreduction of the ketone **39** moiety as shown in Scheme 6.15 [25].

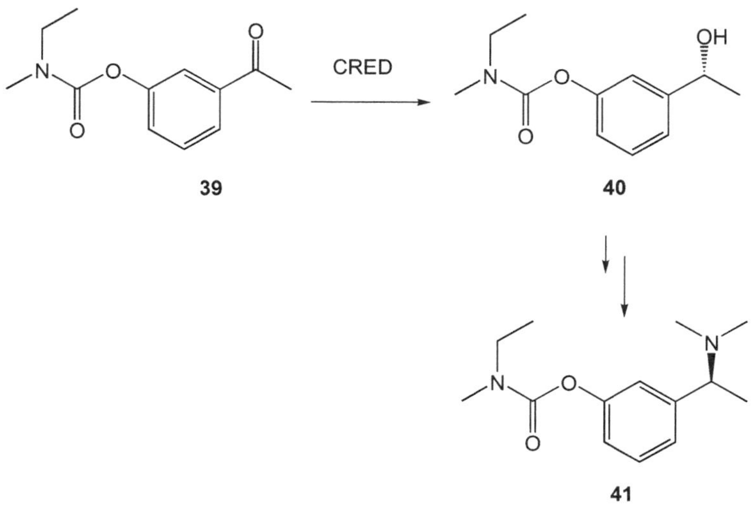

SCHEME 6.15
Bioreduction of ketone 39 en route to rivastigmine 41.

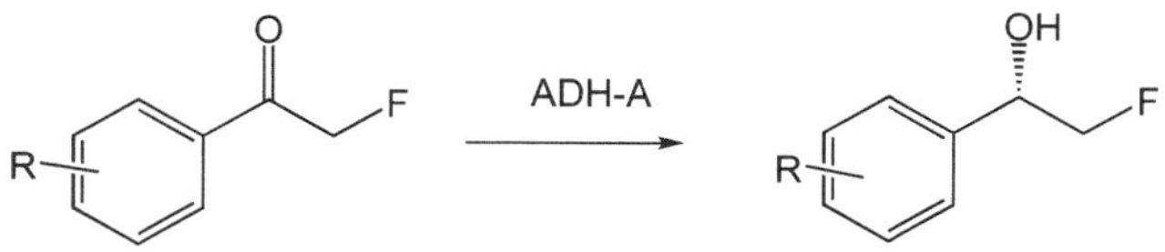

42 43

SCHEME 6.16
Bioreduction of 42 using ADH-A to yield chiral fluorohydrins 43.

Phenacyl halides have featured in many published articles using CRED technology (typically *en route* to chiral epoxides) and the work by Gotor and coworkers demonstrate how enzymes can be used to make fluorohydrins **43** as shown in Scheme 6.16 using the CRED ADH-A from *Rhodococcus ruber*.

The reactions were performed with whole-cell preparations and no external cofactor was added. A 0.5 M charge of ketone was used in a buffered system. Because of the reactive nature of these substrates, a high concentration of hydride donor was not necessary to achieve high-reaction conversion to product [26].

3.4 Aliphatic and Cyclic Ketone Reductions

3-Hydroxythiophene **45** is a relative simple building block and looks like an obvious choice for bioreduction from the corresponding ketone **44** as shown in Scheme 6.17. The original synthesis by Pfizer was a five-step route from L-aspartic acid. Due to its near-symmetric structure, the identification of highly selective CREDs was difficult. At the time of screening, the best *R* selective CRED came from *Lactobacillus kefir*. Directed evolution of the enzyme using gene shuffling, semisynthetic shuffling, random mutagenesis, and ProSAR to deconvolute the contributions of individual mutations resulted in a first tier of HTS screening. The HTS screening involved identification of an active mutant by NADPH assay. Following this, a second tier of screening was performed under preparative conditions using chiral HPLC analysis to detect selective enzymes. A total of eight rounds of mutations were needed to achieve the desirable process conditions (substrate loading >100 g/L, CRED loading less than 1 g/L enzyme, conversion >98%, reaction time ≤24 h, optical purity≥99%). The process was performed at 130 kg scale [27].

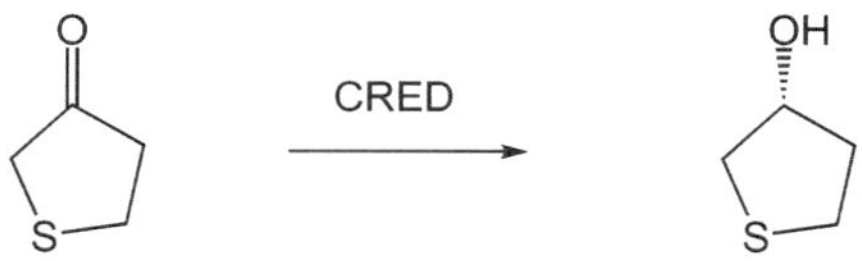

44 45

SCHEME 6.17
Mutated *L. kefir* CRED for the bioreduction to yield 3 hydroxythiophene 45.

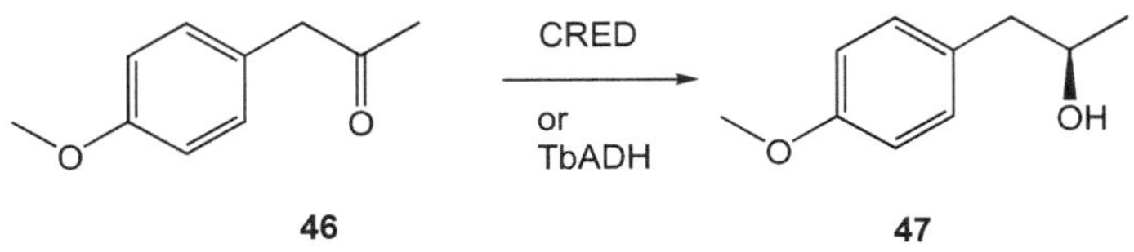

SCHEME 6.18
Enantioselectivity altering of CREDs using cyclodextrins.

TbADH (*Thermaonaerobium brockii* alcohol dehydrogenase) and 12 other commercial CREDs were used to reduce phenylethyl ketone **46** as shown in Scheme 6.18. α-,β- and γ-Cyclodextrins were used as additives to alter the enantioselectivity and decrease the reaction rate. The studies showed that CRED 122, in the presence of increasing concentration of γ-cyclodextrin, led to a decrease in enantiopurity of the desired (*R*) product; however, when the molar ratio of cyclodextrin reached 3.8, the configuration of the alcohol was swapped to (*S*). The decreased activity of the enzyme was attributed to transformation of the enzymes' α-helix to the β-sheet or to a random coil. Maximum activity was associated with maximum amount of α-helix content [28].

The potential to use CREDs for the double reduction of diketones to access potentially eight different stereoisomers in a single reaction step is a powerful tool. Bariotaki et al. demonstrated that the δ,β-diketo esters **48** could be selectively reduced using CRED enzymes as shown in Scheme 6.19. Their work showed that seven of the eight stereoisomers were obtained in pure form using one pot synthesis with yields more than 80% and 95% de [29].

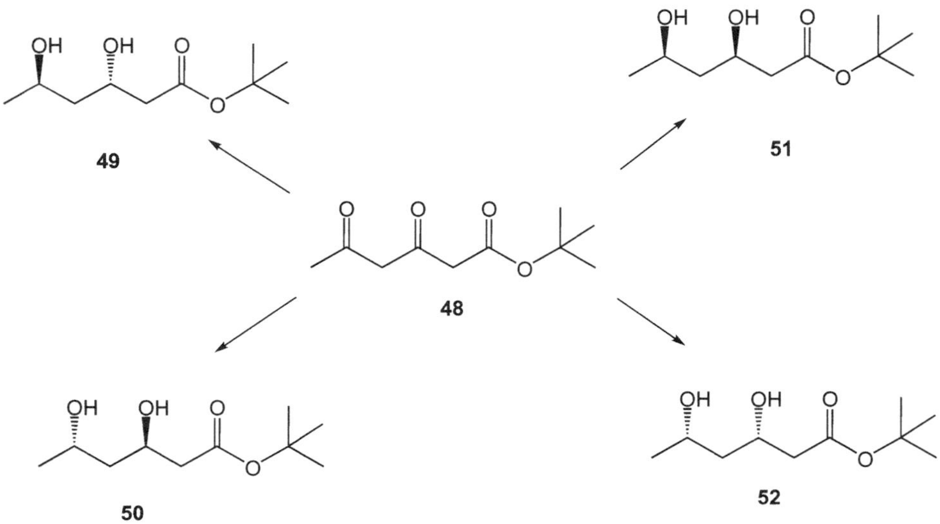

SCHEME 6.19
Bioreductions of diketones.

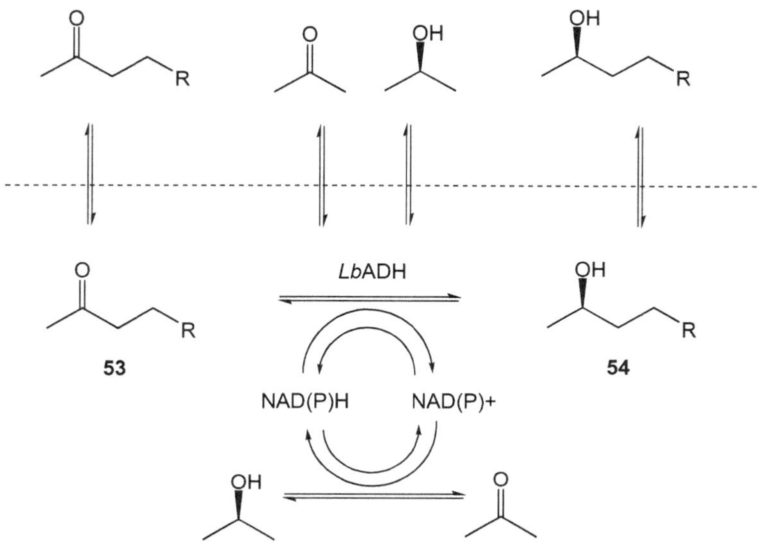

SCHEME 6.20
CSTR using CRED from *L. brevis*.

The continuous production of aliphatic chiral alcohols was demonstrated by Leuchs et al. as shown in Scheme 6.20 by using continuous stirred tank reactor (CSTR). The CRED from *Lactobacillus brevis* was used in a biphasic system with MTBE as cosolvent to reduce aliphatic ketones with the general structure **53**. It was shown that increasing the chain length resulted in a decreased yield due to less available substrate in the aqueous phase. The continuous process was run with a ketone concentration of 100 mmol/L and IPA was used as the reductant (1 mol/L). It was also demonstrated that 200 rpm was ideal for longevity of the enzyme. Doubling the residence time did not lead to a significant increase in the space–time yield [30].

Jakoblinnert et al. showed that the reduction of ketones could be performed in neat substrates with defined water activity (*aw*). Optimum water activity was around an *aw* value of 0.84, nonequilibrated wet cells were active but rapidly deactivated within few hours, whereas lyophilized cells remained stable for several days. A 67.6% conversion was achieved in 24 h, corresponding to 57.4 g/L of optically pure (*S*)-3-butyn-2-ol. This product concentration was already twofold higher than the previously reported maximum value (24.7 g/L). Full conversions could be obtained within 120 h, yielding a product concentration of 87 g/L without acetone removal [31].

An NADH-dependent reductase (ScCR) was identified by genome mining studies of *Streptomyces coelicolor*. This dehydrogenase showed broad tolerance to prochiral ketones including aryl ketones, α- and β-ketoesters and had particularly high activity and excellent enantioselectivity (>99% ee) for the last substrate type. For example, the enzyme reduced ethyl 4-chloro-3-oxobutanoate (COBE) to the (*S*)-alcohol in more than 99% ee in 22 h using 2-propanol as a cosubstrate for NADH regeneration. A very high substrate load was tolerated (600 g/L 3.6 M), and the reaction yield was 93% with a total turnover number (TTN) of 12,100 [5e, 32].

Wang et al. carried out the same reduction using a CRED and obtained the (*S*)-alcohol in 96% yield with more than 99.5% ee. DNA shuffling technology was used to improve the activity of the CRED and associated glucose dehydrogenase (GDH) without decreasing ee. After several generations, GDH activity was improved by a factor of 13 and the ketone reductase activity by a factor of 7. Using the improved enzymes, the reaction performed at 160 g/L of substrate input and the loadings of CRED and GDH could be reduced to 0.57 g/L and 0.38 g/L, respectively [33].

Deracemization of racemic 2-hydroxy acids **55** to optically pure (*R*)-enantiomer was successfully realized *via* concurrent asymmetric oxidation and stereoselective reduction employing two resting-cell biocatalysts. Glycerol was used as the source of hydride. The CRED enzyme from *Pseudomonas aeruginosa* CCTCC M 2011394 catalyzed the stereoselective oxidation of (*S*)-2-hydroxy acids to the corresponding 2-ketoacids **56** and the CRED from *Saccharomyces cerevisiae* ZJB5074 reduced the 2-ketoacids to (*R*)-2-hydroxyacids as shown in Scheme 6.21 [34].

Weuster-Botz and coworkers demonstrated the one-pot reduction of dehydrocholic acid **57** to a series of useful bile acids as shown in Scheme 6.22. The example in this series involved the 100 mM conversion of dehydrocholic acid to more than 99.5 mM of 12-keto- ursodeoxycholic acid **60** within 4.5 h in a simple batch process on a liter scale [35].

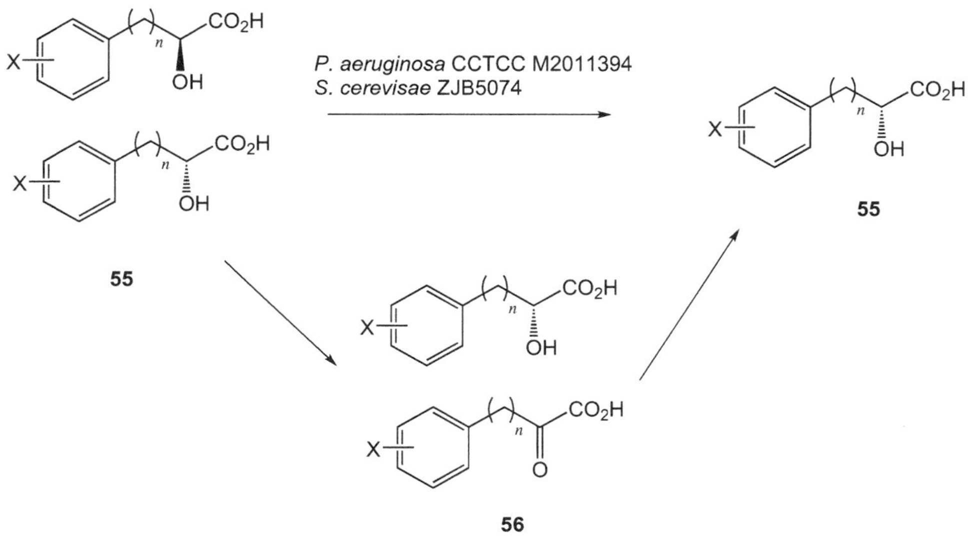

SCHEME 6.21
Concurrent biooxidation and bioreduction using CRED technology.

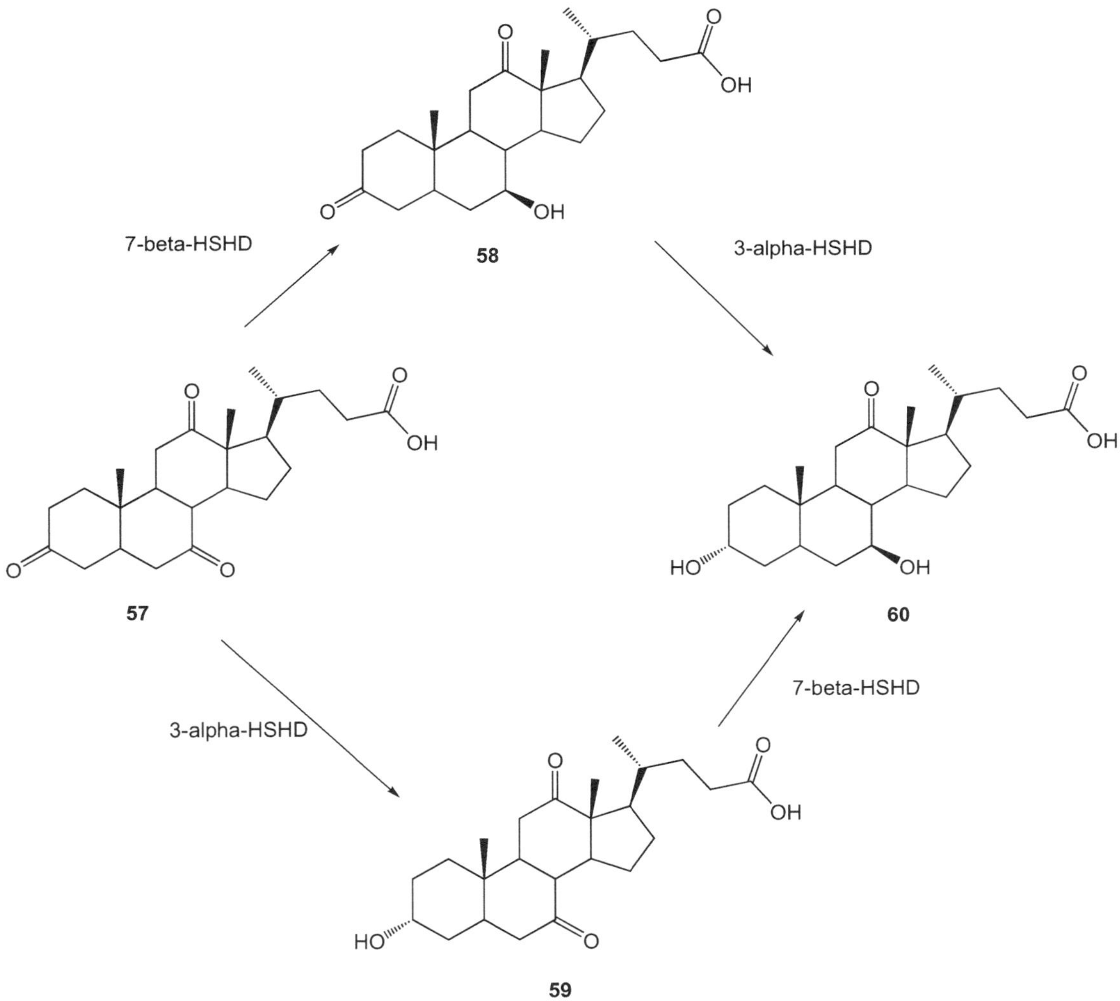

SCHEME 6.22
Bioconversion of dehydrocholic acid to 12-keto-ursodeoxycholic acid.

3.5 Bulky Ketone Reductions

The key challenges with this class of compound come from steric hindrance around the ketone moiety, poor solubility under aqueous conditions, and (often) lack of acceptance by CRED enzymes. Fortunately, more and more enzymes are being identified and evolved in order to tolerate such substrates. The montelukast bioreduction shown earlier in Scheme 6.2 elegantly demonstrates how CREDs can be evolved to accept bulky ketones. Truppo et al. at Merck showed that a series of benzophenones **61** could be selectively reduced using commercially available

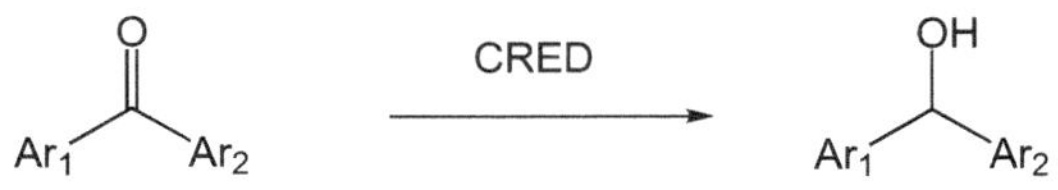

SCHEME 6.23
Benzophenone reductions using commercially available CRED enzymes.

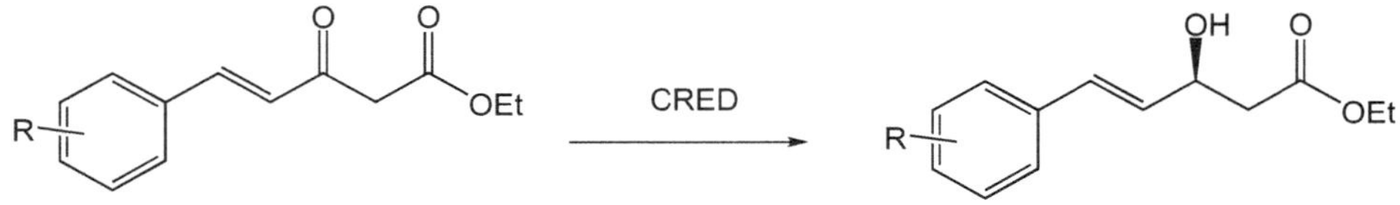

SCHEME 6.24
Chemo- and stereoselective bioreductions.

CRED enzymes with yields more than 90% and ee's up to 99% using GDH as the cofactor recycle system as shown in Scheme 6.23 [36].

An example that highlights the chemo- and stereoselectivity of dehydrogenases was the selective reduction of enone compounds **63** (Scheme 6.24).

Screening 24 commercially available CREDs revealed no competing reduction of the olefin moiety while both alcohol enantiomers were accessible with conversions up to 96% and ee's more than 95% [37].

Chiral diarymethanols **66** were used in synthesis of mGlu2 receptor potentiators. The optically pure compounds were accessed by screening commercial CREDs and both *R* and *S* enantiomers were accessible with high ee as shown Scheme 6.25 [38].

3.6 Reductions Involving Dynamic Kinetic Resolutions

For certain types of molecules, CRED-catalyzed reductions can be employed to generate two chiral centers simultaneously. In such dynamic kinetic reductions, the starting materials contain an easily racemized chiral center that becomes

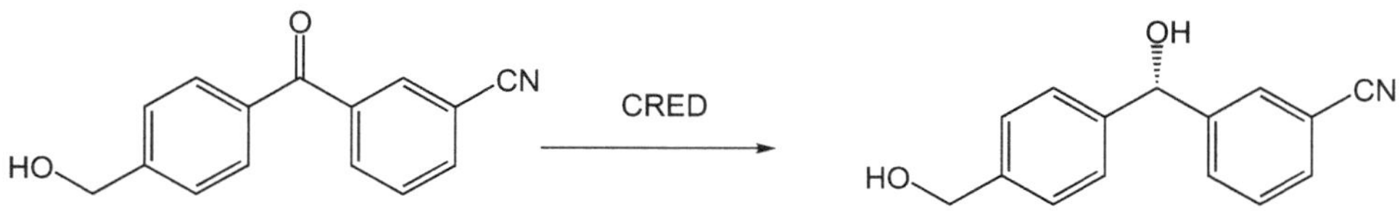

SCHEME 6.25
Chiral access to diarymethanols.

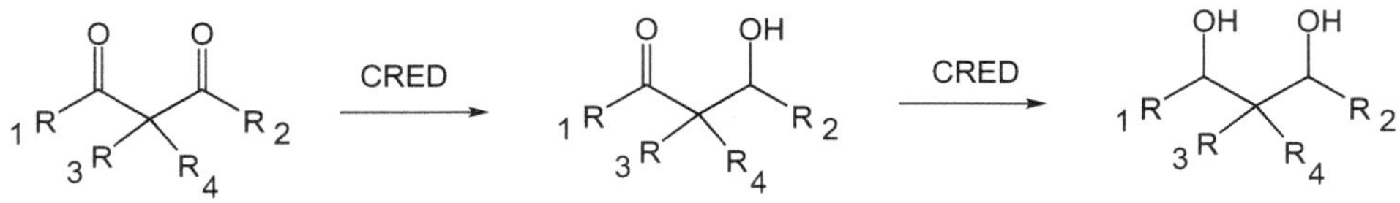

67 68 69

SCHEME 6.26
Synthesis of α-alkyl-β-hydroxy ketones and 1,3-diols via bioreduction.

effectively nonracemizable after ketone reduction. A dehydrogenase selective for one of the two substrate enantiomers thereby "traps" the chiral center by converting it to the alcohol product. The remaining substrate re-establishes the equilibrium racemic mixture and the process continues until the entire starting ketone has been reduced to a single, diastereomerically pure product. This process allows one to obtain one of the four possible diastereomers selectively. α-Substituted β-keto esters are the best-known examples of this phenomenon and highly diastereoselective reductions of such substrates have been described [39].

By screening libraries of CRED enzymes, the most diastereoselective enzymes for each of the four possible diastereomers could be found. Diastereoselectivity often exceeded 99% (no other products could be detected) and the yields typically ranged from 90% to 100%.

In another report using isolated CRED enzymes, Kalaitzakis et al. demonstrated that alkyl-substituted 1,3-diketones **67** of the type in Scheme 6.26 could be reduced with high diastereoselectivities. By screening libraries of CREDs, enzymes capable of producing two or three of the four possible diastereomers were often found. An interesting case was the reduction of compounds such as 3-met hyl-3-allyl-2,4-pentanedione (Scheme 6.26) that gave only a single product in 100% yield. The chiral alcohol was identified as (3R,4R)- 3-methyl-3-allyl-4-hydroxy-pentan-2-one. Other similar examples of this type of reaction were also described [40].

Another example of diastereoselective reduction using CREDs comes from Marocco et al. Using 2-chloro-3-ketophosphonate esters **70** as substrates, both *cis*-(1R,2S)- and *trans*-(1S,2S)-epoxides were prepared according to Scheme 6.27 [41].

Insect hormone sitophilure was synthesized using a practical CRED-catalyzed reaction with high diastereoselectivity as shown in Scheme 6.28 [42]. Screening of a library of CRED enzymes identified a highly diastereoselective enzyme for the reduction of 4-methyl-3,5-heptanedione **74**. A dynamic process was established in which desymmetrization of carbon 4 took place under the conditions of the reaction, with a highly diastereoselective reduction of the ketone leading to a high overall yield of the desired single diastereomer to (4S,5S)-5-hydroxy-4-methyl-3-heptanone **75** (sitophilure) in 87% isolated yield and more than 99% ee.

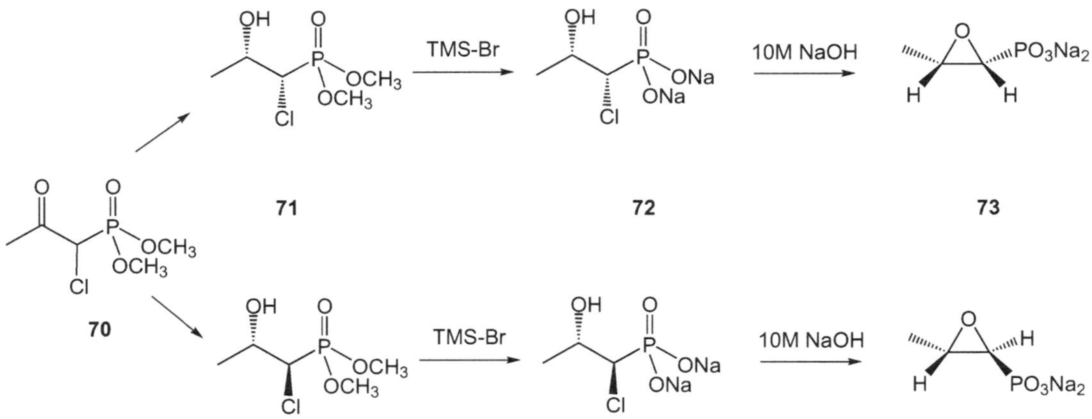

70 71 72 73

SCHEME 6.27
Diastereoselective reductions to access chiral cis and trans epoxides.

74 75

SCHEME 6.28
Diastereoselective reductions to (4S,5S)-5-hydroxy-4-methyl-3-heptanone.

Merck scientists reported a practical enantioselective synthesis of *cis*-2,5-disubstituted pyrrolidines where the key chiral step 3 is a dynamic kinetic bioreduction as shown in Scheme 6.29. The bioreduction is performed at pH 7 and 30°C using GDH as the cofactor recycle system yielding 99% ee and more than 90% assayed yield [43].

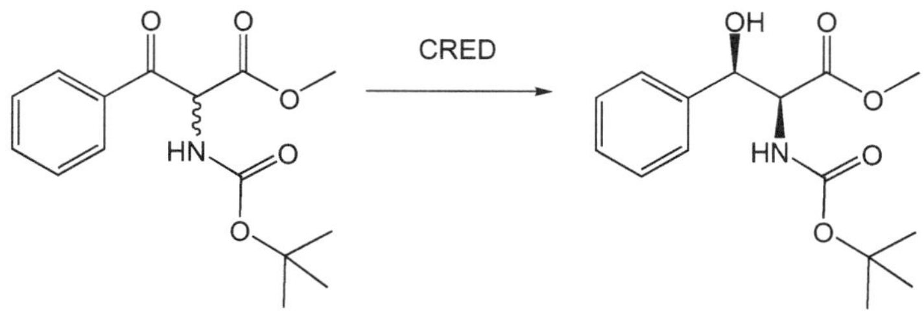

76 77

SCHEME 6.29
Dynamic kinetic bioreduction to access chiral hydroxyl amino esters.

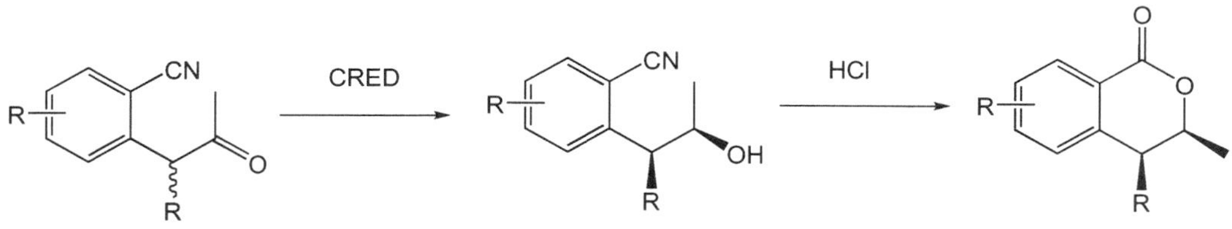

78 79 80

SCHEME 6.30
Dynamic kinetic bioreduction to access 4-alkyl-3-methyl-3,4-dihydroisocoumarins.

A simple, one-pot dynamic kinetic resolution using recombinant *Escherichia coli* cells that overexpressed ADH-A from *R. ruber* provided an enantioselective route to 4-alkyl-3-methyl-3,4-dihydroisocoumarins **80** as shown in Scheme 6.30. Key to the reaction was the introduction of anion exchange resin or triethylamine to promote the desired racemization of the ketone during reduction to yield enantiopure alcohols with excellent selectivities. The reactions were performed in pH 7.5 buffer with hexane cosolvent. The reductant was supplied via isopropyl alcohol cofactor regeneration with excellent conversions [44].

3.7 Aldehyde Reductions

A practical process for the production of (*R*)-2-methylpentanol **82** was described by Gooding et al. (*R*)-2-Methylpentanol is an important chiral intermediate for the synthesis of certain medicinally important compounds, natural products, and liquid crystals.

The process used an evolved, engineered CRED to catalyze the enantioselective reduction of 2-methylvaleraldehyde **81** as shown in Scheme 6.31. The unreacted aldehyde was removed from the process stream, and the *R*-configured product is isolated in high enantiomeric purity. A pilot-scale run was described in which approximately 100 kg of (*R*)-2-methylpentanol was produced with an ee of 98.2% when the reaction was quenched after 45% conversion [45].

An example in which a dynamic racemization was coupled to a stereoselective aldehyde reduction was described by Berkowitz and coworkers for the

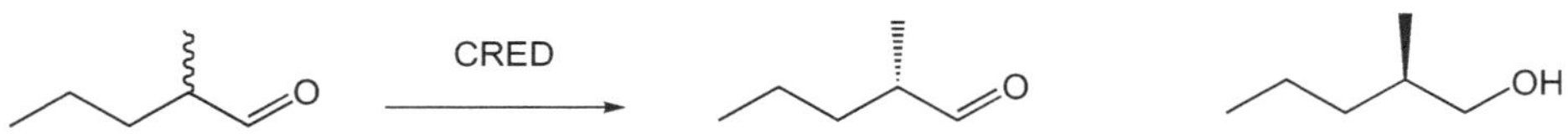

81 82

SCHEME 6.31
Synthesis of (*R*)-2-methylpentanol.

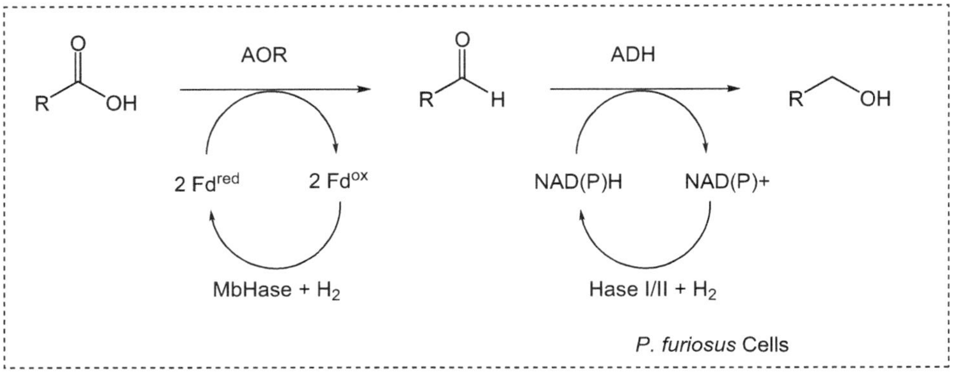

SCHEME 6.32
Acid and aldehyde reduction.

production of optically pure profen drug precursors [46]. Profenols corresponding to naproxen, ibuprofen, flurbiprofen, and ketoprofen were produced in excellent yields and optical purities from the corresponding racemic 2-arylpropionaldehyde precursors. Using the thermostable alcohol dehydrogenase from *Sulfobus solfataricus*, the reduction was carried out at pH of 9 and at 80 °C, with the *S*-form of the aldehyde being selectively reduced to the corresponding alcohol. Since the interconversion of the *R* and *S* stereoisomers of the aldehyde occurred rapidly under the reaction conditions, conversion of the racemic 2-arylpropionaldehydes to the optically pure profenols took place in high yields. The corresponding profen drugs could be obtained by oxidation of the optically pure profenols to the carboxylic acids.

A two-enzyme process to obtain alcohols from carboxylic acids was reported by Ni et al. as shown in Scheme 6.32. The reduction of the acid moiety was demonstrated by the hyperthermophile *Pyrococcus furiosus* using 5 bar of hydrogen as reductant. It was also shown that no reaction occurred in the absence of *P. furiosus* or hydrogen. Both aromatic and aliphatic substrates were used and reduction of the carboxylic acid resulted in the accumulation of aldehyde intermediate. The aldehyde was reduced using a CRED with overall yields of 11–99% [47].

1,4-Butandiol was used as a smart cosubstrate to promote NAD(P)H reductase-dependent reactions. γ-Butyrolactone, the product of two successive essentially irreversible enzyme-catalyzed oxidations, is thermodynamically stable and chemically inert that drives the reaction strongly in the direction of substrate reduction. The authors examined the reduction of $NAD(P)^+$ by 1,4-butanediol with 12 commercially available reductases, 8 of which successfully accepted the diol substrate. In the reduction of cinnamaldehyde with horse liver dehydrogenase, both the initial rate and maximum conversion were higher with 1,4-butandiol than with alternative sacrificial reductants such as *iso*-propanol or ethanol (Scheme 6.33) [48].

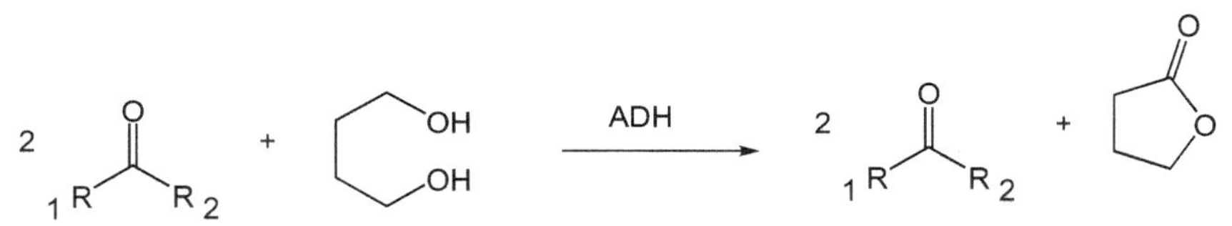

SCHEME 6.33
Smart cosubstrates from CRED reductions.

4 IMMOBILIZATION OF CRED ENZYMES AND MEMBRANE APPLICATIONS

As noted elsewhere, immobilizing enzymes simplify their recovery and reuse. This section considers several examples where this strategy has been applied to dehydrogenases.

Geotrichum candidum is a well-known catalyst for asymmetric ketone reductions, and different cell-disintegration methods were used to liberate intracellular CREDs from these cells. Solid shear (bead milling) proved best. Next, various solid supports were examined for the immobilization of the purified enzyme. The CRED was immobilized on silica with an optimized protein loading of 4 mg/g support. Cross-linking with glutaraldehyde rendered the preparation more stable and suitable for use in consecutive batches. This immobilized enzyme was found to be a highly efficient biocatalyst formulation for the production of *S*(−)-1-(10-naphthyl)ethanol [49].

Acetophenone was reduced by immobilized CRED *R. ruber* DSM 44541 ADA-A to phenylethanol. Immobilization was achieved using a polyphosphazene polymer and it was demonstrated that the enzyme could be used for at least six catalytic cycles without appreciable loss of activity (in all cases 95–100% of activity remained) [50].

Another example of acetophenone reduction was demonstrated by Hildebrand and Lutz by the immobilization of a CRED from *L. brevis* that was used in a plug flow reactor. The enzyme was immobilized onto an amino-epoxy resin from Sepabeads. A fourfold increase in stability was achieved for the CRED by blocking the remaining functional groups on the enzyme-support preparations with glycine or mercaptoethanol. Stabilization of the multipoint covalent attachment could only be achieved by additionally cross-linking the adsorbed proteins with glutaraldehyde. The high stabilization resulted in a half-life time of over 1200 h when stored at 30 °C. A 60-fold increase in stability was achieved when compared to the soluble enzyme [51].

Schroer et al. developed a continuous process for the asymmetric reduction of methyl acetoacetate using *iso*-propanol as a cosubstrate in a continuously stirred tank reactor using an ultrafiltration membrane for retention of cells [52]. The bioreduction was run continuously for 7 weeks with exceedingly high substrate and cosubstrate concentration of up to 2.5 and 2.8 mol/L. Maximal space–time yield of about 700 g/L/day was achieved.

Immobilized resting cells of *G. candidum* were used to reduce a cyclohexanone and *ortho*-fluoroacetophenone in a semicontinuous flow process using supercritical carbon dioxide. Excellent enantioselectivity (ee > 99%) and a higher space–time yields were achieved when compared to that of the corresponding batch process [53].

Mutated *Rhodococcus* phenylacetaldehyde reductase (PAR) or *Leifsonia* alcohol dehydrogenase (LSADH) were applied to water-soluble ketone substrates. For example, 4-hydroxy-2-butanone was reduced to (*S*)/(*R*)-1,3-butanediol, with a high yield and stereoselectivity. Intact *E coli* cells overexpressing mutated PAR (Sar268) or LSADH were directly immobilized with polyethyleneimine or 1,6-hexanediamine and glutaraldehyde and evaluated in a batch reactor. This system produced (*S*)-1,3-butanediol (87% ee) with a space–time yield (STY) of 12.5 mg/h/mL catalyst or (*R*)-1,3-butanediol (99% ee.) with an STY of 60.3 mg/h/mL catalyst. The immobilized cells in a packed bed reactor continuously produced (*R*)-1,3-butanediol with a yield of 99% (about 49.5 g/L) from 5% (w/v) 4-hydroxy-2-butanoate over 500 h. The concentration of PEI used for immobilization influenced the operational stability of immobilized cells, and the cells treated with 3% PEI showed better stability than those treated with lower PEI concentrations The immobilized *E. coli* biocatalyst could be used more than 30 times (for about 500 h) with no decrease in conversion [54].

Ansorge-Schumacher and coworkers used their silCoat method, originally developed to immobilize lipases, as a starting point for immobilizing CRED enzymes [55]. They noted that the direct immobilization of the CRED onto the silCoat bead was not possible; the hydrophobic silicon beads must first be made more hydrophilic. The silicon beads were therefore modified with PEG400 followed by adsorption of the CRED from *C. parapsilosis* (CPCR2). Acetophenone was used as the test substrate and it was found that the relative stability of the CRED was increased by a factor of 120. The same group also investigated the whole-cell entrapment. Once again the silCoat bead was used in the immobilization process. In this case, whole cells of *E. coli* overexpressing the CRED from *C. parapsilosis* (CPCR2) were efficiently adsorbed. Cell loadings up to 400 mg (*E. coli*)/g (carrier) was possible [56]. The work shows that both isolated and whole-cells systems harboring CRED activity can be immobilized with success.

5 IONIC LIQUIDS AND CRED ENZYMES

Water miscible ionic liquids have been used to facilitate stereoselective reductions of poorly water-soluble aliphatic ketones by the *L. brevis* CRED [57]. Ten ionic liquids were screened for activity and solubility. Improved storage stabilities and enzyme activities, along with reduced inhibition by substrate and product were found. The most promising ionic liquid (AMMOENGTM 101) was the subject of more detailed investigations. Batch reactions with cofactor regeneration *via* a GDH strategy showed increased reaction rates, further emphasizing the positive influence of AMMOENGTM 101 ionic liquid.

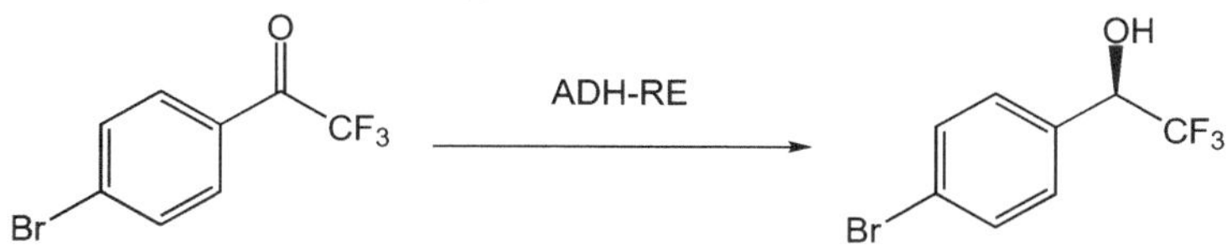

83 84

SCHEME 6.34
Bioreduction of para-bromo-2,2,2-trifluoroacetophenone using ionic liquid additives.

Employing a solvent system containing the ionic liquid [BMIm][BF4] with *Pichia membranaefaciens* Hansen ZJPH07 cells afforded a higher yield and enantiopurity during the (*R*)-selective reduction of ethyl acetoacetate as compared to a monophasic system. It was also demonstrated that the addition of [BMIM][BF4] to the reaction system markedly reduced substrate inhibition and moderately improved the enantioselectivity compared to a monophasic aqueous system [58].

Using traditional organic cosolvents, the bioreduction of *para*-bromo-2,2,2-trifluoroacetophenone **83** (Scheme 6.34) was limited to a maximum product concentration of 10 g/L due to substrate-induced deactivation of the *R. erythropolis* (ADH RE) biocatalyst. However, by employing 10% (v/v) [BMP][NTf2], a water immiscible ionic liquid, more than 5 times the ketone was converted to the chiral alcohol in less than 24 h. The ionic liquid improved the initial reaction rate by more than four times in the presence of ionic liquid compared to an aqueous-only reaction. Moreover, the ionic liquid improved the stability of both the ADH RE and the GDH coenzyme compared to reactions with either organic cosolvents or aqueous buffer systems [59].

Ethyl (R)-3-hydroxybutyrate was prepared from the bioreduction of ethyl acetoacetate using *Acetobacter sp.* (CCTCC M209061) cells in 1-butyl-3-methylimidazolium hexafluorophosphate (C4mim · PF6), an ionic liquid with good biocompatibility. The reaction was performed with a 1:1 ratio of buffer to C4mim · PF6 and a substrate concentration of 55 mmol/L in buffer at pH 5.5. The reaction provided 90.8% conversion and more than 99% ee and was scaled to a 450 mL reaction volume. Also demonstrated in the paper was the successful recycle of the catalyst up to 10 times as an immobilized alginate [60].

6 FUTURE OUTLOOK

The discovery of more and more CREDs has opened many opportunities with regard to substrate tolerance and diversity and scientists are increasingly taking advantage of this first choice technology for asymmetric reductions of prochiral ketones. As the methodology continues to improve further, it becomes even easier to use and fast to scale. Rowan et al. showed that by simply having a small, focused library of CREDs, seven sets of ketones could be reduced to the corresponding *R* and *S* alcohols as summarized in Table 6.3 [61].

Table 6.3 Structural Motifs of the Substrate Classes Involved in the CRED Study

Set 1 (methyl aryl-ketones)	Set 2 (*alpha*-halo aryl-ketones)	Set 3 (fused bicyclics)	Set 4 (*beta*-cyano aryl-ketones)	Set 5 (pyrrolidinone analogues)	Set 6 (*alpha*-branched alkyl aryl-ketones)	Set 7 (*alpha*-trifluoromethyl aryl-ketones)
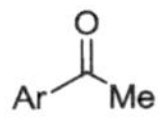	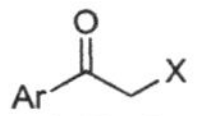 X = Cl or Br	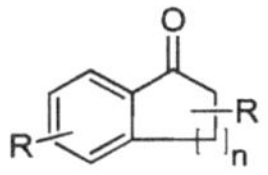	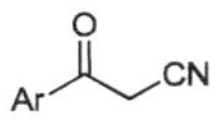	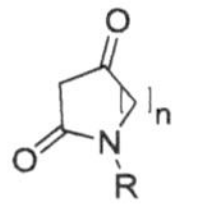	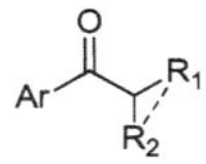	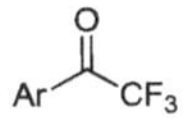
A131 (Prelog) A161 (Anti-Prelog) A601 (Anti-Prelog)	A131 A161 A231 (Anti-Prelog) A601	A131 A161 A231 A281 (Prelog)	A131 A161 A231 A281	A131 A161 A201 (Anti-Prelog)	A131 A161 A231 A601	A131 A161 A231 A601

High-performing CREDs associated with each class are given underneath in blue.

To take maximum advantage of biocatalytic ketone reductions for the synthesis of active pharmaceutical ingredients, it is essential to introduce the strategy into the synthetic process design early so that as the product progresses through the clinical pipeline, the enzymatic step will be there from the start.

CRED research will continue to dominate in the literature and more diverse and active CREDs will be discovered or developed to access bulkier substrates, run under higher concentrations and tolerate higher concentrations of solvents.

Future development of CRED reaction systems could investigate and integrate the use of technologies that are known to speed up catalysis in other systems. One such approach may be to utilize ultrasonication that has been shown to greatly increase the rate of some enzyme reactions, sometimes by an order of magnitude [62]. Ultrasound can induce physical phenomena such as cavitation and acoustic streaming, and these not only lead to extreme conditions of liquid turbulence that can benefit mass transfer but also influence behavior at the molecular level such as protein conformation and secondary structure, and lower catalytic activation energies. If applied carefully, beneficial effects such as increased reaction rates outweigh any negative damaging effects to the cells or enzymes [63].

7 PRACTICAL EXAMPLES

This section gives the organic chemist examples of how to perform the bioreductions in the laboratory using both wild-type whole cell and recombinant systems.

7.1 Screening Conditions

CRED-enzyme screening for identification of hit enzyme for bioreduction [64].

In this section, screening conditions for enzymatic ketone reductions are discussed. A solution of ketone (5 mg) in dimethylsulfoxide (0.2 mL) was added to a mixture of 0.1 M KH_2PO_4 buffer at pH 7 (1 mL) and NADPH (2 mg, or

NADH 2 mg depending on the cofactor requirement of the dehydrogenase). Lyophilized CRED biocatalyst (2 mg) and GDH (2 mg) were added; the vial was sealed and shaken overnight at room temperature. Ethyl acetate (0.5 mL) was added to the vial and the organic layer was separated, filtered through cotton wool and dried ($MgSO_4$). The ethyl acetate was evaporated in an oven at 60°C and the residue was redissolved in the appropriate HPLC solvent. Samples were analyzed by HPLC.

7.2 Whole-Cell Microbial Bioreduction

Whole-cell microbial bioreduction involves both whole cell preparation followed by incubation with ketone substrate [13].

7.2.1 CULTIVATION AND PREPARATION OF THE BIOCATALYST

B. pumilus Phe-C3 (obtained from Institute of Biotechnology, ETH, Zurich) was grown on LB-agar plate for 5 days, and then inoculated into 100 mL M9 medium containing 2% (w/v) glucose in a 500-mL shaking flask. The cells were grown at 30°C at 220 rpm for 38 h. The cells were then harvested by centrifugation at 4°C and washed twice with buffer and stored in a refrigerator until ready for use.

7.2.2 GENERAL PROCEDURE FOR SMALL-SCALE BIOREDUCTION

Ketoesters (3–30 mM) were added to a suspension of *B. pumilus* Phe-C3 cells (3.0 or 12.0 g cdw/L) in 300 mL of phosphate buffer (pH 7.0) containing glucose (2%). The mixture was incubated in a 1000 mL shaking flask at 220 rpm at 30°C. The reaction was stopped by centrifugation, and the product was extracted with solvent. The organic phase was dried over anhydrous Na_2SO_4, filtered, and concentrated under reduced pressure. The residue was purified by column chromatography on silica gel to afford the corresponding hydroxyester.

7.3 Recombinant Bioreduction Using GDH as Recycle System

CRED-mediated biocatalysis for preparation of (*R*)-tetrahydrothiophene-3-ol using GDH as recycle system [27].

To a 3-L three-neck jacketed flask at room temperature fitted with a mechanical stirrer at ~300 rpm, a thermometer, a pH probe, and an 8 N NaOH dosing probe from a Schott Titronic universal titrator (VWR catalogue number 13502-814) was added 100 mM pH 7.0 triethanolamine · HCl with 2 mM $MgSO_4$ buffer ("the buffer" henceforth, 850 mL) and glucose (338 g, 1.88 mol, 1.25 equiv.) in portions over ~5 min under nitrogen. The temperature was lowered to 15°C. To the reaction mixture was added tetrahydrothiophene-3-one (150 g, 1.47 mol) in one portion to give a biphasic reaction mixture followed by KRED CDX-033 (1.5 g, from Codexis) in 50 mL of the buffer, GDH CDX-901 (0.37 g, from Codexis) in 50 mL of the buffer, and NADPNa (0.60 g) in 50 mL of the buffer. The reaction was maintained at pH 7.0 (±0.1) via pH-stat controlled dosing of 8 N NaOH in 20 μL intervals. After stirring at 15°C for 16 h (163 mL of

8 N NaOH was dosed), the circulator temperature was raised to 25 °C (internal temperature reached 25 °C in ~1/2 h). After a total of 23 h (16 h at 15 °C and 7 h at 25 °C; a total of 171 mL of 8 N NaOH was dosed), an aliquot was taken, and HPLC analysis revealed that more than 99% conversion had been achieved. The pH control was stopped, and 5.0 g of $Na_2S_2O_5$ was added. After stirring for an additional 1 h at 25 °C (24 h total), EtOAc (1200 mL) was added. After 15 min, the aqueous and organic phases were separated and the bottom, yellow aqueous phase was returned to the reaction vessel. The top, milky organic phase (containing some emulsion) was passed through 30 g (~3 in. diameter × ~1/2 in. height) of Celite 545 over ~2 min. A clean phase split occurred with both the top organic and the bottom aqueous phases being homogeneous. The bottom aqueous phase was returned to the reaction vessel (combined with the original aqueous phase). The Celite was rinsed with EtOAc (1200 mL). The EtOAc used in the Celite rinse was added to the pooled aqueous phase and stirred at 300 rpm. After 15 min, the phases were separated (clean split). The top organic phase was combined with the organic phase from the first extraction, and to the pooled organic phases was added heptane (600 mL) to phase out and split off additional water. All aqueous waste streams were treated with bleach and disposed of as alkaline aqueous waste. The combined organic phases were concentrated under reduced pressure at 40 °C to yield 136 g (88% yield, 99.3% ee) of the product as a pale-yellow oil.

7.4 Recombinant Bioreduction Using IPA as Recycle System

CRED-mediated biocatalysis for preparation of (*S*)-2-bromo-2-cyclohexen-1-ol) using IPA [6].

CRED A601 cell paste (20 g, from Almac) was resuspended in a solution of pH 7 KH_2PO_4 buffer (3.9 L, 0.1 M) containing $NADP^+$ (1.0 g, 1.3 mmol) and shaken at 20 °C for 30 min. The suspension was transferred to a 5-L reactor to which was added a solution of 2-bromocyclohex-2- enone (100 g, 0.57 mol) dissolved in IPA (800 mL) and DMSO (200 mL). In addition, polypropylene glycol (1.5 mL) was added to prevent excessive foaming. The reaction mixture was stirred mechanically using an overhead stirrer and incubated at 35 °C for 24 h. The depletion of IPA and water was counterbalanced by bubbling a pressurized airflow saturated with 2-propanol and water (1:4 v/v) through the reaction mixture. Upon completion, the resulting mixture was extracted with MTBE (3 × 5 L) and washed with water (2 × 2.5 L) and brine (2 × 2.5 L) and dried over $MgSO_4$. Solvent was removed under reduced pressure to afford the title compound as a colorless oil.

8 CONCLUSIONS

CRED-mediated biocatalysis continues to grow as success has bred success, and recent literature shows how molecular biology is being exploited to facilitate this progress. Established CREDs are being further evolved for new and improved properties, and also being expressed and applied in new ways. New CREDs continue to emerge, realized by genomic-assisted screening technology. This field of biocatalysis will continue to seduce synthetic chemists and can be expected to grow further as mainstream biocatalysis moves into the modern era.

References

[1] (a) R. Noyori, M. Kitamura, T. Ohkuma, Toward efficient asymmetric hydrogenation: architectural and functional engineering of chiral molecular catalysts, Proc. Natl. Acad. Sci USA 101 (2004) 5356. (b) R. Noyori, Chiral metal complexes as discriminating molecular catalysts, Science 248 (1990) 1194. (c) C.A. Sandoval, T. Ohkuma, K. Muniz, R. Noyori, Mechanism of asymmetric hydrogenation of ketones catalyzed by BINAP/1,2-diamine-ruthenium(II) complexes, J. Am. Chem. Soc. 125 (2003) 13490. (d) R.V. Hoffman, N. Maslouh, F. Cervantes-Lee, Highly stereoselective syntheses of syn- and anti-1,2-amino alcohols, J. Org. Chem 67 (2002) 1045. (e) E.J. Corey, J.O. Link, A new chiral catalyst for the synthesis of secondary alcohols and deuterated primary alcohols by carbonyl reduction, Tetrahedron Lett. 30 (1989) 6275.

[2] (a) M. Muller, M. Wolberg, T. Schubert, W. Hummel, Enzyme-catalyzed regio- and enantioselective ketone reductions, Adv. Biochem. Eng. Biotechnol. 92 (2005) 261. (b) J.D. Stewart, S. Rodriquez, M.M. Kayser, Cloning, structure and activity of ketone reductases from Baker's yeast, in: H.A. Kirst, W.-K. Yeh, M.J. Zmijewski (Eds.), Enzyme Technology for Pharmaceutical and Biotechnological Applications, Marcel Dekker, New York, 2001, pp. 175–207. (c) I.A. Kaluzna, A.A. Andrew, M. Bonnilla, M.R. Martzen, J. Stewart, Enantioselective reductions of ethyl 2-oxo-4-phenylbutyrate by *Saccharomyces cerevisiae* dehydrogenases, J. Mol. Catal B Enzymatic 17 (2002) 101. (d) M. Katz, B. Hahn-Hagerdal, M.F. Gorwa-Grauslund, Screening of two complementary collections of *Saccharomyces cerevisiae* to identify enzymes involved in stereoselective reductions of specific carbonyl compounds: an alternative to protein purification, Enz. Microb. Technol. 33 (2003) 163. (e) I.A. Kaluzna, B.D. Feske, W. Wittayanan, I. Ghiviriga, J.D. Stewart, Stereoselective, biocatalytic reductions of alpha-chloro-beta-keto esters, J. Org. Chem. 70 (2005) 342.

[3] (a) T. Johanson, M. Katz, M. Gorwa-Grauslund, Strain engineering for stereoselective bioreduction of dicarbonyl compounds by yeast reductases, FEMS Yeast Res. 5 (2005) 513. (b) M. Kataoka, K. Kita, M. Wada, Y. Yasohara, J. Hasegawa, S. Shimizu, Novel bioreduction system for the production of chiral alcohols, Appl. Microbiol. Biotechnol. 62 (2003) 437.

[4] (a) G. Brown, D. Mangan, I. Miskelly, T.S. Moody, A facile stereoseelctive biocatalytic route to the precursor of woody acetate, Org. Process Res. Dev. 15 (2011) 1036–1039. (b) T. Matsuda, R. Yamanaka, K. Nakamura, Recent progress in biocatalysis for asymmetric reduction and oxidation, Tetrahedron Asymm. 20 (2009) 513–557. (c) K.B. Borges, W. de Souza Borges, R. Durán-Patrón, M.T. Pupo, P.S. Bonato, I.G. Collado, Stereoselective biotransformations using fungi as biocatalysts, Tetrahedron Asymm. 20 (2009) 385–397. (d) I.A. Kaluzna, J.D. Rozzell, S. Kambourakis, Ketoreductases: stereoselective catalysts for the facile synthesis of chiral alcohols, Tetrahedron Asymm. 16 (2005) 3682–3689. (e) A. Weckbecker, W. Hummel, Cloning, expression, and characterization of an R-specific alcohol dehydrogenase from *Lactobacillus kefir*, Biocatal. Biotransform. 24 (2006) 380–389.

[5] (a) S.K. Padhi, I.A. Kaluzna, D. Buisson, R. Azerad, J.D. Stewart, Reductions of cyclic *beta*-keto esters by individual *Saccharomyces cerevisiae* dehydrogenases and a chemo-enzymatic route to (1*R*,2*S*)-2-methyl-1-cyclohexanol, Tetrahedron Asymm. 18 (2007) 2133–2138. (b) P. Panizza, S. Onetto, S. Rodríguez, A recombinant *E. coli* expressing an alfa-alkyl-beta-ketoester reductase with unusual stereoselectivity, Biocatal. Biotransform. 25 (2007) 414–417. (c) D. Zhu, Y. Yang, L. Hua, Stereoselective enzymatic synthesis of chiral alcohols with the use of a carbonyl reductase from *Candida magnoliae* with anti-Prelog enantioselectivity, J. Org. Chem. 71 (2006) 4202–4205. (d) B. Kosjek, J. Nti-Gyabaah, K. Telari, L. Dunne, J.C. Moore, Preparative asymmetric synthesis of 4,4-dimethoxytetrahydro-2H-pyran-3-ol with a ketone reductases

and *in situ* co-factor recycling using glucose dehydrogenase, Org. Process Res. Dev. 12 (2008) 584–588. (e) M. Kataoka, K. Yamamoto, H. Kawabata, M. Wada, K. Kita, H. Yanase, S. Shimizu, Stereoselective reduction of ethyl 4-chloro-3-oxobutanoate by *Escherichia coli* transformant cells coexpressing the aldehyde reductases and glucose dehydrogenase genes, Appl. Microbiol. Biotechnol. 51 (1999) 486–490. (f) T. Matsuda, Y. Yamagishi, S. Koguchi, N. Iwai, T. Kitazume, An effective method to use ionic liquids as reaction media for asymmetric reduction by *Geotrichum candidum*, Tetrahedron Lett. 47 (2006) 4619–4622.

[6] S.J. Calvin, D. Mangan, T.S. Moody, P.J. Stevenson, Overcoming equilibrium issues with carbonyl reductase enzymes, Org. Process Res. Dev. 16 (1) (2012) 82–86.

[7] Y. Ni, P. Hagedoorn, J. Xu, I.W.C.E. Arends, F. Hollmann, A biocatalytic hydrogenation of carboxylic acids, Chem. Comm. 48 (2012) 12056–12058.

[8] Biocatalysis Enzyme Guide, Sustainable chemistry solutions, Burbank California, 2014. Available from: http://www.bio-catalyst.com/biocatalysis-enzyme-guide/).

[9] U.T. Bornsheuer, G.W. Huisman, R.J. Kazlauskas, S. Lutz, J.C. Moore, K. Robbins, Engineering the third wave in biocatalysis, Nature 485 (2012) 185–194.

[10] S.K. Ma, J. Gruber, C. Davis, L. Newman, D. Gray, A. Wang, J. Grate, G.W. Huisman, R.A. Sheldon, A green-by-design biocatalytic process for atorvastatin intermediate, Green Chem. 12 (2010) 81–86.

[11] D. Pollard, M. Truppo, J. Pollard, C.-Y. Chen, Effective synthesis of (s)-3,5- bistrifluoromethylphenyl ethanol by asymmetric enzymatic reduction, Tetrahedron Asymm. 17 (4) (2006) 554–559.

[12] M. Chartrain, C. Roberge, J. Chung, J. McNamara, D. Zhao, R. Olewinski, G. Hunt, P. Salmon, D. Roush, S. Yamazaki, T. Wang, E. Grabowski, B. Buckland, R. Greasham, Asymmetric bioreduction of (2-(4-nitro-phenyl)-*N*-(2-oxo-2-pyridin-3-yl-ethyl)-acetamide) to its corresponding (R) alcohol [(R)-*N*- (2-hydroxy-2-pyridin-3-yl-ethyl)-2-(4-nitro-phenyl)-acetamide] by using *Candida sorbophila* MY 1833, Enzyme Microb. Technol. 25 (1999) 489–496.

[13] C. He, D. Chang, J. Zhang, Asymmetric reduction of substituted α- and β- ketoesters by *Bacillus pumilus* Phe-C3, Tetrahedron Asymm. 19 (2008) 1347–1351.

[14] S. Manner, C. Hansson, J.M. Larsson, V.T. Oltner, T. Frejd, Asymmetric Baker's yeast reductions of bridgehead-substituted bicyclo[2.2.2]octane-2,6-dione derivatives followed by conversion into catalytically active BODOLs for the diethylzinc addition to benzaldehyde, Tetrahedron Asymm. 21 (2010) 1374–1381.

[15] L. Wang, W. Wang, J. Cuui, W. Ren, N. Meng, J. Wang, X. Qian, Preparation of chiral *trans*-5-substituted-acenaphthene-1,2-diols by Baker's yeast-mediated reduction of 5-substituted-acenaphthylene-1,2-diones, Tetrahedron Asymm. 21 (2010) 825–830.

[16] T. Janezko, W. Gladkowski, E. Kostrzewa-Suslow, Microbial transformations of chalcones to produce food sweetener derivatives, J. Mol. Catal. B: Enzym. 98 (2013) 55–61.

[17] T. Kaliperumal, S.N. Gummadi, A. Chadha, Synthesis of both enantiomers of ethyl-4-chloro-3-hydroxbutanoate from a prochiral ketone using *Candida parapsilosis* ATCC 7330, Tetrahedron Asymm. 22 (2011) 1548–1552.

[18] M.J. Homann, R.B. Vail, E. Previte, M. Tamarez, B. Morgan, D.R. Dodds, A. Zaks, Rapid identification of enantioselective ketone reductions using targeted microbial libraries, Tetrahedron 60 (2004) 789–797.

[19] M.M. Musa, N. Lott, M. Laivenieks, L. Watanabe, C. Vieille, R.S. Phillips, A single point mutation reverses the enantiopreference of *Thermoanaerobacter ethanolicus* secondary alcohol dehydrogenase, ChemCatChem 1 (2009) 89–93.

[20] G. Xu, H. Yu, Z. Zhang, J. Xu, Stereocomplementary bioreduction of β-ketonitrile without ethylated byproduct, Org. Lett. 15 (2013) 5408–5411.

[21] R.Z. Zhang, Y. Xu, Y. Sun, W.C. Zhang, R. Xiao, Ser67Asp and His68Asp substitutions in *Candida parapsilosis* carbonyl reductase alter the coenzyme specificity and enantioselectivity of ketone reduction, Appl. Environ. Microbiol. 75 (2009) 2176–2183.

[22] Z. Ou, H. Shi, X. Sun, W. Shem, Synthesis of *S*-licarbazepine by asymmetric reduction of oxcarbazepine with *Saccharomyces cerevisiae* CGMCC No. 2266, J. Mol. Catal. B Enzym. 72 (2011) 294–297.

[23] D. Zhu, Y. Yang, S. Majkowicz, T.H. Pan, K. Kantardjieff, L. Hua, Inverting the enantioselectivity of a carbonyl reductase via substrate-enzyme docking-guided point mutation, Org. Lett. 10 (2008) 525–528.

[24] R. Zhang, B. Zhang, Y. Xu, Y. Li, M. Li, H. Liang, R. Xiao, Efficient (R)-phenylethanol production with enantioselectivity-alerted (*S*)-carbonyl reductase II and NADPH regeneration, PLoS ONE 8 (2013) e83586.

[25] M.K. Sethi, S.R. Bhandya, N. Maddur, R. Shukla, A. Kumar, V.S.N. Jayalakshmi Mittapalli, Asymmetric synthesis of an enantiomerically pure rivastigmine intermediate using ketoreductase, Tetrahedron Asymm. 24 (2013) 374–379.

[26] W. Borzecka, I. Lavandera, V. Gotor, Synthesis of enantiopure fluorohydrins using alcohol dehydrogenases at high substrate concentrations, J. Org. Chem 78 (2013) 7312–7317.

[27] J. Liang, E. Mundroff, E. Volari, R. Jenne, S. Gilson, L. Conway, A. Krebber, J. Wong, G. Huisman, S. Truesdell, J. Lalonde, Highly enantioselective reduction of a small heterocyclic ketone: biocatalytic reduction of tetrahydrothiophene-3-one to the corresponding (R)-alcohol, Org, Proc. Res. Dev. 14 (2010) 188–192.

[28] G.A. Petkova, V. Král, Chiral switch of enzymatic ketone reduction by addition of gamma-cyclodextrin, Bioorg. Med. Chem. 18 (2010) 6651–6656.

[29] A. Bariotaki, D. Kalaitzakis, I. Smonou, Enzymatic reductions for the regio- and stereoselective synthesis of hydroxy-keto esters and dihydroxyesters, Org. Lett. 14 (2012) 1792–1795.

[30] S. Leuchs, T. Nonnen, D. Dechambre, S. Na'amnieh, L. Greiner, Continuous biphasic enzymatic reduction of aliphatic ketones, J. Mol. Cat. B Enz. 88 (2013) 52–59.

[31] A. Jakoblinnert, R. Mladenov, A. Paul, F. Sibilla, U. Schwaneberg, M.B. Ansorge-Schumacher, P. Domínguez de María, Asymmetric reduction of ketones with recombinant *E. coli* whole cells in neat substrates, Chem. Comm. 47 (2011) 12230–12232.

[32] B. Kosjek, D.M. Tellers, M. Biba, R. Farr, J.C. Moore, Biocatalytic and chemocatalytic approaches to the highly stereoselective 1,2-reduction of an α,β-unsaturated ketone, Tetrahedron Asymm. 17 (2006) 2798–2803.

[33] L. Wang, C. Li, Y. Ni, J. Zhang, X. Liu, J. Xu, Highly efficient synthesis of chiral alcohols with a novel NADH-dependent reductase from *Streptomyces coelicolor*, Bioresource Technol. 102 (2011) 7023–7028.

[34] Y. Xue, Y. Zheng, Y. Zhang, J. Sun, Z. Liu, Y. Shen, One-pot, single-step deracemization of 2-hydroxyacids by tandem biocatalytic oxidation and reduction, Chem. Comm. 49 (2013) 10706–10708.

[35] B. Sun, C. Kantzow, S. Bresch, K. Catiglione, D. Weuster-Botz, Multi-enzymatic one-pot reduction of dehydrocholic acid to 12-keto-ursodeoxycholic acid with whole-cell biocatalysts, Biotechnol. Bioeng. 110 (2013) 68–77 1.

[36] M.D. Truppo, D. Pollard, P. Devine, Enzyme-catalyzed enantioselective diaryl ketone reductions, Org. Lett. 9 (2007) 335–338.

[37] Z. Dai, K. Guillemette, T.K. Green, Stereoselective synthesis of aryl γ,δ-unsaturated β-hydroxyesters by ketoreductases, J. Mol. Cat. B Enz. 97 (2013) 264–269.

[38] N.A. Magus, D.S. Coffey, A.C. DeBaille, C.D. Jones, I.A. Kalluzna, S. Kambourakis, Y.J. Pu, L. Wang, J.P. Wepsiec, DIaryl ketone reductase screen and synthesis demonstration to access mGlu2 receptor potentiators, Org. Process. Res. Dev. 15 (2011) 1377–1381.

[39] D. Kalaitzakis, J.D. Rozzell, S. Kambourakis, L. Smonou, Highly Stereoselective Reductions of a-alkyl-1,3-diketones and a- alkyl-b-keto esters catalyzed by isolated NADPH-dependent Ketoreductases, Org. Lett. 22 (2005) 4799–4801.

[40] D. Kalaitzakis, J.D. Rozzell, I. Smonou, S. Kambourakis, Synthesis of valuable chiral intermediates by isolated ketoreductases: application in the synthesis of α-alkyl-β-hydroxy ketones and 1,3-diols, Adv. Synth. Catal. 348 (2006) 1958–1969.

[41] C.P. Marocco, E.V. Davis, J.E. Finnell, P.H. Nguyen, S.C. Mateer, I. Ghiviriga, C.W. Padgett, B.D. Feske, Asymmetric synthesis of (−)-fosfomycin and its trans-(1S,2S)-diastereomer using a biocatalytic reduction as the key step, Tetrahedron Asymm. 22 (2011) 1784–1789.

[42] D. Kalaitzakis, J.D. Rozzell, S. Kambourakis, I. Smonou, A two-step chemoenzymatic synthesis of the natural pheromone (+)-sitophilure utilizing isolated, NADPH-dependent ketoreductases, Eur. J. Org. Chem. (2006) 2309–2313.

[43] F. Xu, J.Y.L. Chung, J.C. Moore, Z. Liu, N. Yoshikawa, R.S. Hoerrner, J. Lee, M. Royzen, E. Cleator, A.G. Gibson, R. Dunn, K.M. Maloney, M. Alam, A. Goodyear, J. Lynch, N. Yasuda, P.N. Devine, Asymmetric synthesis of *cis*-2,5-disubstituted pyrrolidine, the core scaffold of β_3-AR agonists, Org. Lett. 15 (6) (2013) 1342–1345.

[44] J. Mangas-Sanchez, E. Busto, V. Gotor, V. Gotor-Fernandez, One-pot synthesis of enantiopure 3,4-dihydroisocoumarins through dynamic reductive kinetic resolution processes, Org. Lett. 15 (15) (2013) 3872–3875.

[45] O. Gooding, R. Voladri, A. Bautista, T. Hopkins, G. Huisman, S. Jenne, S. Ma, E. Mundorff, M. Savile, Development of a practical biocatalytic process for (R)-2-methylpentanol, Org. Proc. Res. Dev. 14 (2010) 119–126.

[46] J.A. Friest, Y. Maezato, S. Broussy, P. Blum, D.B. Berkowitz, Use of a robust dehydrogenase from an archael hyperthermophile in asymmetric catalysis – dynamic reductive kinetic resolution entry into (*S*)-Profens, J Am. Chem. Soc. 132 (2010) 5930–5931.

[47] Y. Ni, P. Hagedoorn, J. Xu, I.W.C.E. Arends, F. Hollmann, A biocatalytic hydrogenation of carboxylic acids, Chem. Comm. 48 (2012) 12056–12058.

[48] S. Kara, D. Spickermann, J.H. Schrittwieser, C. Leggewie, W.J.H. van Berkel, I.W.C.E. Arend, F. Hollmann, More efficient redox biocatalysis by utilising 1,4-butanediol as a 'smart cosubstrate', Green Chem. 15 (2013) 330–335.

[49] M.S. Bhattacharyya, A. Singh, U.C. Banerjee, Immobilization of intracellular carbonyl reductase from *Geotrichum candidum* for the stereoselective reduction of 1-naphthyl ketone, Bioresource Technol. 101 (2010) 1581–1586.

[50] A. Cuetos, M.L. Valenzuela, L. Lavandera, G.A. Carriedo, Polyphosphazenes as tunable and recyclable supports to immobilize alcohol dehydrogenases and lipases: synthesis, catalytic activity, and recycling efficiency, Biomacromolcules 11 (2010) 1291–1297.

[51] F. Hildebrand, S. Lutz, Immobilisation of alcohol dehydrogenase from *Lactobacillus brevis* and its application in a plug flow reactor, Tetrahedron Asymm. 17 (2006) 3219–3225.

[52] K. Schroer, U. Mackfeld, I.A.W. Tan, C. Wandrey, F. Heuser, S. Bringer-Meyer, A. Weckbecker, W. Hummel, T. Daußmann, R. Pfaller, A. Liese, S. Lütz, Continuous asymmetric ketone reduction processes with recombinant *Escherichia coli*, J. Biotechnol. 132 (2007) 438–444.

[53] T. Matsuda, K. Watanabe, T. Kamitanaka, T. Harada, K. Nakamura, Biocatalytic reduction of ketones by semi-continuous flow process using supercritical carbon dioxide, Chem. Comm. (2003) 1198–1199.

[54] N. Itoh, M. Nakamura, K. Inoue, Y. Makino, Continuous production of chiral 1,3-butanediol using immobilized biocatalysts in a packed bed reactor: promising biocatalysis method with an asymmetric hydrogen-transfer bioreduction, Appl. Microbiol. Biotechnol. 75 (2007) 1249–1256.

[55] A. Scholz, M. Eckstein, M.B. Ansorge-Schumacher, Hydrophilized silicone matrix for the preparation of stable carbonyl reductase immobilizates, ChemCatChem. 5 (2013) 815.

[56] A. Findeisin, O. Thum, M.B. Ansorge-Schumacher, Biocatalytically active silCoat-composites entrapping viable *Escherichia coli*, Appl. Microbiol. Biotechnol. 98 (2014) 1557.

[57] C. Kohlmann, N. Robertz, S. Leuchs, Z. Dogan, S. Lütz, K. Bitzer, S. Na'amnieh, L. Greiner, Ionic liquid facilitates biocatalytic conversion of hardly water soluble ketones, J. Mol. Cat. B Enz. 68 (2011) 147–153.

[58] J. He, L. Zhou, P. Wang, L. Zu, Microbial reduction of acetoacetate to ethyl (R)-3-hydroxybutyrate in an ionic liquid containing system, Process Biochem. 44 (2009) 316.

[59] W. Hussain, D.J. Pollard, M. Truppo, G.J. Lye, Enzymatic ketone reductions with co-factor recycling: Improved reactions with ionic liquid co-solvents, Mol. Cat. B Enz. 55 (2008) 19–29.

[60] X.-T. Wang, D.-M. Yue, M.-H. Zong, W.-Y. Lou, Use of ionic liquid to significantly improve asymmetric reduction of ethyl acetoacetate catalyzed by *Acetobacter sp.* CCTCC M209061 cells, Ind. Eng. Chem. Res. 52 (2013) 12550–12558.

[61] A.S. Rowan, T.S. Moody, R.M. Howard, T.J. Underwood, I.R. Miskelly, Y. He, B. Wang, Preparative access to medicinal chemistry related chiral alcohols using carbonyl reductase technology, Tetrahedron Asymm. 24 (2013) 1369–1381.

[62] (a) E.C.S. Brenelli, Fernandes JLN, Stereoselective acylations of 1,2-azidoalcohols with vinyl acetate, catalyzed by lipase Amano PS, Tetrahedron Asymm. 14 (2003) 1255–1259. (b) G. Lin, H.-S. Liue, Ultrasound-promoted lipase catalyzed reactions, Tetrahedron Lett. 36 (1995) 6067–6068. (c) C.M.R. Ribeiro, E.N. Passaroto, E.C.S. Brenelli, Native enzymes with and without nanoparticles, Tetrahedron Lett. 42 (2001) 6477–6479.

[63] (a) R. Mawson, M. Gamage, N.T. Terefe, K. Knoerzer, Ultrasound in enzyme activation and inactivation. In: H. Feng, G.V. Barbosa-Canovas, J. Weiss (Eds.), Ultrasound Technologies for Food and Bioprocessing, Springer, New York, NY, 2012. 49. (b) M. Souza, E.T. Mezadri, E. Zimmerman, E.X. Leaes, M.M. Bassaco, V. Dal Prá, E. Foletto, A. Cancellier, L.M. Terra, S.L. Jahn, M.A. Mazutti, Evaluation of activity of a commercial amylase under ultrasound-assisted irradiation, Ultrason Sonochem. 20 (2013) 89–94.

[64] Emmanuelle Alanvert, Claire Doherty, Thomas S. Moody, Nicholena Nesbit, Andrew S. Rowan, Stephen J.C. Taylor, Fatima Vaughan, Tony Vaughan, Jonathan Wiffen, Ian Wilson, Highly stereoselective biocatalytic reduction of alpha-halo ketones, Tetrahedron Asymm. 20 (2009) 2462–2466.

CHAPTER 7

Organic Synthesis with Amino Acid Dehydrogenases, Transaminases, Amine Oxidases, and Amine Dehydrogenases

Samantha K. Au*, Jonathan Groover, Brent D. Feske**, Andreas S. Bommarius*,†**

*School of Chemical and Biomolecular Engineering, Georgia Institute of Technology, Parker H. Petit Institute of Bioengineering and Bioscience, Atlanta, GA, USA

**Chemistry and Physics Department, Armstrong State University, Savannah, GA, USA

†School of Chemistry and Biochemistry, Georgia Institute of Technology, Atlanta, GA, USA

1 INTRODUCTION

Enzymes are widely known for rate enhancement, and they also feature high selectivities, including regio-, chemo-, and enantioselectivities. Enzymes operate under mild processing conditions with greener solvents and can have easier product separation compared to heterogeneous catalysts [1]. The American Chemical Society's Green Chemistry Institute Pharmaceutical Roundtable (ACS GCIPR) has supported and continually encouraged the integration of green chemistry such as biocatalysts into the pharmaceutical industry. There is a current focus to identify and expand the limited volume of enzymatic reactions. In 2007, the Pharmaceutical Roundtable identified amination of ketones as a top aspirational reaction [2]. In 2011, the group also identified *key green research areas*, with an emphasis on bioprocess engineering and focus on use of green organic solvents [3].

Organic Synthesis Using Biocatalysis. http://dx.doi.org/10.1016/B978-0-12-411518-7.00007-X

1.1 Enzymatic Production of Enantiomerically Pure Amino Acids

Enantiomerically pure amino acids have been used as pharmaceutical and agrochemical intermediates [4]. The desired amino acids have been further expanded past the natural production of L-amino acids through fermentation [5]. Reduction of keto acids can be achieved through amino acid dehydrogenases (AADHs) or transaminases (TAs).

AADHs have favorable equilibrium toward production of amino acids and use NADH as a cofactor. TAs have also been used for the production of amino acids in recent work, including the production of unnatural amino acids [6] and deracemization of amino acids [7]. However, TAs suffer from the disadvantage of being equilibrium limited and requiring a sacrificial amine donor [8].

1.2 Enzymatic Production of Enantiomerically Pure Amines

Enantiomerically pure amines are commonly used as precursors for active pharmaceutical ingredients (APIs). In 2013, one in four of the top 200 selling drugs contained a chiral amine moiety [9]. In addition, chemical synthesis of these enantiomerically pure amines is critical as 80% of small-molecule pharmaceutical drugs approved by the FDA in 2006 contained chiral centers and 75% were single enantiomers [10]. Chiral amine compounds with high optical purities can be difficult to prepare by many types of traditional catalysts.

The chapter introduces the reader to the use of amine dehydrogenases (AmDHs), derived from AADHs, TAs, and amine oxidases (AO) for the production of chiral amines. It should also be noted that, although not covered in this chapter, imine reductases offer another biocatalytic route to enantiomerically pure amines [11].

2 AMINO ACID DEHYDROGENASES

2.1 Introduction

AADHs (EC 1.4.1.X) reductively aminate α-keto acids to α-amino acids through a reversible reaction; however, production toward the amino acid is thermodynamically favorable (Figure 7.1). One major advantage for AADHs is that ammonia can be directly used as an amine donor and AADHs have been used for incorporation of ammonia into organic compounds. AADHs nearly always form enantiomerically pure amino acids that are important for large-scale commercial applications [12]. Most AADHs are naturally L-selective, with notable

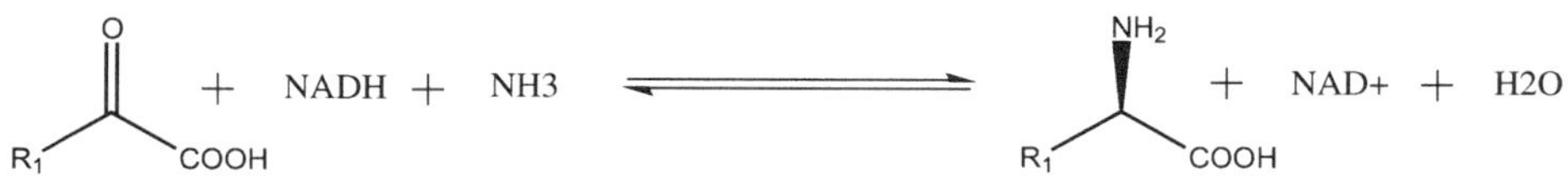

FIGURE 7.1
Reaction scheme of AADHs.

Table 7.1 Amino Acid Dehydrogenase (AADH)

EC Number	Enzyme
1.4.1.1	Alanine DH
1.4.1.2	Glutamate DH
1.4.1.3	Glutamate DH
1.4.1.4	Glutamate DH
1.4.1.7	Serine DH
1.4.1.8	Valine DH
1.4.1.9	Leucine DH
1.4.1.10	Glycine DH
1.4.1.11	3,5-Diaminohexanoate DH
1.4.1.12	2,4-Diaminopentanoate DH
1.4.1.15	Lysine DH
1.4.1.16	Diaminopimelate DH
1.4.1.19	Tryptophan DH
1.4.1.20	Phenylalanine DH

exceptions such as *meso*-diaminopimelate dehydrogenases (DAPDH) [13]. As seen in the two examples described further, the cofactor is constantly recycled with cofactor regeneration enzymes, such as formate dehydrogenase (FDH). Table 7.1 lists common AADHs.

2.2 Classification and Structure

Most AADHs are homo-oligomers and often exist as hexamers, although octamers, tetramers, dimers, and even monomers can also be found. Most AADHs feature the same three-dimensional structure type and superimpose very well. The first crystal structure reported was the hexameric *Clostridium symbiosum* L-glutamate dehydrogenase [14]. Three dimensional structures of many AADHs have since been reported (Table 7.2). Generally, each monomeric subunit consisted of a deep cleft with an *N*- and C-terminal domain (separated by approximately 50 Å). The C-terminal domain is responsible for cofactor binding and contains the "Rossman" fold, while the *N*-terminal domain is responsible for substrate binding of the amino acid.

2.3 Catalytic Mechanism

AADHs produce the enantiomerically pure L-amino acids and additional insight into the mechanism and enzyme–substrate complex [22a,35] further support this observed stereochemistry. The catalytic mechanism of AADHs has been studied most thoroughly with GluDH from *C. symbiosum* [14,22a]

Table 7.2 X-Ray Structures of Amino Acid Dehydrogenases

LeuDH:	*Bacillus sphaericus* (1LEH) [15]
	Sporosarcina psychrophila (3VPX) [16]
PheDH:	*Rh. rhodocrous* (1XBG, 1BW9, 1C1D, 1C1X) [17]
AlaDH:	*Phormidium lapideum* (1PJB, 1PJC, 1SAY) [18]
	Thermus thermophilus (2EEZ) [19]
	Mycobacterium tuberculosis (2VHY, 2VHZ, 2VHV, 2VHW, 2VHX) [20]
GluDH:	*Thermotaga maritama* (1B26, 1B3B, 2TMG) [21]
	C. symbiosum (1BGV, 1K89) [22]
	Thermococcus litoralis (1BVU) [23]
	Thermococcus profundus (1EUZ) [24]
	Pyrococcus furiosus (1GTM, 1HRD) [25]
	Pyrobaculum islandicum (1V9L) [26]
	Plasmodium falciparum (2BMA, 3R3J) [27]
	Peptoniphilus asaccharolyticus (2YFQ) [28]
	Thermus thermophilis (3AOE, 3AOG) [29]
	B. subtilis (3K92, 3K8Z) [30]
	Bovine (1HWY, 1HWZ, 1NR1, 1NQT, 1NR7, 3ETE, 3MW9, 3MVO, 3MVQ) [31]
meso-DAPDH:	*Corynebacterium glutamicum* (1DAP, 2DAP, 3DAP) [32]
	Clostridium tetani E88 (3WGZ, 3WGO, 3WGQ, 3WGY) [33]
	Symbiobacterium thermophilum (3WB9, 3WBB, 3WBF) [13a,34]

and with PheDH from *Rhodococcus* M4 [35a]. The mechanism was found to be remarkably similar in both cases where the first step involves attack of ammonia on the dehydrogenase-keto complex to create a carbinolamine intermediate (Figure 7.2).

2.4 Examples of Amino Acid Dehydrogenases for Synthesis of Amino Acids

Example 1

Synthesis of (*S*)-*tert*-leucine via leucine dehydrogenase-catalyzed reductive amination of trimethylpyruvic acid (TMP or ketopivalic acid)

This case illustrates the leucine dehydrogenase (LeuDH)-catalyzed reductive amination of trimethylpyruvate to (*S*)-*tert*-leucine, a building block in the production of chiral

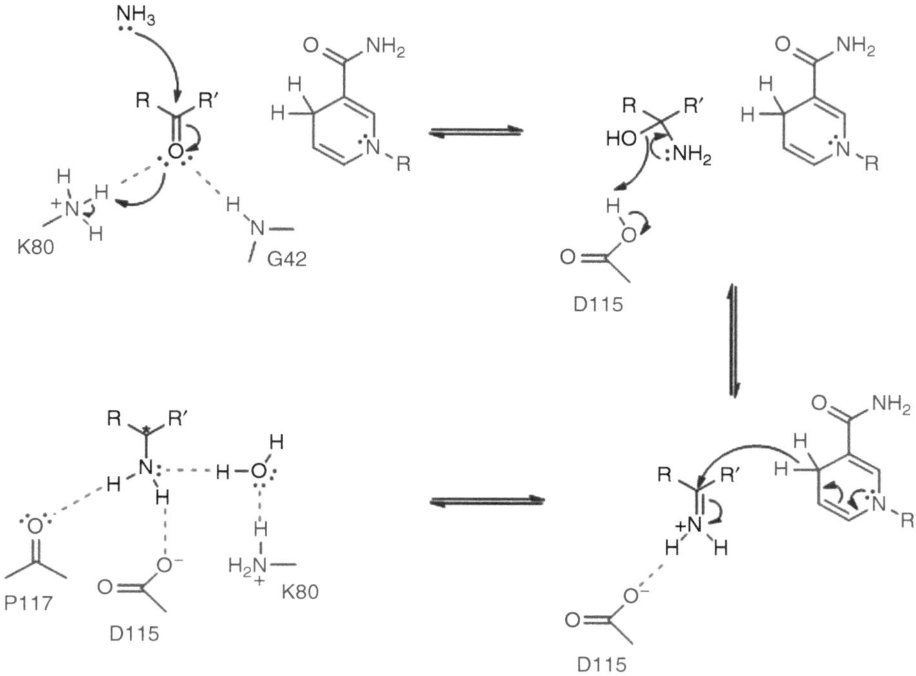

FIGURE 7.2
Proposed reductive amination mechanism for AADH.

ligands as well as in APIs [36]. Despite several reports on the LeuDH-catalyzed synthesis in the 1990s, several reports have appeared recently, mostly from Asia [37] (Figure 7.3).

(*S*)-*tert*-leucine (L-*tert*-leucine, (*S*)-*tert*-butylglycine, 3-methyl-(*S*)-valine, (*S*)-trimethylalanine; Figure 7.4a) is a valuable precursor for a variety of APIs. Examples include the hepatitis C antiviral NS1 protease inhibitors telaprevir [38] (Incivek®, Incivo®; Vertex, Janssen), bocepravir (Victrelis®, Merck), and NS3/4 protease inhibitors [39] asunaprevir (BMS-650032, phase III), faldaprevir (BI 201335, phase III), and vedroprevir (Vertex, phase IIb), or the HIV protease inhibitor atazanavir [40] (Reyataz®; Bristol-Myers-Squibb (BMS)), or peptide deformylase (PDF) inhibitors [41] (such as BB-83698) or matrix metalloprotease inhibitors (MMPI) [42] (Figure 7.4b–h).

The various approaches to synthesize (*S*)-*tert*-leucine have been reviewed [43]. Enantiomeric salt separation is cumbersome and often results in enantiomeric excesses (ees) well short of the 99.5%, regulatory specification limit in many cases. Many biocatalytic approaches failed: neither aminoacylase-based processes, nor hydantoinase/carbamoylase reacted with the respective substrates [36]. By far the most success in the synthesis of (*S*)-Tle was achieved through leucine dehydrogenase (LeuDH)-catalyzed reductive amination of trimethylpyruvate [36,37,44].

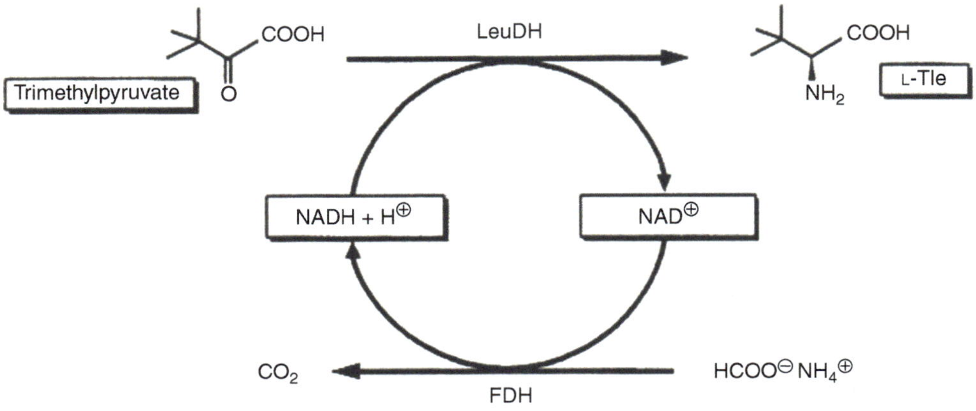

FIGURE 7.3
Scheme adapted from Bommarius [36].

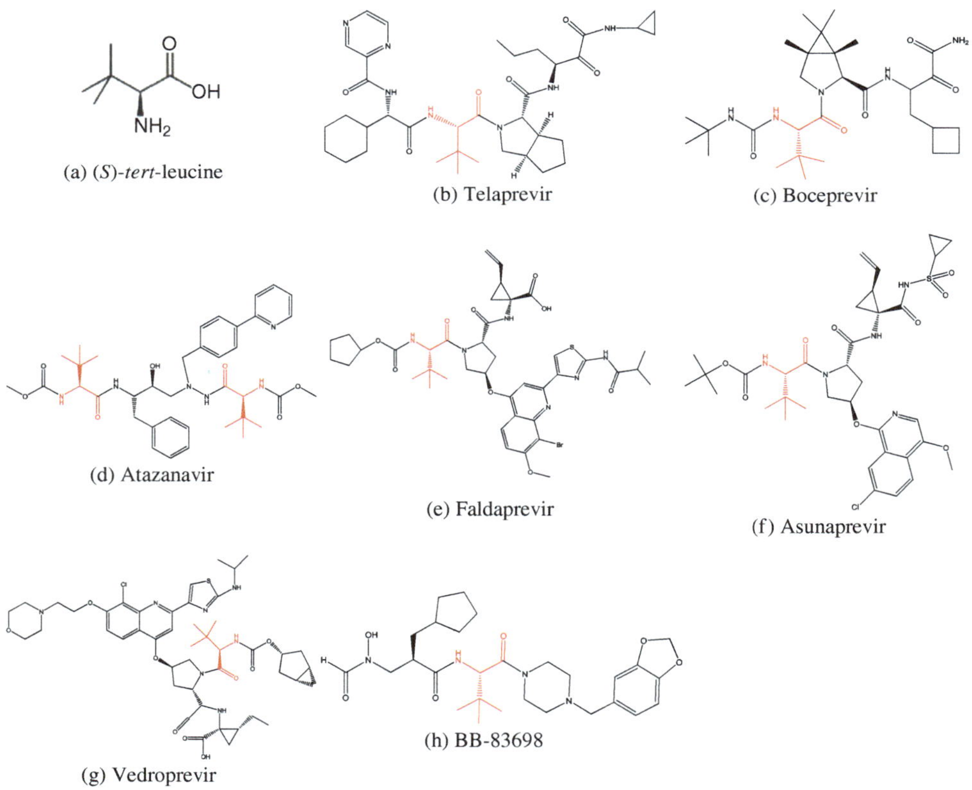

FIGURE 7.4
L-*tert*-leucine and APIs containing L-*tert*-leucine (a) (*S*)-*tert*-leucine (b) telaprevir (c) boceprevir (d) atazanavir (e) faldaprevir (f) asunaprevir (g) vedroprevir (h) BB-83698.

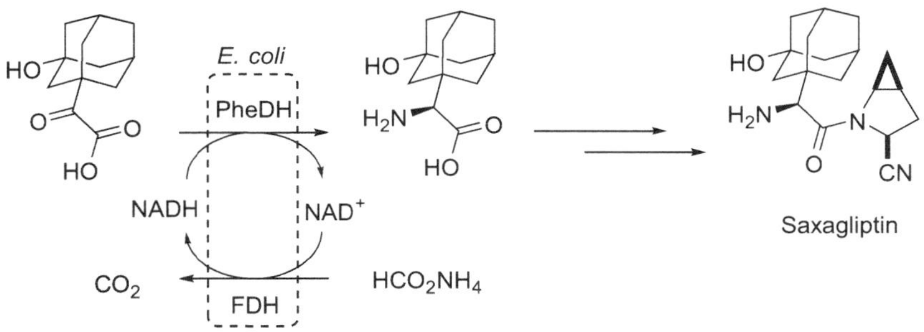

FIGURE 7.5
Reductive amination to (*S*)-3-hydroxyadamantylglycine, a precursor to Saxagliptin, using a PheDH/FDH system [46].

Example 2

Preparation of L-Tle on a 1 mole scale

The frozen wet cells (30 g, prepared as in Ref. [37a], containing the *leudh* gene from *Lysinibacillus sphaericus* CGMCC 1.1677 and the *fdh* gene from *Candida boidinii*) are suspended in deionized water (300 mL) and then disrupted by ultrasonication and centrifuged to obtain cell-free extracts of LeuDH. Ammonium formate (189 g) and NAD^+ (0.50 g) are then added to the cell-free extracts. The pH of the initial reaction mixture is adjusted to 8.5 with NaOH and the volume is adjusted to 500 mL with deionized water. Substrate solution (3 M, pH 8.5, 500 mL) containing 195 g TMP is then fed continuously at a feeding rate of 100 mL/h using a peristaltic pump. The reaction was monitored by HPLC and complete conversion was reached after 6 h. The reaction mixture is heated in a boiling water bath for 30 min to precipitate proteins and then filtered to remove these proteins. The aqueous filtrate is adjusted to pH 3.0 and loaded on a glass column packed with cation-exchange resin (001X7), followed by washing with deionized water, and then elution with 5% ammonia solution. The elution fractions are combined and evaporated to give L-Tle (172 g, 88% yield) with >99% ee (Figure 7.5).

Example 3

Preparation of (S)-3-hydroxyadamantylglycine from 2-(3-hydroxy-1-adamantyl)-2-oxoethanoic acid from a modified PheDH

Saxagliptin is a peptidyl peptidase IV inhibitor that is currently used for the treatment of type-2 diabetes [45]. Modified phenylalanine dehydrogenase (PheDH) from *Thermoactinomyces intermedius* reductively aminates 2-(3-hydroxy-1-adamantyl)-2-oxoethanoic acid to (S)-3-hydroxyadamantylglycine, an intermediate of Saxagliptin. The engineered PheDH contains two mutations and a 12 amino acid extension of the C-terminus.

Escherichia coli cells that coexpressed *Pichia pastoris* PheDH and FDH genes reductively aminated the ketoacid to (S)-3-hydroxyadamantylglycine. Deionized water (196.1 kg), keto acid (40 kg, 178.4 mol), and ammonium formate (22.3 kg, 353.6 mol) were mixed into an 800 L volume vessel and adjusted to pH 8.0. Further, NAD^+ (499.5 g, 0.753 mol), dithiothreitol (111 g, 0.72 mol), deionized water (30.4 kg), and 407 kg of clarified cell lysate (prepared as in Ref. [46]) were added. The reaction mixture was maintained at 40°C and pH 8.0. The reaction reached 99% conversion after 49 h. Protein and insoluble material was removed by filtration after the reaction reached completion. Water and ammonia were removed by distillation to provide 407 kg of amino acid solution (100 g/L) which was converted to its Boc-derivative.

3 TRANSAMINASES

3.1 Introduction

Chiral amine compounds with high optical purities can be difficult to prepare by many types of traditional catalysts. One well-known enzymatic synthesis of chiral amines involves TAs [6,47]. The ω-TA transfers an amine group from an amino donor onto the ketone moiety of the amino acceptor (Figure 7.6). The most popular example, which won the presidential green chemistry award in 2010, has been the development of an ω-TA by Merck and Codexis for an (*R*)-selective TA to synthesize sitagliptin (Januvia®) [48].

Despite many recent advances, TAs inherently possess equilibrium limitations and require the use of an amine donor, which becomes extremely problematic with industrial applications [8,49]. The ketone coproduct must be constantly removed during the reaction to shift the equilibrium and prevent product inhibition to achieve conversion much beyond about 50% [50]. Since TAs exhibiting complementary enantioselectivities have been reported, recent work involves coupling the enzymes to avoid any equilibrium limitations [7]. TAs have traditionally possessed (*S*)-selectivity but recently (*R*)-selective TAs [47c,51] have been developed and have thus expanded the substrate scope. In addition, advances in by-product removal allows for favorable equilibrium and high conversion [52].

Recent work has successfully improved TA stability [53] and has expanded their limited substrate range to include intermediates important for pharmaceuticals [54] such as aminotetraline, aminochromanone [55], and mexiletine derivatives [56], which will be covered in this chapter.

3.2 Classification and Structure

TAs can be divided into two classes: Class I and IV. Both types rely on the cofactor pyridoxal 5′-phosphate (PLP). Fold class I represents the aspartate aminotransferases

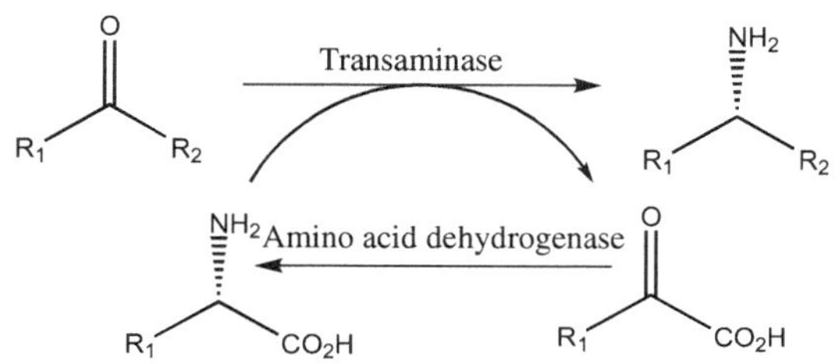

FIGURE 7.6
TA reaction scheme.

Table 7.3 Source and Selectivity of Some TAs

Source	PDB ID	Selectivity
Aspergillus fumigatus	4CHI [57]	(*R*)-selective
Aspergillus terreus	4CE5 [58]	(*R*)-selective
C. violaceum	4A6R, 4A6T, 4A6U, 4A72, 4AH3, 4BA4, 4BA5 [59]	(*S*)-selective
Paracoccus denitrificans	4GRX [60]	(*S*)-selective
Pseudomonas aeruginosa	4BQ0, 4B9B, 4B98 [59a]	(*S*)-selective
Pseudomonas putida	3A8U (unpublished)	(*S*)-selective
Vibrio fluvialis	3NUI, 4E3Q and 4E3R [61]	(*S*)-selective

(the majority of known TAs, 18 out of 27), while class IV comprises D-amino acid aminotransferases, of which only a very few are known. Three dimensional structures of some TAs were determined by X-ray crystallography. Table 7.3 lists sources, protein data base (PDB) ID and selectivity of some TAs.

3.3 Catalytic Mechanism

TAs transfer an amino group from an amine donor to a carbonyl acceptor with PLP as a cofactor. The reaction scheme can be simply divided into two parts (Figure 7.7). First, the amino group from the donor is transferred to PLP to create pyridoxamine phosphate (PMP) and then the amino group is transferred from PMP to the acceptor carbonyl group converting it into amine [62].

TAs exhibit a dual binding pocket, with a larger pocket accommodating bulkier substituents and a smaller pocket for smaller substituents [63]. The smaller binding pocket limits larger compounds as substrates, which is often a problem for industrial applications [64]. Recent protein-engineering efforts in the active site of TAs have increased the enzyme's activity toward bulky substrates [65].

3.4 Examples of TAs

Example 4

Preparation of tetralone and chromanone derivatives via ω-TA

This case describes the amination of tetralone and chromanone derivatives via TA [55]. Aminotetralines and aminochromanes are common scaffolds to pharmaceutical compounds (Figure 7.8). For example, 1-aminotetraline (Figure 7.9, 2d) core is present in the pharmaceutical drugs sertraline (Zoloft®) [66] and norsertraline [67]. The 2-aminotetraline (Figure 7.9, 2a) and 2-aminochromane (Figure 7.9, 2f) are the backbones of rotigotine (Neupro®) [68] and the antidepressant drug alnespirone [69], respectively.

Arthrobacter citreus (S)-selective ω-TA (*ArS*-ωTA, prepared as in Ref. [70]) crude powder extract (270 mg, 1.26 U m/g) is resuspended in phosphate buffer (13.5 mL, 100 mM

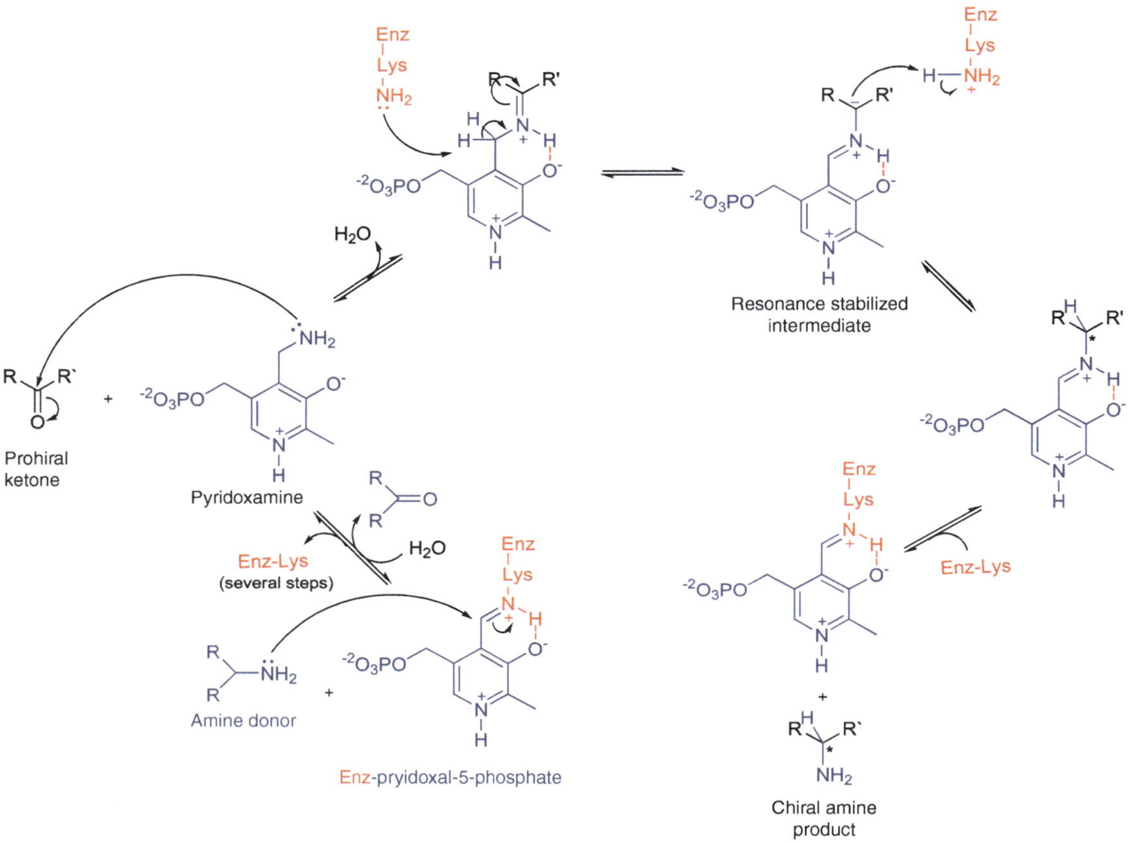

FIGURE 7.7
Mechanism of enzymatic transamination.

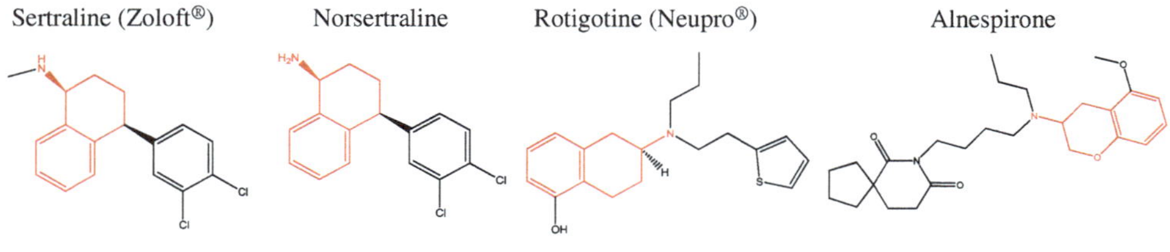

FIGURE 7.8
Pharmaceutically relevant structures containing aminotetraline and aminochromane.

pH 7). L-Alanine (301 mg, 3.375 mmol), lactate dehydrogenase (90 U m/L), glucose dehydrogenase (GDH) (30 U m/L), NADH (1 mM), and PLP (1 mM) are added to the solution. The ketone substrate (1f) (100 mg, 0.675 mmol) is dissolved in DMSO (135 μL, 10% v/v) and added to the reaction mixture. The flask is shaken on an orbital shaker at 150 rpm, 30°C for 24 h. The reaction is stopped by adding 10 N NaOH (2.7 mL), extracted with EtOAc (3 × 5 mL) and dried with anhydrous Na_2SO_4. The organic phase is evaporated and hydrogen chloride is added (2 N, 1 mL). The residual oil is dried under

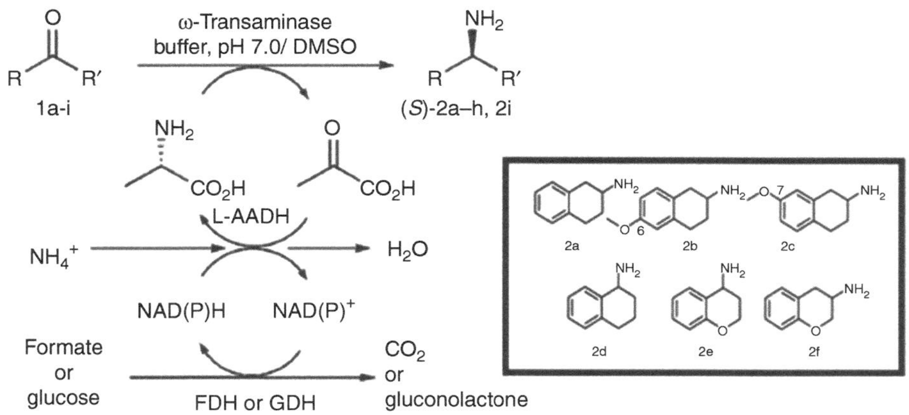

FIGURE 7.9
Use of ω-TA for aminotetralines and –chromanones. *(Adapted from Pressnitz, 2013 [55].)*

vacuum and the product is recrystallized in Et_2O (2 mL). (*R*)-3-chromanone (2f) is isolated as a yellow solid (98.2 mg, 78% yield, >99% ee).

The co-product pyruvate of the TA reaction is constantly removed by lactate dehydrogenase (LDH) or recycled back to alanine by alanine AADH (AlaDH). This is done to push the equilibrium toward the amine. L-Alanine is added for the amination with (*S*)-ωTAs and D-alanine is added for amination with (*R*)-ωTAs. The cofactor NADH is regenerated by 11 U of *C. boidinii* formate dehydrogenase (*Cb*FDH, Codexis).

Example 5

Deracemization of (R/S)-mexiletine [1-2,6-dimethylphenoxy)-2-propanamine] via ω-TA

This case describes the deracemization of mexiletine (dimethyl-phenoxyisopropylamine) by both (*R*) and (*S*)- TAs [56]. The procedure utilizes the (*R*) and (*S*)-selective TAs simultaneously to achieve high yield and selectivity. (*R/S*)-mexiletine is an antiarrhythmic and analgesic drug [71] but the (*R*)-mexiletine is more potent and exhibits fewer side effects, so there is a high demand for the enantiomerically pure (*R*)-compound [72].

Deracemization of racemic mexiletine (di-*o*-methyl-phenoxyisopropylamine) is successfully achieved in a simple two-step procedure. This work extends the utility of the TA to include deracemization of amines and the asymmetric synthesis of ketones to bypass the limitations of kinetic resolution. The cofactor is recycled by amino acid oxidase [73] in the first deracemization step and by dehydrogenases in the second asymmetric synthesis step [47a,74]. In addition, both enantiomers of mexiletine can be obtained by simply switching the order of the addition of stereospecific TAs.

To achieve kinetic resolution of mexiletine (di-*o*-methyl-phenoxyisopropylamine), lyophilized *E.coli* cells containing *Chromobacterium violaceum* (S)-selective ω-TA (100 mg, prepared as described in Ref. [75]) are resuspended in phosphate buffer (20 mL, 100 mM, 0.9 mM sodium pyruvate, 1 mM PLP). *Crotalus adamanteus* L-amino acid oxidase (30 mg, 9 U, Sigma) is rehydrated in the buffer. (*R/S*)-mexiletine (100 mg, 23 mM)

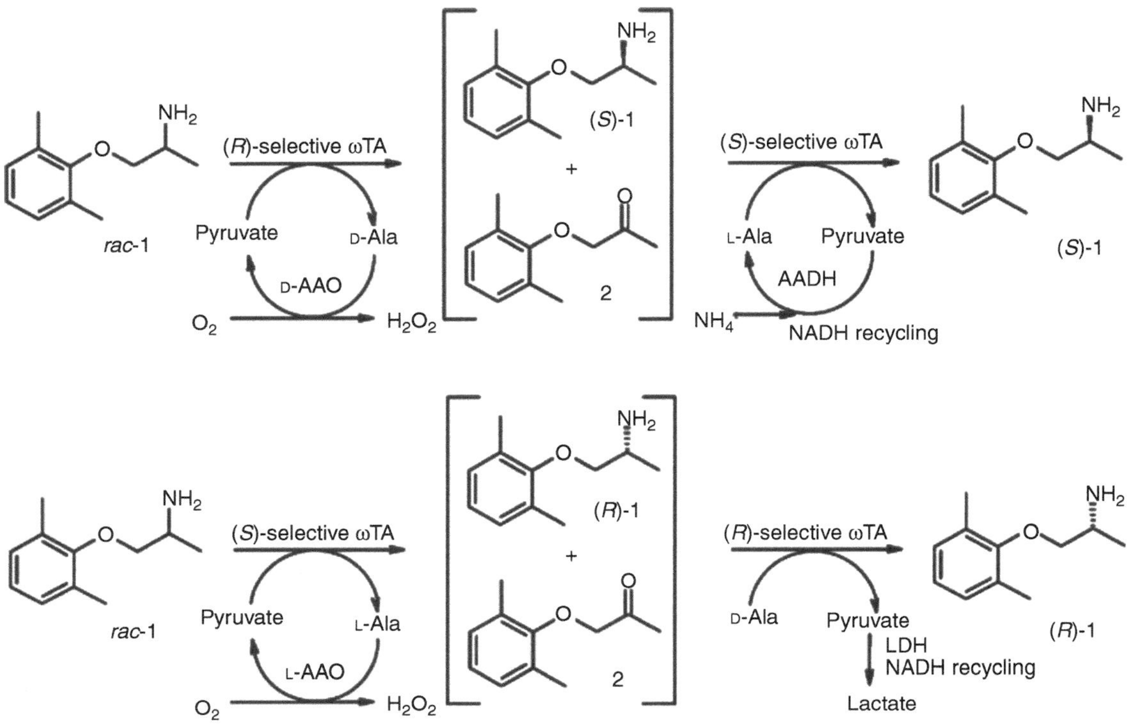

FIGURE 7.10
Deracemization reaction scheme to create both (*R*) and (*S*)-mexiletine. *(Adapted from Koszelewski, 2009 [56].)*

is added to the reaction. The mixture is allowed to proceed at 30°C for 24 h and is shaken at 300 rpm.

Upon completion of the first step, the reaction mixture was heated to 75°C for 30 min to denature the biocatalysts. After cooling to 30°C, Lactate dehydrogenase (50 μL, 100 U), glucose (3 equiv.), NAD^+ (1 mM), GDH (100 μL, 47 U) is added to the reaction. This mixture was allowed to incubate for 1 h, and then (*R*)-selective ATA-117 (30 mg, Codexis) is added to the reaction. The mixture is again reacted at 30°C for 24 h and shaken at 300 rpm. After 24 h, the pH is adjusted to 1 and the ketone is extracted with dichloromethane (5 × 10 mL). The pH is then adjusted to 12 and the amine is extracted with dichloromethane (4 × 10 mL). (*R*)-Mexiletine was isolated (99% yield, >99% ee) (Figure 7.10).

Example 6

Synthesis of Sitagliptin via ω-TA

The most important application of enzymatic transamination has recently been demonstrated for the synthesis of Sitagliptin (Januvia®) [48], a commercial DPP-4 inhibitor from Merck for the treatment of diabetes. Merck and Codexis started from a TA enzyme with

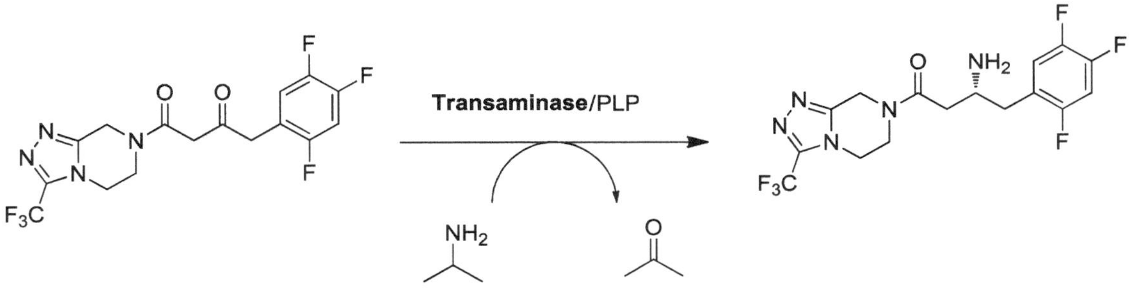

FIGURE 7.11
Enzymatic transamination for the synthesis of Sitagliptin.

barely any activity for the corresponding ketone. They utilized substrate walking and carried out a series of enzyme evolution ultimately leading to an efficient TA enzyme for the conversion of the ketone to Sitagliptin. Isopropyl amine was used as the donor and the product acetone was removed from the reaction mixture to drive the forward reaction (Figure 7.11). The optimum reaction condition utilized 200g/L of Sitagliptin ketone, 6 g/L enzyme and 50% DMSO yielding the desired (*R*)-amine, Sitagliptin in 99.95% ee and 92% yield.

4 AMINE OXIDASE

4.1 Introduction

Amine oxidases (AOs) can be used to synthesize primary, secondary and tertiary amines. Deracemization of chiral amines can be achieved by an enantioselective oxidation by an AO followed by a reduction using a nonselective reducing agent (Figure 7.12) [47d,76]. There are two types of AOs: Type I (Cu/TOPA dependent, CAOs, EC.1.4.3.6) and Type II (Flavin dependent, EC 1.4.3.4). This chapter will solely focus on the flavin dependent monoamine oxidases.

The major disadvantages of the AOs are that the reaction must be coupled with a reducing agent in order to achieve the pure chiral amine and these enzymes have limited substrate ranges. Through protein engineering of the *Aspergillus niger* AO, the substrate range has been expanded and chemo-enzymatic routes to cyclic secondary [77] and tertiary amines [78] have been achieved. In addition,

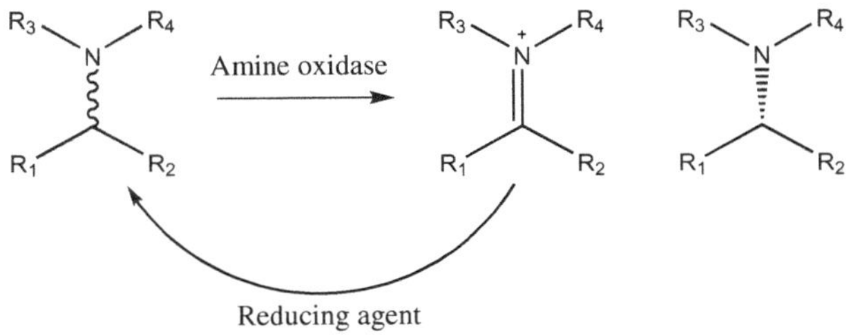

FIGURE 7.12
AO reaction scheme.

Table 7.4 Crystal Structures of Variants of MAO-N

Variants of *A. niger*	Crystal Structure
MAO-N-D3	2VVL [83]
MAO-N-D5	2VVM [83]
MAO-N-D11	3ZDN [76b]

the substrate ranges of AOs have been extended to allow the deracemization of alkaloid natural products [76b,79] and substituted pyrrolidines [80] (as described further). Development of a high throughput colorimetric assay based on identification of the by-product hydrogen peroxide was crucial in order to screen libraries of MAO-N variants [81].

The known monoamine oxidases, including *A. niger* (MAO-N), are (*S*)-selective. Recently, an (*R*)-selective AO from *Arthrobacter nicotinovorans* was characterized and engineered to broaden its substrate specificity [82]. The crystal structures of MAO-N variants have been solved (Table 7.4). These enzymes have two channels that intersect at the active site and seem to play a role in both the substrate selectivity and enantioselectivity [79a].

4.2 Catalytic Mechanism

MAO is a flavin-dependent enzyme in which the cofactor is covalently bound to the enzyme via a cysteine residue. Four mechanisms have been proposed for this complicated oxidation reaction. A hydride mechanism, an addition-elimination mechanism, a carbanion mechanism, and a radical mechanism have all been proposed [70].

The most popularly accepted mechanism is the radical pathway that was put forth by Silverman [75]; however, this mechanism remains controversial since the single electron transfer from the amine to the flavin is thermodynamically unfavorable. Silverman argues that this crucial step could be favored by the enzyme-substrate intrinsic binding energy and this could affect the redox potential of the amine and flavin. This mechanism (Figure 7.13) has been further

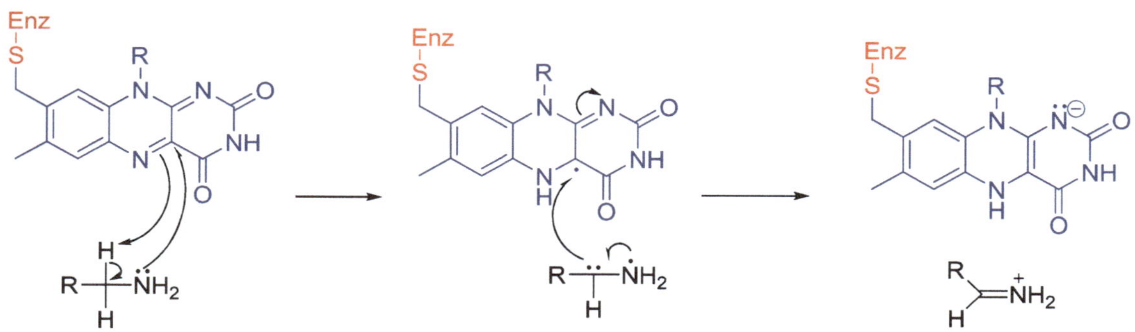

FIGURE 7.13
Proposed radical mechanism for MAO [70].

supported by spectroscopic studies that discovered stable radical intermediates in human MAO [84].

4.3 Examples of Amine Oxidases

Example 7

Preparation of 2,5-disubstituted pyrrolidines via TA/monoamine oxidase

This case discusses the preparation of 2,5-disubstituted pyrrolidines via TA/monoamine oxidase [85]. Disubstituted pyrrolidines are important scaffolds in pharmaceutical compounds, such as (+)-anatoxin-a [86] and are also common intermediates for the synthesis of many drug candidates [87,88]. In addition, disubstituted pyrrolidines are present in natural products [89].

ω-TAs have been successful in the regioselective amination of diketones to 2,6- disubstituted piperidines when followed by hydrogenation using Pd/C [90]. However, poor diastereoselectivity in the reduction step is observed in this chemoenzymatic route when applied to 2,5-disubstituted pyrrolidines. The previous achievements of these two enzymes led to the efficient multienzyme coupling of ω-TA (TA) with monoamine oxidase [91]. This novel synthesis of chiral 2,5-disubstituted pyrrolidines [80b] achieved excellent enantio- and disastereoselectivities (Figure 7.14).

Monoamination of diketone **(1a–g)** *to pyrroline* **((S)-3, (R)-3)**, *preparative scale*: TA ATA-113 (Codexis, 125 mg) is reconstituted in HEPES buffer (50 mL, 100 mM, pH 7.5). The buffer contains PLP (2.02 mM), NAD^+ (1.5 mM), glucose (55.5 mM), GDH (50 U), lactate dehydrogenase (LDH, 113 U), both commercially available, and alanine (500 mM). Diketone (25 mM, **(1a–g)**) is dissolved in DMSO and added to the reaction. The mixture is reacted at 30°C, 250 rpm for 24 h. Completion of the reaction is monitored by GC-MS.

Deracemization: Cells expressing MAO-N D5 variant (50 mg/mL, prepared as described in Ref. [85]) are added with reducing agent NH_3BH_3 (25 mM, 5 eq.) and the previous reaction mixture containing pyrroline **((S)-3, (R)-3)**. The mixture is reacted at 30°C, 250 rpm for 24 h. The completed reaction is centrifuged at 6000 rpm for 10 min. The

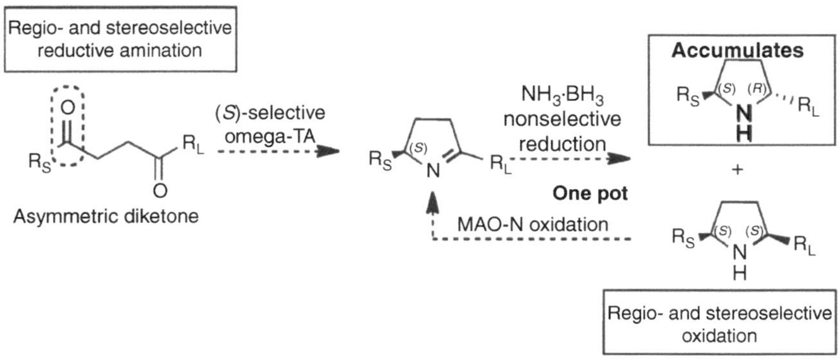

FIGURE 7.14
Chemoenzymatic synthesis of 2,5-disubstituted pyrrolidines involving ω-TA and monoamine oxidase [85].

supernatant is adjusted to pH 2 and extracted with EtOAc (150 mL) to remove any unreacted diketone **(1a–g)**. Next, the solution is adjusted to pH 12 and extracted with EtOAc (3 × 150 mL). The solvent is removed *in vacuo*, dissolved in Et_2O and washed with basified water (pH 12) to remove residual DMSO. The organic layer is dried over $MgSO_4$, filtered and the solvent is again evaporated. The resulting pyrrolines (**(2*S*, 5*R*)-4a–g**) are filtered through a short pad of silica and eluted in cyclohexane/EtOAc (90/10). GC-FID analysis is performed on Agilent 6850 with a Varian CP CHIRASIL-DEX CB 25 m × 0.25 mm DF = 0.25 column. 2,5- disubstituted pyrrolidine (**(2*S*, 5*R*)-4a**) is isolated as a clear oil (82% yield, >99% ee).

5 AMINE DEHYDROGENASES

5.1 Introduction

Amine dehydrogenases (AmDHs) were recently developed through protein engineering using existing AADH as scaffolds, with the first example involving the leucine AmDH (L-AmDH) appearing in 2012 [92]. The AmDH converts the analogous ketone (as opposed to the α-keto acid, the substrate of the AADH) to its respective amine (Figure 7.15). Reductive amination of a ketone by AmDHs offer an advantage over transamination by avoiding equilibrium limitations and requiring a sacrificial amine donor [47b]. On the other hand, cofactor regeneration is critical to achieve favorable equilibrium with AmDHs.

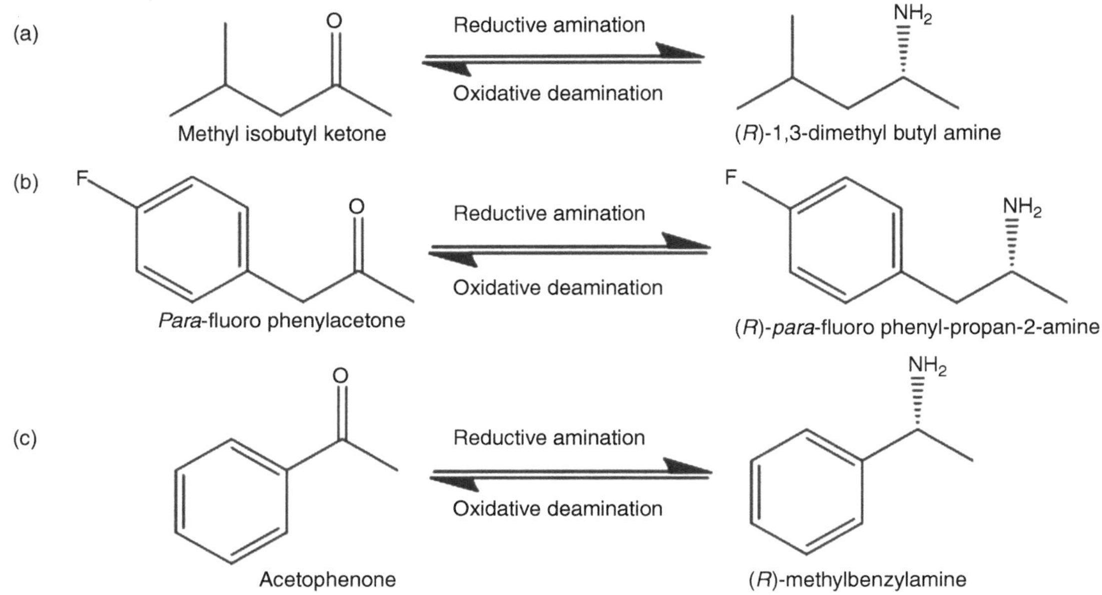

FIGURE 7.15
Reaction scheme for a typical substrate of leucine AmDH (L-AmDH) (a), phenylalanine AmDH (F-AmDH) (b), and chimeric AmDH (cFL1-AmDH) (c).

L-AmDH [92] was the first AmDH developed that was quickly followed by the F-AmDH (phenylalanine AmDH) [93] and cFL1-AmDH (chimeric AmDH) [94]. The same corresponding mutations to two residues in the active site were critical to create an AmDH. In addition, biphasic reaction systems have been incorporated with AmDHs to improve the solubility of substrates (up to 200 mM) and the substrate range has been expanded [95]. The AmDH family has been further extended to cover a wide range of substrate selectivity with pharmaceutical intermediates as targets.

5.2 Examples of Amine Dehydrogenases

Example 8

Preparation of (*R*)-para-fluorophenylisopropylamine from p-fluorophenylacetone (pFPA) via Phenylalanine DH-derived AmDH (F-AmDH)

Phenylalanine AmDH (F-AmDH) catalyzes the reaction (*R*)-*para*-fluorophenylisopropylamine from *p*-fluorophenyl-acetone (*p*FPA) via phenylalanine DH-derived AmDH (F-AmDH) (Figure 7.16). Although (*R*)-*para*-fluorophenylisopropylamine is a model target amine, its unsubstituted cousin (*R*)-phenyl-isopropylamine, obtained from phenylacetone, is a valuable precursor for pharmaceuticals; however, both substrate and product are restricted compounds in the United States: (*R*)-phenyl-isopropylamine or (*R*)-amphetamine is a Schedule-3 controlled substance.

Large-scale reactions involving the AmDH are paired with a cofactor regeneration enzyme, such as GDH or *C. boidinii* *cb*FDH. The reaction is performed in NH_4Cl/OH buffer (150 mL, 225 mM pH 9.6). *p*FPA (20 mM), NAD^+ (2 mM), F-AmDH (17.5 mg, prepared as described in Ref. [93]) and *Bacillus subtilis* GDH (200 U) are added to the reaction. The mixture is reacted at 25°C for 48 h. After completion, the reaction is adjusted to pH 13 with 5 mL of 10 N NaOH. The amine is extracted with methyl *tert*-butyl ether. The sample is dried with anhydrous $MgSO_4$. (*R*)- *p*FPA was isolated (73.9% yield, >99.8% ee).

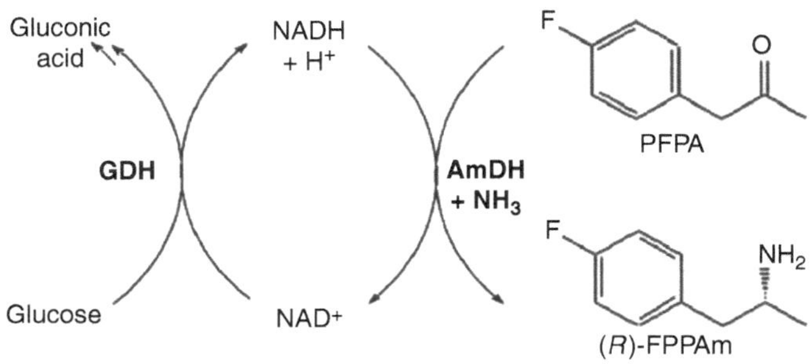

FIGURE 7.16
(*R*)-*p*-F-phenylisopropylamine from *p*-F-phenylacetone [93].

Example 9

Preparation of (*R*)-methylbenzylamine from acetophenone via chimeric DH-derived AmDH (cFL1-AmDH)

The chimeric AmDH (cFL1-AmDH) was created by combining the substrate binding domain of the F-AmDH (residues 1 through 151 of the *Bacillus badius* PheDH) and the cofactor binding domain of the L-AmDH (residues 141 through 368 of the *Bacillus stearothermophilus* [94]. The model reaction of the cFL1-AmDH is acetophenone to methylbenzylamine, a reaction that is not catalyzed by either of its parent enzymes. Surprisingly, cFL1-AmDH has a high temperature optimum of 70°C and a melting temperature of 72°C.

Large-scale conversion of 20 mM acetophenone to methylbenzylamine was achieved by pairing 2 U of cFL1-AmDH (prepared as in Ref. [94]) with a thermostable *B. subtilis* GDH [96] for cofactor regeneration. The reaction was performed at 40°C for 24 h in a 5 M NH_4OH/Cl buffer pH 9.6 providing 91% conversion of acetophenone to methylbenzylamine.

Example 10

Development of a biphasic organic solvent system to allow for conversion of hydrophobic substrates

Many pharmaceutical targets of AmDH have little or no solubility in aqueous media, including the model substrate of the F-AmDH, phenylacetone. Reactions in organic solvents have numerous advantages beyond increased solubility of substrates, including fewer hydrolytic side reactions, easy product removal, and the potential for overcoming unfavorable equilibria [97]. Water-miscible organic solvents led to AmDH deactivation, which is commonly observed for dehydrogenases [98]. This led to the development of a biphasic organic solvent system in which AmDHs, along with the cofactor and cofactor regeneration enzyme such as FDH, remain in the aqueous phase (Figure 7.17). A large amount of hydrophobic substrate can be introduced into the system by incorporating organic solvents and substrate and product inhibition is avoided. This system allows for the conversion of previously unattainable substrates, such as the adamantyl methyl ketone and 3-methyl-1-phenyl-2-butanone [95].

To the reaction, 150 mM of substrate was added. A phase ratio of 1:4 organic solvent heptane to aqueous is used for all reactions. 0.1 mg and 0.25 mg of F-AmDH (prepared as described in Ref. [93]) was added for the *p*FPA and 3-methyl-1-phenyl-2-butanone reactions, respectively. The aqueous reaction medium was 5 M NH_4HCO_2/OH pH 9.6 and the reaction was performed at 35°C. 0.25 mg of cFL1-AmDH was added to the reactions involving acetophenone and 1-adamantyl methyl ketone. The aqueous reaction medium was 5 M NH_4Cl/OH pH 9.6 and the reaction was performed at 50°C. Twice as much regeneration enzyme activity (*Cb*-FDH, *Bs*-GDH) was added to ensure that the reversible reaction would not occur and that enough co-substrate NADH was always present during the reaction. 1 mM NADH/NAD^+ was added to the reactions. All reactions were stopped by adding methanol and analyzed by HPLC to detect amine peaks, except the adamantyl methyl ketone was analyzed by GC due to absorbance issues. Full conversion was not attempted; however, 12% conversion was achieved after 24 h and (*R*)-amine was made (99.67% ee in organic layer, 99.99% ee in aqueous layer).

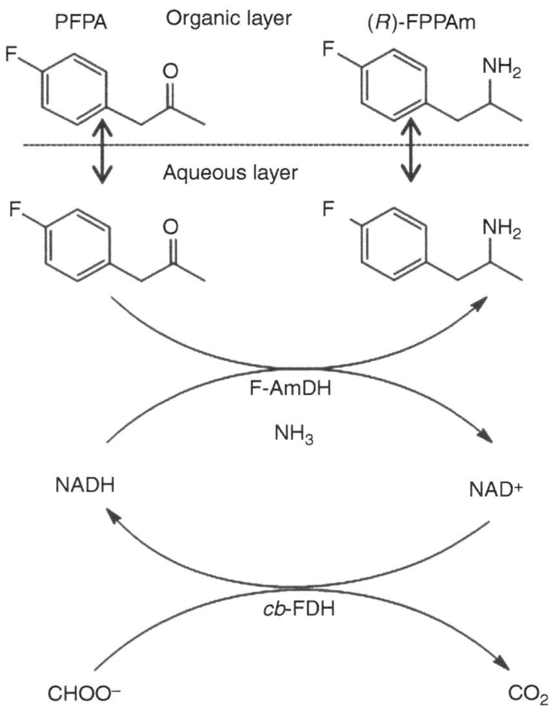

FIGURE 7.17
Biphasic system reaction involving F-AmDH and *p*FPA.

Acknowledgment

Samantha Au and Andreas Bommarius acknowledge support from the NSF I/UCRC Center for Pharmaceutical Development (grant 0969003).

References

[1] C.C.C.R.d. Carvalho, Enzymatic and whole cell catalysis: finding new strategies for old processes, Biotechnol. Adv. 29 (2011) 75–83.

[2] J.C. David, P.J. Dunn, J.D. Hayler, G.R. Humphrey, Johnnie L. Leazer Jr., R.J. Linderman, K. Lorenz, J. Manley, B.A. Pearlman, A. Wells, A. Zaksh, T.Y. Zhang, Key green chemistry research areas – a perspective from pharmaceutical manufacturers, Green Chem. 9 (2007).

[3] J-G. Concepcion, Q.B. Broxterman, B-S Yang, D. Ende, J. Baird, C. Bertsch, R.E. Hannah Phil Dell'Orco, H. Noorman, S. Yee, R. Reintjens, A. Wells, V. Massonneau, J. Manley, Key green engineering research areas for sustainable manufacturing: a perspective from pharmaceutical and fine chemicals manufacturers, Org. Process Res. Dev. (15) (2011) 900–911.

[4] (a) V. Gotor-Fernandez, V. Gotor, Biocatalytic routes to chiral amines and amino acids, Curr. Opin. Drug Disc. 12 (6) (2009) 784–797. (b) M. Breuer, K. Ditrich, T. Habicher, B. Hauer, M. Kesseler, R. Sturmer, T. Zelinski, Industrial methods for the production of optically active intermediates, Angew. Chem. Int. Ed. 43 (7) (2004) 788–824.

[5] W. Leuchtenberger, K. Huthmacher, K. Drauz, Biotechnological production of amino acids and derivatives: current status and prospects, Appl. Microbiol. Biotechnol. 69 (1) (2005) 1–8.

[6] S. Mathew, H. Yun, Omega-transaminases for the production of optically pure amines and unnatural amino acids, Acs Catal. 2 (6) (2012) 993–1001.

[7] E.S. Park, J.S. Shin, Deracemization of amino acids by coupling transaminases of opposite stereoselectivity, Adv. Synth. Catal. 356 (17) (2014) 3505–3509.

[8] P. Tufvesson, J. Lima-Ramos, J.S. Jensen, N. Al-Haque, W. Neto, J.M. Woodley, Process considerations for the asymmetric synthesis of chiral amines using transaminases, Biotechnol. Bioeng. 108 (7) (2011) 1479–1493.

[9] F. Weber, G. Sedelmeier, 200 High-selling drugs, Nachr. Chem. 61 (5) (2013) 528–529.

[10] A.M. Thayer, Centering on chirality, Chem. Eng. News 85 (32) (2007) 11–19.

[11] (a) F. Leipold, S. Hussain, D. Ghislieri, N.J. Turner, Asymmetric reduction of cyclic imines catalyzed by a whole-cell biocatalyst containing an (*S*)-imine reductase, Chemcatchem 5 (12) (2013) 3505–3508. (b) M. Gand, H. Muller, R. Wardenga, M. Hohne, Characterization of three novel enzymes with imine reductase activity, J. Mol. Catal. B Enzym. 110 (2014) 126–132. (c) P.N. Scheller, S. Fademrecht, S. Hofelzer, J. Pleiss, F. Leipold, N.J. Turner, B.M. Nestl, B. Hauer, Enzyme toolbox: novel enantiocomplementary imine reductases, Chembiochem. 15 (15) (2014) 2201–2204.

[12] (a) G. Krix, A.S. Bommarius, K. Drauz, M. Kottenhahn, M. Schwarm, M.R. Kula, Enzymatic reduction of alpha-keto acids leading to L-amino acids, D- or L-hydroxy acids, J. Biotechnol. 53 (1) (1997) 29–39. (b) N.M.W. Brunhuber, J.S. Blanchard, The biochemistry and enzymology of amino-acid dehydrogenases, Crit. Rev. Biochem. Mol. 29 (6) (1994) 415–467.

[13] (a) W.D. Liu, Z. Li, C.H. Huang, R.T. Guo, L.M. Zhao, D.L. Zhang, X. Chen, Q.Q. Wu, D.M. Zhu, Structural and mutational studies on the unusual substrate specificity of *meso*-diaminopimelate dehydrogenase from symbiobacterium thermophilum, Chembiochem. 15 (2) (2014) 217–222. (b) K. Vedha-Peters, M. Gunawardana, J.D. Rozzell, S.J. Novick, Creation of a broad-range and highly stereoselective D-amino acid dehydrogenase for the one-step synthesis of D-amino acids, J. Am. Chem. Soc. 128 (33) (2006) 10923–10929.

[14] P.J. Baker, K.L. Britton, P.C. Engel, G.W. Farrants, K.S. Lilley, D.W. Rice, T.J. Stillman, Subunit assembly and active-site location in the structure of glutamate-dehydrogenase, Protein Struct. Funct. Genet. 12 (1) (1992) 75–86.

[15] P.J. Baker, A.P. Turnbull, S.E. Sedelnikova, T.J. Stillman, D.W. Rice, A role for quaternary structure in the substrate-specificity of leucine dehydrogenase, Structure 3 (7) (1995) 693–705.

[16] Y. Zhao, T. Wakamatsu, K. Doi, H. Sakuraba, T. Ohshima, A psychrophilic leucine dehydrogenase from *Sporosarcina psychrophila:* purification, characterization, gene sequencing and crystal structure analysis, J. Mol. Catal. B Enzym. 83 (2012) 65–72.

[17] (a) A. Pasquo, K.L. Britton, P.J. Baker, G. Brearley, R.J. Hinton, A.J.G. Moir, T.J. Stillman, D.W. Rice, Crystallization of NAD(+)-dependent phenylalanine dehydrogenase from Nocardia sp239, Acta Crystallogr. D 54 (1998) 269–272. (b) J.L. Vanhooke, J.B. Thoden, N.M.W. Brunhuber, J.S. Blanchard, H.M. Holden, Phenylalanine dehydrogenase from *Rhodococcus* sp. M4: high-resolution X-ray analyses of inhibitory ternary complexes reveal key features in the oxidative deamination mechanism, Biochem. 38 (8) (1999) 2326–2339.

[18] P.J. Baker, Y. Sawa, H. Shibata, S.E. Sedelnikova, D.W. Rice, Analysis of the structure and substrate binding of *Phormidium lapideum* alanine dehydrogenase, Nat. Struct. Biol. 5 (7) (1998) 561–567.

[19] T.S. Kumarevel, P. Karthe, S. Kuramitsu, S. Yokoyama, Crystal structure of alanine dehydrogenase from themus thermophilus. Only published as: http://www.metalife.com/PDB/2EEZ_35769.

[20] D. Agren, M. Stehr, C.L. Berthold, S. Kapoor, W. Oehlmann, M. Singh, G. Schneider, Three-dimensional structures of apo- and holo-L-alanine dehydrogenase from *Mycobacterium tuberculosis* reveal conformational changes upon coenzyme binding, J. Mol. Biol. 377 (4) (2008) 1161–1173.

[21] S. Knapp, W.M. deVos, D. Rice, R. Ladenstein, Crystal structure of glutamate dehydrogenase from the hyperthermophilic eubacterium *Thermotoga maritima* at 3.0 angstrom resolution, J. Mol. Biol. 267 (4) (1997) 916–932.

[22] (a) T.J. Stillman, P.J. Baker, K.L. Britton, D.W. Rice, Conformational flexibility in glutamate-dehydrogenase – role of water in substrate recognition and catalysis, J. Mol. Biol. 234 (4)

(1993) 1131–1139. (b) T.J. Stillman, A.M.B. Migueis, X.G. Wang, P.J. Baker, K.L. Britton, P.C. Engel, D.W. Rice, Insights into the mechanism of domain closure and substrate specificity of glutamate dehydrogenase from *Clostridium symbiosum*, J. Mol. Biol. 285 (2) (1999) 875–885.

[23] K.L. Britton, K.S.P. Yip, S.E. Sedelnikova, T.J. Stillman, M.W.W. Adams, K. Ma, D.L. Maeder, F.T. Robb, N. Tolliday, C. Vetriani, D.W. Rice, P.J. Baker, Structure determination of the glutamate dehydrogenase from the hyperthermophile *Thermococcus litoralis* and its comparison with that from *Pyrococcus furiosus*, J. Mol. Biol. 293 (5) (1999) 1121–1132.

[24] M. Nakasako, T. Fujisawa, S. Adachi, T. Kudo, S. Higuchi, Large-scale domain movements and hydration structure changes in the active-site cleft of unligated glutamate dehydrogenase from *Thermococcus profundus* studied by cryogenic X-ray crystal structure analysis and small-angle X-ray scattering, Biochem. 40 (10) (2001) 3069–3079.

[25] K.S.P. Yip, T.J. Stillman, K.L. Britton, P.J. Artymiuk, P.J. Baker, S.E. Sedelnikova, P.C. Engel, A. Pasquo, R. Chiaraluce, V. Consalvi, R. Scandurra, D.W. Rice, The structure of pyrococcus-furiosus glutamate-dehydrogenase reveals a key role for ion-pair networks in maintaining enzyme stability at extreme temperatures, Structure 3 (11) (1995) 1147–1158.

[26] M.W. Bhuiya, H. Sakuraba, T. Ohshima, T. Imagawa, N. Katunuma, H. Tsuge, The first crystal structure of hyperthermostable NAD-dependent glutamate dehydrogenase from *Pyrobaculum islandicum*, J. Mol. Biol. 345 (2) (2005) 325–337.

[27] (a) K. Zocher, K. Fritz-Wolf, S. Kehr, M. Fischer, S. Rahlfs, K. Becker, Biochemical and structural characterization of *Plasmodium falciparum* glutamate dehydrogenase 2, Mol. Biochem. Parasit. 183 (1) (2012) 52–62. (b) C. Werner, M.T. Stubbs, R.L. Krauth-Siegel, G. Klebe, The crystal structure of plasmodium falciparum glutamate dehydrogenase, a putative target for novel antimalarial drugs, J. Mol. Biol. 349 (3) (2005) 597–607.

[28] T. Oliveira, S. Panjikar, J.B. Carrigan, M. Hamza, M.A. Sharkey, P.C. Engel, A.R. Khan, Crystal structure of NAD(+)-dependent *Peptoniphilus asaccharolyticus* glutamate dehydrogenase reveals determinants of cofactor specificity, J. Struct. Biol. 177 (2) (2012) 543–552.

[29] T. Tomita, T. Kuzuyama, M. Nishiyama, Structural basis for leucine-induced allosteric activation of glutamate dehydrogenase, J. Biol. Chem. 286 (43) (2011) 37406–37413.

[30] K. Gunka, J.A. Newman, F.M. Commichau, C. Herzberg, C. Rodrigues, L. Hewitt, R.J. Lewis, J. Stulke, Functional dissection of a trigger enzyme: mutations of the *Bacillus subtilis* glutamate dehydrogenase RocG that affect differentially its catalytic activity and regulatory properties, J. Mol. Biol. 400 (4) (2010) 815–827.

[31] (a) P.E. Peterson, J. Pierce, T.J. Smith, Crystallization and characterization of bovine liver glutamate dehydrogenase, J. Struct. Biol. 120 (1) (1997) 73–77. (b) T.J. Smith, P.E. Peterson, T. Schmidt, J. Fang, C.A. Stanley, Structures of bovine glutamate dehydrogenase complexes elucidate the mechanism of purine regulation, J. Mol. Biol. 307 (2) (2001) 707–720. (c) S. Banerjee, T. Schmidt, J. Fang, C.A. Stanley, T.J. Smith, Structural studies on ADP activation of mammalian glutamate dehydrogenase and the evolution of regulation, Biochem. 42 (12) (2003) 3446–3456.

[32] (a) G. Scapin, S.G. Reddy, J.S. Blanchard, Three-dimensional structure of *meso*-diaminopimelic acid dehydrogenase from *Corynebacterium glutamicum*, Biochem. 35 (42) (1996) 13540–13551. (b) G. Scapin, M. Cirilli, S.G. Reddy, Y. Gao, J.C. Vederas, J.S. Blanchard, Substrate and inhibitor binding sites in *Corynebacterium glutamicum* diaminopimelate dehydrogenase, Biochem. 37 (10) (1998) 3278–3285. (c) M. Cirilli, G. Scapin, A. Sutherland, J.C. Vederas, J.S. Blanchard, The three-dimensional structure of the ternary complex of *Corynebacterium glutamicum* diaminopimelate dehydrogenase-NADPH-L-2-amino-6-methylene-pimelate, Protein Sci. 9 (10) (2000) 2034–2037.

[33] X.Z. Gao, X. Chen, W.D. Liu, J.H. Feng, Q.Q. Wu, L. Hua, D.M. Zhu, A novel *meso*-diaminopimelate dehydrogenase from *Symbiobacterium thermophilum*: overexpression, characterization, and potential for D-amino acid synthesis, Appl. Environ. Microb. 78 (24) (2012) 8595–8600.

[34] X.Z. Gao, F. Huang, J.H. Feng, X. Chen, H.L. Zhang, Z.X. Wang, Q.Q. Wu, D.M. Zhu, Engineering the *meso*-diaminopimelate dehydrogenase from *Symbiobacterium thermophilum* by site saturation mutagenesis for D-phenylalanine synthesis, Appl. Environ. Microb. 79 (16) (2013) 5078–5081.

[35] (a) N.M.W. Brunhuber, J.B. Thoden, J.S. Blanchard, J.L. Vanhooke, Rhodococcus L-phenylalanine dehydrogenase: kinetics, mechanism, and structural basis for catalytic specificity, Biochem. 39 (31) (2000) 9174–9187. (b) K.L. Britton, P.J. Baker, P.C. Engel, D.W. Rice, T.J. Stillman, Evolution of substrate diversity in the superfamily of amino-acid dehydrogenases – prospects for rational chiral synthesis, J. Mol. Biol. 234 (4) (1993) 938–945. (c) J.E. Rife, W.W. Cleland, Determination of the chemical mechanism of glutamate-dehydrogenase from Ph studies, Biochem. 19 (11) (1980) 2328–2333. (d) T. Sekimoto, T. Matsuyama, T. Fukui, K. Tanizawa, Evidence for Lysine-80 as general base catalyst of Leucine dehydrogenase, J. Biol. Chem. 268 (36) (1993) 27039–27045.

[36] A.S. Bommarius, M. Schwarm, K. Drauz, Biocatalysis to amino acid-based chiral pharmaceuticals – examples and perspectives, J. Mol. Catal. B Enzym. 5 (1-4) (1998) 1–11.

[37] (a) W. Liu, H. Ma, J. Luo, W. Shen, X. Xu, S. Li, Y. Hu, H. Huang, Efficient synthesis of L-*tert*-leucine through reductive amination using leucine dehydrogenase and formate dehydrogenase coexpressed in recombinant *E. coli*, Biochem. Eng. J. 91 (2014) 204–209. (b) W.M. Liu, J.X. Luo, X.J. Zhuang, W.H. Shen, Y. Zhang, S. Li, Y. Hu, H. Huang, Efficient preparation of enantiopure L-*tert*-leucine through immobilized penicillin G acylase catalyzed kinetic resolution in aqueous medium, Biochem. Eng. J. 83 (2014) 116–120.

[38] J. Iqbal, V.H. Dahanukar, S. Oruganti, B. Kandagatla, S.K.U. Kunhimon, R.R. Sagyam, B. Ireni, V.V.S.M. Kondepudi, M.M. Muttavarapu, R. Bucchikonda, Process for preparation of intermediates of telaprevir, WO 2014,045,263 A2 20140327, 2014.

[39] K. Bjornson; E. Canales; J.J. Cottell; K.K. Karki; A.A. Katana; D. Kato; T. Kobayashi; J.O. Link; R. Martinez; B.W. Phillips, Preparation of *N*-(3-alkyl- and 3-carbocyclyl)prolyl-1-aminocyclopropanecarboxylic acid peptides as inhibitors of hepatitis C virus WO 2014008285 A1 20140109. 2014.

[40] S. Simhadri, Y. Mohammad, V.S.K. Indukuri, S.R. Gorantla, Preparation of atazanavir bisulfate, WO 2014,030,173 A2 20140227, 2014.

[41] D.R.P. Guay, Drug forecase – the peptide deformylase inhibitors as antibacterial agents, Ther. Clin. Risk Manag. 3 (4) (2007) 513–525.

[42] D.E. Levy, F. Lapierre, W.S. Liang, W.Q. Ye, C.W. Lange, X.Y. Li, D. Grobelny, M. Casabonne, D. Tyrrell, K. Holme, A. Nadzan, R.E. Galardy, Matrix metalloproteinase inhibitors: a structure-activity study, J. Med. Chem. 41 (2) (1998) 199–223.

[43] (a) N.J. Turner, J.R. Winterman, R. Mccague, J.S. Parratt, S.J.C. Taylor, Synthesis of homochiral L-(S)-*tert*-Leucine via a lipase-catalyzed dynamic resolution process, Tetrahedron Lett. 36 (7) (1995) 1113–1116. (b) J.Z. Jin, D.L. Chang, J. Zhang, Discovery and application of new bacterial strains for asymmetric synthesis of L-*tert*-butyl leucine in high enantioselectivity, Appl. Biochem. Biotech. 164 (3) (2011) 376–385. (c) J. Li, J. Pan, J. Zhang, J.H. Xu, Stereoselective synthesis of L-*tert*-leucine by a newly cloned leucine dehydrogenase from *Exiguobacterium sibiricum*, J. Mol. Catal. B Enzym. 105 (2014) 11–17. (d) A.S. Bommarius, M. Schwarm, K. Stingl, M. Kottenhahn, K. Huthmacher, K. Drauz, Synthesis and use of enantiomerically pure *tert*-leucine, Tetrahedron Asymmetry 6 (12) (1995) 2851–2888.

[44] U. Kragl, D. VasicRacki, C. Wandrey, Continuous production of L-*tert*-leucine in series of two enzyme membrane reactors – modelling and computer simulation, Bioprocess Eng. 14 (6) (1996) 291–297.

[45] (a) L.L. Nielsen, Incretin mimetics and DPP-IV inhibitors for the treatment of type 2 diabetes, Drug Discov. Today 10 (10) (2005) 703–710. (b) D.J. Augeri, J.A. Robl, D.A. Betebenner, D.R. Magnin, A. Khanna, J.G. Robertson, A.Y. Wang, L.M. Simpkins, P. Taunk, Q. Huang, S.P. Han, B. Abboa-Offei, M. Cap, L. Xin, L. Tao, E. Tozzo, G.E. Welzel, D.M. Egan, J. Marcinkeviciene, S.Y. Chang, S.A. Biller, M.S. Kirby, R.A. Parker, L.G. Hamann, Discovery and preclinical profile of saxagliptin (BMS-477118): A highly potent, long-acting, orally active dipeptidyl peptidase IV inhibitor for the treatment of type 2 diabetes, J. Med. Chem. 48 (15) (2005) 5025–5037.

[46] R.L. Hanson, S.L. Goldberg, D.B. Brzozowski, T.P. Tully, D. Cazzulino, W.L. Parker, O.K. Lyngberg, T.C. Vu, M.K. Wong, R.N. Patel, Preparation of an amino acid intermediate for the

dipeptidyl peptidase IV inhibitor, saxagliptin, using a modified phenylalanine dehydrogenase, Adv. Synth. Catal. 349 (8-9) (2007) 1369–1378.

[47] (a) D. Koszeleivski, I. Lavandera, D. Clay, G.M. Guebitz, D. Rozzell, W. Kroutil, Formal asymmetric biocatalytic reductive amination, Angew. Chem. Int. Edit 47 (48) (2008) 9337–9340. (b) D. Koszelewski, K. Tauber, K. Faber, W. Kroutil, Omega-transaminases for the synthesis of non-racemic alpha-chiral primary amines, Trends Biotechnol. 28 (6) (2010) 324–332. (c) S. Schatzle, F. Steffen-Munsberg, A. Thontowi, M. Hohne, K. Robins, U.T. Bornscheuer, Enzymatic asymmetric synthesis of enantiomerically pure aliphatic, aromatic and arylaliphatic amines with (*R*)-selective amine transaminases, Adv. Synth. Catal. 353 (13) (2011) 2439–2445. (d) M. Hohn, U.T. Bornscheuer, Biocatalytic routes to optically active amines, Chemcatchem. 1 (1) (2009) 42–51. (e) D. Koszelewski, K. Tauber, K. Faber, W. Kroutil, Omega-transaminases for the synthesis of non-racemic alpha-chiral primary amines, Trends Biotechnol. 28 (6) (2010) 324–332. (f) W. Kroutil, E.M. Fischereder, C.S. Fuchs, H. Lechner, F.G. Mutti, D. Pressnitz, A. Rajagopalan, J.H. Sattler, R.C. Simon, E. Siirola, Asymmetric preparation of prim-, sec-, and *tert*-amines employing selected biocatalysts, Org. Process Res. Dev. 17 (5) (2013) 751–759. (g) C.E. Paul, M. Rodriguez-Mata, E. Busto, I. Lavandera, V. Gotor-Fernandez, V. Gotor, S. Garcia-Cerrada, J. Mendiola, O. de Frutos, I. Collado, Transaminases applied to the synthesis of high added-value enantiopure amines, Org. Process Res. Dev. 18 (6) (2014) 788–792.

[48] C.K. Savile, J.M. Janey, E.C. Mundorff, J.C. Moore, S. Tam, W.R. Jarvis, J.C. Colbeck, A. Krebber, F.J. Fleitz, J. Brands, P.N. Devine, G.W. Huisman, G.J. Hughes, Biocatalytic asymmetric synthesis of chiral amines from ketones applied to Sitagliptin manufacture, Science 329 (5989) (2010) 305–309.

[49] M.S. Malik, E.S. Park, J.S. Shin, Features and technical applications of omega-transaminases, Appl. Microbiol. Biotechnol. 94 (5) (2012) 1163–1171.

[50] K.T. Dominik Koszelewski, Kurt Faber, Wolfgang Kroutil, W-transaminases for the synthesis of non-racemic a-chiral primary amines, Trends Biotechnol. 28 (2010) 324–332.

[51] F.G. Mutti, C.S. Fuchs, D. Pressnitz, J.H. Sattler, W. Kroutil, Stereoselectivity of four (*R*)-selective transaminases for the asymmetric amination of ketones, Adv. Synth. Catal. 353 (17) (2011) 3227–3233.

[52] A.P. Green, N.J. Turner, E. O'Reilly, Chiral amine synthesis using omega-transaminases: an amine donor that displaces equilibria and enables high-throughput screening, Angew. Chem. Int. Edit 53 (40) (2014) 10714–10717.

[53] K. Deepankumar, M. Shon, S.P. Nadarajan, G. Shin, S. Mathew, N. Ayyadurai, B.G. Kim, S.H. Choi, S.H. Lee, H. Yun, Enhancing thermostability and organic solvent tolerance of w-transaminase through global incorporation of fluorotyrosine, Adv. Synth. Catal. 356 (5) (2014) 993–998.

[54] (a) E. Busto, R.C. Simon, B. Grischek, V. Gotor-Fernandez, W. Kroutil, Cutting short the asymmetric synthesis of the ramotroban precursor by employing w-transaminases, Adv. Synth. Catal. (2014). (b) R.C. Simon, J.H. Sattler, J.E. Farnberger, C.S. Fuchs, N. Richter, F. Zepeck, W. Kroutil, Enzymatic asymmetric synthesis of the silodosin amine intermediate, Tetrahedron Asymmetry 25 (2014) 284–288.

[55] D. Pressnitz, C.S. Fuch, J.H. Sattler, T. Knaus, P. Macheroux, F.G. Mutti, W. Kroutil, Asymmetric amination of tetralone and chromanone derivatives employing w-transaminase, ACS Catal. 3 (2013) 555–559.

[56] D. Koszelewski, D. Pressnitz, D. Clay, W. Kroutil, Deracemization of mexiletine biocatalyzed by omega-transaminases, Org. Lett. 11 (21) (2009) 4810–4812.

[57] M. Thomsen, L. Skalden, G.J. Palm, M. Hohne, U.T. Bornscheuer, W. Hinrichs, Crystallization and preliminary X-ray diffraction studies of the (*R*)-selective amine transaminase from *Aspergillus fumigatus*, Acta Crystallogr. F. 69 (2013) 1415–1417.

[58] A. Lyskowski, C. Gruber, G. Steinkellner, M. Schurmann, H. Schwab, K. Gruber, K. Steiner, Crystal structure of an (*R*)-selective omega-transaminase from *Aspergillus terreus*, PloS ONE 9 (1) (2014).

[59] (a) C. Sayer, M.N. Isupov, A. Westlake, J.A. Littlechild, Structural studies of *Pseudomonas* and Chromobacterium-aminotransferases provide insights into their differing substrate specificity, Acta Crystallogr. D 69 (2013) 564–576. (b) M.S. Humble, K.E. Cassimjee, M. Hakansson, Y.R. Kimbung, B. Walse, V. Abedi, H.J. Federsel, P. Berglund, D.T. Logan, Crystal structures of the *Chromobacterium violaceum*? – Transaminase reveal major structural rearrangements upon binding of coenzyme PLP, Febs J. 279 (5) (2012) 779–792.

[60] C. Rausch, A. Lerchner, A. Schiefner, A. Skerra, Crystal structure of the omega-aminotransferase from *Paracoccus denitrificans* and its phylogenetic relationship with other class III aminotransferases that have biotechnological potential, Proteins 81 (5) (2013) 774–787.

[61] T.H. Jang, B. Kim, O.K. Park, J.Y. Bae, B.G. Kim, H. Yun, H.H. Park, Crystallization and preliminary X-ray crystallographic studies of omega-transaminase from Vibrio fluvialis JS17, Acta Crystallogr. F 66 (2010) 923–925.

[62] K.E. Cassimjee, M.S. Humble, V. Miceli, C.G. Colomina, P. Berglund, Active site quantification of an omega-transaminase by performing a half transamination reaction, ACS Catal. 1 (9) (2011) 1051–1055.

[63] F. Steffen-Munsberg, C. Vickers, A. Thontowi, S. Schatzle, T. Meinhardt, M.S. Humble, H. Land, P. Berglund, U.T. Bornscheuer, M. Hohne, Revealing the structural basis of promiscuous amine transaminase activity, Chemcatchem. 5 (1) (2013) 154–157.

[64] (a) J.S. Shin, B.G. Kim, Exploring the active site of amine: pyruvate aminotransferase on the basis of the substrate structure-reactivity relationship: how the enzyme controls substrate specificity and stereo selectivity, J. Org. Chem. 67 (9) (2002) 2848–2853. (b) E.S. Park, M. Kim, J.S. Shin, Molecular determinants for substrate selectivity of omega-transaminases, Appl. Microbiol. Biotechnol. 93 (6) (2012) 2425–2435.

[65] (a) A. Nobili, S-M Fabian, H. Kohls, I. Trentin, C. Schulzke, M. Hohne, U.T. Bornscheuer, Engineering the active site of the amine transaminase from *Vibrio fluvialis* for the asymmetric synthesis of aryl-alkyl amines and amino alcohols, ChemCatChem. 7 (5) (2015) 757–760. (b) E.S. Park, S.R. Park, S.W. Han, J.Y. Dong, J.S. Shin, Structural determinants for the non-canonical substrate specificity of the omega-transaminase from *Paracoccus denitrificans*, Adv. Synth. Catal. 356 (1) (2014) 212–220.

[66] (a) M.F. Flament, R.M. Lane, R. Zhu, Z. Ying, Predictors of an acute antidepressant response to fluoxetine and sertraline, Int. Clin. Psychopharm. 14 (5) (1999) 259–275. (b) J.R.T. Davidson, B.O. Rothbaum, B.A. van der Kolk, C.R. Sikes, G.M. Farfel, Multicenter, double-blind comparison of sertraline and placebo in the treatment of posttraumatic stress disorder, Arch. Gen. Psychiat. 58 (5) (2001) 485–492.

[67] Z.X. Han, S.G. Koenig, H. Zhao, X.P. Su, S.P. Singh, R.P. Bakale, Development of a large-scale stereoselective process for (1R,4S)-4-(3,4-dichlorophenyl)-1 2,3,4-tetrahydronaphthalen-1-amine hydrochloride, Org. Process Res. Dev. 11 (4) (2007) 726–730.

[68] R.L. Watts, J. Jankovic, C. Waters, A. Rajput, B. Boroojerdi, J. Rao, Randomized, blind, controlled trial of transdermal rotigotine in early Parkinson disease, Neurology 68 (4) (2007) 272–276.

[69] G. Griebel, R. Misslin, M. Pawlowski, B.G. Lemaitre, G. Guillaumet, J. Bizotespiard, Anxiolytic-like effects of a selective 5-Ht1a agonist, S20244, and its enantiomers in mice, Neuroreport 3 (1) (1992) 84–86.

[70] S.S. Erdem, O. Karahan, I. Yildiz, K. Yelekci, A computational study on the amine-oxidation mechanism of monoamine oxidase: insight into the polar nucleophilic mechanism, Org. Biomol. Chem. 4 (4) (2006) 646–658.

[71] (a) M.M. Cavalluzzi, A. Catalano, C. Bruno, A. Lovece, A. Carocci, F. Corbo, C. Franchini, G. Lentini, V. Tortorella, Synthesis of (*R*)-, (*S*)-, and (*RS*)-hydroxymethylmexiletine, one of the major metabolites of mexiletine, Tetrahedron Asymmetry 18 (20) (2007) 2409–2417. (b) A. De Luca, F. Natuzzi, J.F. Desaphy, G. Loni, G. Lentini, C. Franchini, V. Tortorella, D.C. Camerino, Molecular determinants of mexiletine structure for potent and use-dependent block of skeletal muscle sodium channels, Mol. Pharmacol. 57 (2) (2000) 268–277. (c) A. De-Luca, F. Natuzzi, S. Pierno, H. Jockusch, A. Duranti, G. Lentini, C. Franchini, V. Tortorella, D.C.

Camerino, Use-dependent block of skeletal muscle sodium current by chiral derivatives of mexiletine and tocainide and antimyotonic activity on ADR mouse, Biophys. J. 72 (2) (1997) Thp72-Thp72. (d) P.E. Fenster, K.A. Comess, Pharmacology and clinical use of mexiletine, Pharmacotherapy 6 (1) (1986) 1–9.

[72] (a) A. Deluca, F. Natuzzi, G. Lentini, C. Franchini, V. Tortorella, D.C. Camerino, Stereoselective effects of mexiletine enantiomers on sodium currents and excitability characteristics of adult skeletal-muscle fibers, N-S Arch. Pharmacol. 352 (6) (1995) 653–661. (b) C. Franchini, A. Carocci, A. Catalano, M.M. Cavalluzzi, F. Corbo, G. Lentini, A. Scilimati, P. Tortorella, D.C. Camerino, A. De Luca, Optically active mexiletine analogues as stereoselective blockers of voltage-gated Na^+ channels, J. Med. Chem. 46 (24) (2003) 5238–5248.

[73] M.D. Truppo, N.J. Turner, J.D. Rozzell, Efficient kinetic resolution of racemic amines using a transaminase in combination with an amino acid oxidase, Chem. Commun. (16) (2009) 2127–2129.

[74] D. Koszelewski, I. Lavandera, D. Clay, D. Rozzell, W. Kroutil, Asymmetric synthesis of optically pure pharmacologically relevant amines employing omega-transaminases, Adv. Synth. Catal. 350 (17) (2008) 2761–2766.

[75] (a) R.B. Silverman, Radical ideas about monoamine-oxidase, Accounts Chem. Res. 28 (8) (1995) 335–342. (b) A.P.B. Vintem, N.T. Price, R.B. Silverman, R.R. Ramsay, Mutation of surface cysteine 374 to alanine in monoamine oxidase A alters substrate turnover and inactivation by cyclopropylamines, Bioorg. Med. Chem. 13 (10) (2005) 3487–3495.

[76] (a) M. Alexeeva, R. Carr, N.J. Turner, Directed evolution of enzymes: new biocatalysts for asymmetric synthesis, Org. Biomol. Chem. 1 (23) (2003) 4133–4137. (b) D. Ghislieri, A.P. Green, M. Pontini, S.C. Willies, I. Rowles, A. Frank, G. Grogan, N.J. Turner, Engineering an enantioselective amine oxidase for the synthesis of pharmaceutical building blocks and alkaloid natural products, J. Am. Chem. Soc. 135 (29) (2013) 10863–10869. (c) R. Carr, M. Alexeeva, A. Enright, T.S.C. Eve, M.J. Dawson, N.J. Turner, Directed evolution of an amine oxidase possessing both broad substrate specificity and high enantioselectivity, Angew. Chem. Int. Edit 42 (39) (2003) 4807–4810.

[77] R. Carr, M. Alexeeva, M.J. Dawson, V. Gotor-Fernandez, C.E. Humphrey, N.J. Turner, Directed evolution of an amine oxidase for the preparative deracemisation of cyclic secondary amines, Chembiochem. 6 (4) (2005) 637–639.

[78] C.J. Dunsmore, R. Carr, T. Fleming, N.J. Turner, A chemo-enzymatic route to enantiomerically pure cyclic tertiary amines, J. Am. Chem. Soc. 128 (7) (2006) 2224–2225.

[79] (a) D. Ghislieri, D. Houghton, A.P. Green, S.C. Willies, N.J. Turner, Monoamine oxidase (MAO-N) catalyzed deracemization of tetrahydro-beta-carbolines: substrate dependent switch in enantioselectivity, ACS Catal. 3 (12) (2013) 2869–2872. (b) I. Rowles, K.J. Malone, L.L. Etchells, S.C. Willies, N.J. Turner, Directed evolution of the enzyme monoamine oxidase (MAO-N): highly efficient chemo-enzymatic deracemisation of the alkaloid (+/-)-Crispine A, Chemcatchem. 4 (9) (2012) 1259–1261.

[80] (a) V. Kohler, K.R. Bailey, A. Znabet, J. Raftery, M. Helliwell, N.J. Turner, Enantioselective biocatalytic oxidative desymmetrization of substituted pyrrolidines, Angew. Chem. Int. Edit 49 (12) (2010) 2182–2184. (b) E. O'Reilly, C. Iglesias, D. Ghislieri, J. Hopwood, J.L. Galman, R.C. Lloyd, N.J. Turner, A regio- and stereoselective w-transaminase/monoamine oxidase cascade for the synthesis of chiral 2,5-disubstituted pyrrolidines, Angew. Chem. Int. Edit 53 (2014) 1–5.

[81] M. Alexeeva, A. Enright, M.J. Dawson, M. Mahmoudian, N.J. Turner, Deracemization of alpha-methylbenzylamine using an enzyme obtained by in vitro evolution, Angew. Chem. Int. Edit 41 (17) (2002) 3177-+.

[82] R.S. Heath, M. Pontini, B. Bechi, N. Turner, Development of an *R*-selective amine oxidase with broad substrate specificity and high enantioselectivity, ChemCatChem. 000 (2014) 1–8.

[83] K.E. Atkin, R. Reiss, V. Koehler, K.R. Bailey, S. Hart, J.P. Turkenburg, N.J. Turner, A.M. Brzozowski, G. Grogan, The structure of monoamine oxidase from *Aspergillus niger* provides a molecular context for improvements in activity obtained by directed evolution, J. Mol. Biol. 384 (5) (2008) 1218–1231.

[84] S.E.J. Rigby, R.M.G. Hynson, R.R. Ramsay, A.W. Munro, N.S. Scrutton, A stable tyrosyl radical in monoamine oxidase A, J. Biol. Chem. 280 (6) (2005) 4627–4631.

[85] E. O'Reilly, C. Iglesias, D. Ghislieri, J. Hopwood, J.L. Galman, R.C. Lloyd, N.J. Turner, A regio- and stereoselective w-transaminase/monoamine oxidase cascade for the synthesis of chiral 2,5-disubstituted pyrrolidines, Angew. Chem. Int. Edit 53 (9) (2014) 2447–2450.

[86] (a) J.B. Brenneman, R. Machauer, S.F. Martin, Enantioselective synthesis of (+)-anatoxin-a via enyne metathesis, Tetrahedron 60 (34) (2004) 7301–7314. (b) J.B. Brenneman, S.F. Martin, Application of intramolecular enyne metathesis to the synthesis of aza[4.2.1]bicyclics: Enantiospecific total synthesis of (+)-anatoxin-a, Org. Lett. 6 (8) (2004) 1329–1331.

[87] F. Xu, J.Y.L. Chung, J.C. Moore, Z.Q. Liu, N. Yoshikawa, R.S. Hoerrner, J. Lee, M. Royzen, E. Cleator, A.G. Gibson, R. Dunn, K.M. Maloney, M. Alam, A. Goodyear, J. Lynch, N. Yasuda, P.N. Devine, Asymmetric synthesis of *cis*-2,5-disubstituted pyrrolidine, the core scaffold of beta(3)-AR agonists, Org. Lett. 15 (6) (2013) 1342–1345.

[88] (a) S. Hanessian, M. Bayrakdarian, X.H. Luo, Total synthesis of A-315675: a potent inhibitor of influenza neuraminidase, J. Am. Chem. Soc. 124 (17) (2002) 4716–4721. (b) E. Lorthiois, W. Breitenstein, F. Cumin, C. Ehrhardt, E. Francotte, E. Jacoby, N. Ostermann, H. Sellner, T. Kosaka, R.L. Webb, D.F. Rigel, U. Hassiepen, P. Richert, T. Wagner, J. Maibaum, The discovery of novel potent trans-3,4-disubstituted pyrrolidine inhibitors of the human aspartic protease renin from in silico three-dimensional (3D) pharmacophore searches, J. Med. Chem. 56 (6) (2013) 2207–2217. (c) S.W. Wright, M.J. Ammirati, K.M. Andrews, A.M. Brodeur, D.E. Danley, S.D. Doran, J.S. Lillquist, L.D. McClure, R.K. McPherson, S.J. Orena, J.C. Parker, J. Polivkova, X. Qiu, W.C. Soeller, C.B. Soglia, J.L. Treadway, M.A. VanVolkenburg, H. Wang, D.C. Wilder, T.V. Olson, *cis*-2,5-dicyanopyrrolidine inhibitors of dipeptidyl peptidase IV: synthesis and in vitro, in vivo, and X-ray crystallographic characterization, J. Med. Chem. 49 (11) (2006) 3068–3076.

[89] V.K. Aggarwal, C.J. Astle, M. Rogers-Evans, A concise asymmetric route to the bridged bicyclic tropane alkaloid ferruginine using enyne ring-closing metathesis, Org. Lett. 6 (9) (2004) 1469–1471.

[90] R.C. Simon, B. Grischek, F. Zepeck, A. Steinreiber, F. Belaj, W. Kroutil, Regio- and stereoselective monoamination of diketones without protecting groups, Angew. Chem. Int. Edit 51 (27) (2012) 6713–6716.

[91] E. O'Reilly, C. Iglesias, N.J. Turner, Monoamine oxidase–w-transaminase cascade for the deracemisation and dealkylation of amines, ChemCatChem. 00 (1-4) (2014).

[92] M.J. Abrahamson, E. Vazquez-Figueroa, N.B. Woodall, J.C. Moore, A.S. Bommarius, Development of an amine dehydrogenase for synthesis of chiral amines, Angew. Chem. Int. Edit 51 (16) (2012) 3969–3972.

[93] M.J. Abrahamson, J.W. Wong, A.S. Bommarius, The evolution of an amine dehydrogenase biocatalyst for the asymmetric production of chiral amines, Adv. Synth. Catal. 355 (9) (2013) 1780–1786.

[94] B. Bommarius, M. Schurmann, A. Bommarius, A chimeric amine dehydrogenase offers surprising properties not found in either parent, ChemComm. 50 (95) (2014) 14953–14955.

[95] K.A. Samantha, R.B. Bettina, S.B. Andreas, Biphasic reaction system allows for conversion of hydrophobic substrates by amine dehydrogenases, ACS Catal. 4 (2014) 4021–4026.

[96] E. Vazquez-Figueroa, J. Chaparro-Riggers, A.S. Bommarius, Development of a thermostable glucose dehydrogenase by a structure-guided consensus concept, Chembiochem. 8 (18) (2007) 2295–2301.

[97] (a) A.M. Klibanov, Improving enzymes by using them in organic solvents, Nature 409 (6817) (2001) 241–246. (b) A. Zaks, A.M. Klibanov, Enzyme-catalyzed processes in organic-solvents, P. Natl. Acad. Sci. USA 82 (10) (1985) 3192–3196.

[98] (a) H. Ogino, H. Ishikawa, Enzymes which are stable in the presence of organic solvents, J. Biosci. Bioeng. 91 (2) (2001) 109–116. (b) G. Cainelli, P.C. Engel, P. Galletti, D. Giacomini, A. Gualandi, F. Paradisi, Engineered phenylalanine dehydrogenase in organic solvents: homogeneous and biphasic enzymatic reactions, Org. Biomol. Chem. 3 (24) (2005) 4316–4320.

CHAPTER 8
Biocatalysis for Organic Chemists: Hydroxylations

Laila Roper, Gideon Grogan
Department of Chemistry, University of York, Heslington, York, UK

1 INTRODUCTION
1.1 Hydroxylations in Biocatalysis

Of all the biotransformation enzymes studied, oxidases and oxygenases offer perhaps the most promise, because they are able to catalyze highly regio- and enantioselective oxidations of organic molecules without recourse to toxic heavy metal reagents and high temperatures or pressures [1]. This is especially true for hydroxylation reactions, where a range of enzymes is available for the selective oxyfunctionalization of nonactivated carbon atoms in a selective and environmentally benign manner. This chapter serves as an introduction to biocatalytic hydroxylations, reviewing the major enzymes of interest, their mechanisms, examples of reactions catalyzed, and some experimental procedures that illustrate the benefits and also the complications involved in their application. The activation of oxygen for the hydroxylation of organic substrates is achieved in many ways by enzymes, depending on their type, structure, and mechanistic characteristics. In most examples, the activity of hydroxylases is often dependent on both external, nonproteinaceous cofactors and proteins that assist in electron transfer. Although these aspects necessarily introduce complications into the application of hydroxylases over and above that of say, hydrolase enzymes, many recent advances have been made that seek to address these complications and maximize the potential of these useful activities. In this chapter, we will address the three major classes of hydroxylating enzyme that are of interest for synthetic applications: cytochrome P450s, which use oxygen species bound to an iron atom coordinated within a heme prosthetic group; flavin-dependent hydroxylases that use oxidizing species formed on the flavin coenzyme, and nonheme iron hydroxylases that are dependent on the cofactor 2-oxoglutarate. In each example, biological context is provided, including details on structure and mechanism, as these can have profound consequences for the strategy to be employed in creating and applying the biocatalyst of choice.

Organic Synthesis Using Biocatalysis. http://dx.doi.org/10.1016/B978-0-12-411518-7.00008-1

2 CYTOCHROME P450s

2.1 Reactions Catalysed by P450s

Cytochrome P450s (P450s or CYPs) are iron-dependent, heme-containing oxygenases that constitute a superfamily present in all domains of life [2]. In addition to being studied for their important roles in drug metabolism in higher eukaryotes, and in biodegradation and secondary metabolite production in prokaryotes, they have also been investigated extensively for their ability to catalyze oxidative reactions of unactivated C—H bonds in a highly regio- and stereoselective manner [3–5]. The general reaction catalyzed by P450s is summarized in Scheme 8.1.

During the reaction, one atom of oxygen is incorporated into the substrate as a hydroxyl group; the other atom of oxygen is reduced to water. In order to accomplish this, two electrons are required to reduce the iron as part of the catalytic cycle. Hence, in addition to heme and oxygen, catalysis is dependent on a reduced nicotinamide cofactor, such as NAD(P)H that does not usually interact with the part of the protein that contains the heme domain. Rather, the electrons from NAD(P)H are transferred to heme through one or two "electron transport proteins" or "redox partners" which are described within. P450s have been described that are capable of hydroxylating a vast and diverse range of organic molecules, from terpenes and antibiotic compounds in bacteria, through to secondary metabolites in plants and drug compounds in mammals, where they are the major agent of first-phase metabolism, rendering hydrophobic compounds more hydrophilic as a first step toward their excretion from the body. Some examples of P450-catalysed reactions are shown in Figure 8.1. P450s are also capable of catalyzing numerous other oxidative reactions including heteroatom dealkylation, heteroatom deoxygenation, carbon–carbon bond cleavage, nitration and de-nitration reactions, and Baeyer–Villiger reactions. Those enzymes that are most valued in biocatalysis, however are those that catalyze the regio- and stereo-selective hydroxylation of nonactivated carbon atoms, exemplified by the prototypical P450, P450cam from *Pseudomonas putida* [6] that catalyzes the selective 5-*exo* hydroxylation of (1*R*)-camphor **1** shown in Figure 8.1. This figure also illustrates other significant examples of hydroxylations catalyzed by bacterial, plant, and mammalian P450s [7–10].

2.2 Distribution in Nature and Classification of P450s

Since their first description, P450s have been discovered in most if not all phyla [11], and gene sequences either known or suspected to encode P450s now number in the tens of thousands [11]. Their roles in different organisms vary from detoxification of organic molecules such as drugs in humans, through to

$$RH + O_2 + NAD(P)H + H^+ \rightarrow ROH + H_2O + NAD(P)^+$$

SCHEME 8.1
Overall Reaction Catalyzed by a Cytochrome P450.

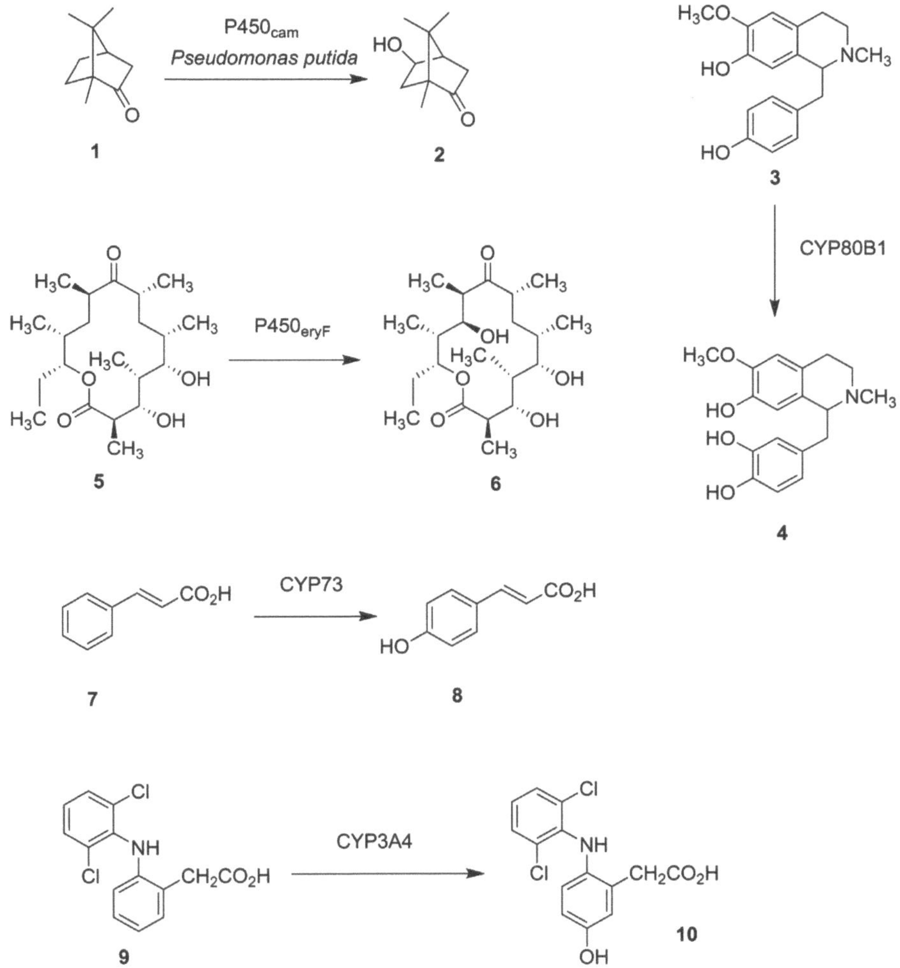

FIGURE 8.1
Some hydroxylation reactions catalyzed by cytochromes P450. Hydroxylation of camphor 1 by $P450_{cam}$ from *P. putida* [6]; Hydroxylation of (*S*)-*N*-methylcoclaurine 3 by CYP80B1 in morphine biosynthesis [7]; Hydroxylation of 6-deoxyerythronolide B 5 by $P450_{Eryf}$ from *Saccharopolyspora erythraea* [8]; Hydroxylation of cinnamic acid 7 by CYP73 from *Arabidopsis thaliana* [9]; Hydroxylation of diclofenac 9 by human CYP3A4 [10].

biosynthesis in plants, and biodegradation and secondary metabolism in bacteria. However, they are all related phylogenetically, and a system of nomenclature has been applied to make sense of the large numbers and, if possible, to allow the activity of new P450s to be predicted from sequence [12]. P450s are usually grouped into families, as designated by a number such as CYP101, CYP51, and so on. P450s are then designated a subfamily through the addition of a capital letter. Finally, the individual P450 is assigned a number that is specific to the gene itself. An illustrative example is provided by $P450_{cam}$ CYP101A1, introduced previously. An additional example is CYP3A4, one of the most important P450s in human metabolism.

2.3 Structure of P450s

The protein components of P450s that carry the heme – the "heme domains" – are highly conserved structurally, with a well-known P450 fold that is made up of characteristic secondary structural elements [13]. Each heme domain is prism shaped, with the heme porphyrin ring parallel to the triangular faces of the protein (Figure 8.2a). The prism has a "proximal" face, corresponding to the area of protein beneath the heme and containing the cysteine thiolate residue that is bound to iron (Figure 8.2b). This is also the face that interacts most extensively with redox partners. The "distal" face is above the heme, and this is the direction from which substrates approach and are accepted through active site channels to the heme for oxygenation. The heme domain is typically described as having two domains overall: a small beta-strand domain (at the top of the picture in Figure 8.2a) and a larger helical domain. Each P450 has of the order of 12–14 helices that are designated A–L. Characteristic helical features include the I-helix, which traverses the heme over the edge of its distal face, and contains a pocket that is used to bind molecular oxygen, and the F and G helices at the distal surface of the protein. The F and G helices, together with the intervening loop, and the "B"-region, which is close in space and includes the B′ helix (Figure 8.2b), are variable in length and sequence among P450s and have been termed "substrate recognition regions" (SRSs). It has been shown that open–close motions of these regions are involved in substrate recognition and binding.

The heme is bound in a hydrophobic pocket in the heme domain, usually accessed from the surface by a channel that allows the transfer of protons, crucial

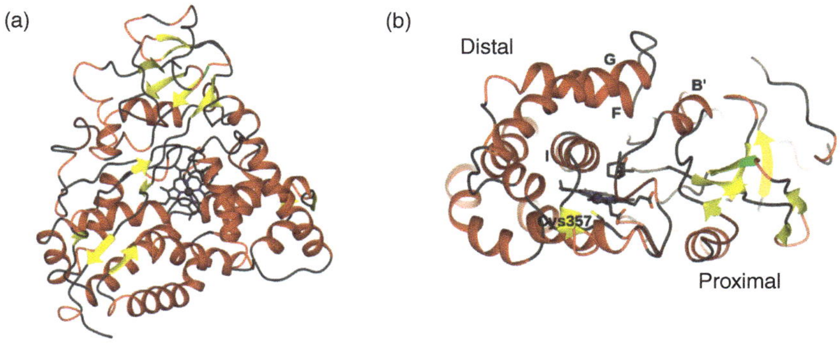

FIGURE 8.2
Structure of $P450_{cam}$ as a model of the general P450 fold. (a) View from distal face showing prism-shaped structure, with heme shown in cylinder format in gray. (b) View from side, with distal face at the top of the page. Helices **I**, **F**, **G** and **B′** are highlighted: the last three constitute "substrate recognition sites" in P450s. The Cys357 residue that coordinates to iron from the proximal face is also labeled.

to the mechanism of the enzyme. Throughout the cycle, the iron is bound within the heme cofactor, coordinated in a plane by four nitrogen atoms that form the porphyrin ring. Beneath the plane of the heme, the fifth ligand is almost always a cysteine residue (Cys) with a deprotonated thiol side-chain (Cys-thiolate, Figure 8.2b). In the "resting" state, that is, in the absence of a substrate, the sixth ligand to iron, above the heme, is usually a water molecule, but this will be displaced by different oxygen species throughout the catalytic cycle. If examined using UV-spectrophotometry, the Cys-thiolate bound to iron with a water ligand gives rise to a characteristic peak at 420 nm. When the heme is reduced, with a chemical reagent such as sodium dithionite, then reacted with carbon monoxide, a characteristic shift to 450 nm is seen in the absorption maximum (hence **Pigment 450**).

2.4 Catalytic Cycle of P450s

The catalytic cycle of a P450 is shown in more detail in Figure 8.3. During this cycle, two electrons, usually derived from NAD(P)H, and two protons, ultimately from bulk solvent, are delivered to the P450 heme iron. Initial binding of the substrate (R-H) leads to the displacement of the sixth water ligand in **I** to give **II**. This results in the subsequent electronic reorganisation in the *d*-orbitals of the heme iron and ultimately a shift in the spin-state equilibrium, from low-spin

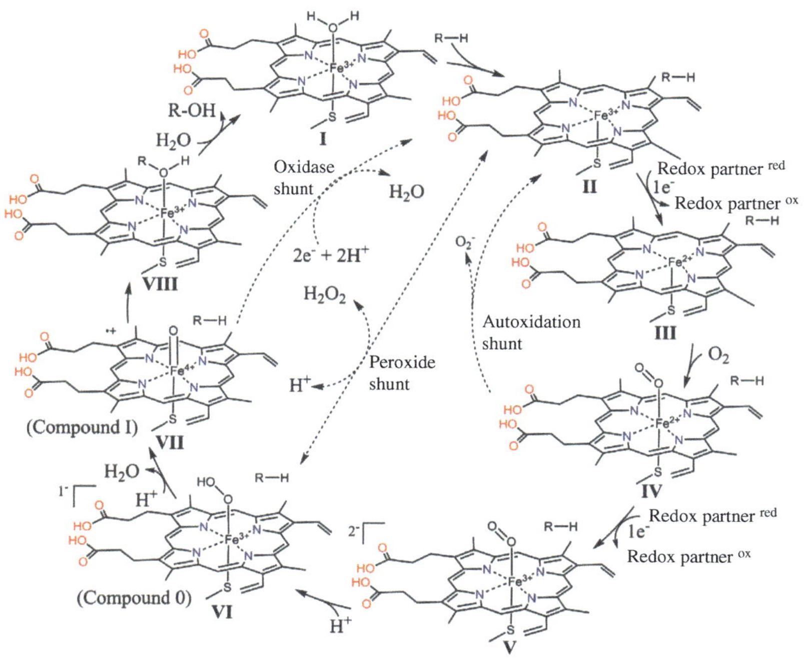

FIGURE 8.3
Catalytic cycle of cytochromes P450. In addition to the catalytic cycle described in the text, "shunt" pathways that are responsible for the nonproductive breakdown of iron-oxygen species throughout the cycle are labeled.

(S =1/2), toward high-spin (S = 5/2). The transition to a high-spin increases the reduction potential of the heme iron, thus favoring the reduction of the heme iron to the ferrous Fe^{2+} state **III** *via* electron transfer from the redox partner. Dioxygen is now bound to form the ferrous-oxy intermediate **IV** that is reduced by the redox partner in the second electron transfer resulting in the ferric peroxy state **V**. Protonation generates the ferric hydroperoxo species **VI**, also known as "compound 0." A second protonation causes the loss of water and the formation of the ferryl-oxo species **VII**, also known as "compound I" that possesses a porphyrin-radical cation. It is this oxidant species that is thought to attack the substrate, resulting in hydroxylation of a C—H bond (**VIII**). Evidence for intermediate **VII** has recently been obtained using Mossbauer spectroscopic analysis of the iron-bound heme treated with *meta*-chloroperbenzoic acid [14]. Finally, the hydroxylated product dissociates, and water again assumes the sixth iron coordinating position (**I**), completing the cycle.

2.5 Types of Electron Delivery System or Redox Partners

The electrons that must be delivered to the heme iron to accomplish catalysis are transferred from the reduced cofactor NADPH by the redox partners. Nature uses a variety of strategies to accomplish this, almost all of which rely on intermediary electron transfer proteins that themselves require nonproteinaceous cofactors. There are currently at least 10 different systems or classes known [15], but for the purposes of applied biocatalysis, it is important to be aware of the four most commonly encountered in the biocatalysis literature (Figure 8.4).

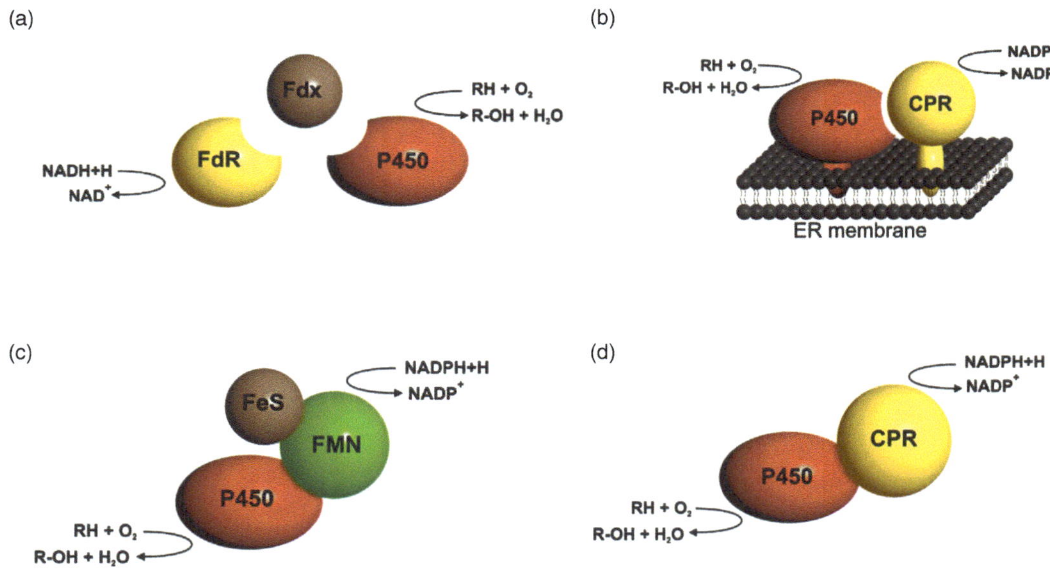

FIGURE 8.4
Some classes of P450 as defined by their redox partners. (a) Class I is typified by $P450_{cam}$; (b) Class II by plant and human enzymes, such as CYP3A4; (c) Class VII by $P450_{Rhf}$ from *Rhodococcus* sp. NCIMB 9784; (d) Class IV by $P450_{BM3}$ from *B. megaterium*.

In bacteria, most P450s systems are of the "Class I" variety (Figure 8.4a), in which electrons are removed from NAD(P)H by an "NADPH dependent ferredoxin reductase" (FdR), a flavin-dependent enzyme. Nicotinamide electrons are then transferred to the coenzyme flavin adenine dinucleotide (FAD) that reduces the next partner in the chain, an iron–sulfur cluster containing protein called the ferredoxin (Fd). The reduced ferredoxin can then transfer two electrons, sequentially, to the heme domain, with which it interacts through the proximal face. It is found that, while some P450s appear to have their own "cognate" electron transport proteins, some Fdr/Fd couples can reduce numerous P450 partners. This has proven useful in constructing biocatalytic systems using P450s as we shall see later. The model enzyme of this class and probably the best-characterized bacterial P450 is again CYP101A1 ($P450_{cam}$).

Class II systems (Figure 8.4b), operate in eukaryotic species such as plants, fungi, and humans, These consist of a two-protein chain in which the first enzyme is an flavin mononucleotide (FMN)- plus FAD-dependent P450 reductase which is capable of accepting electrons from NADPH and transferring these to the heme domain. In eukaryotes, both P450 reductase and heme domain are usually secured within the membrane of the endoplasmic reticulum, and as such, eukaryotic P450s have proved challenging to study and to apply in synthetic reactions. Good examples of these "Class II" systems include human P450s such as CYP3A4, and all known plant P450s.

Clearly, the requirement for additional proteins dictates that extra effort must be expended in gene cloning, expression and, if necessary, purification, when considering the use of P450s for biocatalysis, either in whole-cell preparations or as pure enzymes. However, there do exist in nature "fusion" systems of P450 in which the electron transfer proteins and heme domain are fused together in one genetic product, meaning that all that is required to effect the reaction is the protein, NADPH and the substrate. The most studied of these is the model Class VIII (Figure 8.4d), P450 from *Bacillus megaterium*, $P450_{BM3}$ (CYP102A), a fatty acid hydroxylase for which the assumed natural role is the terminal and subterminal hydroxylation of fatty acid substrates [16]. $P450_{BM3}$ is composed of an FAD-containing, NADPH-dependent reductase domain, an FMN containing domain, and a heme domain. This P450 (CYP102A) is also the most active P450 known, with turnover rates for the substrate arachidonic acid of 283 s^{-1} [16]. The excellent activity of $P450_{BM3}$ has resulted in its being the focus of intensive research by many groups involved in biocatalysis, seeking to engineer or evolve the enzyme for improved or altered activity. It is not the only class of P450 fusion however. An additional system, $P450_{Rhf}$ from *Rhodococcus* sp. NCIMB 9784, a Class VII P450 (Figure 8.4c), features an alternative fusion, of a *C*-terminal iron-sulfur protein and an FMN domain fused to an *N*-terminal heme domain [17]. Although the native substrate of $P450_{Rhf}$ has not been discovered, the reductase domain "RhfRED" has proved useful as a promiscuous agent for the delivery of electrons from NADPH to foreign heme domains, as part of nonnatural fusion enzymes [18]. Thus, RhfRED has been shown to work in concert with a range of both bacterial and plant P450s.

2.6 Summary

The complex nature of electron transport in P450s dictates that, for the most part, these enzymes have been applied within whole-cell systems in which both electron transfer, and indeed the recycling of the cofactor NAD(P)H can be accomplished by the cell's biochemistry. Such whole-cell systems used to constitute the natural or "wild-type" strain of the organism of interest, but, since the advent of routine gene cloning and construction of recombinant strains of bacteria expressing foreign genes, these have been replaced by cells of the workhorse laboratory bacterium *Escherichia coli*. In the examples that follow, we will illustrate the use of a wild-type strain of fungus that contains many Class II P450s, for a hydroxylation reaction, but also some examples of how Class I, Class VII and Class VIII P450s have been exploited in biocatalysis in *E.coli*, and how, more recently, this bacterium has also been used to express genes encoding natural and nonnatural "fusions" of reductase and heme domains of P450 to achieve selective oxidations.

3 EXAMPLES OF P450-CATALYSED BIOTRANSFORMATIONS

3.1 Hydroxylations by Whole-Cell Preparations of Filamentous Fungi

In the 1950s, some of the first hydroxylation reactions using microbes can be found in patents filed by researchers such as Murray and Peterson at Upjohn, who discovered that filamentous fungi such as the *Mucorales* group could catalyze the highly regio- and stereoselective hydroxylation of important steroids such as progesterone [19]. The interest in filamentous fungi as hydroxylation catalysts continued in the 1960s and 1970s, and the Upjohn company published a series of papers related to the use of a fungus, *Sporotrichum sulfurescens*, for hydroxylating a range of compounds including protected *N*-heterocycles and related structures [20–25]. *S. sulfurescens*, later designated *Beauveria sulfurescens* and finally *Beauveria bassiana*, was extensively investigated as a hydroxylation catalyst for this and a wider range of compounds until the early 2000s, as it displayed an unusually active repertoire of enzymes for the selective hydroxylation of nonactivated carbon atoms [26]. The following example details an experimental procedure for the hydroxylation of *N*-Cbz-substituted piperidines (Figure 8.5) by whole-cell preparations of *B. bassiana* ATCC 7159 [27].

3.2 Hydroxylation of *N*-Benzyloxycarbonyl-4-Methylpiperidine by *B. bassiana*

The fungus was obtained from the American type culture collection (ATCC) as a freeze-dried ampoule containing dried cell material. The culture was revived by suspension of this material, under aseptic conditions, in sterile water, and its dispersion onto a solid growth medium composed of malt extract agar. The agar

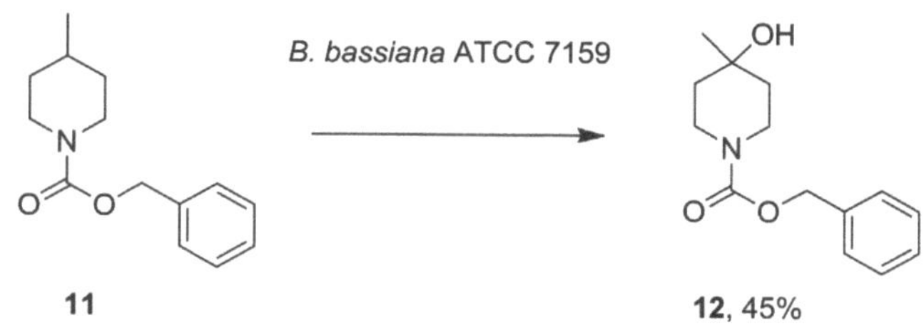

FIGURE 8.5
Hydroxylation of *N*-benzyloxycarbonyl-4-methylpiperidine 11 by *B. bassiana*.

plate was grown for 3–7 days to allow a powdery, white mycelium to develop on the surface of the plate.

A loop full of the organism was scraped from the plate and used to inoculate, using sterile technique, 60 mL of a medium containing 7.5 g/L corn steep solids and 10 g/L glucose in water, which had been adjusted to a pH of 4.85. After 3 days growth at 28 °C with shaking in an orbital incubator at 150 rpm, the mature culture was used to inoculate a larger, 600 mL volume of medium inside a 2 L Erlenmeyer flask. This was again grown at 28 °C for 3 days in an orbital shaker. After 72 h growth, 66 mg of **11**, dissolved in 1 mL of ethanol, was added to the flask, and the progress of the reaction monitored by extracting, at intervals, 0.5 mL of culture broth, followed by extraction into 0.5 mL of ethyl acetate.

Once TLC analysis had indicated that no substrate remained, the fungus was removed from the culture by either centrifugation or by filtering through a sheet of muslin. The supernatant, saturated with sodium chloride (which helped to eliminate emulsions in the extraction process), was extracted with ethyl acetate. The combined organic phases were dried over anhydrous $MgSO_4$, and filtered. The solvent was removed by rotary evaporation and the residue subjected to flash column chromatography. Hydroxylated product 12 was recovered in 45 % yield overall.

Although P450s from *Beauveria* have not been isolated, enzyme inhibition experiments have shown that it is almost certainly P450s that are responsible for the hydroxylation reactions in these instances [28]. A recent genome sequence of *B. bassiana* suggested that there are as many as 83 P450 genes in the organism [29], and any one, or several, may be responsible for the reaction catalyzed for a given substrate. These enzymes are almost certain to be predominantly of Class II, with a fungal P450 reductase serving a P450 in the endoplasmic reticulum of the cells. Although these reactions are certainly useful for generating small amounts of metabolites for toxicology studies, for example, it is recognized that heterologous systems that produce only one active P450 will be more efficient for the production of single hydroxylated products. Some examples of these "designer" biocatalysts are given in the next examples.

3.3 Hydroxylation of *N*-Protected Heterocycles by a "Class I" Cytochrome P450 Expressed in a Heterologous Host

Although wild-type organisms, such as *B. bassiana* proved very useful in industry for hydroxylation reactions, from the late 1990s, the number of available P450 gene sequences expanded dramatically. Advances in molecular biology have allowed these genes to be expressed in large amounts in easy-to-handle organisms such as *E. coli*, which has meant that new biocatalysts can contain P450s originally derived from a vast array of organisms, but expressed in bacteria or yeast. Of course, the situation is complicated by the need for redox partner proteins. We will examine the application of natural and nonnatural P450 fusion systems later, but heterologous systems have also been developed for the expression of classical bacterial "Class I" P450 systems that require a ferredoxin reductase and a ferredoxin to transfer electrons from NADPH.

An excellent example of this is provided by cytochrome $P450_{pyr}$, an enzyme from *Sphingomonas* sp. HXN-200, which was shown to have good hydroxylation activity toward substrates such as *N*-benzyl heterocycles and terpenes in the wild-type strain [30,31]. One major disadvantage of the *Sphingomonas* is that it needs to be grown on the flammable hydrocarbon octane as the sole carbon source in order to induce expression of the P450 enzyme.

In the following example, a strain of *E. coli* that has been engineered to express both the Type I P450 domain, and a ferredoxin reductase and ferredoxin, is applied to the hydroxylation of *N*-benzylpyrrolidin-2-one **13** (Figure 8.6) [32]. The requirement for redox partner proteins was solved by using a two-plasmid system for expression of the required genes in *E. coli*. Hence, ferredoxin and ferredoxin reductases genes were cloned into a pETDuet vector that allows the co-transcription of both genes from the same plasmid. The gene encoding $P450_{pyr}$ was cloned into another expression plasmid, pRSFDuet, and a strain of *E. coli* DH5a transformed with both plasmids to yield a recombinant biocatalyst competent for the hydroxylation of a range of substrates.

3.4 Hydroxylation of *N*-Benzylpyrrolidinone by Recombinant *E. coli* Expressing Cytochrome $P450_{pyr}$

The strain of DH5α expressing the $P450_{pyr}$, ferredoxin and ferredoxin reductases was employed in the hydroxylation (Figure 8.6) of *N*-benzylpyrrolidin-2-one **13**

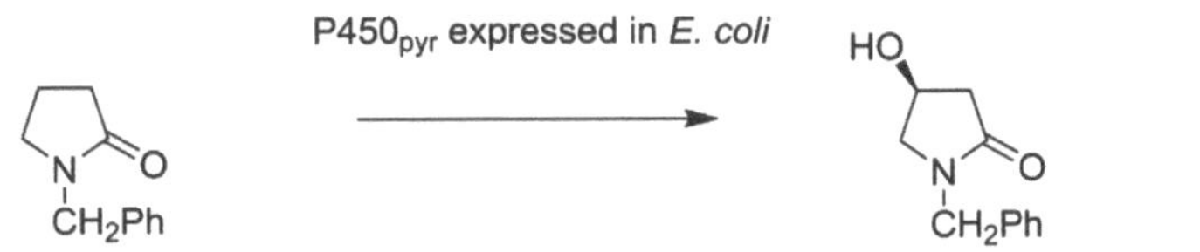

13, 25 mM   (*S*)-**14**, 10.8 mM, 99% e.e.

FIGURE 8.6
Hydroxylation of *N*-benzylpyrrolidinone 13 by recombinant *E. coli* expressing cytochrome $P450_{pyr}$.

to 4*S*-hydroxy-*N*-benzoylpyrrolidin-2-one **14** in high regio- and stereoselectivity (ee 99%). The *E. coli* strain was maintained on LB agar containing ampicillin and kanamycin as antibiotic resistance markers. A single colony was used to inoculate a 30 mL starter culture that was grown for 12 h at 37 °C. A 2 mL portion was added to 100 mL of LB medium containing antibiotics and grown with shaking for 2 h, after which 0.5 mM of IPTG was added as an inducer. After 3 h further growth, the cells were harvested by centrifugation and resuspended in potassium phosphate buffer pH 7.5 to a density of 1.5 g dry cell weight per liter. To 10 mL of cell suspension were then added 400 μL of a glucose solution (50% w/v) and 43.8 mg (a final concentration of 25 mM) of *N*-benzylpyrrolidin-2-one. The reaction was maintained near pH of 8.0 through the addition of 2M sodium hydroxide. The reaction was monitored by HPLC after removing 100 μL of reaction mixture, mixing with 100 μL cold methanol and filtration of samples. A 10.8 mM final concentration of product was observed, although it was not isolated.

3.5 Hydroxylations by Class VIII Cytochromes P450, Such as $P450_{BM3}$

The coexpression of redox partners is one way of achieving a functional P450-catalyzed hydroxylation system in recombinant *E. coli*, but there are other strategies that take inspiration from natural systems that consist of P450 heme domains fused to their redox partners. $P450_{BM3}$ is the best example of this arrangement, displaying the highest turnover numbers for a P450 catalyzed reaction in part, it is thought, because of the fusion of the P450 reductase component to the heme domain. Given the high activity of $P450_{BM3}$ toward its natural substrates (fatty acids), it is no surprise that it has been the target of engineering studies that look to expand its substrate specificity to include, for example, terpenes, drug molecules, aromatic compounds, and also alkanes, even those as small as ethane and methane.

The group of Frances Arnold has pioneered methods for the mutation of $P450_{BM3}$ for the hydroxylation of alternative target compounds, including alkanes [33,34]. The mutants that are employed result from *in vitro* evolution experiments that generate random diversity within the $P450_{BM3}$ gene sequence. Vast libraries of variants are then screened for improved activity. A description of *in vitro* evolution techniques is outside the remit of his chapter but certainly, for the organic chemist who finds a $P450_{BM3}$ mutant in the literature that has a possible application of interest, it is now a simple task to order the gene of interest in an expression plasmid from a synthetic gene company, and to carry out standard transformation and expression protocols to furnish a recombinant strain of the required activity.

3.6 Hydroxylation of Octane by a Mutant of Cytochrome $P450_{BM3}$

In the forthcoming example, whole cells of *E. coli* (DH5α) expressing mutants of $P450_{BM3}$ were used to hydroxylate alkanes such as octane **15** (Figure 8.7) [35].

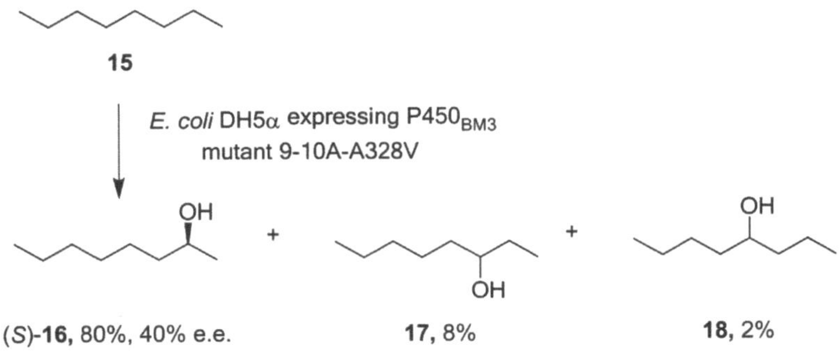

FIGURE 8.7
Hydroxylation of octane 15 by a mutant of Cytochrome $P450_{BM3}$.

Starter (3 mL) cultures were inoculated from colonies on an agar plate and grown overnight on LB medium containing 100 μg/mL ampicillin. These were then used to inoculate 75 mL volume cultures in "M9" medium; a simple growth medium for *E. coli* that consists of 0.5% (w/v) glucose; 0.2 mM $CaCl_2$, 4 mM $MgSO_4$ and 100 μg/mL ampicillin. After 24 h growth at 37 °C, the cells were harvested by centrifugation and suspended in potassium phosphate buffer (20 mL) that also contained 0.5% (w/v) glucose, 1 mM IPTG, 0.5 mM aminolevulinic acid (this is often added to growth media for recombinant P450 expression as it is a precursor in heme biosynthesis) and octane **15** as substrate (5 mM from a 500 mM stock of alkane in dimethyl sulfoxide). After 8-h reaction, the reactions were extracted into dichloromethane and the mixtures analyzed by both nonchiral and chiral gas chromatography. In one example, the strain of DH5α expressing the "9-10A-A328V" variant was used to transform octane providing 80% of (*S*)-2-octanol **16** with 40% e.e. Small amounts of 3-hydroxylated (**17**) and 4-hydroxylated (**18**) products were also obtained.

3.7 Hydroxylations Using Nonnatural Fusions Consisting of the Reductase of P450Rhf, a Class VII P450, With "Foreign" Heme Domains

One of the benefits of using the fusion system of $P450_{BM3}$ is that it allows the expression of a single polypeptide that is a complete biohydroxylation catalyst. The discovery of natural P450 fusions led quickly to the realization that the redox partner domains of such enzymes might be promiscuous for electron delivery from NAD(P)H to other heme domains that had substantially different, desirable substrate specificities compared to the wild-type fusion enzyme. Indeed, the reductase system of $P450_{BM3}$ has been used in nonnatural fusions with human P450s such as $P450_{3A4}$ [36] and recently, P450153A1 from *Marinobacter aqueoli* [37] for the terminal hydroxylation of fatty acid substrates. However, one of the other natural Class VII P450 fusion enzymes has also demonstrated its worth as a promiscuous reductase for transferring electrons from NADPH to foreign heme domains. $P450_{Rhf}$ is a cytochrome P450, initially described in

Rhodococcus sp. NCIMB 9784, which consists of an *N*-terminal heme domain fused to an FMN-containing domain and a *C*-terminal iron–sulfur protein; a redox system which has something in common with bacterial phthalate dioxygenases. Although the natural substrate for the $P450_{Rhf}$ remains unknown, a number of groups have been successful in removing the wild-type heme domain and replacing it with useful Type I P450 heme domains, including, for example, pikC, a P450 involved in pikromycin biosynthesis [38], $P450_{cam}$ [39], the 5-*exo* camphor hydroxylase, and CYP153A13a from *Alcanivorax borkumensis* SK2, an alkane hydroxylase [40]. A straightforward example is illustrated by the fusion of the P450 CYP154 from *Nocardia farcinica* IFM10152 to the $P450_{Rhf}$ reductase domain [18].

3.8 Hydroxylation of Testosterone by a Nonnatural Fusion Enzyme Consisting of the Reductase of P450Rhf With Heme Domain CYP154 from *N. farcinica*

The hydroxylation of testosterone and other steroids presents a special challenge to abiotic chemistry as there are so many chemically equivalent carbon atoms on the steroid skeleton. Now that individual P450 genes can be cloned and expressed individually, the selectivity of these enzymes can be exploited. Class I P450 genes from *B. megaterium* for example, encoding CYP106A2 homologs, have been shown to be selective for the 15β-hydroxylation of testosterone [41]. CYP154 from *N. farcinica* IFM 10152 has been shown to be selective for the 16α position of the steroid **19** [42] (Figure 8.8). Fusion of the CYP154 gene to the $P450_{Rhf}$ reductases, and cloning and expression of the fusion gene in *E. coli* resulted in a recombinant strain that was able to perform this hydroxylation as a whole-cell catalyst [18].

The gene fusion was created using techniques detailed in Ref. [18]. A single colony of a recombinant strain of *E. coli* Rosetta 2 (DE3) that had been transformed with an expression plasmid containing the RhfRed-CYP154 gene was used to inoculate a 5-mL culture in LB medium containing 100 μg/mL kanamycin and chloramphenicol each as antibiotic markers. After shaking overnight at 37 °C, the cell culture was centrifuged and resuspended in 500 μL of sterile M9 medium. A 250 μL of this suspension was then used to inoculate 250 mL M9 medium containing 100 μg/mL kanamycin and chloramphenicol,

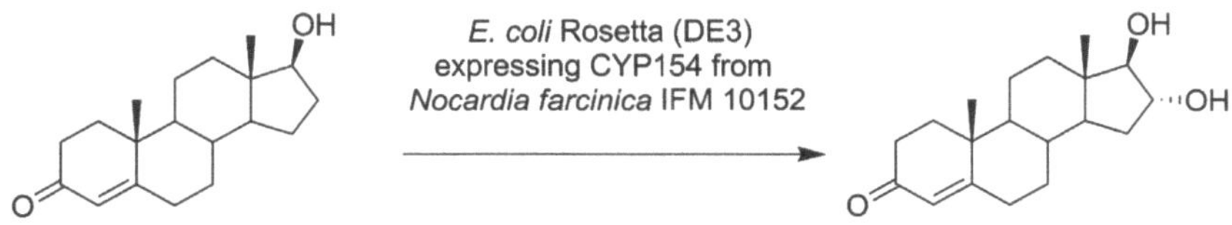

19 **20**

FIGURE 8.8
Hydroxylation of testosterone 19 by a nonnatural fusion enzyme consisting of the reductase of P450Rhf with heme domain CYP154 from *N. farcinica*.

glucose (0.4 g/L), and 0.05% $FeCl_3$ (w/v) in a 1 L Erlenmeyer flask. The culture was grown until it had reached an optical density of A_{600} = 0.8 at which point 0.5 mM aminolevulinic acid and 0.4 mM IPTG were added. The culture was then grown for a further 18 h with shaking at 25 °C, after which the cells were harvested by centrifugation. A portion of the resultant wet cell paste (300 mg) was suspended in 3 mL of a potassium phosphate buffer at pH 7.0 in a sterile 15-mL plastic tube. To this was added 0.5-mM testosterone (from a DMSO stock solution) along with 25 μL of a 50% solution of aqueous glycerol. The tube was incubated at 4 °C with shaking. Samples were extracted at intervals for analysis by GC/GC-MS after extraction and derivatization with *N,O*-bis(trimethylsilyl) trifluoroacetamide (BSTFA) using an established procedure [19].

3.9 Summary

The application of cytochromes P450 in biocatalysis in the organic chemistry laboratory is much less straightforward than, say the application of hydrolases such as proteases or lipases, as the latter are stable, mechanistically simple, and can be produced in freeze-dried powders that act as off-the-shelf reagents. However, as the extraordinary benefits of P450-catalysed hydroxylations become more appreciated, many more groups are developing either their own facilities to manufacture biological catalysts, or are working in collaboration with partners in molecular biology and biotechnology to create catalysts that allow them to realize synthetic chemistry objectives.

4 FLAVIN-DEPENDENT HYDROXYLASES

4.1 Hydroxylations by Flavoprotein Monooxygenases

The activation of molecular oxygen for oxygenation reactions can also be achieved within enzymes by flavin coenzymes such as FAD **21** and flavin mononucleotide (FMN) **22** (Figure 8.9). Flavins, derived from riboflavin (vitamin B_5) possess a characteristic tricyclic isoalloxazine structure which, when reduced to $FADH_2$ **23**, usually by NAD(P)H, can react with molecular oxygen to form

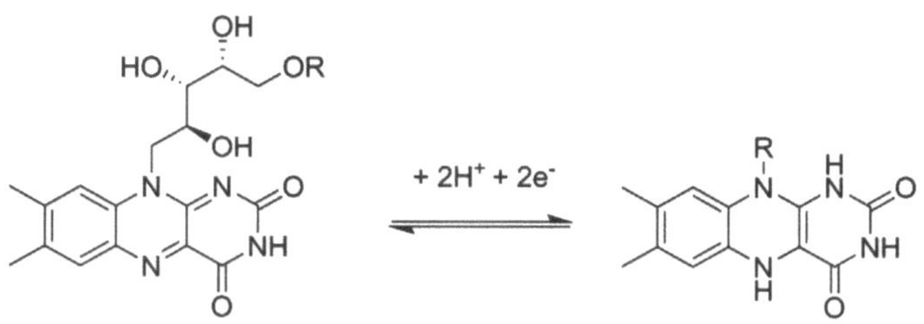

21, R = ADP = Flavin adenine dinucleotide (FAD)
22, R = PO_3^{2-} = Flavin mononucleotide (FMN)

23, $FADH_2$ (or $FMNH_2$)

FIGURE 8.9
Structure of flavin coenzymes FAD 21 and FMN 22 and their reduction to FAD(FMN)H_2 23.

(a) $NAD(P)H + H^+ + FAD \rightarrow NADP^+ + FADH_2$

(b) $FADH_2 + O_2 \rightarrow FADH\text{-}O\text{-}O\text{-}H$

(c) $FADH\text{-}O\text{-}O\text{-}H + SH \rightarrow S\text{-}OH + FAD + H_2O$

SCHEME 8.2
Overall Reaction Catalyzed by a Flavin-Dependent Hydroxylase.

a (hydro)peroxy flavin intermediate suitable for the oxidation of organic substrates. These reactions are summarized in Scheme 8.2.

In addition to Baeyer–Villiger oxidation and the electrophilic oxidation of heteroatoms such as sulfur, some enzymes that form this (hydro)peroxy flavin can also catalyze the hydroxylation of aromatic substrates, particularly where the aromatic ring is already activated with, for example, a phenol group. Flavoprotein monooxygenases (FPMOs) have been divided up into six classes on the basis of structure, mechanism, and other catalytic characteristics [43]. FPMO classes that catalyze aromatic hydroxylation reactions include Class A FPMOs, which are single polypeptides capable of both the oxidation of NADPH to give $FADH_2$, and of the oxygenation of an organic substrate using the (hydroperoxy)flavin species. Examples of Class A FPMOs include *para*-hydroxybenzoate hydroxylase (PHBH) and hydroxybiphenyl monooxygenase (HbpA), described further. In Class E FPMOs, the reduction of flavin is accomplished by a separate flavin reductase, which then transfers the reduced flavin to a separate protein that is responsible for forming the (hydroperoxy)flavin and oxygenating the hydrocarbon substrate. The application of Class E FPMOs raises similar issues to those of cytochromes P450 in respect of the need to express more than one gene to reconstitute the hydroxylation system. For the purposes of discussion of FPMOs therefore, we will focus on Class A enzymes, which are able to catalyze the flavin reduction step and the substrate oxygenation reaction within one polypeptide.

4.2 Class A FPMOs: Mechanism and Structure: *Para*-hydroxy Benzoate Hydroxylase (PHBH)

Much of the work into the structure and mechanism of aromatic hydroxylation by Class A FPMOs has come from studies of *para*-hydroxybenzoate hydroxylase (PHBH), the structure of which was first determined in 1979 [44]. PHBH catalyses not only the hydroxylation of 4-hydroxybenzoate **24** but also 2,4-dihydroxybenzoate **26**, 4-mercaptobenzoate **27**, and fluorinated and chlorinated derivatives of the parent substrate (Figure 8.10).

The mechanism of PHBH activity is illustrated in Figure 8.11 [45]. Initially, the flavin FAD in **I** must be reduced by NAD(P)H and H^+ to give $FADH_2$ (**II**). Reaction with molecular oxygen through one single electron transfer step from

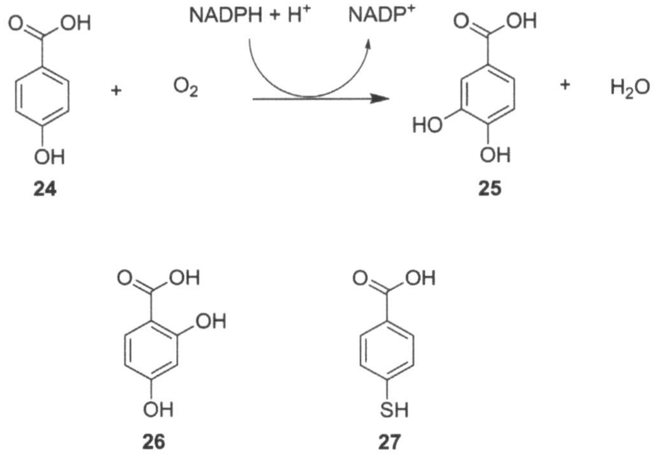

FIGURE 8.10
Aromatic hydroxylation reactions catalysed by *p*-hydroxybenzoate hydroxylase (PHBH).

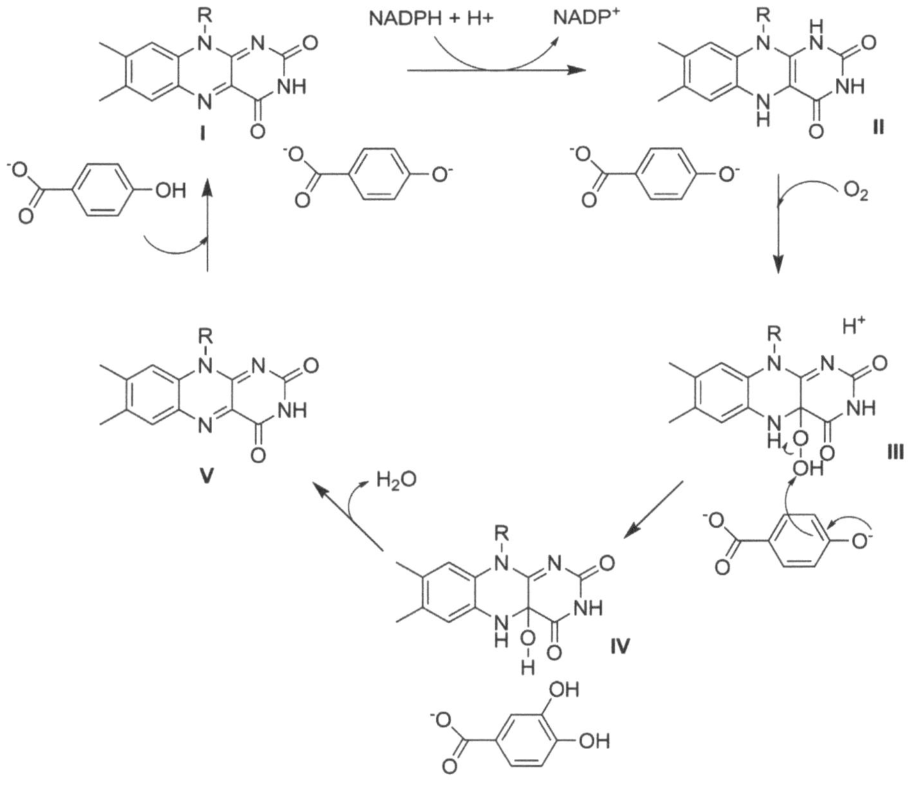

FIGURE 8.11
Catalytic cycle of PHBH reaction. FAD (**I**) is reduced to $FADH_2$ using NADPH (**II**). This then reacts with oxygen to form a (hydroperoxy)flavin **III** that hydroxylates the aromatic ring. This generates the hydroxyflavin (**IV**), which loses water to regenerate FAD (**V**).

reduced FAD to molecular oxygen, results in an FAD semiquinone radical and a superoxide radical that recombine to form the (hydro)peroxyflavin (**III**). It is this intermediate that acts as the oxidant in the aromatic hydroxylation reaction. Following the electrophilic substitution reaction that introduces the hydroxyl group into the aromatic ring, the FAD is left as a C_{4a} FAD hydroxide (**IV**), from which water is eliminated to yield oxidized flavin suitable for another catalytic cycle (**V**).

The structure of PHBH is shown in Figure 8.12. It is an enzyme of 394 amino acids, with a three domain structure composed of an *N*-terminal "Rossman fold" domain that binds the ADP of the FAD coenzyme, a second domain that binds the *para*-hydroxybenzoate substrate and, a third, largely helical domain that is involved in dimerization with another subunit. The tricyclic isoalloxazine ring of the flavin is bound at the interface between the first two major domains, and there is known to be movement of flavin associated with the accommodation of substrate [46]. The structure of PHBH in the presence of NADPH has not been determined, but it is thought that there is only a transient association of the two cofactors to effect flavin reduction after which the $NADP^{+}$ dissociates to permit substrate binding [46]. Although PHBH itself has not been the focus of studies in applications in synthesis, other Class A FPMO aromatic hydroxylases, such as hydroxybiphenyl monooxygenase from *Pseudomonas azelaica* have been used for the preparation of hydroxylated compounds.

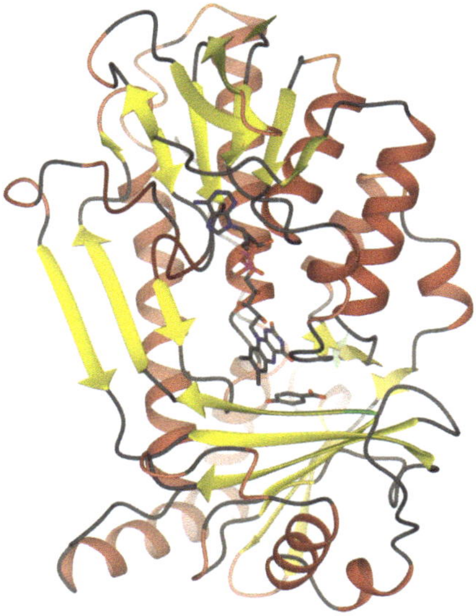

FIGURE 8.12
Structure of PHBH [44,46] shown in ribbon format, with FAD, in cylinder format and carbon atoms in gray, shown bound at the domain interface. The substrate, *p*-hydroxybenzoate can also be seen beneath the FAD molecule.

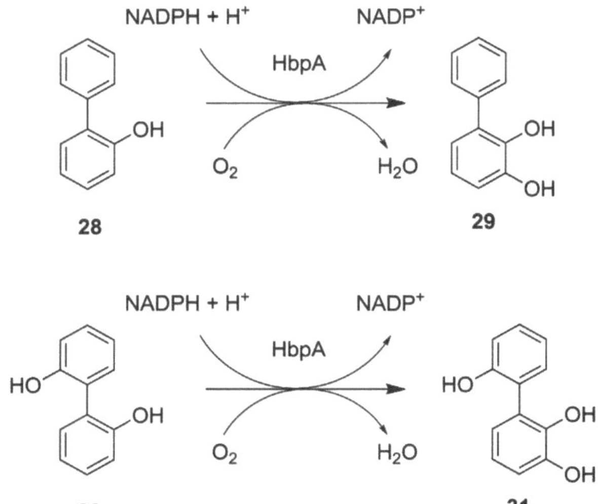

FIGURE 8.13
Reactions catalyzed by hydroxybiphenyl monooxygenase (HbpA) from *P. azelaica*.

4.3 Hydroxybiphenyl Monooxygenase (HbpA) from *Pseudomomas azelaica*

Hydroxybiphenyl monooxygenase enzyme "HbpA" from *P. azelaica* is a larger Class A FPMO than PHBH, of 586 amino acids. The enzyme was originally isolated from the named organism after discovery of its ability to grow on hydroxybiphenyl as a sole carbon source [47]. HbpA catalyses the hydroxylation of 2-hydroxybiphenyl **28** to 2,3-dihydroxy biphenyl **29** (Figure 8.13), but has also been shown to hydroxylate a range of 3-aryl, 3-alkyl- and 3-halo catechol substrates [48]. The gene encoding HbpA was cloned and expressed in *E. coli* JM109 and both the wild-type and mutant strains have been employed for whole cell biotransformations of a number of aromatic compounds [48]. In the following example, HbpA expressed in *E. coli* JM109 is used to hydroxylate *ortho*-substituted phenols.

4.4 Hydroxylation of 2,2′-Dihydroxybiphenyl by HbpA Expressed in *E. coli* JM109

In this example, HbpA expressed in *E. coli* JM109 is used to hydroxylate 2,2′-dihydroxybiphenyl **30** to 2,2′3-trihydroxybiphenyl **31** (Figure 8.13). A plasmid pHBP461 was first generated in which the gene encoding HbpA was cloned and able to be expressed under the control of the *lac* promoter, permitting chemical induction using IPTG. A single colony that had been transformed with the plasmid was used to inoculate 500 mL of LB broth containing 100 μgm/L as antibiotic marker. The cells were then grown on an orbital shaker at 30 °C to an optical density (A_{546}) of 0.6. The cells were harvested by centrifugation, washed in LB and then transferred to a fermenter containing LB medium with 100 μg/mL ampicillin, 5-mM sodium succinate and 0.1-mM IPTG as inducer. The cells were grown until an optical density (A_{546}) of 7.0 had been reached, after which 0.85 g of 2,2′-dihydroxybiphenyl was added from a 1 M stock solution in methanol. After 10 min and also after 30 min, additional 0.85 g amounts of the substrate were added, and after 90 min a further amount of sodium succinate was added, to a final concentration of 5 mM, in order to assist nicotinamide cofactor regeneration within the reaction mixture. The reaction was stopped after 170 min by lowering the pH to 2.0 using perchloric acid. The cells were removed by centrifugation and the 2,2′,3-trihydroxybiphenyl product was extracted by adsorption to a resin; Serdolit A400, from which it was eluted subsequently with acidic methanol to give a final product yield of 70%. As part of this study, in addition to hydroxybiphenyl it was also established that HbpA could effectively hydroxylate 2-propyl phenol, 2-methyl phenol and 2-halophenols, with typically 20–35% relative activity when compared to that observed for **28**.

4.5 Summary

Flavin-dependent hydroxylases are attractive biocatalysts in that they are simpler to apply than cytochrome P450s. Certainly studies of scaled-up reactions have illustrated that they can be applied at above 1-L scale for the generation of gram quantities of products. However their substrate range is more limited than those of P450s, and appears at this stage to be limited to activated aromatic nuclei. However, niche applications of these enzymes to certain synthetic problems could prove valuable, and certainly related flavoprotein monooxygenases also have great potential in, for example, asymmetric Baeyer–Villiger reactions and the epoxidations of alkenes.

5 2-OXOGLUTARATE DEPENDENT HYDROXYLASES

5.1 Introduction

A further class of monooxygenase enzymes that has been employed in biocatalytic applications for the selective introduction of hydroxyl groups is the 2-oxo glutarate (2-OG) dependent hydroxylases [49]. These enzymes, which use non-heme iron to catalyze oxidations, are also dependent for activity upon the cofactor 2-OG, which participates in the oxygen activation reaction. Ascorbic acid is also thought to improve 2-OG dependent reactions, although is not involved in the mechanism of the enzyme. The general reaction catalyzed by 2-OG dependent hydroxylases is shown in Scheme 8.3.

The reaction catalyzed is therefore an oxidative decarboxylation of 2-OG, in which one atom of oxygen is transferred to succinate and the other to the substrate, through an iron (IV) oxo intermediate that is formed as part of the catalytic

$$\text{R-H} + O_2 + \text{2-oxo-glutarate} \rightarrow \text{R-OH} + CO_2 + \text{succinate}$$

SCHEME 8.3
Overall Reaction Catalyzed by a 2-Oxoglutarate Dependent Hydroxylase.

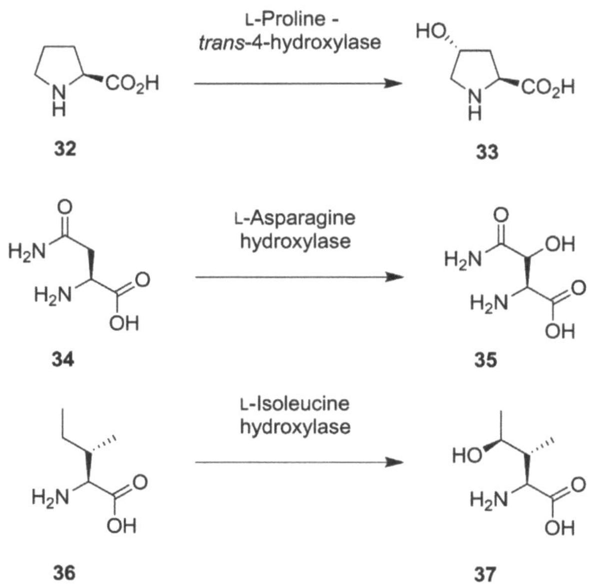

FIGURE 8.14
Hydroxylation of amino acids catalyzed by 2-oxoglutarate dependent hydroxylases.

mechanism. 2-OG dependent enzymes in general are a vast family, and include enzymes that catalyze, for example, the ring expansion reaction in cephalosporin biosynthesis (deacetoxycephalosporin C synthase); the desaturation of the carbapenem nucleus (CarC – carbapenem synthase), and numerous hydroxylation reactions [49]. Of the latter, the enzymes that catalyze the hydroxylation of amino acids have been the most intensively studied from the perspective of preparative biocatalysis, and include enzymes that catalyze the regio- and enantioselective hydroxylation of proline **32** [50], asparagine **34** [51], and isoleucine **36** [52]. Some of these reactions are illustrated in Figure 8.14.

5.2 Structure and Mechanism

2-OG dependent hydroxylases are attractive catalysts from the perspective of applied biocatalysis as they do not require additional redox partners for catalysis (in contrast to P450s), although their substrate ranges are much narrower than those of P450s. Crucial to the oxygenation reaction is an iron atom, which is octahedrally coordinated by three amino acid side chains (three histidines, or two histidines and an acidic residue, glutamate or aspartate) that are to be found within a fairly well-conserved motif usually described as HXD/E…H [49]. In complexes of 2-OG dependent hydroxylases whose structures have been determined by X-ray crystallography, two further ligand positions are occupied by the 2-oxo group of the cofactor and one of the 2-OG carboxylate oxygens, leaving a sixth ligand position for either water or oxygen. The overall structure of the 2-OG dependent hydroxylases is a double-stranded β-helix (DSBH) fold,

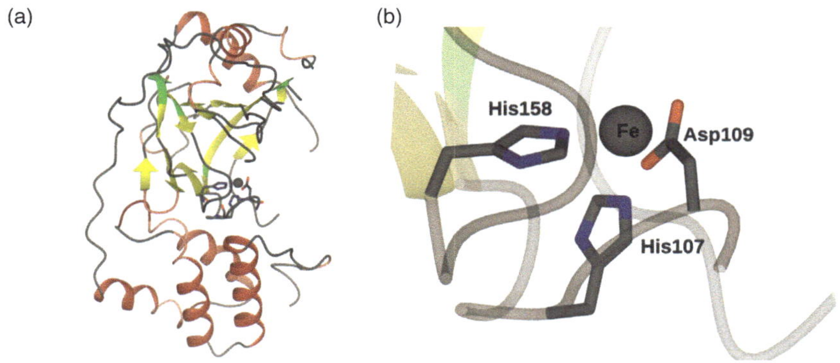

FIGURE 8.15
(a) Structure of proline-*cis*-3 hydroxylase in ribbon forma showing two-domain structure with the active site at the domain interface [53]. (b) Detail of active site showing His158, His107 and Asp109 coordinating to the catalytic iron atom (**Fe**).

in which the core of the protein usually consists of eight β-strands divided into two four-stranded antiparallel β-sheets, forming a β-sandwich [49]. An example is provided by proline-*cis*-3-hydroxylase from *Streptomyces* sp. [53]. In this enzyme, the protein main chain is folded into two domains. The *N*-terminal domain contains ten β-strands, eight of which form the β-sandwich and two α-helices (Figure 8.15a). The *N*-terminal domain is connected to a *C*-terminal helical domain by a linker region containing several proline residues. The *C*-terminal domain is unusual for a 2-OG dependent hydroxylase, and is a helical bundle, containing five α-helices, terminating in a loop close to the active site. At the domain interface, residues His158, Asp109 and His107 coordinate the catalytic iron atom (Figure 8.15b). Residues Ser170 and Arg168 are part of an "RXS" motif proposed to bind the 5-carboxylate of 2-oxoglutarate. Also within the active site are four basic residues His43, Arg95, Arg97 and Arg122 (not shown), the side-chains of which all point directly toward the active site iron. It has been hypothesized that some or all of these residues may form direct or indirect electrostatic interactions with the carboxylate of the proline substrate and in doing so position the proline C_3-H bond correctly for oxidation.

5.3 Catalytic Cycle

2-OG-dependent hydroxylases are thought to employ the ordered sequential mechanism shown in Figure 8.16 using a high valent iron-oxo species to effect hydroxylation [54]. The reaction cycle initiates with the binding of the cosubstrate, 2-oxoglutarate. Two water molecules are displaced as 2-OG is coordinated to the iron centre in a bidentate manner (**II**). It is proposed that binding occurs at the 5-carboxylate of 2-OG, although rotation of the His 135 side chain may allow for binding to the 1-carboxylate and/or 2-oxo binding (**III**). Subsequent rearrangement is followed by binding of the primary substrate, such as proline, to the active site (**IV**). Spectroscopic studies with clavaminate synthase

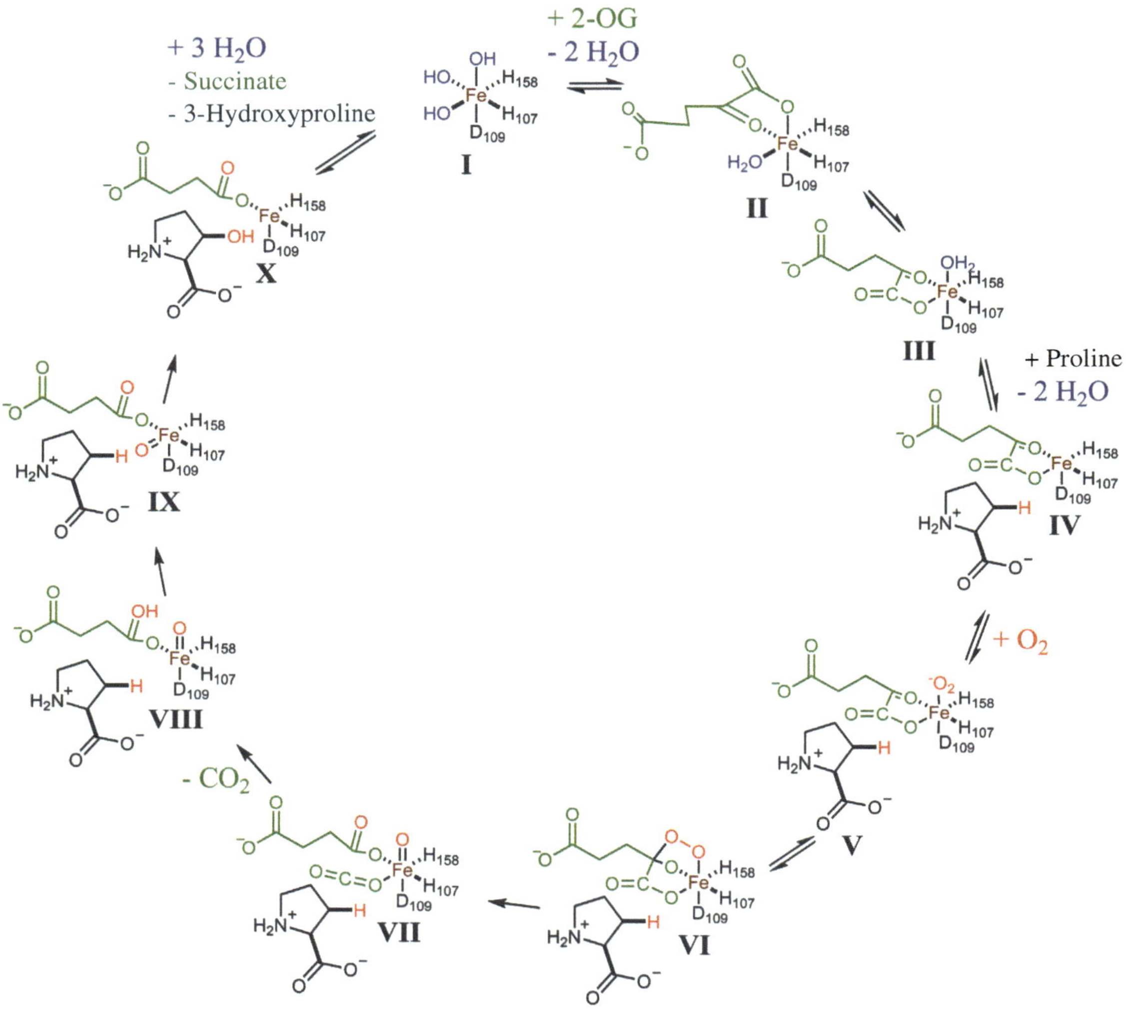

FIGURE 8.16
Catalytic cycle proposed for proline-3-hydroxylase [53].

(another 2-OG dependent enzyme) indicate a change in geometry on substrate binding, from 6-coordinate octahedral to 5-coordinate square pyramidal, with the displacement of another water molecule (**IV**). This geometry change activates the iron for dioxygen binding (**V**). Once bound, reaction of dioxygen with the 2-oxo group of 2-OG may lead to the formation of a six-membered cyclic peroxide (**VI**). The cyclic peroxide collapses to form a persuccinate species and the proposed ferryl (FeIV = O) reactive intermediate (**VII**). Carbon dioxide is released from the active site and it is at this point that the pentacoordinate iron coordination sphere (**VIII**) is thought to rearrange so that the oxygen atom is positioned *trans* to His 158 (**IX**). The ferryl reactive species then reacts with the proline substrate (possibly through a radical) (X) and both succinate and *cis*-3-hydroxyproline are released from the active site.

5.4 Applications of 2-OG-Dependent Hydroxylases in Biocatalysis

2-OG dependent hydroxylases have attracted the most attention for preparative biocatalysis because of their ability to catalyze the hydroxylation of amino acids. Although eukaryotic enzymes only catalyze the hydroxylation of peptidic substrates, homologous enzymes also display the ability to hydroxylate isolated amino acids including proline, isoleucine, asparagine, and lysine. The proline hydroxylases have attracted most attention, both because regio- and enantiocomplementary enzymes that allow access to *cis*- and *trans*- forms of both 3-hydroxy and 4-hydroxy proline have been found, but also because of their wider substrate specificities, which encompass non-canonical amino acids such as pipecolic acid. Some experimental examples are given further.

5.5 Production of *cis*-3-Hydroxyproline Using Recombinant Cells of *E. coli* Expressing Proline-3-Hydroxylase from *Streptomyces* sp.

In this example, *cis*-3-hydroxyproline **38**, an intermediate in the synthesis of a Bristol–Myers–Squibb pharmaceutical candidate BMS-564929 **40** (Figure 8.17) was produced from L-proline with a yield of 85% [55]. The gene encoding proline 3-hydroxylase from *Streptomyces* sp. was cloned into a proprietary vector by the investigators, which put the expression under the control of the *lac* promoter in much the same way as for commercially available pET plasmids. A recombinant strain of *E. coli* BL21 (DE3) was created by transforming the bacteria with the recombinant plasmid. Whole-cell biotransformations were performed using pregrown cells in buffer, which were prepared in the following

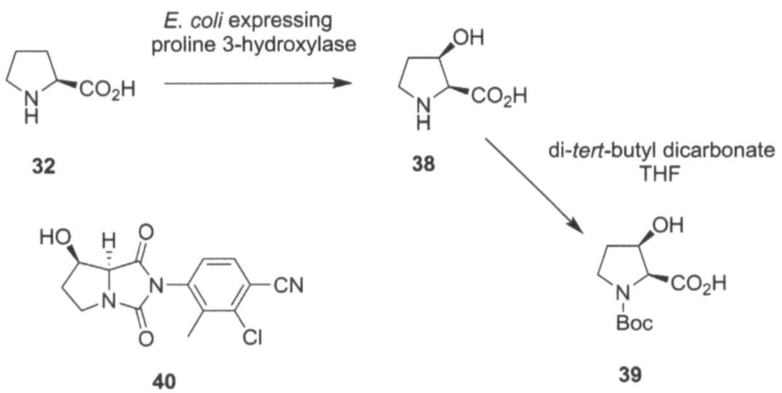

FIGURE 8.17
Hydroxylation of L-proline 32 by recombinant cells of *E. coli* expressing proline 3-hydroxylase [54]. *cis*-3-Hydroxy proline was used to make the pharmaceutical target **40**.

way: Cells were grown on "MT5" medium that contained 4% (v/v) glycerol; 2% (w/v) yeastamin; 0.6% (w/v) disodium hydrogen phosphate; 0.3% (w/v) potassium dihydrogen phosphate; 0.125% (w/v) ammonium sulfate and 0.0246% (w/v) magnesium sulfate heptahydrate, with 50 mg/L kanamycin as antibiotic marker. Cells were grown at 30 °C with shaking until an optical density (A_{600}) of 1.0, at which point 0.1 mM IPTG was added to effect induction. Cells were grown at 30 °C with shaking for 24 h at which point they were harvested by centrifugation and washed with 50 mM phosphate buffer, ready for application in resting whole-cell biotransformation reactions.

A suspension of these cells (10–20% in the phosphate buffer) was placed in a 2-L bioreactor. To this was added 0.88 g ascorbic acid, 0.28 g iron sulfate heptahydrate, 17.1 g 2-oxoglutarate, and 10 g L-proline as substrate. The reaction was stirred at 30 °C and at pH 7.0 for 24 h. It was found that the yields of the reaction were improved through the addition of both 2-OG (8 g/h) and ascorbate (3 g/h) during the reaction. The reaction was monitored by HPLC.

In order to recover the hydroxylated product, it was converted to its Boc-protected derivative **39** after an aqueous work up. The *E. coli* cells were first separated by centrifugation. The supernatant was heated to 90 °C for 10 min and the solution then cooled, resulting in a white precipitate that was filtered off. The resultant solid was washed with water, and the combined wash and filtrate fractions were adjusted to pH 2.5 and kept for 18 h at 5 °C. A resultant precipitate was filtered off and washed, and the wash and filtrate combined. The aqueous phase was washed with ethyl acetate (1.5 L) and the pH of the aqueous fraction was adjusted to pH 12.5 using sodium hydroxide. The solution was stirred at room temperature and 26.6 g of di-*tert*-butyl dicarbonate (corresponding to two equivalents of hydroxyproline), dissolved in 375 mL of tetrahydrofuran was added. After 3 h, 850 mL of isopropyl acetate was added, and the mixture stirred for 5 min. The pH of the aqueous fraction was adjusted to pH of 2.2, and the solution extracted with isopropyl acetate. The combined organic extracts were dried and filtered, and the solvent removed to yield the Boc-protected *cis*-3-hydroxy proline that could be further purified by recrystallization.

5.6 Hydroxylation of Pipecolic Acid Using Recombinant Cells of *E. coli* Expressing Proline-*trans*-4-hydroxylase from *Dactylsporangium* sp.

On a smaller scale, a similar, but regio- and enantiocomplementary 2-OG dependent hydroxylase was used to hydroxylate pipecolic acid **41** (Figure 8.18) [56]. This enzyme can also be used to generate *trans*-4-hydroxy proline. The gene encoding proline *trans*-4-hydroxylase from *Dactylsporangium* sp. was synthesized with codon expression suitable for optimal expression in *E. coli*, and the gene was cloned into the Novagen vector pET-28b. In an effort to improve expression further, the proline hydroxylase gene was in this case coexpressed with genes encoding chaperones: pL1SL2 or pGKJE8 in the strain *E. coli* Codon Plus RP. Cells were grown in 80 mL of medium consisting of 1% (v/v) glycerol,

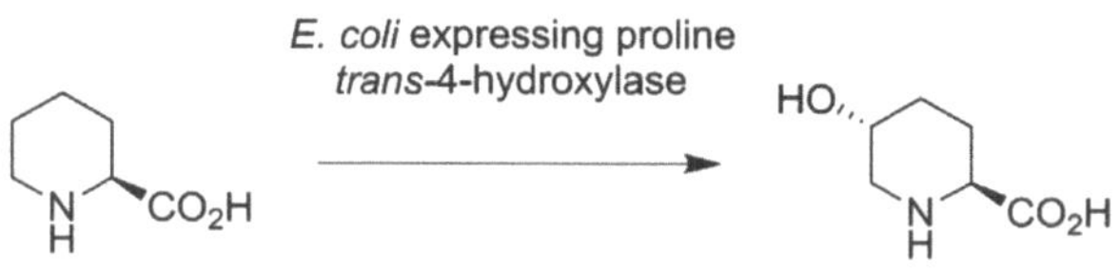

FIGURE 8.18
Hydroxylation of L-pipecolic acid 41 by recombinant cells of *E. coli* expressing proline *trans*-4-hydroxylase [55].

10 g/L ammonium sulfate, 2 g/L sodium chloride, 1 g/L dipotassium hydrogen phosphate, 1 g/L magnesium sulfate, 5 g/L ammonium chloride, 5 g/L sodium acetate, and 0.01% (w/v) arabinose, with both chloramphenicol (34 μg/mL) and kanamycin (50 μg/mL) as antibiotic markers. The cultures were incubated at 37°C with shaking, and when the optical density (A_{600}) had reached 1.1 to 1.4, 0.2 mM IPTG was added to effect induction. L-Pipecolic acid **41** (52 mg), iron (II) sulfate heptahydrate (11.1 mg) and 2-OG (0.145 g) were then added, and the cells incubated for 72 h at 28°C. During the reaction, at 24 and 48 h, 0.4 mL of glycerol was added to the reactions. After 72 h, the cells were removed by centrifugation and the pH of the supernatant adjusted to pH 1.6. The hydroxylated product was initially isolated by cation-exchange chromatography on a Dowex W × 8, 50–100 mesh (H^+ form). Fractions containing the hydroxylated product were then further purified using an anion exchange column Dowex 1 × 8, 20-50 mesh (HO^- form). From 52 mg pipecolic acid, 35 mg (61% yield) of *trans*-4-hydroxy pipecolic acid **42** were isolated.

5.7 Summary

2-OG dependent hydroxylases are beginning to gain more attention as biocatalysts for the production of hydroxylated amino acids, and also for other compounds, as the diversity of genes and knowledge of their natural substrate specificity increases. They are more readily usable than P450s, but their narrower substrate specificity dictates that applications at this stage have been limited. These applications are sure to increase however as we gain more knowledge of their substrate specificity, catalytic characteristics, and process suitability.

6 CONCLUSIONS

The examples of enzyme-catalyzed hydroxylation given within this chapter reveal the potential value of these catalysts in generating selectively oxygenated intermediates for organic synthesis. Despite the advantages of biohydroxylations over abiotic catalysis in terms both of selectivity and green chemistry that are illustrated previously, it is clear that many challenges remain to be overcome before these catalysts can be applied on large scales with the same confidence as hydrolase, reductase, and lyase enzymes that are already frequently encountered in industrial biocatalysis. In addition to problems of poor process stability and, with some exceptions, low catalytic rates of hydroxylation enzymes, there are also technical and facilities considerations associated with the requirements for cofactors and redox partners, which make these enzymes, much less user-friendly than other enzyme classes. However, advances in gene synthesis, cloning, and expression technology are making recombinant catalysts more accessible to synthetic chemistry research groups, and the growing appreciation of the potential value of biohydroxylation catalysts will ensure that research that seeks to overcome obstacles to application will continue in the future.

References

[1] Z. Li, J.B. van Beilen, W.A. Duetz, A. Schmid, A. de Raadt, H. Griengl, B. Witholt, Oxidative biotransformations using oxygenases, Curr. Opin. Chem. Biol. 6 (2) (2002) 136–144.

[2] P.R. Ortiz de Montellano (Ed.), Cytochrome P450: Structure, mechanism, and biochemistry, third ed., Kluwer Academic/Plenum Publishers, New York, 2005.

[3] G. Grogan, Cytochromes P450: exploiting diversity and enabling application as biocatalysts, Curr. Opin. Chem. Biol. 15 (2) (2011) 241–248.

[4] R. Fasan, Tuning P450 enzymes as oxidation catalysts, ACS Catal. 2 (4) (2012) 647–666.

[5] V.B. Urlacher, M. Girhard, Cytochrome P450 monooxygenases: an update on perspectives for synthetic application, Trends Biotechnol. 30 (1) (2012) 26–36.

[6] J. Hedegaard, I.C. Gunsalus, Mixed function oxidation: IV. An induced methylene hydroxylase in camphor oxidation, J. Biol. Chem. 240 (10) (1965) 4038–4043.

[7] H.H. Pauli, T.M. Kutchan, Molecular cloning and functional heterologous expression of two alleles encoding (*S*)-*N*-methylcoclaurine 3′-hydroxylase (CYP80B1), a new methyl jasmonate-inducible cytochrome P-450-dependent mono-oxygenase of benzylisoquinoline alkaloid biosynthesis, Plant J. 13 (6) (1998) 793–801.

[8] J.F. Andersen, C.R. Hutchinson, Characterization of *Saccharopolyspora erythraea* cytochrome P-450 genes and enzymes, including 6-deoxyerythronolide B hydroxylase, J. Bacteriol. 174 (3) (1992) 725–735.

[9] M.A. Pierrel, Y. Batard, M. Kazmaier, C. Mignotte-Vieux, F. Durst, D. Werck-Reichhart, Catalytic properties of the plant cytochrome P450 CYP73 expressed in yeast, Eur. J. Biochem. 224 (3) (1994) 835–844.

[10] J.S. Ngui, W. Tang, R.A. Stearns, M. Shou, R.R. Miller, Y. Zhang, J.H. Lin, T.A. Baillie, Cytochrome P450 3A4-mediated interaction of diclofenac and quinidine, Drug Metab. Dispos. 28 (9) (2000) 1043–1050.

[11] D. Nelson, The cytochrome P450 homepage, Hum. Genom. 4 (1) (2009) 59–65.

[12] D. Nelson, Cytochrome P450 nomenclature, in: I. Phillips, E. Shephard (Eds.), Cytochrome P450 protocols, vol. 320, Humana Press, New York, USA, 2006, pp. 1–10.

[13] A.W. Munro, H.M. Girvan, A.E. Mason, A.J. Dunford, K.J. McLean, What makes a P450 tick?, Trends Biochem. Sci. 38 (3) (2013) 140–150.

[14] J. Rittle, M.T. Green, Cytochrome P450 compound I: capture, characterization, and C-H bond activation kinetics, Science 330 (6006) (2010) 933–937.

[15] F. Hannemann, A. Bichet, K.M. Ewen, R. Bernhardt, Cytochrome P450 systems – biological variations of electron transport chains, Biochim. Biophys. Acta (BBA) – General Subjects 1770 (3) (2007) 330–344.

[16] A.W. Munro, D.G. Leys, K.J. McLean, K.R. Marshall, T.W.B. Ost, S. Daff, C.S. Miles, S.K. Chapman, D.A. Lysek, C.C. Moser, C.C. Page, P.L. Dutton, P450 BM3: the very model of a modern flavocytochrome, Trends Biochem. Sci. 27 (5) (2002) 250–257.

[17] G.A. Roberts, G. Grogan, A. Greter, S.L. Flitsch, N.J. Turner, Identification of a new class of cytochrome P450 from a *Rhodococcus* sp, J. Bacteriol. 184 (14) (2002) 3898–3908.

[18] F. Sabbadin, R. Hyde, A. Robin, E.-M. Hilgarth, M. Delenne, S. Flitsch, N. Turner, G. Grogan, N.C. Bruce, LICRED: A versatile drop-in vector for rapid generation of redox-self-sufficient cytochrome P450s, Chembiochem 11 (7) (2010) 987–994.

[19] H.C. Murray, D.H. Peterson, Oxygenation of steroids by *Mucorales* fungi. US Patent 2,602,769 A, 1952.

[20] G.S. Fonken, M.E. Herr, H.C. Murray, L.M. Reineke, Microbiological hydroxylation of monocyclic alcohols, J. Am. Chem. Soc. 89 (3) (1967) 672–675.

[21] G.S. Fonken, M.E. Herr, H.C. Murray, L.M. Reineke, Microbiological oxygenation of alicyclic amides, J. Org. Chem. 33 (8) (1968) 3182–3187.

[22] R.A. Johnson, M.E. Herr, H.C. Murray, L.M. Reineke, G.S. Fonken, Microbiological oxygenation of some azabicycloalkanes, J. Org. Chem. 33 (8) (1968) 3195–3201.

[23] R.A. Johnson, H.C. Murray, L.M. Reineke, G.S. Fonken, Stereochemistry of microbiological hydroxylation. II. Oxygenation of 1-benzoylalkylpiperidines, J. Org. Chem. 34 (8) (1969) 2279–2284.

[24] R.A. Johnson, M.E. Herr, H.C. Murray, G.S. Fonken, Stereochemistry of microbiological hydroxylation. IV. Stereochemistry of the microbiological oxygenation of *N*-acylcyclohexylamines, J. Org. Chem. 35 (3) (1970) 622–626.

[25] R.A. Johnson, H.C. Murray, L.M. Reineke, Stereochemistry of microbiological hydroxylation. VI. Microbiological oxygenation of acyclic *N*-alkylbenzamides, J. Am. Chem. Soc. 93 (19) (1971) 4872–4880.

[26] G.J. Grogan, H.L. Holland, The biocatalytic reactions of *Beauveria* spp, J. Mol. Catal. B Enzym. 9 (1–3) (2000) 1–32.

[27] S.J. Aitken, G. Grogan, C.S.Y. Chow, N.J. Turner, S.L. Flitsch, Biohydroxylations of Cbz-protected alkyl substituted piperidines by *Beauveria bassiana* ATCC 7159, J. Chem. Soc. Perkin Trans. 1 (1998) 3365–3370.

[28] H.L. Holland, T.A. Morris, P.J. Nava, M. Zabic, A new paradigm for biohydroxylation by *Beauveria bassiana* ATCC 7159, Tetrahedron 55 (24) (1999) 7441–7460.

[29] G. Xiao, S.-H. Ying, P. Zheng, Z.-L. Wang, S. Zhang, X.-Q. Xie, Y. Shang, R.J. St. Leger, G.-P. Zhao, C. Wang, M.-G. Feng, Genomic perspectives on the evolution of fungal entomopathogenicity in *Beauveria bassiana*, Sci. Rep. 2 (2012) 483.

[30] Z. Li, H.-J. Feiten, D. Chang, W.A. Duetz, J.B. van Beilen, B. Witholt, Preparation of (*R*)- and (*S*)-*N*-protected 3-hydroxypyrrolidines by hydroxylation with *Sphingomonas* sp. HXN-200, a highly active, regio- and stereoselective, and easy to handle biocatalyst, J. Org. Chem. 66 (25) (2001) 8424–8430.

[31] D. Chang, H.-J. Feiten, B. Witholt, Z. Li, Regio- and stereoselective hydroxylation of N-substituted piperidin-2-ones with *Sphingomonas* sp. HXN-200, Tetrahedron Asymmetry 13 (19) (2002) 2141–2147.

[32] W. Zhang, W.L. Tang, Z. Wang, Z. Li, Regio- and stereoselective biohydroxylations with a recombinant *Escherichia coli* expressing P450pyr monooxygenase of *Sphingomonas* sp. HXN-200, Adv. Synth. Catal. 352 (18) (2010) 3380–3390.

[33] A.M. Sawayama, M.M.Y. Chen, P. Kulanthaivel, M.-S. Kuo, H. Hemmerle, F.H. Arnold, A panel of cytochrome P450 BM3 variants to produce drug metabolites and diversify lead compounds, Chem. Eur. J. 15 (43) (2009) 11723–11729.

[34] P. Meinhold, M.W. Peters, A. Hartwick, A.R. Hernandez, F.H. Arnold, Engineering cytochrome P450 BM3 for terminal alkane hydroxylation, Adv. Synth. Catal. 348 (6) (2006) 763–772.

[35] M.W. Peters, P. Meinhold, A. Glieder, F.H. Arnold, Regio- and enantioselective alkane hydroxylation with engineered cytochromes P450 BM-3, J. Am. Chem. Soc. 125 (44) (2003) 13442–13450.

[36] V. Dodhia, A. Fantuzzi, G. Gilardi, Engineering human cytochrome P450 enzymes into catalytically self-sufficient chimeras using molecular Lego, J. Biol. Inorg. Chem. 11 (7) (2006) 903–916.

[37] D. Scheps, S. Honda Malca, S.M. Richter, K. Marisch, B.M. Nestl, B. Hauer, Synthesis of ω-hydroxy dodecanoic acid based on an engineered CYP153A fusion construct, Microb. Biotechnol. 6 (6) (2013) 694–707.

[38] S. Li, L.M. Podust, D.H. Sherman, Engineering and analysis of a self-sufficient biosynthetic cytochrome P450 PikC fused to the RhFRED reductase domain, J. Am. Chem. Soc. 129 (43) (2007) 12940–12941.

[39] M. Nodate, M. Kubota, N. Misawa, Functional expression system for cytochrome P450 genes using the reductase domain of self-sufficient P450RhF from *Rhodococcus* sp. NCIMB 9784, Appl. Microbiol. Biotechnol. 71 (4) (2006) 455–462.

[40] M. Bordeaux, A. Galarneau, F. Fajula, J. Drone, A regioselective biocatalyst for alkane activation under mild conditions, Angew. Chem. Int. Ed. 50 (9) (2011) 2075–2079.

[41] M. Lisurek, M.-J. Kang, R.W. Hartmann, R. Bernhardt, Identification of monohydroxy progesterones produced by CYP106A2 using comparative HPLC and electrospray ionisation collision-induced dissociation mass spectrometry, Biochem. Biophys. Res. Commun. 319 (2) (2004) 677–682.

[42] H. Agematu, N. Matsumoto, Y. Fujii, H. Kabumoto, S. Doi, K. Machida, J. Ishikawa, A. Arisawa, Hydroxylation of testosterone by bacterial cytochromes P450 using the *Escherichia coli* expression system, Biosci. Biotechnol. Biochem. 70 (1) (2006) 307–311.

[43] W.J.H. van Berkel, N.M. Kamerbeek, M.W. Fraaije, Flavoprotein monooxygenases, a diverse class of oxidative biocatalysts, J. Biotechnol. 124 (4) (2006) 670–689.

[44] R.K. Wierenga, R.J. de Jong, K.H. Kalk, W.G.J. Hol, J. Drenth, Crystal structure of *p*-hydroxybenzoate hydroxylase, J. Mol. Biol. 131 (1) (1979) 55–73.

[45] S. Montersino, D. Tischler, G.T. Gassner, W.J.H. van Berkel, Catalytic and structural features of flavoprotein hydroxylases and epoxidases, Adv. Synth. Catal. 353 (13) (2011) 2301–2319.

[46] B. Entsch, W.J. van Berkel, Structure and mechanism of *para*-hydroxybenzoate hydroxylase, FASEB J. 9 (7) (1995) 476–483.

[47] H.P. Kohler, D. Kohler-Staub, D.D. Focht, Degradation of 2-hydroxybiphenyl and 2,2′-dihydroxybiphenyl by *Pseudomonas* sp. strain HBP1, Appl. Environ. Microbiol. 54 (11) (1988) 2683–2688.

[48] A. Schmid, H.-P.E. Kohler, K.-H. Engesser, *E. coli* JM109 pHBP461, a recombinant biocatalyst for the regioselective monohydroxylation of ortho-substituted phenols to their corresponding 3-substituted catechols, J. Mol. Catal. B: Enzym. 5 (1–4) (1998) 311–316.

[49] I.J. Clifton, M.A. McDonough, D. Ehrismann, N.J. Kershaw, N. Granatino, C.J. Schofield, Structural studies on 2-oxoglutarate oxygenases and related double-stranded β-helix fold proteins, J. Inorg. Biochem. 100 (4) (2006) 644–669.

[50] T. Shibasaki, H. Mori, S. Chiba, A. Ozaki, Microbial proline 4-hydroxylase screening and gene cloning, Appl. Environ. Microbiol. 65 (9) (1999) 4028–4031.

[51] J.M. Neary, A. Powell, L. Gordon, C. Milne, F. Flett, B. Wilkinson, C.P. Smith, J. Micklefield, An asparagine oxygenase (AsnO) and a 3-hydroxyasparaginyl phosphotransferase (HasP) are involved in the biosynthesis of calcium-dependent lipopeptide antibiotics, Microbiology 153 (3) (2007) 768–776.

[52] T. Kodera, S.V. Smirnov, N.N. Samsonova, Y.I. Kozlov, R. Koyama, M. Hibi, J. Ogawa, K. Yokozeki, S. Shimizu, A novel L-isoleucine hydroxylating enzyme, L-isoleucine dioxygenase from *Bacillus thuringiensis*, produces (2*S*,3*R*,4*S*)-4-hydroxyisoleucine, Biochem. Biophys. Res. Commun. 390 (3) (2009) 506–510.

[53] H. Mori, T. Shibasaki, K. Yano, A. Ozaki, Purification and cloning of a proline 3-hydroxylase, a novel enzyme which hydroxylates free L-proline to *cis*-3-hydroxy-L-proline, J. Bacteriol. 179 (18) (1997) 5677–5683.

[54] I.J. Clifton, L.-C. Hsueh, J.E. Baldwin, K. Harlos, C.J. Schofield, Structure of proline 3-hydroxylase, Eur. J. Biochem. 268 (24) (2001) 6625–6636.

[55] R.M. Johnston, L.N. Chu, M. Liu, S.L. Goldberg, A. Goswami, R.N. Patel, Hydroxylation of L-proline to *cis*-3-hydroxy-L-proline by recombinant *Escherichia coli* expressing a synthetic L-proline-3-hydroxylase gene, Enzym. Microb. Technol. 45 (6-7) (2009) 484–490.

[56] C. Klein, W. Hüttel, A simple procedure for selective hydroxylation of L-proline and L-pipecolic acid with recombinantly expressed proline hydroxylases, Adv. Synth. Catal. 353 (8) (2011) 1375–1383.

CHAPTER 9

Miscellaneous Key Non-C—C Bond Forming Enzyme Reactions

Florian Rudroff, Michael J. Fink, Marko D. Mihovilovic
Vienna University of Technology, Institute of Applied Synthetic Chemistry, Vienna, Austria

1 NITRILE HYDRATASES

Nitrile hydratases (NHases) catalyze the hydration of organic nitriles to amides under very benign reaction conditions (neutral aqueous environment and room temperature) and therefore offer a chemoselective alternative to classical approaches, where functional group compatibility is often limited due to the harsh acidic or basic solutions used [1]. Starting with their application in acrylamide production [2,3], this enzyme class is one of the most prominent in industrial processes with respect to production volume ($>3 \times 10^7$ kg/a for acrylonitrile hydration) [4]. Hence, Lonza (Switzerland) uses a nitrile hydratase to convert 3-cyanopyridine into nicotinamide (6×10^3 tons/year). Very recently, a one-pot industrial protocol for the synthesis of a chiral intermediate for cilastatin was published that employed a nitrile hydratease/amidase approach [5].

NHases are widely found in bacteria, which thereby make use of nitriles as sole source of nitrogen by coupling with an amidase (hydrolases acting on amides that yield the corresponding carboxylic acid). This feature provides a means for new biocatalyst discovery via nitrile-selective screening [6,7]. An iron-based colorimetric assay for NHase/amidase-activity was recently developed and optimized for high throughput discovery [8].

All NHases are either Fe- or Co-dependent, with mechanistic evidence pointing toward a direct activation of the nitrile by the metal ion [9,10]. The metal binding site requires two cysteine residues in higher oxidation states (sulfinic and sulfenic acid) by posttranslational modification, and the smallest functional quarternary structure is a $\alpha\beta$ dimer that often forms larger multimers in solution [11]. Hence, most applications on laboratory and production scale use wild-type strains or recombinant hosts, as isolated enzyme formulations usually suffer from low activity and operational instability.

Organic Synthesis Using Biocatalysis. http://dx.doi.org/10.1016/B978-0-12-411518-7.00009-3

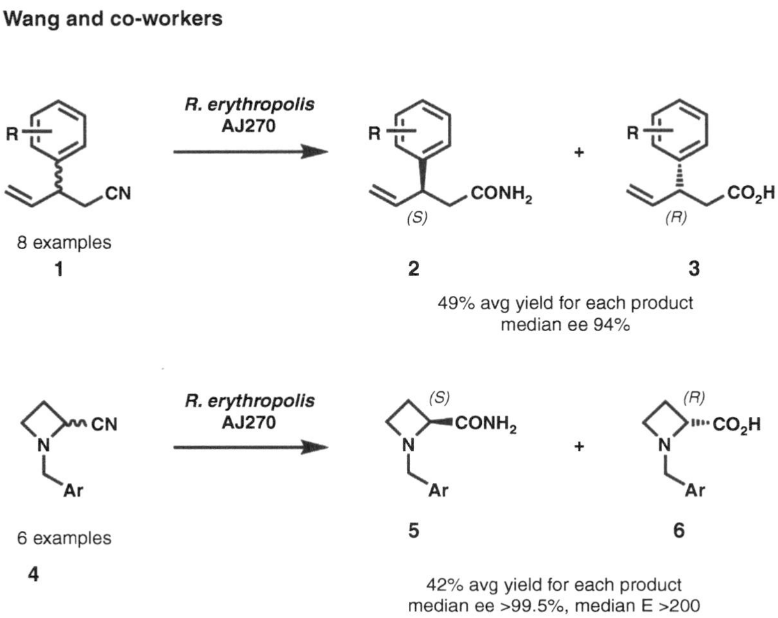

SCHEME 9.1
Kinetic resolutions of aliphatic nitriles toward enantiomerically enriched amides and carboxylic acids catalyzed by wild-type strain *R. erythropolis* AJ270 [13].

In general, NHases show high substrate promiscuity paired with high activity (e.g., >2 kU/mg enzyme for the NHase *Rhodococcus rhodochrous* J1) [12], making them ideal catalysts for chemoselective transformations. Enantioselectivity at the nitrile hydratase step is usually poor to nondetectable, but by combining with enantioselective amidases, optically pure carboxylic acids can ultimately be obtained. This was recently demonstrated by Wang and co-workers. First, a phenomenological mechanistic study showed that a β-vinyl substituent dramatically increased stereoselectivity in biotransformations using wild-type strain *Rhodococcus erythropolis* AJ270 (when compared to the saturated ethyl analog; Scheme 9.1, top) [13]. They were thus able to produce 3-phenyl-4-pentenoic (*S*)-amides **2** and (*R*)-acids **3** with optimum yields (49% average yield for each product) and good optical purities (median ee 94%). The subsequent application of these chiral building blocks in the synthesis of chiral γ-amino acids as well as γ- and ε-lactams was also demonstrated. In a later study, the same organism was used to resolve racemic mixtures of azetidine-2-carbonitriles **4** to enantiomerically pure amides **5** and acids **6**, which serve as useful intermediates in pharmaceutical and agrochemical fields, and are unique amino acid structures in various alkaloids (Scheme 9.1, bottom) [14]. Here too, the natural *in vivo* cascade reaction yielded both products with excellent yield and optical purity (42% average yield, >99.5% median ee, *E* >200).

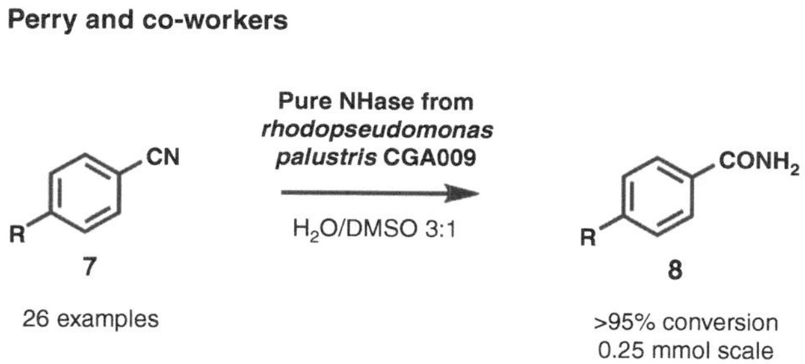

SCHEME 9.2
Chemoselective hydration of various (hetero-)aromatic nitriles to amides using NHase from *R. palustris* CGA009 [15].

Whenever this double hydrolytic activity is not desirable, and an amide (rather than the carboxylic acid) is the target, process development mostly relies on screening and optimization of culture conditions for wild-type strains rather than genetic engineering. There are, however, exceptions to this trend: Perry and coworkers recently reported the isolation and application of a solvent-tolerant NHase from *Rhodopseudomonas palustris* CGA009 (Scheme 9.2) [15]. A series of *p*-substituted benzonitriles and unsubstituted furan-, thiophene-, and pyridine-carbonitriles were efficiently hydrated to the corresponding carboxamides in aqueous buffer/DMSO solvent mixtures on analytical scales. The authors further probed the enzyme with sterically congested 2- and 2,6-susbtituted nitriles **7**, showing that 2-fluoro-, 2-chloro-, and 2,6-difluorobenzonitrile are well accepted without side reactions; unfortunately, larger substituents could not be converted with appreciable efficiencies.

Martinkova and coworkers efficiently suppressed amidase activity to less than 20% by immobilizing *Rhodococcus equi* A4 cells in polyvinyl alcohol/polyethylene glycol beads (LentiKats®); the formulation was tested with four small aromatic and aliphatic nitriles, and good recyclability and operational stability was demonstrated [16].

Medium and culture engineering by the group of Zheng provided a good example of a common approach: by medium-enrichment screening of soil samples they were able to identify nine potential strains for the hydration of 2-amino-2,3-dimethylbutyronitrile **9**; the resulting amide **10** is an important intermediate of highly potent imidazolinone herbicides and can be used in racemic form (Scheme 9.3) [17]. The main challenges in this investigation were to eliminate amidase activity and to increase cyanide tolerance (cyanide inhibition is a known issue with NHases and α-aminonitriles spontaneously decompose to aldehydes, ammonia, and hydrogen cyanide) [18]. Optimizing reaction temperature for both catalyst and starting material stability, organic cosolvent addition (hexane), and fed-batch operation with the best strain candidate *Rhodococcus boritolerans* CCTCC M 20818 led to a lab-scale method for producing α-aminoamide **10**. The purity (98.5% after recrystallization), yield (91%), and

Zheng and co-workers

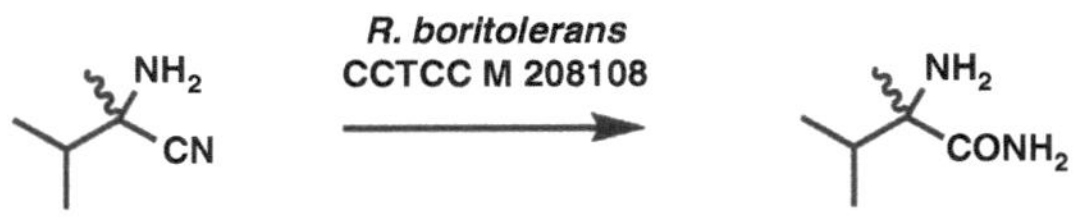

9

10

12 g scale, 50 g/L
91% yield

SCHEME 9.3
Process optimization for NHase-catalyzed production of 2-amino-2,3-dimethylbutyramide using wild-type strain *R. boritolerans* CCTCC M 208108 [17].

productivity (50 g/L product concentration, 12.3 g product g^{-1} catalyst; two reaction cycles) were all sufficient to meet targets.

Lastly, an elegant approach to circumventing amidase activity and improve general performance was presented by Payne's group at Du Pont. The thermostable NHase from *Comamonas testosteroni* 5-MGAM-4D was recombinantly overexpressed in *Escherichia coli*, increasing biocatalyst activity almost threefold (7752 U/g dry cell weight vs. 2577 U/g in the wild-type strain) [19]. Functional biosynthesis of this cobalt-dependent enzyme required adaptations to the growth medium (cobalt chloride and sodium citrate). More importantly, an expression construct including an additional DNA sequence approximately 1 kb downstream of the NHase β-subunit gene did not change protein production level, but increased activity by more than 20-fold. This region encodes a potential accessory protein (P7K). Although this peptide seems not to be a factor for NHase solubility and no catalytic motif was found, the authors hypothesized that it serves as a chaperone or facilitates cobalt incorporation. This finding underlines the aforementioned complexity of functional quaternary structure and required posttranslational modifications of NHases, when taken out of their natural host.

With the final system, small aliphatic nitriles (C_3–C_6, also hydroxy-functionalized) could be hydrated at 0.5 M substrate concentrations with high efficiencies (1–20 h, 0.2–0.4% w/v dry cell weight) and quantitative yields of amide products were obtained. Using an alginate-immobilized cell formulation, the biocatalyst showed no significant loss of activity after three recycling steps even at elevated substrate concentration (16% acrylonitrile, 3.0 M solution). This example strongly demonstrates the capabilities of this enzyme class.

2 NITRILASES

Nitrilases (EC 3.5.5.1) catalyze the direct hydrolysis of organic cyanides or nitriles into the corresponding carboxylic acids without the release of intermediate amides. These enzymes are generally believed to contain a conserved Glu-Lys-Cys catalytic motif [20]. Although yet unproven due to a lack of crystal structures, the

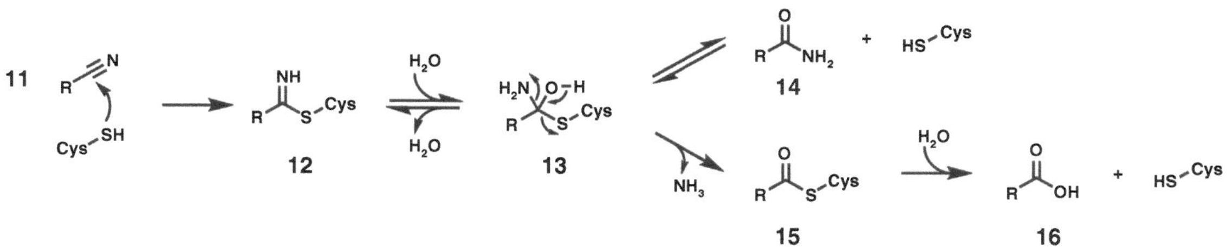

SCHEME 9.4
Generally accepted mechanism for nitrilase-catalyzed hydration of organic nitriles to carboxylic acids (bottom right, 16) and possible side reaction leading to amides (top right, 14); Cys-SH: active site cysteine residue [21,22].

mechanism is supposed to start with a nucleophilic attack of the cysteine thiol to give a thioimidate **12**, which is in equilibrium with a tetrahedral intermediate **13** upon addition/elimination of water. General base action of the catalytically active glutamate and charge stabilization of the neighboring lysine leads to collapse of this intermediate, mainly by expulsion of ammonia and subsequent hydrolysis of the enzyme-bound thioester **15** (Scheme 9.4, bottom) to give carboxylic acid **16**; alternatively, cleavage of the C—S bond forms an amide **14** that was observed to occur with few substrates (Scheme 9.4, top) [21,22].

Nitrilases are widespread in plants, fungi, archaea, and bacteria [23–25], possibly owing to their central role in metabolizing organocyanides as carbon and/or nitrogen sources [26]. The enzyme family can be divided into three groups according to substrate specificity [27]:

1. Aromatic nitrilases, which mainly act on (hetero-)aromatic nitriles, for example, benzonitrile
2. Aliphatic nitrilases
3. Nitrilases, which preferentially hydrolyze arylacetonitriles

Some enzymes show extended substrate range and span more than one of these defined ranges. Similar to nitrile hydratases, nitrilases must associate as oligomers to form the active catalysts. This process may be triggered by various events (substrate, high ionic strength, organic solvents, heat, or high enzyme concentration) [27,28].

Nitrilases are known to be highly regioselective, often exclusively hydrolyzing only one particular nitrile group in a molecule [29]. This desymmetrization potential was employed in synthesizing the important building block (*R*)-4-cyano-3-hydroxybutyric acid **18**. This molecule is the key intermediate for the chiral side chain of Atorvastatin **19**, but may also be converted to L-carnitine **20** or (*R*)-4-amino-3-hydroxybutyric acid (GABOB, **21**), an anticonvulsant drug, in only three steps (Scheme 9.5, right) [30].

Initially, DeSantis et al. discovered nitrilase activity on **17** in a library screening, identifying many capable enzymes for the synthesis of both optical antipodes with appreciable enantiomeric purity (>90% ee) on laboratory scale

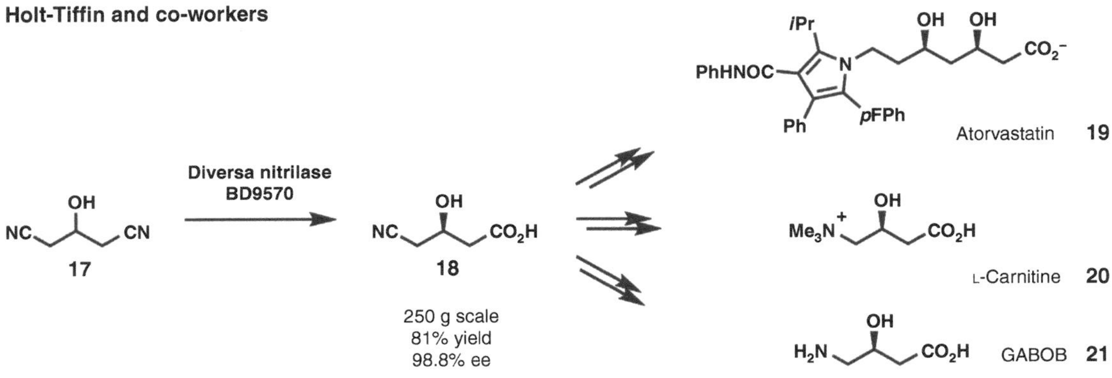

SCHEME 9.5
Desymmetrization of 3-hydroxyglutaronitrile 17 toward (*R*)-4-cyano-3-hydroxybutyric acid 18, a key intermediate in the synthesis of Atorvastatin 19, L-Carnitine 20 and (*R*)-4-amino-3-hydroxybutyric acid (GABOB, 21) [29,30].

(1 g, 0.24 M substrate concentration) [31]. Unfortunately, chiral induction deteriorated when the substrate titer was increased to production level (88% ee at 2.25 M). In a follow-up study, the researchers developed a HTS protocol using the pseudo-asymmetric ^{15}N-monolabeled (*R*)-enantiomer **17** as substrate. This way, all possible single-site mutants (>10k genetic variants, >30k clones) were screened and mass-spectrometric detection could be correlated to enantiomeric excess. A single-site mutant was eventually capable of producing the desired building block **18** with 98.5% ee and a productivity of 619 g/L/d [32]. This enzyme was later used by Holt-Tiffin and coworkers in combination with an improved and scalable synthesis for the symmetric starting material **17** on 250 g scale (81% yield, 98.8% ee) [33].

Even more striking is the ability to discriminate diastereomers of α,β-unsaturated nitriles: nitrilase AtNIT1 from *Arabidopsis thaliana* resolved a series of approximately 60:40 *E*/*Z* mixtures of **22** into configurationally pure *Z*-nitriles **23** and *E*-acids **24** (Scheme 9.6) [34]. This synthetic method thus represents a

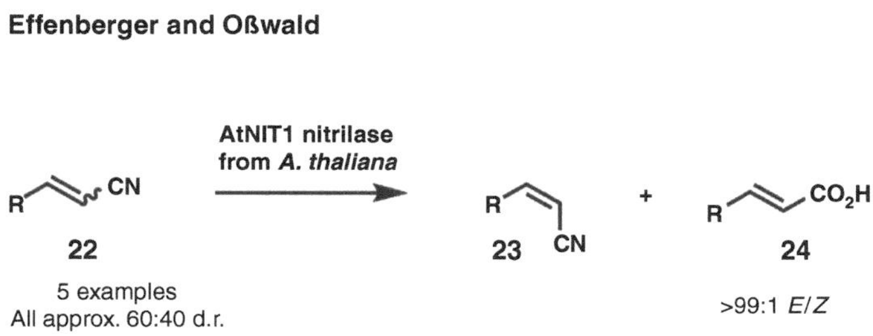

SCHEME 9.6
Resolution of diastereomeric vinyl nitriles using a nitrilase from *A. thaliana* [34].

DeSantis and co-workers

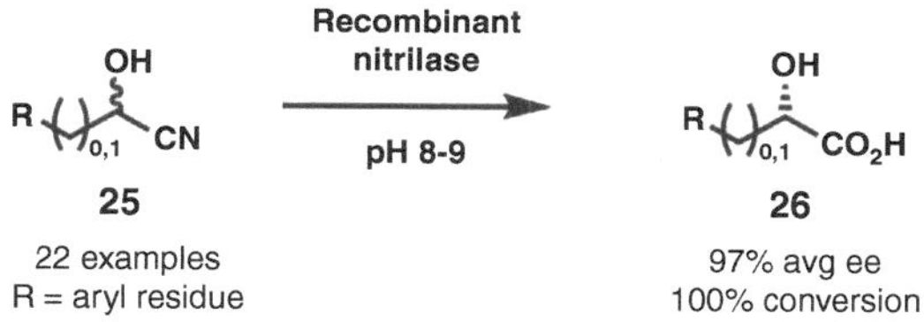

SCHEME 9.7
Dynamic kinetic resolution of racemic cyanohydrins 25 using recombinant nitrilases toward enantiopure mandelic and lactic acid derivatives 26 [31].

reactive purification method for product mixtures from standard condensation reactions.

Enantioselective transformations catalyzed by nitrilases often suffer from poor chiral recognition. Exceptions from this trend are benzaldehyde and phenylacetaldehyde cyanohydrins. As an additional advantage, these substrates racemize readily at near-neutral pH via reversible loss of hydrogen cyanide representing good starting materials for dynamic kinetic resolution processes. This was demonstrated using 22 substituted phenyl and heteroaryl derivates **25** with two recombinant nitrilases; a preparative biotransformation yielded (*S*)-phenyllactic acid **26** in 84% yield and 96% ee on 1 g scale (Scheme 9.7) [31].

Nitrilases are already well-established industrial biocatalysts with more than five known processes, for example, nicotinic and (*R*)-mandelic acids [35–38]. The aforementioned examples and the fact that many taxa offer an abundant source of still undiscovered wild-type enzymes may make this class of biocatalysts even more attractive to synthetic organic chemists in the future.

3 HYDANTOINASES AND CARBAMOYLASES

Both hydantoinases (the traditional name for a subgroup of dihydropyrimidinases EC 3.5.2.2) and carbamoylases (EC 3.5.1.77, EC 3.5.1.87) belong to the class of hydrolytic enzymes acting on carbon-nitrogen bonds other than peptide bonds (linear EC 3.5.1.x and cyclic EC 3.5.2.x). Although biological function and natural substrates in their fungal or bacterial origins remain mostly unclear to date [39,40], many such biocatalysts are well characterized for their application in enantioselective α-amino acid (AA) synthesis using the hydantoinase process [39,41]. Optically pure D- and L-AAs are widely used as additives in human and animal nutrition, in the pharmaceutical and cosmetics industries or as chiral building blocks for organic synthesis.

This multi-enzyme process (Scheme 9.8) starts with the enantioselective but often-reversible hydrolysis of one hydantoin (HYD, **28**) enantiomer to the open-chain *N*-carbamoyl amino acid **29** by a hydantoinase, followed by stereoselective hydrolysis of the carbamate (CARB, **29**), liberating ammonia, and carbon

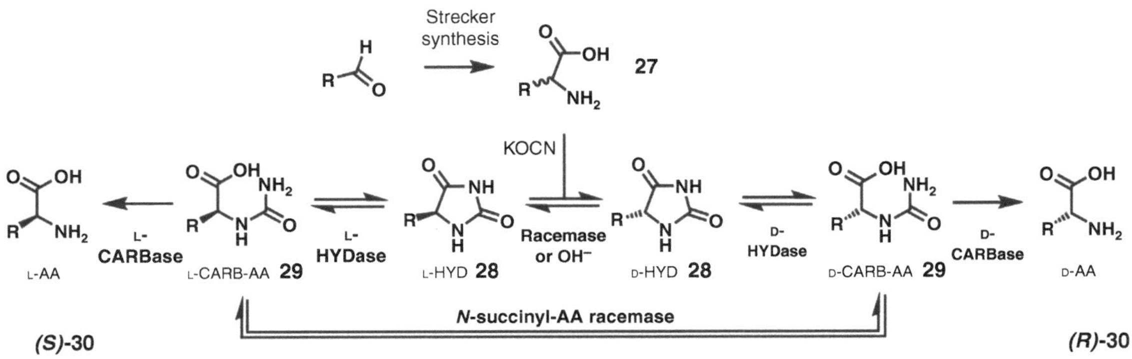

SCHEME 9.8
Basic principle for the synthesis of optically pure L- and D-amino acids via Strecker synthesis of racemic hydantoins and their (dynamic) kinetic resolution and stepwise hydrolysis toward the amino acids. KOCN, potassium cyanate; HYD, hydantoin **28**; CARB-AA, *N*-carbamoyl amino acid **29**; AA, amino acid **30**.

dioxide. In order to overcome the 50% yield limit set by the racemic nature of the starting material, either a chemical, base catalyzed (for a compilation of reaction rates, see Ref. [42]) or enzymatic racemization (by means of a hydantoin racemase, EC 5.1.99.5) [43–45] system is introduced to the process. Only recently, a different dynamic protocol was developed on the *N*-CARB-AA **29** stage, making use of the substrate promiscuity of *N*-succinyl amino acid racemase (EC 5.1.1.10) [46,47].

Both hydantoinases and carbamoylases share broad substrate acceptance toward substituted starting materials, although polar residues are only poorly tolerated. Hydantoinases only accept 5-monosubstituted HYD substrates **28**, and more commonly convert D- rather than L-enantiomers [48,49]. Successful mutagenesis approaches to overcome this limitation are known [47,50,51]. As a benefit from sequential double stereocontrol, it is possible to combine a poorly selective hydantoinase with an L-selective carbamoylase to produce L-AAs **(*S*)-30** [52]. For the latter enzymes, both types of stereoinduction are known from numerous microorganisms. Interestingly, whereas D-carbamoylases are inhibited by several metal ions, their enantiocomplementary biocatalysts actually *require* divalent cations (particularly Mn^{2+} and Co^{2+}), indicating a completely different mode of action. Both enzyme types are fairly thermostable (maximum reaction rates at 40–70 °C, although optimum stabilities are typically below 50 °C) and tolerate reasonably broad pH ranges (6.0–9.0). Still, most research in this area is focused on improvement of process stability in terms of higher operational temperature and also higher pH. Higher reaction pH values eliminate the need for an additional enzyme to racemize 5-substituted hydantoins.

Other efforts deal with immobilization [53–56] of enzyme formulations or genetic engineering of recombinant host organisms for D-selective hydrolases, as the products from this sequence are naturally degraded [57,58].

4 EPOXIDE HYDROLASE FOR EPOXIDE AND DIOL FORMATION

Epoxide hydrolases (EH, EC.3.3.2.3) were discovered decades ago and have been found in almost every organism examined, from microorganisms to plants and animals. The role of EHs in nature is detoxification, catabolism, and regulation of signaling molecules. In many higher organisms, EHs have very special tasks and thus have very limited substrate scope. Although EHs from higher organisms (mammalian, insects, and plants) were mainly investigated to gain insights into their biological roles, microbial enzymes were exploited in organic synthesis for the production of optical pure epoxides and *vicinal* diols. Attention in biocatalysis was gained after several EHs became available in recombinant form in *E. coli*. Key features that make this enzyme class especially useful in preparative synthesis are cofactor independence, decent solvent stabilities, substrate-dependent but generally very high regio- and enantioselectivities and the ability to purify and lyophilize the enzyme without significant loss of activity. Owing to these properties, a remarkable number of applications in synthetic chemistry were investigated and published for the preparation of Ibuprofen [59], α-Bisabolol [60], (*R*)-(−)-Mevalonolactone [61,62] and other chiral compounds [63,64].

Most EHs have α/β-hydrolase fold topology and consist of a core and a lid domain [65,66]. The lid domain is mainly α-helical and contains two tyrosine residues that point toward the catalytic triad and cover the core domain. Both tyrosine residues are involved in substrate binding, transition-state stabilization, and activation of the epoxide by protonation. The catalytic center is composed of two aspartate and one histidine residue. The first crystal structure of an epoxide hydrolase was solved for the enzyme from *Agrobacterium radiobacter* AD1 (EchA) [67]. The reaction mechanism of EHs is depicted in Scheme 9.9. First, a nucleophilic attack of the aspartic residue on the epoxide ring of the substrate **31** takes place and a covalently bound ester **32** is formed. This intermediate is subsequently hydrolyzed by a so-called charge relay system (general base catalysis) and the diol **33** is released from the active site. Key reaction partners are a histidine residue and a water molecule. It is worth mentioning that a limonene epoxide hydrolase discovered by Arand et al. displayed a different crystal structure and catalytic cycle that is discussed elsewhere [68].

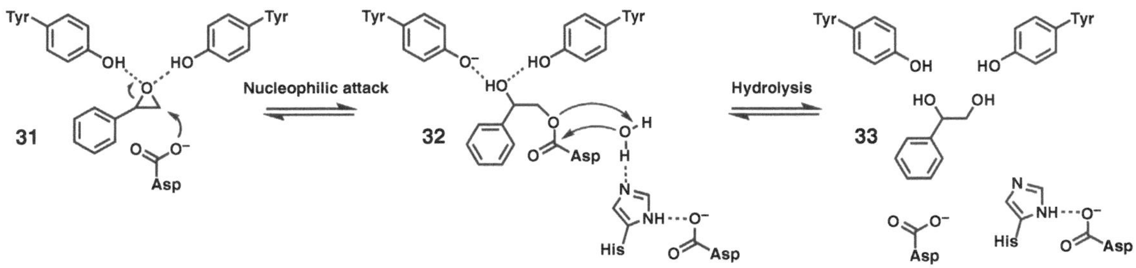

SCHEME 9.9
Reaction mechanism of epoxide hydrolases.

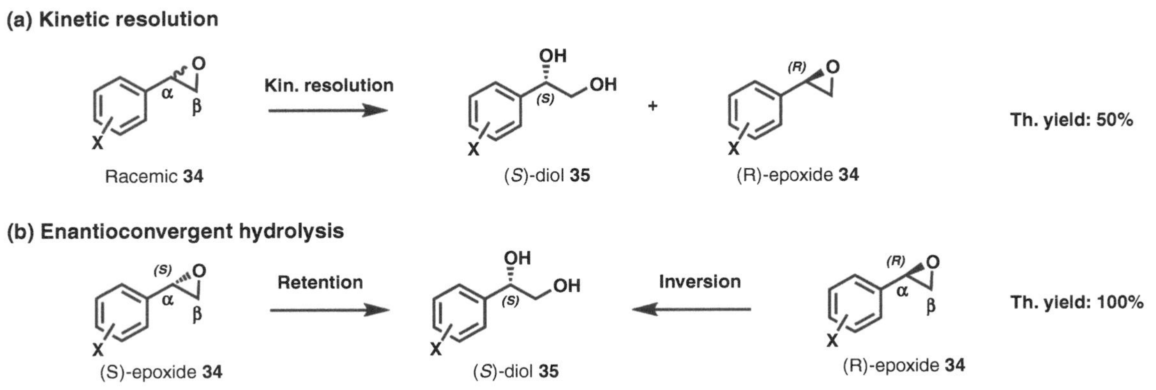

SCHEME 9.10
Three different concepts for the use of EHs in synthetic chemistry.

In the last decade, several examples of valuable hydrolysis of racemic epoxides have been described. Due to the nature of the epoxide substrates, three distinct synthetic strategies can be performed (Scheme 9.10). The first and the most widely used is a simple kinetic resolution (Scheme 9.10a) in which the reaction stops after the fast reacting enantiomer **(*S*)-34** has been consumed and the remaining epoxide **(*R*)-35** can be isolated in high optical purity. Reactions with high *E*-values provide enantiopure epoxides and a theoretical yield of 50%. The stereochemistry of the optically active diol **35** depends on whether the attack occurs at the α- or β-position, leading either to inversion or retention of the stereocenter, respectively. Due to the loss of 50% of material in classic kinetic resolutions, the enantioconvergent hydrolysis is industrially as well as synthetically more appealing. As depicted in Scheme 9.10b, a different alternative scenario is possible: first, by employing a single, highly selective EH, optically pure (*S*)-diol **35** is formed and the remaining epoxide (*R*)-**34** is hydrolyzed chemically with inversion of the stereogenic center, thus providing 100% final product yield (*S*)-**35**. This pathway strongly depends on a stereoselective chemical hydrolysis of the epoxide and often results in decrease of the overall optical purity of the diol.

A similar effect can be achieved by using two enantiocomplementary epoxide hydrolases that attack the opposite oxirane carbon, thereby providing 100% yield of optically enriched diol (*S*)-**35** (Scheme 9.10b).

Over the past 15 years, several examples were investigated in which optically pure epoxides have been prepared by kinetic resolutions [69–71]. Because many epoxides are poorly water-soluble, they tend to hydrolyze spontaneously and the products can inhibit the enzyme, leading to the investigation of aqueous/organic two-phase systems. Li and coworkers described a stereoselective epoxide hydrolase from *Sphingomonas* sp. HXN-200 in a biphasic system (water/*n*-hexane) for producing enantiomerically pure diols. Several different styrene oxides were hydrolyzed in very high concentrations (up to 200 mM) up to 99% ee [72].

Li and co-workers

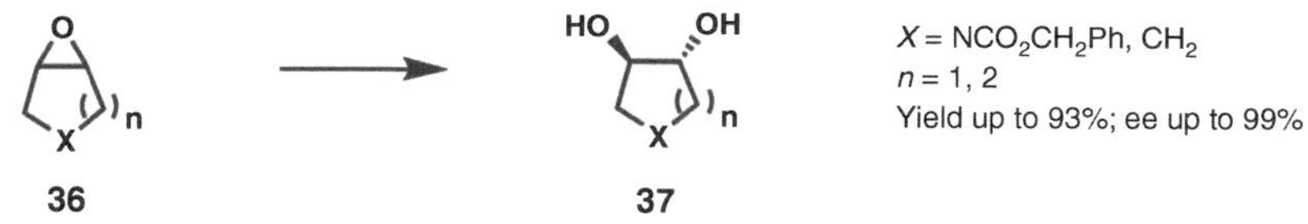

SCHEME 9.11
Enantioselective hydrolysis of meso-epoxides 36 with resting cells of *E. coli* expressing SpEH [73].

Furthermore, various *meso*-epoxides **36** were tested and gave *trans*-diols **37** in good to very good optical purities and excellent yields (Scheme 9.11) [73].

Recently, the same group reported two different epoxide hydrolases from *Sphingomonas* sp. and *Solanum tuberosum* with distinct stereopreferences that enabled enantiodivergent synthesis of diols in excellent yields and selectivities (Scheme 9.12) [74].

Another example of an enantioconvergent transformation of racemic epoxides was reported by Faber and coworkers who employed a chemoenzymatic approach for the synthesis of optically enriched diols. They used an (*S*)-selective epoxide hydrolase from *Nocardia* sp. in connection with an acid catalyzed hydrolytic step to convergently prepare (*S*)-diols **41** in good to excellent yields (71–98%) with enantiomeric excess values up to 99% (Scheme 9.13) [75].

Overall, epoxide hydrolases are very important enzymes for the asymmetric synthesis of optically pure epoxides and 1,2-diols. The biphasic application of EHs and the development of continuous processes have made them industrially attractive. In spite of that, their application in synthetic chemistry is still limited due to low regioselectivity, chemical instability, low solubility of epoxides in water and available alternative chemical routes. Nevertheless, it is important to exploit and investigate new epoxide hydrolases for synthetic routes to chiral epoxides and diols.

Li and co-workers

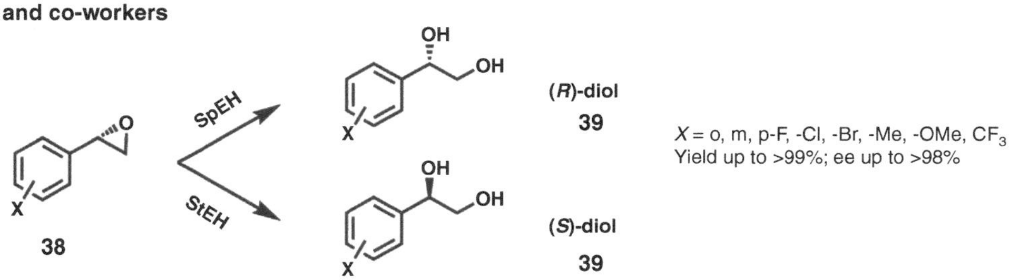

SCHEME 9.12
Enantioselective hydrolysis of styrene epoxides 38 by two different epoxide hydrolases from *Sphingomonas sp.* HXN-200 (SpEH) and *S. tuberosum* (StEH) [74].

Faber and co-workers

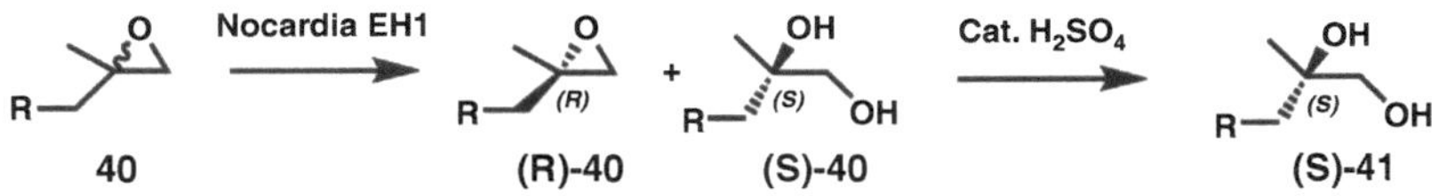

R = C_4H_9, $(CH_2)_2CH{=}CH_2$, $(CH_3)_3Br$, C_6H_5
Yield: 71-98%, 92-98% ee

SCHEME 9.13
A chemoenzymatic enantioconvergent synthetic route toward (*S*)-diols 41 [75].

5 DEHALOGENASE

Dehalogenases (EC 4.5.1.X) are lyases that catalyse the cleavage of carbon–halogen bonds by the elimination of hydrogen halide or the reverse reaction. The halide can be replaced by a suitable anionic nucleophile such as cyanide, azide, and nitrite ions for the reverse reaction resulting in the formation of corresponding derivatives. Based on the primary substrate preference, dehalogenases are roughly divided into three groups: halohydrin dehalogenses (HHDH) converting vicinal halogenated alcohols to epoxides and *vice versa*, haloalkane dehalogenases showing activity toward halogenated alkanes, and haloacid dehalogenases converting halogenated alkanoic acids [76–78]. Halogenated organic compounds are a large class of substances that are widely used in the production of herbicides and fungicides. Many can be harmful to human health and the environment due to their toxicity and chemical stability. The possibility of biodegrading such compounds has attracted great interest during the past decade. Microorganisms contain almost limitless source of carbon–halogen bond cleaving enzymes and hundreds of bacteria and/or the corresponding enzymes have been successfully implemented in industry for the degradation of these compounds. In organic synthesis, the most prominent representatives are the halohydrin dehalogenases (HHDHs) that satisfy synthetic needs by having a very broad substrate scope, high stereoselectivities and no need for expensive and/or unstable cofactors.

HHDHs can catalyze ring closure of halo alcohols **42** to the corresponding chiral epoxides **43**. In the presence of an appropriate nucleophile, they can also catalyze epoxide ring opening to yield the α-hydroxy/Nuc compounds **44** (Scheme 9.14). This class of enzymes has been divided into three distinct groups, namely A, B, and C, based on their sequence homology [79]. Two representatives from group A – HheA from *Athrobacter* sp. AD2 and H-lyase A – show 97.7% sequence identity, whereas sequence homology between groups is limited to *ca*. 25%. To date, the best-investigated HHDH (HheC from *A. radiobacter* AD1) belongs to group C and differs from HheA and HheB by having high enantioselectivity toward a range of predominantly aromatic substrates. Based on the crystal structure published by de Jong et al., the catalytic mechanism was elucidated [80]. The active site contains three conserved amino acid residues (Ser132, Tyr145, and Arg149) that play key roles (Scheme 9.15).

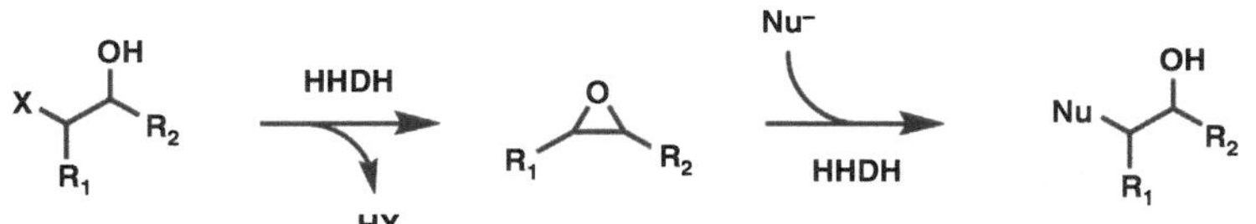

42 43 44

X = Cl, Br, I
Nu^- = CN^-, N_3^-, NO_2^-, NCS^-, NCO^-

SCHEME 9.14
General reaction scheme of halohydrin dehaloenases.

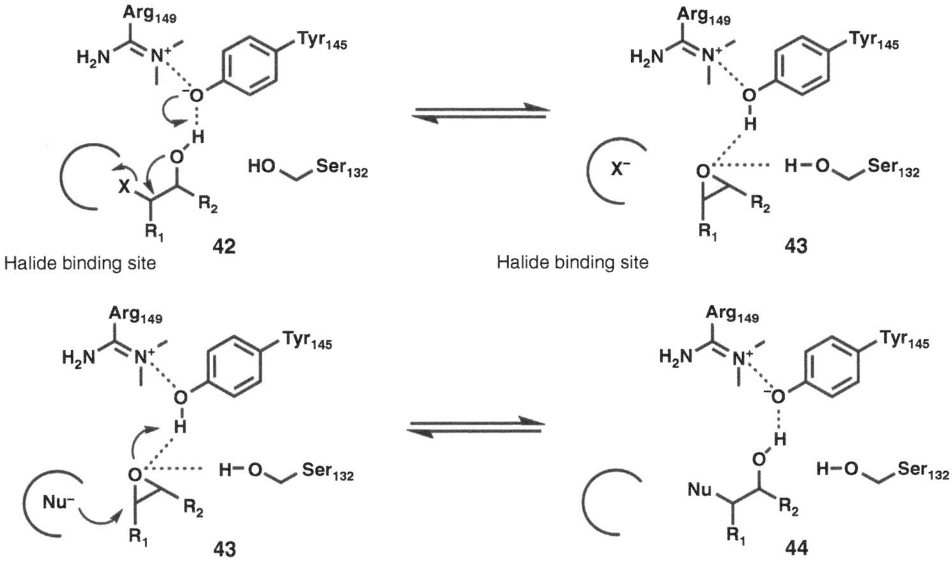

SCHEME 9.15
Catalytic cycle of HheC [80].

In the first step, Tyr 145 acts as a base that deprotonates the substrate hydroxyl **42**. This is followed by an intramolecular nucleophilic attack of the oxygen on the halogen-bearing carbon atom. Subsequently, the halide is released, "trapped" in the halide-binding site and the epoxide **43** is formed. A nucleophile such as CN^- that occupies the halide binding site can attack the epoxide ring to form an α-hydroxynitrile, for example., **44**. Although Tyr 145 is directly involved in the catalytic cycle, Arg 149 polarizes the hydroxyl group of Tyr 145 and lowers its pK_a value. Ser132 interacts with the substrate hydroxyl group and keeps this moiety in close proximity to Tyr 145.

One remarkable feature of HHDHs is their complementary enantiopreferences: HheC favors (*R*)-enantiomers while HheA-AD2 preferentially produces (*S*)-enantiomers.

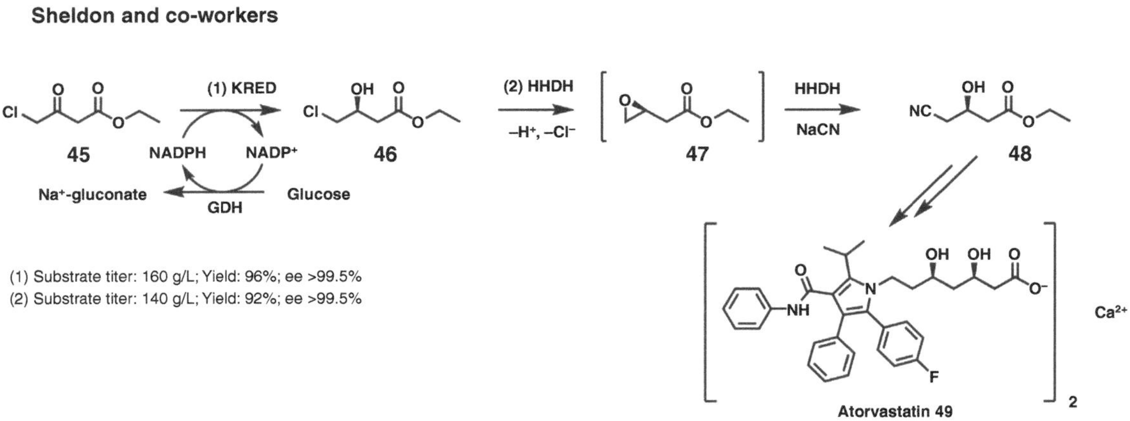

SCHEME 9.16
Industrial application of HheC for the synthesis of Artorvastatin side chain [81].

In order to apply a biocatalyst to industrial processes, it must meet specific requirements with respect to catalytic efficiency, stability, and easy preparation. Wild-type enzymes are often lacking one or the other of these aspects, and for this reason, several protein-engineering strategies – both rational and randomized – have been developed over the past decade in order to circumvent these drawbacks. The most impressive modification of a HHDH was performed by Codexis (USA) for the synthesis of (*R*)-4-cyano-3-hydroxybutyrate (HN, **48**), the side chain of the cholesterol-lowering drug Atorvastatin (Lipitor/Pfizer; Scheme 9.16) [81]. This study was based on a three-enzyme and two-step process in which a ketoreductase (KRED) and a glucose dehydrogenase (GDH) were used for the enantioselective reduction of ethyl-4-chloroacetoacetate **45** to produce **46**. Subsequently, HheC was applied in the presence of NaCN at pH 7 for the formation of HN **48**, via the intermediate epoxide **47**. A proprietary mutagenesis strategy (ProSAR) that combined techniques such as error-prone PCR, site-directed mutagenesis, rational design, and gene shuffling was used in a multivariate approach to improve the catalytic properties of the enzyme.

After 18 rounds of evolution and the introduction of at least 35 mutations, the volumetric activity of the mutant was increased ~4000-fold and this allowed a process having a space–time yield greater than 360 $g_{product}$L/d. The optical purity of the desired product was higher than 99.5%. As a result of this extensive protein engineering study, several new and useful HHDHs became commercially accessible. The synthetic applicability of these HHDHs was demonstrated on several examples in the past [30,82–85]. Kosjek and coworkers published the resolution of 2,2-disubstituted epoxides **50** via biocatalytic azidolysis and subsequent synthesis of chiral amino alcohols **52** and aziridines **53** containing a tertiary center (Scheme 9.17) [86].

Optimized reaction conditions resulted in a procedure for the preparation of azido arylpropanols **51** used for 10 examples with very good yields (40%) and

Kosjek and co-workers

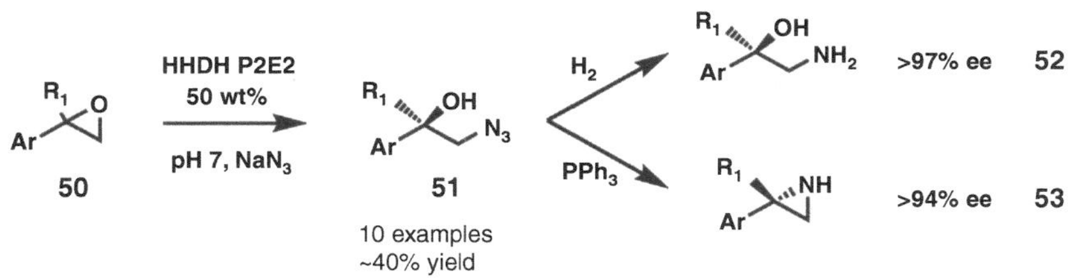

SCHEME 9.17
Synthesis of 1-azido-2-arylpropan-2-ols 51 via kinetic resolution of 2,2-disubstituted epoxides 50 [86].

excellent stereoselectivities (>97%) on 100 mg scales. These novel 2,2-disubstituted azido alcohols **51** are synthetically valuable building blocks and can be converted either into enantiomerically enriched unprotected aziridines **53** through a Staudinger reaction or into 2,2-disubstituted amino alcohols **52** via a simple azide reduction in the presence of hydrogen and Pd/C in ethanol.

Overall, HHDHs represent a very mild and selective alternative method for the introduction of azides and other nucleophiles into organic compounds, which is in sharp contrast to chemical methods. Investigations to expand the scope of such enzymes, especially for new nucleophiles, would be highly desirable.

6 ENOATE REDUCTASES

Enoate reductases (EREDs), also referred also as "alkene" or "ene" reductases, are NAD(P)H-dependent flavoenzymes that catalyze the conjugate 1,4-addition of hydride to an electron-deficient C—C double bond (most commonly α,β-unsaturated carbonyl or nitro compounds). The electron-withdrawing group is further activated by hydrogen bonding to enzyme amino acid side chains such as histidine or asparagine. Subsequent protonation of the resulting enol(ate) by an active site proton donor (Tyr) or a water molecule completes the formal addition of H_2 across the double bond. This mechanism was established for the family of old yellow enzymes (OYEs) and is likely applicable to most, if not all, ERED-catalyzed double bond reductions [87–89]. The active site geometry usually dictates *anti* configuration of the new hydrogen atoms in the products and this is obtained in most cases with high specificity and enantioselectivity [90–93].

Recently, deviations from this trend were investigated on tri- and tetra-substituted olefins and *Saccharomycete* OYEs, leading to the identification of a *flipped* binding mode (turning the substrate by 180° along a virtual axis through the carbonyl-olefin plane), which may be an influential factor in poorly selective transformations [94–96]. There is now a well-established portfolio of EREDs, including extensive structural information from X-ray crystal structures in a variety of forms (flavin-reduced and -oxidized forms, with and without bound

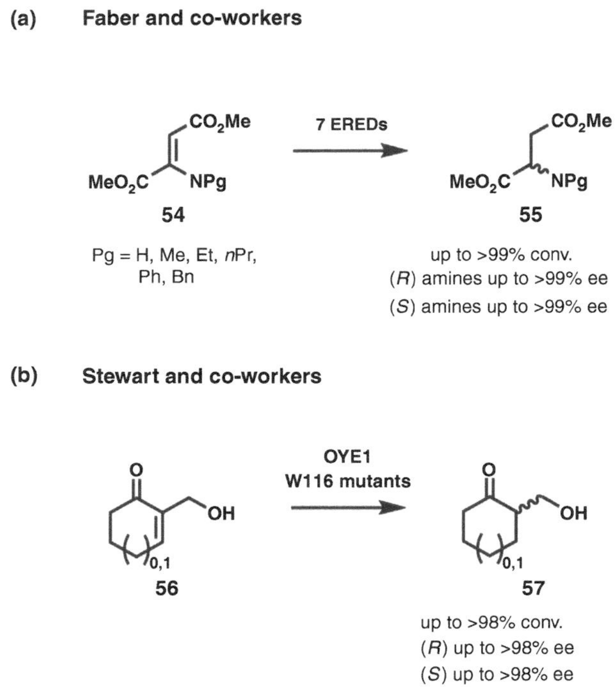

SCHEME 9.18
Synthesis of N-protected (*R*)- and (*S*)-amino acids 55 using various EREDs (a) [100] and reduction of cyclic Baylis–Hillman adducts using OYE1 mutants (b) to obtain α-hydroxy methyl-cycloketones 57 [101].

substrate and cofactor, etc.). Several novel enzymes in this family have been uncovered, often along with their three-dimensional structures over the past several years. Characterization and structural elucidation of nicotinamide-dependent cyclohex-2-en-1-one reductase (NCR) from *Zymomonas mobilis*, a typical OYE, provided new insights on the stereochemical influence of loop regions of the common $\alpha_8\beta_8$ barrel fold [97]. An interesting discovery was made via BLAST search of a leukotriene B_4 dehydrogenase subfamily [98], revealing a double-bond reductase from *Nicotiana tabacum*. This NADPH-dependent enzyme does not contain any flavin cofactor and also shows opposite stereopreference in some cases when compared to OYE family members. It also complemented the typical ERED substrate profile by reducing exocyclic C—C double bonds such as that found in (*R*)-pulegone, which are often difficult for "standard" EREDs [99].

Screening of a library of seven EREDs enabled the enantioselective synthesis of *N*-acyl protected amino acids **55** with excellent optical purities for both antipodes (Scheme 9.18a). Although no single enzyme could be established as perfect catalyst, stereoselectivity could be effectively tuned by size variation of the protective group (>99% ee for (*S*) using ethyl carbamate **54**, 92% ee for (*R*) employing benzyl carbamate **54**). The authors attributed the tendency of

(*R*)-preference with larger acyl residues to the aforementioned presence of an alternative-binding mode, corroborated by ^{2}H-labeling experiments [100].

Access to both enantiomers of valuable α-hydroxy methyl-cycloketones **57** by reducing easily accessible Baylis–Hillman adducts **56** was reported using a focused library of OYE1 W116 mutants that were previously identified and created in a study on terpenones (Scheme 9.18b) [101]. Careful analysis of X-ray crystal structure data of selected mutants revealed key substrate–protein interactions which dictate the orientation toward the cofactor, thereby explaining the formidable improvement in both activity and selectivity when compared to the wild type: hydroxymethylcyclohexenone **56** ($n = 1$) was converted quantitatively with >98% ee and the homologous cyclopentenone **56** ($n = 0$) could now be converted to the (*S*) product **57** enantiomer with synthetically useful activity and selectivity (>98% conversion, >98% ee) [102].

Another member of the OYE superfamily, OYE2.6 from *Pichia stipitis* CBS 6054 (also known as Scheffersomyces stipitis), was used in the gram-scale synthesis of (*R*)-citronellal from geranial [103]. This biocatalyst proved to be superior to OYE1 in terms of stereoselectivity and rate acceleration, but most notably also with regard to operational stability; structural, and biochemical properties were later derived from a newly solved X-ray crystal structure. Comparative analysis against OYE1 showed that amino acid position 113 (Trp in the equivalent position of OYE1, Ile in OYE 2.6) is equally important for substrate orientation, but the overall molecular mechanism seems to differ significantly in these enzymes [104].

A novel ERED from *Shewanella oneidensis* (SYE-4) is a potent recent addition to the ERED family. This enzyme reduces both cyclic and linear unsaturated carbonyl compounds as well as substituted maleimides [105]. In particular, the fast and selective transformation of 2-methyl maleates **56** (99% conversion after 6 h, 98–99% ee) by SYE-4 may provide a well-defined alternative to the undisclosed ERED used in the pilot-scale production (70 g, 35 g/L) of dimethyl (*R*)-2-methylmaleate **57** employed by Almac (Scheme 9.19) [106]. The Almac group further combined this redox enzyme with a hydrolase in a one-pot reaction,

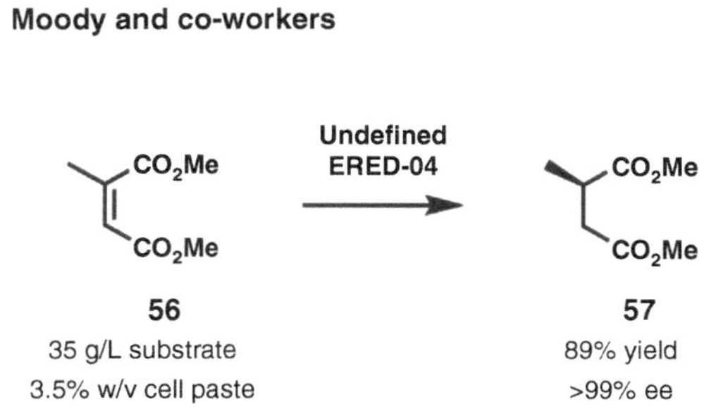

SCHEME 9.19
Pilot-scale preparation of a chiral diester building block using an undefined ERED [106].

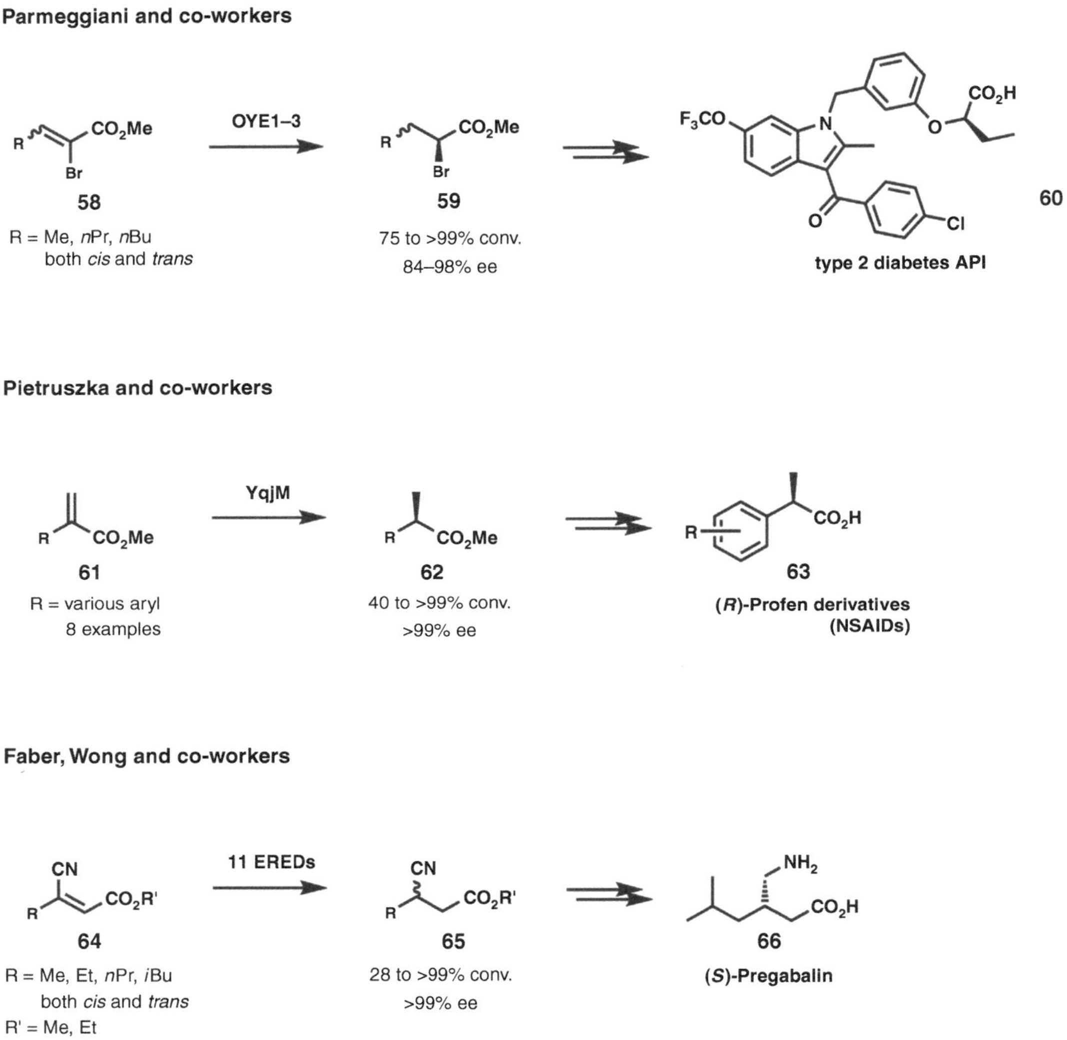

SCHEME 9.20
Selective reduction toward chiral α-bromoesters 58 for type 2 diabetes API 60 functionalization using OYEs (top) [94]; synthesis of various (*R*)-Profen derivatives 63 using YqjM (center) [107]; application of 11 EREDs in the preparation of (*S*)-Pregabalin 66 and other γ-aminobutyric acid (bottom) analogs [108].

furnishing the desired chiral compound methyl (*R*)-2-methylsuccinate in high yield and perfect optical purity (89% yield, >99% ee).

Precursors for chiral ether side chains of a type-2 diabetes API **60** could be synthesized by OYE-mediated reduction of α-bromoacrylic acid derivatives **58** (Scheme 9.20, top) [94]. Three enzymes were tested on a library of seven unsaturated acids. Compared to classic bakers' yeast systems, significant improvement

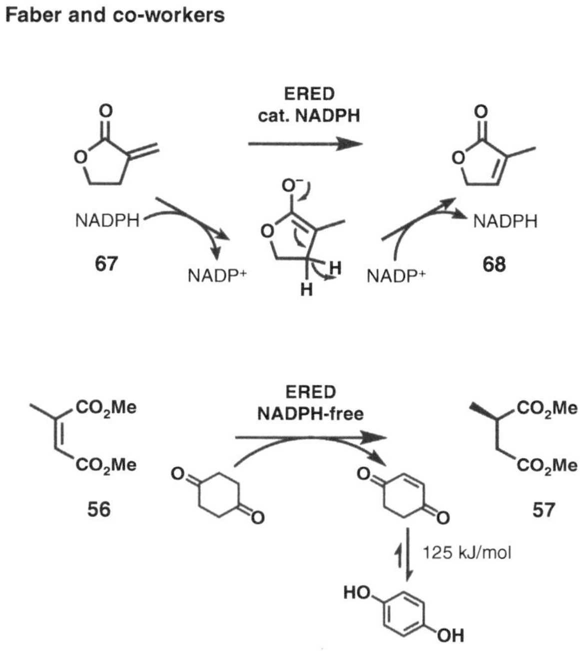

SCHEME 9.21
Gross redox-neutral processes catalyzed by EREDs: isomerization of α-methylene-γ-butyrolactone 67 to the thermodynamically favored olefin 68 (top) [109]; NADPH-free reduction of a dimethyl maleate 56 by concomitant oxidation and aromatization of a sacrificial substrate (bottom) [110].

in both activity (75 to >99% conversion) and selectivity toward the desired (*S*)-bromoacids **59** (84–98% ee) were obtained. Using the very versatile enzyme YqjM, a facile approach toward (*R*)-Profen derivatives **63** was presented (Scheme 9.20, center): seven such nonsteroidal anti-inflammatory drugs **63** (NSAIDs) could be prepared with excellent optical purities (>99% ee). Optimization of the reaction buffer and organic cosolvent provided the basis of an alternative chemo-enzymatic pathway toward (*R*)-flurbiprofen methyl ester (enzymatic step: 68% yield, >99% ee, 1 g/L substrate, 0.04% w/v biocatalyst) [107]. The possibility for ERED-catalyzed reduction of β-cyanoacrylates **64** was explored by Pfizer, testing a library of 11 enzymes on 12 substrates of this class (Scheme 9.20, bottom) [108]. This way, gram-scale reductions with excellent selectivity toward GABA analogues such as Pregabalin **66** were reported using wild-type and evolved mutant biocatalysts. Although the authors of this industrial contribution admit that volumetric productivity of the employed enzymes is still too low, this route still has high scalability potential in comparison to other chemical means due to the high selectivity and cheap cofactor regeneration system utilized (ADH/isopropanol).

Generally, in direct comparison with ADHs, ERED-mediated processes still lag behind in terms of productivity and industrial applicability, but the increasing output in basic and applied research in this area surely points the way for a promising future in application for this enzyme class.

6.1 Redox-Neutral and Oxidative Processes Catalyzed by EREDs

Recently, several unusual reaction types were reported in context with various EREDs. OYEs catalyze a C—C double bond isomerization of α-methylene-γ-butyrolactone **67** to the corresponding α,β-unsaturated lactone **68**, thus enriching the thermodynamically more stable internal double bond with favorable kinetics (Scheme 9.21, top) [109]. An enzymatic disproportionantion reaction catalyzed by the same enzyme produces equimolar amounts of cyclohexanone and phenol from 2-cyclohexenone or substituted analogues thereof [110]. This technology was also put to synthetically practical use in the NADPH-free, stereoselective reduction of dimethyl 2-methylmaleate **56** producing 2-methylmaleate **57**, using cyclohexane-1,4-dione as electron source. The redox equilibrium is mainly driven by the tautomerization of the coproduct to *p*-cresol, liberating approximately 125 kJ/mol (Scheme 9.21, bottom).

An intriguing finding was made while investigating ERED-catalyzed reductions of vinylnitro compounds **69**: after successful hydride addition to the Michael acceptor, the resulting nitroalkane **70** was also prone to enzymatic reduction. Spontaneous protonation of the nitroso intermediate **71** and subsequent oxime **72** reduction leads to aldehyde **74** after hydrolysis of an intermediate imine **73**, thus representing a biological equivalent of a catalyzed Nef reaction.

This mechanistic pathway is in direct competition with the reverse order of reductions (first nitro-reduction), as the resulting nitrosovinyl species **75** may isomerize to **76**, cyclize to **77** and cyclorevert orthogonally. This liberates hydrogen cyanide in a C—C-bond cleavage reaction and amounts to a net oxidation to give **78** (Scheme 9.22) [111].

Last, the characterization of a thermostable ERED from *Geobacillus kaustophilus* led to the discovery of a reverse enzymatic process, namely catalytic α,β-dehydrogenation of carbonyl compounds under dioxygen consumption at 70 °C [112]. This reaction is highly unfavored at ambient temperature and has so far

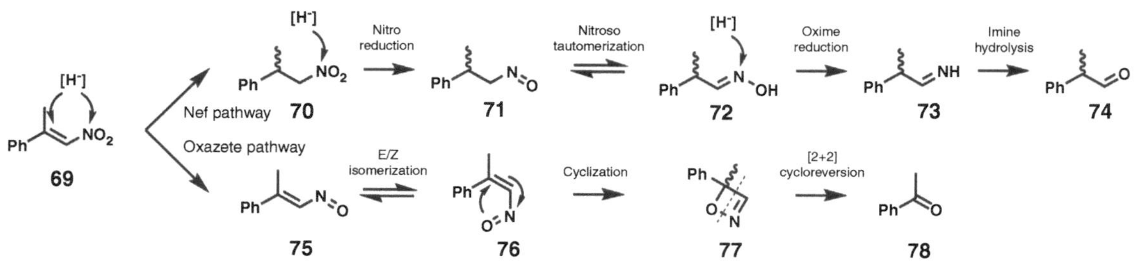

SCHEME 9.22
Possible cascade reaction pathways in ERED-catalyzed reactions with nitrovinyl compounds.

only been achieved by OYEs bearing an artificial cofactor with increased redox potential or chemically by highly reactive reagents such as IBX. These atypical ERED reaction types were recently reviewed in detail [113].

7 BAEYER–VILLIGER MONOOXYGENASES

Baeyer–Villiger monoxygenases (BVMOs) are intensively studied NADPH- and flavin-dependent biocatalysts, which so far represent the only available catalytic system for asymmetric Baeyer–Villiger oxidations with synthetically useful enantioselectivities [114–116]. Mechanistically, enzymatic Baeyer–Villiger oxidations (BVOx) closely resemble their chemical counterparts as determined by Criegee: the prosthetic flavin group is first reduced by NADPH and subsequently activated by dioxygen to form a peroxy anion. This species can perform a nucleophilic attack on a carbonyl carbon atom, leading to rearrangement of one of the α-substitutents on the substrate and its cleavage from the flavin. Elimination of water from the latter species completes the catalytic cycle. This well-accepted mechanism was postulated almost 40 years ago [117,118], but a lack of structural data hindered further progress. Recently, several X-ray crystal structures from two prominent BVMOs (the substrate-promiscuous CHMO from *Rhodococcus* sp. [119] and the thermostable PAMO from *Thermobifida fusca* [120]) were solved. These data and the subsequent analyses provided detailed insights into the mechanism of oxygen activation [121,122], substrate binding [123,124] and cofactor specificity [125]. Together, this structural knowledge provided a fruitful starting ground for evolving and optimizing BVMOs to harness one of their key advantages over chemical reagents or catalysts: BVMOs use aerial oxygen rather than hazardous peroxides as the primary oxidant. This gives them a unique advantage for large-scale process development [126]. Additionally, BVMOs display interesting behavior in the Baeyer–Villiger rearrangement process by imposing certain stereoelectronic effects that enable access to regioisomeric lactone products [127–129]. Only recently, the first industrial application using a BVMO [130] reinvigorated applied research in this catalyst class: a highly evolved mutant CHMO from *Acinetobacter* sp. NCIMB 9871 was used for enantioselective sulfur oxidation in the final step toward active pharmaceutical ingredient Esomeprazole. This also stimulated work for improved operational stability of such biocatalysts by enzyme engineering [131] and whole-cell immobilization [132,133].

Extending recent discoveries for enantiocomplementary enzymes of this class [134], cycloketone-converting BVMOs from *Xanthobacter* sp. ZL5 ($CHMO_{Xantho}$) and *Comamonas* sp. NCIMB 9872 ($CPMO_{Coma}$) were used in the key asymmetric step *en route* to both enantiomers of carba-Showdomycin **81**, a carbon-analog to the potent antiproliferative *C*-nucleoside Showdomycin, with excellent optical purity and good yield (89 and >99% ee, 63 and 71% yield, **80**; Scheme 9.23, top) [135]. Another set of enantiodivergent BVMOs (CHMOs from *Arthrobacter* or *Brachymonas* sp., and a cyclododecanone monooxygenase from *Rhodococcus ruber* SC1, $CDMO_{Rhodo}$) catalyzed the formation of optically pure β-amino acids **84** (e.g., (*R*)-β-leucine, 33% yield, >99% ee) and β-amino alcohols, after enzymatic cleavage of the intermediate esters **83** (Scheme 9.23, bottom) [136,137]. Apart

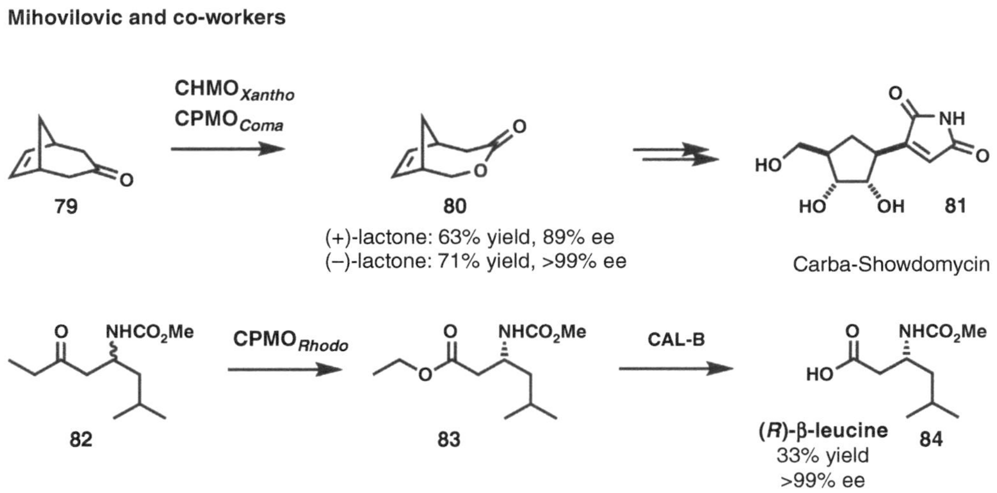

SCHEME 9.23
Desymmetrization of bicyclo[3.2.1]oct-6-en-3-one 79 toward both enantiomers of carba-analogous Showdomycin ((+)-81 (-)-81) using two BVMOs (top) [135]; regioselective kinetic resolution of β-aminoketones 82 toward β-amino acids 84 (bottom) [136,137].

from the well-documented enantioselective properties of this catalyst class, a novel diastereoselective desymmetrization was recently reported: 4-methylene-cyclohexanones **85** could be selectively converted to *E*-configured 5-methylene-ε-caprolactones **86** (>96:4 *E*:*Z*, four examples) using a wild-type cyclohexanone monooxygenase from *Acinetobacter* sp. NCIMB 9871 (CHMO$_{Acineto}$, Scheme 9.24) [138]. Additionally, the authors created efficient variants of this catalyst over three rounds of mutation, thereby completely inverting diastereoselectivity toward the corresponding *Z*-lactones **86** (4:96 d.r.); the effects of the five modified amino acids in the final variant were rationalized by MM/QM studies.

Kinetic resolutions catalyzed by BVMOs are usually limited to the theoretical 50% yield threshold. In 2002 a dynamic protocol mediated by base-catalyzed

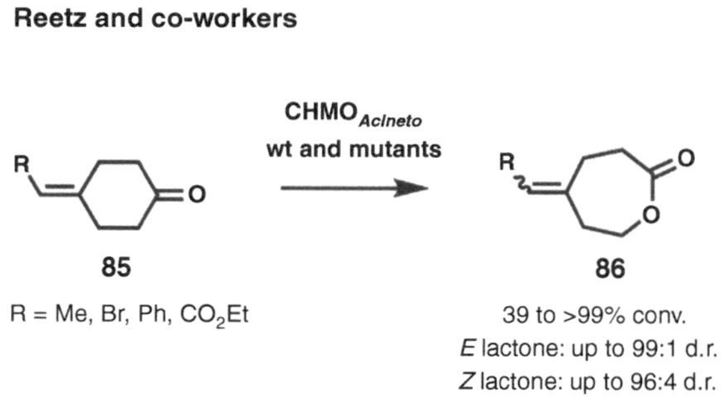

SCHEME 9.24
Diastereoselective desymmetrization of 4-methylenecyclohexanones by CHMO$_{Acineto}$ and evolved mutants [138].

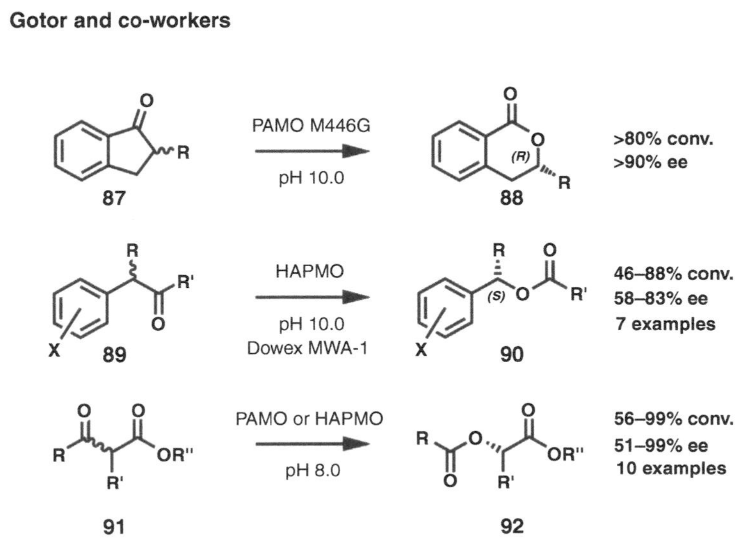

SCHEME 9.25
Dynamic kinetic resolutions catalyzed by BVMOs at elevated pH [139–142].

racemization of the ketone starting material was published [139] and then further investigated [140]. The main drawback was that the conditions were limited to a single compound. Gotor and coworkers later extended this technology to additional substrates bearing more activated α-C—H bonds (homo-benzylic **87** [141], benzylic **89** [141] and 1,3-dicarbonyl **91** [142]), thereby producing the desired lactone or ester products with significantly more than 50% yield and excellent optical purities (Scheme 9.25).

Recent efforts in BVMO biocatalyst discovery were marked by the complete analysis of camphor metabolism in *Pseudomonas putida* NCIMB 10007, particularly the characterization of the very interesting 3,6-diketocamphane 1,6-monooxygenase (3,6-DKCMO). This enzyme was the missing counterpart to 2,5-DKCMO, each acting specifically on regioisomeric diketocamphanes. Although the respective native lactone products are unstable under physiological conditions and spontaneously convert to a single homochiral cyclopentenone (which is further converted by another BVMO in the pathway), their high specificity for such similar compounds makes these BVMOs promising candidates for difficult catalytic tasks in terms of isomer discrimination [143].

A surprising new class of BVMOs was found during the characterization of eight flavin monooxygenases (FMOs) from *Rhodococcus jostii* RHA1: whereas FMOs in general lack the ability to perform Baeyer–Villiger oxidations (BVOx), three biocatalysts of the test set stood out from this trend [144]. Not only do these enzymes readily perform BVOx on a range of small-ring ketones, but they also display a heretofore indifference toward the nicotinamide cofactor and use NADH and NADPH equally well. This feature may prove to be highly useful in

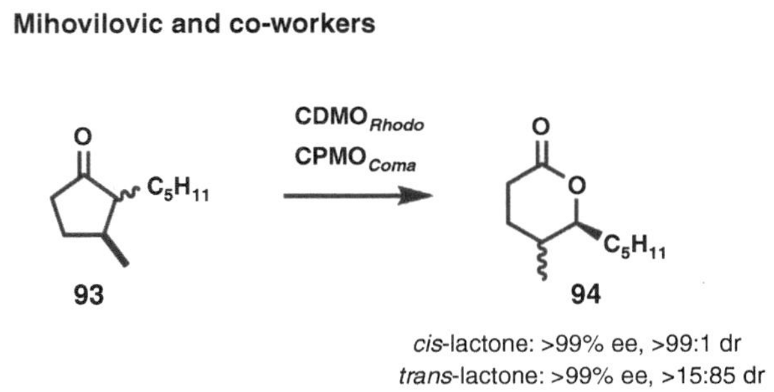

SCHEME 9.26
Kinetic resolution of *cis*- and *trans*-tetrahydrojasmones 93 toward natural *Aerangis* lactone and its epimer 94 using two BVMOs [147].

enzymatic redox cascades, circumventing the need for complex cofactor shuttling. It may also be highly beneficial in whole-cell applications, where metabolic redox burden can theoretically be halved as both the intracellular NADH and NADPH pools can be tapped [145].

Meanwhile, given the diverse available information on BVMOs, their substrate promiscuity, variable activity and selectivity over a structurally diverse spectrum of ketones, it has become increasingly difficult to set a newly discovered catalyst into context or compare its performance to existing BVMOs. A system to counter this problem was published in combination with an elaborate substrate profile of large cycloketone-converting $CDMO_{Rhodo}$: the authors developed a statistical scoring algorithm for BVMOs based on the application-oriented properties activity and selectivity. Nine enzymes were ranked according to their performance in various substrate classes and their overall versatility. The model is complemented by a clear graphical output for facile interpretation and operates with minimal bias; moreover, the algorithm design should allow application to other catalyst classes, but requires largely comparable reaction conditions to provide a meaningful outcome [146].

The enzyme $CDMO_{Rhodo}$ was used in a chemo-enzymatic synthesis toward the natural isomer of *Aerangis* lactone **94**, an important aroma compound from African orchids. The proposed setup combines a diastereoselective Rh-catalyzed continuous flow hydrogenation with the Baeyer–Villiger reaction in a single operation, yielding the desired fragrance lactone **94** in perfect optical purity with high scalability potential (>99% ee, >99% de, 3.4 g/L, <60 min; Scheme 9.26) [147]. Strikingly, the diastereomeric ratio obtained from the chemical step (>90%, **93**) is augmented to the final value by $CDMO_{Rhodo}$, another example for diastereoselectivity among BVMOs. Additionally, $CPMO_{Coma}$ was employed to gain access to the epimeric *Aerangis* lactone with comparable, but not perfect diastereomeric purity. Nevertheless, a 7-step stereoselective synthesis could be shortened to a single operation process by applying BVMO catalysis. A few remarkable examples of the use of BVMOs in enzymatic cascade type reactions published very recently substantiate the great potential of this enzyme class in complex organic syntheses [148–150].

8 DIHYDROXYLATION OF AROMATIC COMPOUNDS

Aromatic ring hydroxylating diooxygenases are another very attractive class of enzymes for biocatalysis [151]. These enzymes, also known as Rieske nonheme iron dioxygenases, are multicomponent enzyme systems that add both atoms of molecular oxygen across an arene bond (EC 1.14.12.-). Such transformations are a key step in the microbial degradation pathway of aromatic compounds. In the normal metabolic pathway, addition of oxygen to yield a *cis*-dihydrodiol is followed by an enzyme-catalyzed oxidation to the corresponding catechol. Catechol formation can be blocked by employing mutant strains that lack the *cis*-dihydrodiol dehydrogenase and these organisms allow the accumulation of chiral *cis*-diols in the medium, from which they can be isolated in good yields. More than 100 such enzymes have been identified based on their biological activities and DNA sequences [152–155]. Many of these biocatalysts display a broad substrate scope (more than 400 *cis*-diol products) and are often highly regio-, stereo-, and enantioselective (Scheme 9.27). Furthermore, this process has no equivalent in synthetic chemistry, which makes this class of enzymes very appealing for the preparation of new chiral building blocks.

Ramaswamy and coworkers carried out a detailed mechanistic study of the toluene 2,3-dioxygenase system (TDO), which consists of a reductase, a ferredoxin and a terminal dioxygenase, all of whose structures have been determined by X-ray crystallography [156]. The dioxygenase, which was cocrystallized with toluene, is a heterohexamer containing a catalytic and a structural subunit. The catalytic subunit contains a Rieske [2Fe–2S] cluster and mononuclear iron at the active site. Another Rieske [2Fe–2S] type cluster is located close to the protein surface of the ferredoxin, allowing it to act as an electron transport protein. The reductase belongs to the glutathione reductase family of flavoproteins and is comprised of three domains: an NADH binding domain, an FAD binding domain and a *C*-terminal domain. The orientation of the toluene substrate within the active site is consistent with the regiospecificity of oxygen incorporation found in the product. In 1993, Boyd and coworkers published [157] a robust and predictive model for arene *cis* dihydroxylations with respect to enantiopreference across organisms and substrates. In the last two decades, biocatalytic processes have been developed that use either

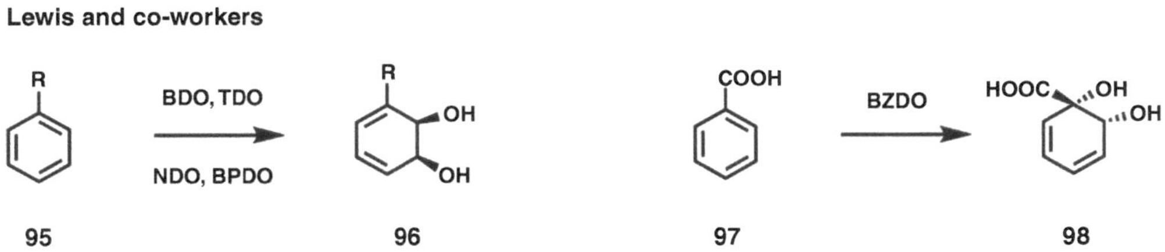

SCHEME 9.27
Regio- and stereoselectivity of biocatalytic arene cis dihydroxylation mediated by dioxygenases. BDO, benzene dioxygenase; TDO, toluene dioxygenase; NDO, naphthalene dioxygenase; BPDO, biphenyl dioxygenase; BZDO, benzoate dioxygenase [151].

dehydrogenase-deficient mutant strains or recombinant cells expressing complete dioxygenase systems. This work has been carried out on both laboratory and industrial scales that have yielded gram to multi kilogram amounts of *cis*-cyclohexadiene diols. Since these chiral building blocks became accessible, a vast number of synthetic applications have been published [151,154,157,158]. In the following, only two examples with significant impact on the scientific community will be discussed.

A general problem in the development and discovery of new antibiotics are the limitations in chemical synthesis, which become even more evident with increased complexity of natural products. Myers and coworkers reported a convergent enantioselective route to structurally diverse 6-deoxytetracycline antibiotics by exploiting a key building block from the transformation of benzoic acid **97** by benzoic acid dioxygenase (BZDO, **98**, Scheme 9.28) [159,160]. The first milestone in this synthesis was the preparation of intermediates A **100** and B **101** in Scheme 9.28. Both precursors are the core structure of the AB-ring of tetracycline and could be obtained in 11% yield for A **100** after 12 steps while B **101** was prepared in 12% yield after 15 steps (Scheme 9.28, a). The missing C ring was constructed by a Michael–Dieckmann cascade cyclization of the AB-ring system and the D-ring precursor **102** which yielded (−)-doxycycline **103** in 8.3% yield after 18 steps (Scheme 9.28). The overall yield of this approach represents an improvement of orders of magnitude over those of all previously published total syntheses of tetracycline (0.002% vs. 8.3%). Because of the modularity of the approach, novel tetracycline analogues became accessible for

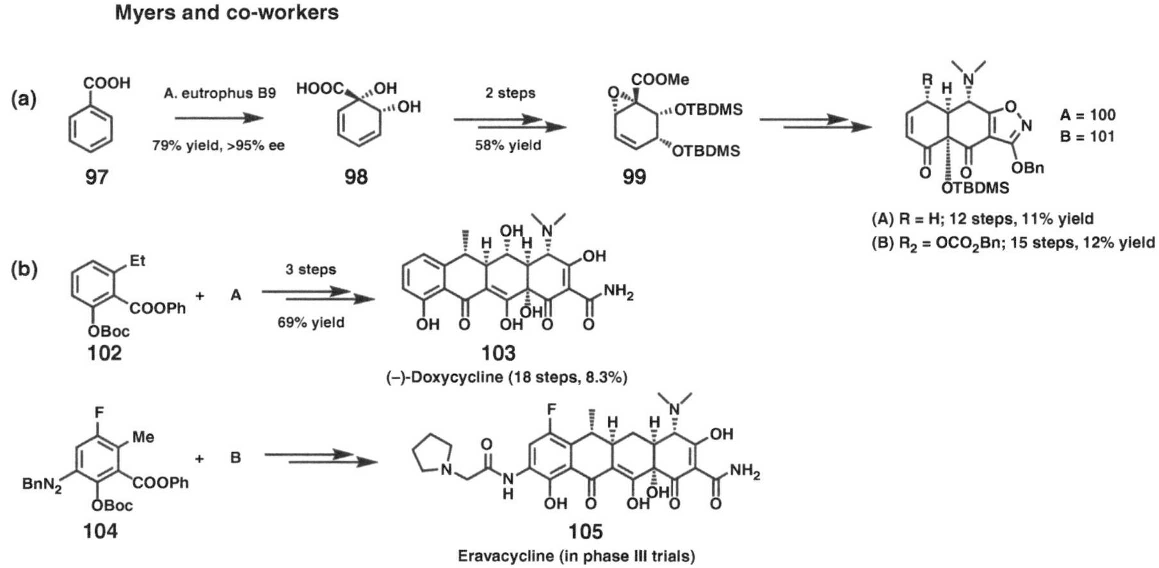

SCHEME 9.28
Myers' synthetic approach for the preparation of tetracycline derivatives starting from benzoic acid 97 [159,160].

Hudlicky and co-workers

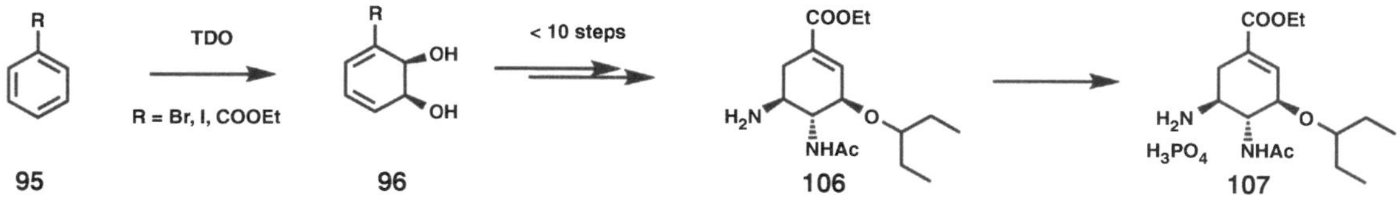

95 96 106 Oseltamivir 107 Tamiflu

SCHEME 9.29
Synthetic strategies for the synthesis of oseltamivir 106, a precursor for the protein neuraminidase inhibitor Tamiflu 107 [154].

the first time, including Eravacycline **105**, which is in Phase III clinical trials (Scheme 9.28, b).

Several different important milestones in applications to synthesis of arene dioxygenases have been reported by the groups of Banwell [161], Hudlicky and others [154]. Key contributions have been the synthesis of $PGE_{2\alpha}$, (+) and (−)-pinitol, pancratistatin, lycoricidine, narciclasine, (+)-codeine, and oseltamivir (Scheme 9.29) [154].

With the synthesis of oseltamivir **106**, Hudlicky and coworkers again demonstrated the importance and versatility of dihydroxylated aromatic compounds obtained from toluene dioxygense [162]. Ethylbenzoate **95** is a very good substrate for TDO and the resulting diol product **96** has been exploited for a very efficient chemoenzymatic synthesis of Tamiflu **107**. This example also utilized the so-called "plane of latent symmetry" concept [163]. By proper considerations of symmetry, both enantiomers may become available from a single isomer generated by biocatalytic means (Scheme 9.30). Oseltamivir **106** contains such a latent symmetry axis through the C_1 and C_4 atoms, which was used for a flexible synthetic route. In more detail, if substituents at C_3, C_4, and C_5 are not specifically defined, then configurations of these carbon atoms describe an enantiomeric switch (Scheme 9.30) that could be controlled by a simple flip of the double bond.

Hudlicky and co-workers

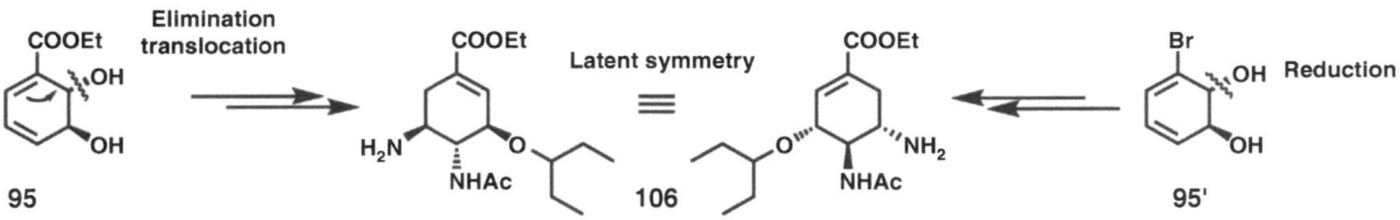

95 106 Oseltamivir 95'

SCHEME 9.30
Symmetry-based design for oseltamivir 106 from cis-dihydrodiols 95 and 95' [163].

One appealing aspect of using ethyl benzoate **95** (rather than bromobenzene **95′**) is to avoid the Pd-catalyzed introduction of the carboxylic ester functionality in the later stage of the synthesis. The direct oxidation of ethyl benzoate **95** thus provided a very efficient and scalable synthesis without the need for elaborate chromatographic purification steps. They reduced the number of key synthetic steps from 10 to 5 and provided an overall yield of 24% [164].

Dihydroxylation of aromatic compounds by various enzymes (TDO, BZDO, NDO, etc.) yields very interesting and valuable starting materials for synthesis. Due to the complexity of the hydroxylating enzymes, most studies are performed on wild type enzymes. While some dioxygenases such as TDO show very broad substrate acceptance, others such as BZDO are much less tolerant. These cases might be improved by protein engineering and directed evolution strategies as these arene *ipso-* and *ortho-*diols provide new access to uncharted chemical space.

9 EXPERIMENTAL PROCEDURES

9.1 Enantioselective Hydrolysis of Cyclohexene Oxide 37 Using Epoxide Hydrolase SpEH From *Sphingomonas* sp. HXN-200 in *E. coli* on Multigram Scale

E. coli (SpEH) was grown in Lysogeny broth or Luria broth (LB) medium (2 mL) containing kanamycin (50 mg/L) at 37 °C for 7–10 h and then inoculated into TB medium (50 mL) containing kanamycin (50 mg/L) (Scheme 9.31). When the OD_{600} reached 0.6 (around 2 h), isopropylthiogalactoside (IPTG, 0.5 mM) was added to induce the expression of protein. The cells continued to grow for 10–12 h at 25 °C and the cell density reached >4 g (cell dry weight) cdw/L. Cells were harvested by centrifugation (5,000 × *g*, 5 min) and resuspended in Tris buffer (200 mL, 50 mM, pH 7.5) at a cell density of 10 g cell dry weight (cdw)/L. Cyclohexene oxide (100 mmol, 9.82 g, **36**) was added to the reaction mixture, which was shaken (250 rpm) in an incubator at 30 °C for 120 min. The reaction was monitored by GC, and terminated by cooling on ice once the conversion of *meso*-epoxides **36** reached 99%. The reaction mixture was immediately saturated with NaCl and extracted four times with EtOAc (200 mL). The organic phases were combined, dried over Na_2SO_4 and the solvents were removed by evaporation. The crude product was purified by crystallization in ethyl acetate (dissolved

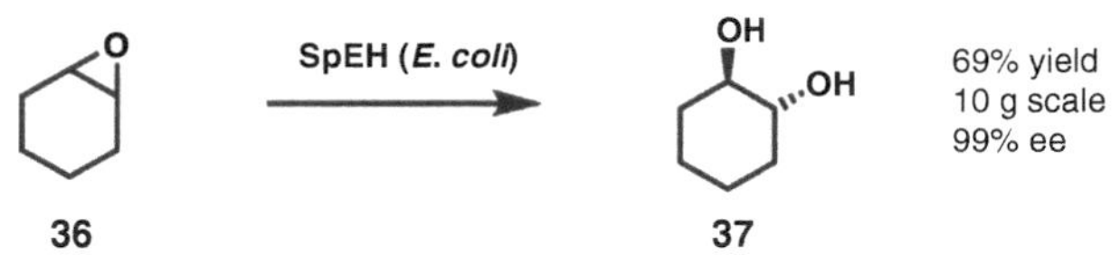

SCHEME 9.31
Enantioselective hydrolysis of cyclohexene oxide 37 using epoxide hydrolase SpEH from *Sphingomonas* sp. HXN-200 in *E. coli* on multigram scale [72].

at 60 °C and slowly cooled to 4 °C), yielding pure *trans*-diol **37** in 69% yield after recrystallization (99% ee) [72].

9.2 Multigram Synthesis of an Atorvastatin Intermediate 48 Using a Mutant Halohydrin Dehalogenase

A 1 L jacketed three-neck round bottom flask equipped with a rubber septum, a pH electrode connected to a pH stat, and a mechanical stirrer was charged with 400 mL 500 mM NaCN (Scheme 9.32). The whole system was airtight since gaseous HCN is generated. *Caution: HCN is highly toxic and only should be handled with adequate training and safety precautions.* The pH of the solution was approx. 11.2. The pH electrode was calibrated at room temperature 20–23 °C before use. The vessel was sealed and the pH was adjusted to approx. 7.5 using conc. H_2SO_4 (approx. 7 mL). HHDH (obtained as described in Ref 81) was charged as an aqueous solution (1.05 g in 20 mL deionized water) and the reaction mixture was heated to 40 °C. Then, the pH stat control was started and 70 g (0.42 moles) of the chlorohydrin **46** was added with a syringe over approximately 10 min. The pH stat maintained the pH at 7.3 ± 0.05 with the addition of a 25% NaCN solution containing 0.25% NaOH. After approx. 18 h, 73 mL of NaCN/NaOH solution was consumed. At that point, the progress of the reaction was checked by GC and no starting material was detected. The pH of the reaction mixture was adjusted to approx. 3 with conc. H_2SO_4 (approx. 2.5 mL). For safety reasons, the reaction was cooled to 25 °C. The pH electrode and the rubber septum were removed (Caution: HCN gas was present). The flask was equipped with a condenser and nitrogen dip tube extending into the liquid. Vacuum (approximately 100 mm Hg) was applied from a diaphragm pump (a caustic trap containing approximately 4 M NaOH was used to collect the HCN in the off gas). The reaction mixture was heated to 40 °C. A nitrogen bleed was used to facilitate the removal of HCN. After 2.5 h, HCN removal was complete (<5 ppm in the off gas). The mixture was cooled and treated with 3.5 g of Celite and 3.5 mL bleach (containing 6.1% sodium hypochlorite). This mixture was filtered through a Celite pad. The filtrate was extracted four times with 280 mL of ethyl acetate. The combined ethyl acetate phases were concentrated *in vacuo* to give 61.6 g (93%) of the cyanohydrin **48**. The chemical purity as determined by GC was 99.5% and the chiral purity was >99.5% ee for the *R*-enantiomer (no *S*-enantiomer was detected) [81].

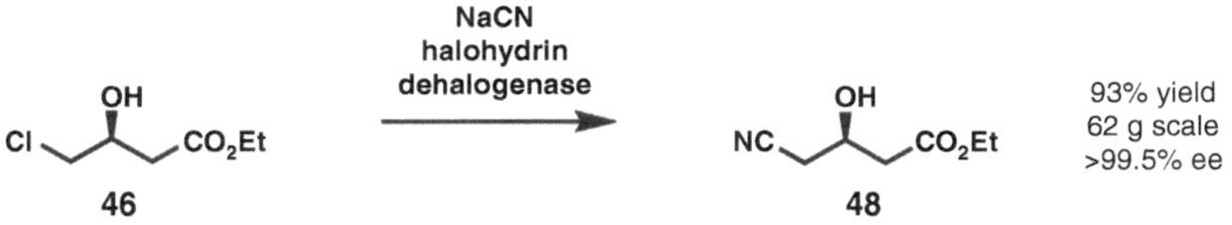

SCHEME 9.32
Multigram synthesis of an atorvastatin intermediate 48 using a mutant halohydrin dehalogenase [81].

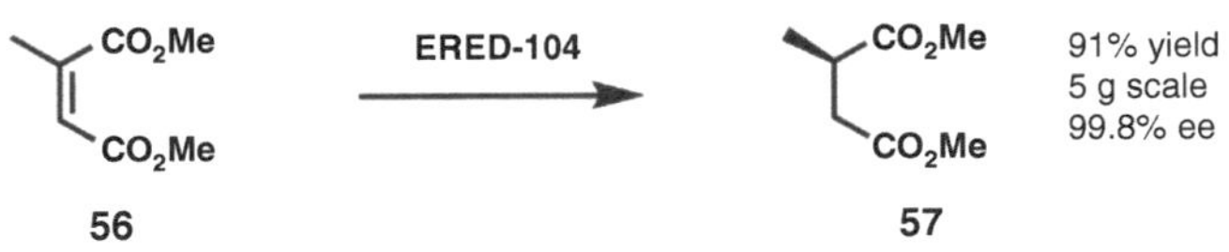

SCHEME 9.33
Reduction of dimethyl methylmaleate using mutant ERED-104 [106].

9.3 Reduction of Dimethyl Methylmaleate Using Mutant ERED-104

A solution of dimethyl methylmaleate (5.00 g, 31.65 mmol, **56**) in toluene (40 mL) was added to a suspension of ERED-04 cell paste (5 g, obtained as described in Ref [106]) in pH 8 Tris HCl buffer (100 mL, 0.1M) containing NAD^+ (250 mg, 0.38 mmol) while stirring at 30 °C (Scheme 9.33). Isopropyl alcohol (2.60 mL, 2.03 g, 33.78 mmol) was added. In addition, polypropylene glycol (1 mL) was added to prevent excessive foaming. The reaction mixture was stirred mechanically using an overhead stirrer at 30 °C for 59 h while sparging at 2 L/min to remove acetone formed. The depletion of 2-propanol was counterbalanced by passing the pressurized airflow through a solution of 2-propanol in water (3% v/v) before passing through the reaction mixture. The reaction was monitored by *in situ* ^1H NMR sampling. Complete conversion was observed after 59 h. The mixture was filtered through Celite and the phases separated. The aqueous phase was extracted with MTBE (3 × 50 mL). The combined organic extracts were washed with brine (150 mL) and dried over Na_2SO_4. Solvent was removed under reduced pressure to provide (*R*)-dimethyl 2-methylsuccinate **57** as an oil with an enantiomeric excess of 99.8% as determined by HPLC analysis; yield: 4.61 g (91%) [106].

9.4 Diastereoselective Baeyer–Villiger Oxidation of a 4-Methylenecyclohexanone 87 Using Cyclohexanone Monooxygenase from *Acinetobacter* on Lab Scale (300 mg)

The wild-type CHMO or mutant CHMO were stored in the form of glycerol frozen stocks of *E. coli* BL21-Gold (DE3) (Stratagene) harboring the pET-22b (+) expression vector (Scheme 9.34). Fresh LB medium (50 mL) containing carbenicillin (100 mg/mL) was inoculated with 1% of an overnight preculture of the CHMO-expressing *E. coli* strains in 300-mL Erlenmeyer flasks. The culture

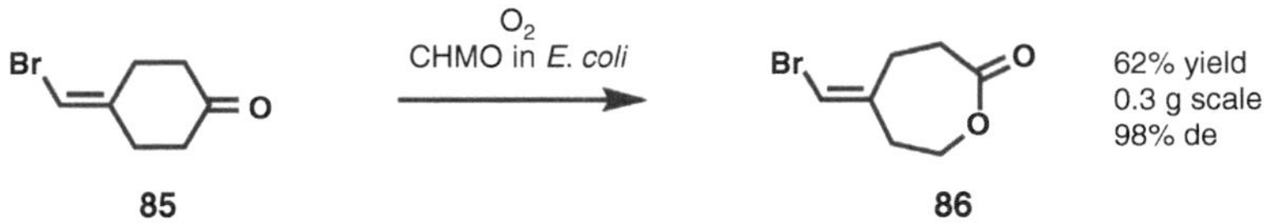

SCHEME 9.34
Diastereoselective Baeyer–Villiger oxidation of a 4-methylenecyclohexanone 87 using cyclohexanone monooxygenase from *Acinetobacter* on Lab Scale (300 mg) [138].

was incubated at 37 °C with shaking at 200 rpm until the OD_{600} reached 0.7. Then, IPTG was added to a final concentration of 0.2 mM, followed by adding 4-bromomethylenecyclonexanone **85** in acetonitrile with the final concentration of 20 mM. The final reaction mixture was shaken at 30 °C and 200 rpm. The reaction was monitored by GC analysis of the aliquot samples every 2 h until the reaction completion. Then, the reaction was stopped and biotransformation mixtures were extracted with ethyl acetate. The organic phase was dried with Na_2SO_4, evaporated and the residue subjected to column chromatography (ethyl acetate/petroleum ether 1:2), yielding the lactone product **86** as white solid in 62% yield (98% de) [138].

9.5 Asymmetric Dihydroxylation of Cinnamonitrile 108 Using Chlorobenzene Dioxygenase in *E. coli* on Multigram Scale

The 30-L biotransformations were performed in a 42-L stirred tank reactor with regulated temperature, pH, stirrer speed, and internal pressure (New MBR, Zurich, Switzerland) (Scheme 9.35). The pH was kept at 7.2 by the addition of 25% w/v aqueous ammonia or 30% phosphoric acid. Mixing was achieved with two 6-bladed Rushton turbine impellers. Before sterilization, the reactor contained salts equivalent to 30 L of M9 medium dissolved in 20 L of water. After sterilization at 121 °C for 40 min, the reactor contents were supplemented aseptically with glucose solution (0.5%), US* (for preparation of salt mix US* see Ref. [165]) (0.1%), magnesium sulfate (according to M9 recipe), thiamine (10 ppm), kanamycin (50 ppm), and antifoaming agent. The volume was adjusted to 27 L with water. The reactor was inoculated with 3 L of M9 preculture (Freshly transformed E. coli JM101 cells were plated on LB agar plates. After incubation overnight, 5 mL LB medium was inoculated with a single colony and incubated for 10 h. A part (1 mL) of this LB preculture was used to inoculate 100 mL of M9 medium and grown overnight) and batch cultivation was performed at a stirring speed of 500 rpm and an aeration of 15 L/min for about 10 h. All incubation and growth steps were carried out at 30 °C. Silicone oil-based antifoam agent (Sigma) was used to control foaming. Fed-batch was started with pumping of the feed consisting of 46% glucose and 1% $MgSO_4$ with a rate of 10 g/L/h into the reactor. Stirring speed was increased to 700 rpm. After 1 h of fed-batch growth, additional kanamycin (1.5 g), US* (30 mL) and thiamine (0.3 g) were added and the expression of CDO genes was induced by the addition of 0.05% dicyclopropylketone. After 3 h the biotransformation was

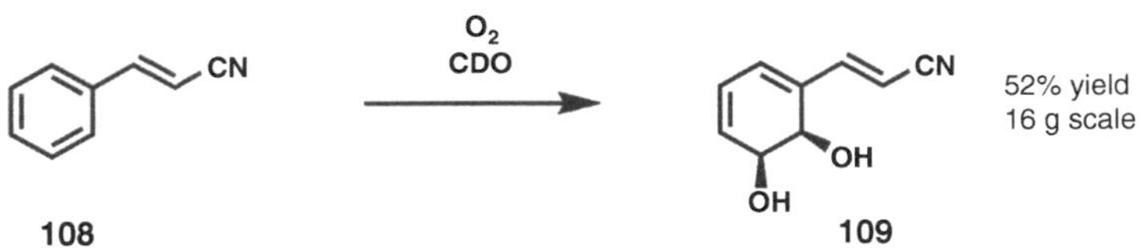

SCHEME 9.35
Asymmetric dihydroxylation of cinnamonitrile 108 using chlorobenzene dioxygenase in *E. coli* on multigram scale [166].

started by the addition of substrate **108**. Substrate was added repeatedly to keep the concentrations below 1.5 mM. Samples were taken at different time points and diluted 10 times with ice-cold acetonitrile and centrifuged at 5000 rpm at 4°C. Supernatants were filtered through a cellulose membrane filter (Spartan, Schleicher & Schuell GmbH, Germany) and analyzed by HPLC.

After 20 h, the reaction mixture (30 L) was centrifuged at 5,000 rpm and at 4°C for 15 min. The supernatant was removed and evaporated on a rotary evaporator at room temperature to a volume of approximately 1 L. The dissolved salts and proteins were precipitated by adding 4 L of acetonitrile. The mixture was centrifuged again under the same conditions and the supernatant was evaporated down to 20 mL. The product was isolated after silica gel chromatography using 60% ethyl acetate and 40% hexane as the eluent. *trans*-3-[(5*S*,6*R*)-5,6-Dihydroxycyclohexa-1,3-dienyl]-acrylonitrile **109** was obtained in 52% yield [166].

References

[1] S. Prasad, T.C. Bhalla, Nitrile hydratases (NHases): at the interface of academia and industry, Biotechnol. Adv. 28 (2010) 725–741.

[2] H. Yamada, M. Kobayashi, Nitrile hydratase and its application to industrial production of acrylamide, Biosci. Biotechnol. Biochem. 60 (1996) 1391–1400.

[3] H. Yamada Y. Tani, Microbial production of amides from nitriles WO 8,301,784A1, 1983.

[4] A. Liese, K. Seelbach, A. Buchholz, J. Haberland, Processes: Lyases EC 4, in: A. Liese, K. Seelbach, C. Wandrey (Eds.), Industrial Biotransformations, Wiley-VCH Verlag GmbH & Co, KGaA, Weinheim, 2006.

[5] R.-C. Zheng, Z.-Y. Yang, C.-C. Li, Y.-G. Zheng, Y.-C. Shen, Industrial production of chiral intermediate of cilastatin by nitrile hydratase and amidase catalyzed one-pot, two-step biotransformation, J. Mol. Catal. B Enzym. 102 (2014) 161–166.

[6] L. Martinkova, V. Mylerova, Synthetic applications of nitrile-converting enzymes, Curr. Org. Chem. 7 (2003) 1279–1295.

[7] L. Martinkova, V. Vejvoda, V. Kren, Selection and screening for enzymes of nitrile metabolism, J. Biotechnol. 133 (2008) 318–326.

[8] Z.-J. Lin, R.-C. Zheng, L.-H. Lei, Y.-G. Zheng, Y.-C. Shen, Ferrous and ferric ions-based high-throughput screening strategy for nitrile hydratase and amidase, J. Microbiol. Methods 85 (2011) 214–220.

[9] Y. Sugiura, J. Kuwahara, T. Nagasawa, H. Yamada, Nitrile hydratase. The first non-heme iron enzyme with a typical low-spin iron(III)-active center, J. Am. Chem. Soc. 109 (1987) 5848–5850.

[10] K. Hashimoto, H. Suzuki, K. Taniguchi, T. Noguchi, M. Yohda, M. Odaka, Catalytic mechanism of nitrile hydratase proposed by time-resolved X-ray crystallography using a novel substrate, tert-butylisonitrile, J. Biol. Chem. 283 (2008) 36617–36623.

[11] M. Nakasako, M. Odaka, M. Yohda, N. Dohmae, K. Takio, N. Kamiya, I. Endo, Tertiary and quaternary structures of photoreactive Fe-type nitrile hydratase from *Rhodococcus* sp. N-771: roles of hydration water molecules in stabilizing the structures and the structural origin of the substrate specificity of the enzyme, Biochemistry 38 (1999) 9887–9898.

[12] T. Nagasawa, K. Takeuchi, H. Yamada, Characterization of a new cobalt-containing nitrile hydratase purified from urea-induced cells of *Rhodococcus rhodochrous* J1, Eur. J. Biochem. 196 (1991) 581–589.

[13] M. Gao, D.-X. Wang, Q.-Y. Zheng, M.-X. Wang, An unusual β-vinyl effect leading to high efficiency and enantioselectivity of the amidase, nitrile biotransformations for the preparation

of enantiopure 3-arylpent-4-enoic acids and amides and their applications in synthesis, J. Org. Chem. 71 (2006) 9532–9535.

[14] D.-H. Leng, D.-X. Wang, J. Pan, Z.-T. Huang, M.-X. Wang, Highly efficient and enantioselective biotransformations of racemic azetidine-2-carbonitriles and their synthetic applications, J. Org. Chem. 74 (2009) 6077–6082.

[15] G.W. Black, T. Gregson, C.B. McPake, J.J. Perry, M. Zhang, Biotransformation of nitriles using the solvent-tolerant nitrile hydratase from *Rhodopseudomonas palustris* CGA009, Tetrahedron Lett. 51 (2010) 1639–1641.

[16] D. Kubac, A. Cejkova, J. Masak, V. Jirku, M. Lemaire, E. Gallienne, J. Bolte, R. Stloukal, L. Martinkova, Biotransformation of nitriles by *Rhodococcus* equi A4 immobilized in LentiKats, J. Mol. Catal. B Enzym. 39 (2006) 59–61.

[17] Z.-J. Lin, R.-C. Zheng, Y.-J. Wang, Y.-G. Zheng, Y.-C. Shen, Enzymatic production of 2-amino-2,3-dimethylbutyramide by cyanide-resistant nitrile hydratase, J. Ind. Microbiol. Biotechnol. 39 (2012) 133–141.

[18] P.F.B. Brandao, S. Verseck, C. Syldatk, Bioconversion of D,L-tert-leucine nitrile to D-tert-leucine by recombinant cells expressing nitrile hydratase and D-selective amidase, Eng. Life Sci. 4 (2004) 547–556.

[19] K.L. Petrillo, S. Wu, E.C. Hann, F.B. Cooling, A. Ben-Bassat, J.E. Gavagan, R. DiCosimo, M.S. Payne, Over-expression in *Escherichia coli* of a thermally stable and regio-selective nitrile hydratase from *Comamonas testosteroni* 5-MGAM-4D, Appl. Microbiol. Biotechnol. 67 (2005) 664–670.

[20] D.E. Stevenson, R. Feng, F. Dumas, D. Groleau, A. Mihoc, A.C. Storer, Mechanistic and structural studies on *Rhodococcus* ATCC 39484 nitrilase, Biotechnol. Appl. Biochem. 15 (1992) 283–302.

[21] M. Kobayashi, M. Goda, S. Shimizu, Nitrilase catalyzes amide hydrolysis as well as nitrile hydrolysis, Biochem. Biophys. Res. Commun. 253 (1998) 662–666.

[22] B.C.M. Fernandes, C. Mateo, C. Kiziak, A. Chmura, J. Wacker, F. van Rantwijk, A. Stolz, R.A. Sheldon, Nitrile hydratase activity of a recombinant nitrilase, Adv. Synth. Catal. 348 (2006) 2597–2603.

[23] D.E. Robertson, J.A. Chaplin, G. DeSantis, M. Podar, M. Madden, E. Chi, T. Richardson, A. Milan, M. Miller, D.P. Weiner, K. Wong, J. McQuaid, B. Farwell, L.A. Preston, X. Tan, M.A. Snead, M. Keller, E. Mathur, P.L. Kretz, M.J. Burk, J.M. Short, Exploring nitrilase sequence space for enantioselective catalysis, Appl. Environ. Microbiol. 70 (2004) 2429–2436.

[24] J.L. Seffernick, S.K. Samanta, T.M. Louie, L.P. Wackett, M. Subramanian, Investigative mining of sequence data for novel enzymes: a case study with nitrilases, J. Biotechnol. 143 (2009) 17–26.

[25] P. Mueller, K. Egorova, C.E. Vorgias, E. Boutou, H. Trauthwein, S. Verseck, G. Antranikian, Cloning, overexpression, and characterization of a thermoactive nitrilase from the hyperthermophilic archaeon *Pyrococcus abyssi*, Protein Expression Purif. 47 (2006) 672–681.

[26] T. Janowitz, I. Trompetter, M. Piotrowski, Evolution of nitrilases in glucosinolate-containing plants, Phytochemistry (Elsevier) 70 (2009) 1680–1686.

[27] R.N. Thuku, D. Brady, M.J. Benedik, B.T. Sewell, Microbial nitrilases: versatile, spiral forming, industrial enzymes, J. Appl. Microbiol. 106 (2009) 703–727.

[28] S. Oßwald, A. Yanenko, Hydrolysis of Nitriles to Carboxylic Acids, in: K. Drauz, H. Gröger, O. May (Eds.), Enzyme Catalysis in Organic Synthesis, Wiley-VCH, Weinheim, 2012, pp. 545–559.

[29] F. Effenberger, Acids Selective hydrolysis of aliphatic dinitriles to monocarboxylic acids by a nitrilase from *Arabidopsis thaliana*, S. Oßwald, Synthesis 12 (2001) 1866–1872.

[30] Z.-Y. You, Z.-Q. Liu, Y.-G. Zheng, Chemical and enzymatic approaches to the synthesis of optically pure ethyl (R)-4-cyano-3-hydroxybutanoate, Appl. Microbiol. Biotechnol. 98 (2014) 11–21.

[31] G. DeSantis, Z. Zhu, W.A. Greenberg, K. Wong, J. Chaplin, S.R. Hanson, B. Farwell, L.W. Nicholson, C.L. Rand, D.P. Weiner, D.E. Robertson, M.J. Burk, An enzyme library approach to biocatalysis: development of nitrilases for enantioselective production of carboxylic acid derivatives, J. Am. Chem. Soc. 124 (2002) 9024–9025.

[32] G. DeSantis, K. Wong, B. Farwell, K. Chatman, Z. Zhu, G. Tomlinson, H. Huang, X. Tan, L. Bibbs, P. Chen, K. Kretz, M.J. Burk, Creation of a productive, highly enantioselective nitrilase through gene site saturation mutagenesis (GSSM), J. Am. Chem. Soc. 125 (2003) 11476–11477.

[33] S. Bergeron, D.A. Chaplin, J.H. Edwards, B.S.W. Ellis, C.L. Hill, K. Holt-Tiffin, J.R. Knight, T. Mahoney, A.P. Osborne, G. Ruecroft, Nitrilase-catalyzed desymmetrization of 3-hydroxyglutaronitrile: preparation of a statin side-chain intermediate, Org. Process Res. Dev. 10 (2006) 661–665.

[34] F. Effenberger, S. Osswald, (E)-Selective hydrolysis of (E,Z)-α,β-unsaturated nitriles by the recombinant nitrilase AtNIT1 from *Arabidopsis thaliana*, Tetrahedron Asymmetry 12 (2001) 2581–2587.

[35] A. Schmid, J.S. Dordick, B. Hauer, A. Kiener, M. Wubbolts, B. Witholt, Industrial biocatalysis today and tomorrow, Nature (London) 409 (2001) 258–268.

[36] N.M. Shaw, K.T. Robins, A. Kiener, Lonza: 20 years of biotransformations, Adv. Synth. Catal. 345 (2003) 425–435.

[37] D. Brady, A. Beeton, J. Zeevaart, C. Kgaje, F. van Rantwijk, R.A. Sheldon, Characterisation of nitrilase and nitrile hydratase biocatalytic systems, Appl. Microbiol. Biotechnol. 64 (2004) 76–85.

[38] P. Kaul, A. Banerjee, U.C. Banerjee, Nitrile hydrolyses, in: J. Polaina, A.P. MacCabe (Eds.), Industrial Enzymes, Springer, India, 2007, pp. 531–547.

[39] S. Martinez-Rodriguez, A.I. Martinez-Gomez, F. Rodriguez-Vico, J.M. Clemente-Jimenez, F.J. Las Heras-Vazquez, Carbamoylases: characteristics and applications in biotechnological processes, Appl. Microbiol. Biotechnol. 85 (2010) 441–458.

[40] J. Ogawa, N. Horinouchi, S. Shimizu, Hydrolysis and Formation of Hydantoins, in: K. Drauz, H. Gröger (Eds.), Enzyme Catalysis in Organic Synthesis, Wiley-VCH, Weinheim, 2012, pp. 651–674.

[41] S.G. Burton, R.A. Dorrington, Hydantoin-hydrolysing enzymes for the enantioselective production of amino acids: new insights and applications, Tetrahedron Asymmetry 15 (2004) 2737–2741.

[42] M. Pietzsch, C. Syldatk, in: K. Drauz, H. Waldmann (Eds.), Enzyme Catalysis in Organic Synthesis, Wiley-VCH, Weinheim, 2002, pp. 761–799.

[43] S. Martinez-Rodriguez, F.J. Las Heras-Vazquez, J.M. Clemente-Jimenez, L. Mingorance-Cazorla, F. Rodriguez-Vico, Complete conversion of D,L-5-monosubstituted hydantoins with a low velocity of chemical racemization into D-amino acids using whole cells of recombinant *Escherichia coli*, Biotechnol. Prog. 18 (2002) 1201–1206.

[44] F.J. Las Heras-Vazquez, S. Martinez-Rodriguez, L. Mingorance-Cazorla, J.M. Clemente-Jimenez, F. Rodriguez-Vico, Overexpression and characterization of hydantoin racemase from *Agrobacterium tumefaciens* C58, Biochem. Biophys. Res. Commun. 303 (2003) 541–547.

[45] S. Suzuki, Y. Takenaka, N. Onishi, K. Yokozeki, Molecular cloning and expression of the hyu genes from *Microbacterium liquefaciens* AJ 3912, responsible for the conversion of 5-substituted hydantoins to α-amino acids, in Escherichia coli, Biosci. Biotechnol. Biochem 69 (2005) 1473–1482.

[46] S.-K. Hsu, H.-H. Lo, W.-D. Lin, I.C. Chen, C.-H. Kao, W.-H. Hsu, Stereoselective synthesis of L-homophenylalanine using the carbamoylase method with in situ racemization via *N*-acylamino acid racemase, Process Biochem. (Amsterdam, Neth) 42 (2007) 856–862.

[47] C.-K. Lo, C.-H. Kao, W.-C. Wang, H.-M. Wu, W.-H. Hsu, L.-L. Lin, H.-Y. Hu, Engineering of the critical residues at the stereochemistry-gate loops of *Brevibacillus agri* dihydropyrimidinase

for the production of L-homophenylalanine, Process Biochem. (Amsterdam, Neth) 44 (2009) 309–315.

[48] H.S. Nandanwar, R.M. Vohra, G.S. Hoondal, Trimeric L-N-carbamoylase from newly isolated *Brevibacillus reuszeri* HSN1: a potential biocatalyst for production of L-α-amino acids, Biotechnol. Appl. Biochem. 60 (2013) 219–230.

[49] Y. Mei, B. He, N. Liu, P. Ouyang, Screening and distributing features of bacteria with hydantoinase and carbamoylase, Microbiol. Res. 164 (2009) 322–329.

[50] C.-H. Kao, H.-H. Lo, S.-K. Hsu, W.-H. Hsu, A novel hydantoinase process using recombinant *Escherichia coli* cells with dihydropyrimidinase and L-*N*-carbamoylase activities as biocatalyst for the production of L-homophenylalanine, J. Biotechnol. 134 (2008) 231–239.

[51] H. Groeger, K. Drauz, Methods for the Enantioselective Biocatalytic Production of L-amino Acids on an Industrial Scale, Wiley-VCH Verlag GmbH & Co, KGaA, 2004, pp. 131–147.

[52] G.F. Matcher, S.G. Burton, R.A. Dorrington, Mutational analysis of the hydantoin hydrolysis pathway in *Pseudomonas putida* RU-KM3S, Appl. Microbiol. Biotechnol. 65 (2004) 391–400.

[53] I. Aranaz, N. Acosta, A. Heras, Encapsulation of an *Agrobacterium radiobacter* extract containing D-hydantoinase and D-carbamoylase activities into alginate-chitosan polyelectrolyte complexes, J. Mol. Catal. B Enzym. 58 (2009) 54–64.

[54] H.S. Nandanwar, R.M. Vohra, G.S. Hoondal, Enhanced stability of newly isolated trimeric L-methionine-N-carbamoylase from *Brevibacillus reuszeri* HSN1 by covalent immobilization, Biotechnol. Appl. Biochem. 60 (2013) 305–315.

[55] M.B. Arcuri, O.A.C. Antunes, S.P. Machado, C.H.F. Almeida, E.G. Oestreicher, Stability of immobilized D-hydantoinase from *Vigna angularis* and repeated cycles of highly enantioenriched production of *N*-carbamoyl- D-phenylglycines, Amino Acids 27 (2004) 69–74.

[56] B.T. Bulawayo, R.A. Dorrington, S.G. Burton, Enhanced operational parameters for amino acid production using hydantoin-hydrolysing enzymes of *Pseudomonas putida* strain RUKM3s immobilised in Eupergit C, Enzym. Microb. Technol. 40 (2007) 533–539.

[57] R.J. Turner, J. Aikens, S. Royer, L. DeFilippi, A. Yap, D. Holzle, N. Somers, I.G. Fotheringham, D-amino acid tolerant hosts for D-hydantoinase whole cell biocatalysts, Eng. Life Sci. 4 (2004) 517–520.

[58] H. Yu, S. Yang, W. Jiang, Y. Yang, Efficient biocatalytic production of D-4-hydroxyphenylglycine by whole cells of recombinant *Ralstonia pickettii*, Folia Microbiol. (Prague, Czech Repub) 54 (2009) 509–515.

[59] M. Cleij, A. Archelas, R. Furstoss, Microbiological transformations 43. Epoxide hydrolases as tools for the synthesis of enantiopure alpha-methylstyrene oxides: a new and efficient synthesis of (*S*)-ibuprofen, J. Org. Chem. 64 (1999) 5029–5035.

[60] X.J. Chen, A. Archelas, R. Furstoss, Microbiological transformations 27. The 1st examples for preparative-scale enantioselective or diastereoselective epoxide hydrolyzes using microorganisms – an unequivocal access to all 4 bisabolol stereoisomers, J. Org. Chem. 58 (1993) 5528–5532.

[61] R.V.A. Orru, I. Osprian, W. Kroutil, K. Faber, An efficient large-scale synthesis of (*R*)-(-)-mevalonolactone using simple biological and chemical catalysts, Synthesis-Stuttgart 9 (1998) 1259–1263.

[62] R.V.A. Orru, S.F. Mayer, W. Kroutil, K. Faber, Chemoenzymatic deracemization of (+/-)-2,2-disubstituted oxiranes, Tetrahedron 54 (1998) 859–874.

[63] N. Bala, S.S. Chimni, Recent developments in the asymmetric hydrolytic ring opening of epoxides catalysed by microbial epoxide hydrolase, Tetrahedron Asymmetry 21 (2010) 2879–2898.

[64] E.Y. Lee, M.L. Shuler, Molecular engineering of epoxide hydrolase and its application to asymmetric and enantioconvergent hydrolysis, Biotechnol. Bioeng. 98 (2007) 318–327.

[65] J.K. Beetham, D. Grant, M. Arand, J. Garbarino, T. Kiyosue, F. Pinot, F. Oesch, W.R. Belknap, K. Shinozaki, B.D. Hammock, Gene evolution of epoxide hydrolases and recommended nomenclature, DNA Cell Biol. 14 (1995) 61–71.

[66] M. Nardini, B.W. Dijkstra, Alpha/beta hydrolase fold enzymes: the family keeps growing, Curr. Opin. Struct. Biol. 9 (1999) 732–737.

[67] M. Nardini, I.S. Ridder, H.J. Rozeboom, K.H. Kalk, R. Rink, D.B. Janssen, B.W. Dijkstra, The X-ray structure of epoxide hydrolase from *Agrobacterium radiobacter* AD1 – an enzyme to detoxify harmful epoxides, J. Biol. Chem. 274 (1999) 14579–14586.

[68] M. Arand, B.M. Hallberg, J.Y. Zou, T. Bergfors, F. Oesch, M.J. van der Werf, J.A.M. de Bont, T.A. Jones, S.L. Mowbray, Structure of *Rhodococcus erythropolis* limonene-1,2-epoxide hydrolase reveals a novel active site, EMBO J. 22 (2003) 2583–2592.

[69] J.-H. Woo, E.Y. Lee, Enantioselective hydrolysis of racemic styrene oxide and its substituted derivatives using newly-isolated *Sphingopyxis* sp. exhibiting a novel epoxide hydrolase activity, Biotechnol. Lett. 36 (2014) 357–362.

[70] A. Duarah, A. Goswami, T.C. Bora, M. Talukdar, B.K. Gogoi, Enantioconvergent biohydrolysis of racemic styrene oxide to *R*-phenyl-1, 2-ethanediol by a newly isolated filamentous fungus *Aspergillus tubingensis* TF1, Appl. Biochem. Biotechnol. 170 (2013) 1965–1973.

[71] W-J Chen, W-Y Lou, M-H Zong, Asymmetric hydrolysis of racemic styrene oxide to (R)-1-phenyl-1,2-ethanediol catalyzed by Mung bean epoxide hydrolase, Am. Chem. Soc., pp. ORGN-667, 2011.

[72] Y. Xu, X. Jia, S. Panke, Z. Li, Asymmetric dihydroxylation of aryl olefins by sequential enantioselective epoxidation and regioselective hydrolysis with tandem biocatalysts, Chem. Commun. 12 (Cambridge, U K) (2009) 1481–1483.

[73] S. Wu, A. Li, Y.S. Chin, Z. Li, Enantioselective hydrolysis of racemic and *meso*-epoxides with recombinant *Escherichia coli* expressing epoxide hydrolase from *Sphingomonas* sp. HXN-200: preparation of epoxides and vicinal diols in high ee and high concentration, ACS Catal. 3 (2013) 752–759.

[74] S.K. Wu, Y.Z. Chen, Y. Xu, A.T. Li, Q.S. Xu, A. Glieder, Z. Li, Enantioselective *trans*-dihydroxylation of aryl olefins by cascade biocatalysis with recombinant *Escherichia coli* coexpressing monooxygenase and epoxide hydrolase, ACS Catal. 4 (2014) 409–420.

[75] R.V.A. Orru, W. Kroutil, K. Faber, Deracemization of (+/-)-2,2-disubstituted epoxides via enantioconvergent chemoenzymatic hydrolysis using Nocardia EH1 epoxide hydrolase and sulfuric acid, Tetrahedron Lett. 38 (1997) 1753–1754.

[76] P.E. Swanson, Dehalogenases applied to industrial-scale biocatalysis, Curr. Opin. Biotechnol. 10 (1999) 365–369.

[77] S. Fetzner, Bacterial dehalogenation, Appl. Microbiol. Biotechnol. 50 (1998) 633–657.

[78] S. Fetzner, F. Lingens, Bacterial dehalogenases – biochemistry, genetics, and biotechnological applications, Microbiol. Rev. 58 (1994) 641–685.

[79] J.E.T.V. Vlieg, L.X. Tang, J.H.L. Spelberg, T. Smilda, G.J. Poelarends, T. Bosma, A.E.J. van Merode, M.W. Fraaije, D.B. Janssen, Halohydrin dehalogenases are structurally and mechanistically related to short-chain dehydrogenases/reductases, J. Bacteriol. 183 (2001) 5058–5066.

[80] R.M. de Jong, K.H. Kalk, L. Tang, D.B. Janssen, B.W. Dijkstra, The X-ray structure of the haloalcohol dehalogenase HheA from *Arthrobacter* sp. strain AD2: insight into enantioselectivity and halide binding in the haloalcohol dehalogenase family, J. Bacteriol. 188 (2006) 4051–4056.

[81] S.K. Ma, J. Gruber, C. Davis, L. Newman, D. Gray, A. Wang, J. Grate, G.W. Huisman, R.A. Sheldon, A green-by-design biocatalytic process for atorvastatin intermediate, Green Chem. 12 (2010) 81–86.

[82] S.-P. Zou, E.-H. Du, Z.-C. Hu, Y.-G. Zheng, Enhanced biotransformation of 1,3-dichloro-2-propanol to epichlorohydrin via resin-based in situ product removal process, Biotechnol. Lett. 35 (2013) 937–942.

[83] M. Schallmey, R.J. Floor, B. Hauer, M. Breuer, P.A. Jekel, H.J. Wijma, B.W. Dijkstra, D.B. Janssen, Biocatalytic and structural properties of a highly engineered halohydrin dehalogenase, Chembiochem. 14 (2013) 870–881.

[84] E.M. Majeric, I. Primozic, T. Hrenar, A. Smolko, I. Dokli, B. Salopek-Sondi, L. Tang, Catalytic activity of halohydrin dehalogenases towards spiroepoxides, Org. Biomol. Chem. 10 (2012) 5063–5072.

[85] W. Szymanski, C.P. Postema, C. Tarabiono, F. Berthiol, L. Campbell-Verduyn, W.S. de, V.J.G. de, B.L. Feringa, D.B. Janssen, Combining designer cells and click chemistry for a one-pot four-step preparation of enantiopure β-hydroxytriazoles, Adv. Synth. Catal. 352 (2010) 2111–2115.

[86] C. Molinaro, A.-A. Guilbault, B. Kosjek, Resolution of 2,2-disubstituted epoxides via biocatalytic azidolysis, Org. Lett. 12 (2010) 3772–3775.

[87] K.M. Fox, P.A. Karplus, Old yellow enzyme at 2 Å resolution: overall structure, ligand binding, and comparison with related flavoproteins, Structure (London) 2 (1994) 1089–1105.

[88] B.J. Brown, Z. Deng, P.A. Karplus, V. Massey, On the active site of old yellow enzyme. Role of histidine 191 and asparagine 194, J. Biol. Chem. 273 (1998) 32753–32762.

[89] R.M. Kohli, V. Massey, The oxidative half-reaction of old yellow enzyme. The role of tyrosine 196, J. Biol. Chem. 273 (1998) 32763–32770.

[90] H.S. Toogood, J.M. Gardiner, N.S. Scrutton, Biocatalytic reductions and chemical versatility of the old yellow enzyme family of flavoprotein oxidoreductases, ChemCatChem. 2 (2010) 892–914.

[91] G. Oberdorfer, G. Steinkellner, C. Stueckler, K. Faber, K. Gruber, Stereopreferences of old yellow enzymes: structure correlations and sequence patterns in enoate reductases, ChemCatChem. 3 (2011) 1562–1566.

[92] G. Oberdorfer, K. Gruber, K. Faber, M. Hall, Stereocontrol strategies in the asymmetric bioreduction of alkenes, Synlett. 23 (2012) 1857–1864.

[93] C.K. Winkler, G. Tasnadi, D. Clay, M. Hall, K. Faber, Asymmetric bioreduction of activated alkenes to industrially relevant optically active compounds, J. Biotechnol. 162 (2012) 381–389.

[94] E. Brenna, F.G. Gatti, A. Manfredi, D. Monti, F. Parmeggiani, Enoate reductase-mediated preparation of methyl (S)-2-bromobutanoate, a useful key intermediate for the synthesis of chiral active pharmaceutical ingredients, Org. Process Res. Dev. 16 (2012) 262–268.

[95] E. Brenna, S.L. Cosi, E.E. Ferrandi, F.G. Gatti, D. Monti, F. Parmeggiani, A. Sacchetti, Substrate scope and synthetic applications of the enantioselective reduction of a-alkyl-b-arylenones mediated by Old Yellow Enzymes, Org. Biomol. Chem. 11 (2013) 2988–2996.

[96] Y.A. Pompeu, B. Sullivan, J.D. Stewart, X-ray crystallography reveals how subtle changes control the orientation of substrate binding in an alkene reductase, ACS Catal. 3 (2013) 2376–2390.

[97] S. Reich, H.W. Hoeffken, B. Rosche, B.M. Nestl, B. Hauer, Crystal structure determination and mutagenesis analysis of the ene reductase NCR, ChemBioChem. 13 (2012) 2400–2407.

[98] E. Nordling, H. Jornvall, B. Persson, Medium-chain dehydrogenases/reductases (MDR). Family characterizations including genome comparisons and active site modelling, Eur. J. Biochem. 269 (2002) 4267–4276.

[99] D.J. Mansell, H.S. Toogood, J. Waller, J.M.X. Hughes, C.W. Levy, J.M. Gardiner, N.S. Scrutton, Biocatalytic asymmetric alkene reduction: crystal structure and characterization of a double bond reductase from *Nicotiana tabacum*, ACS Catal. 3 (2013) 370–379.

[100] C. Stueckler, C.K. Winkler, M. Hall, B. Hauer, M. Bonnekessel, K. Zangger, K. Faber, Stereo-controlled asymmetric bioreduction of α,β-dehydroamino acid derivatives, Adv. Synth. Catal. 353 (2011) 1169–1173.

[101] S.K. Padhi, D.J. Bougioukou, J.D. Stewart, Site-saturation mutagenesis of Tryptophan 116 of *Saccharomyces pastorianus* old yellow enzyme uncovers stereocomplementary variants, J. Am. Chem. Soc. 131 (2009) 3271–3280.

[102] A.Z. Walton, W.C. Conerly, Y. Pompeu, B. Sullivan, J.D. Stewart, Biocatalytic reductions of Baylis–Hillman Adducts, ACS Catal. 1 (2011) 989–993.

[103] D.J. Bougioukou, A.Z. Walton, J.D. Stewart, Towards preparative-scale, biocatalytic alkene reductions, Chem. Commun. (Cambridge, UK) 46 (2010) 8558–8560.

[104] Y.A. Pompeu, B. Sullivan, A.Z. Walton, J.D. Stewart, Structural and catalytic characterization of *Pichia stipitis* OYE 2.6, a useful biocatalyst for asymmetric alkene reductions, Adv. Synth. Catal. 354 (2012) 1949–1960.

[105] N. Iqbal, F. Rudroff, A. Brige, B.J. Van, M.D. Mihovilovic, Asymmetric bioreduction of activated carbon-carbon double bonds using Shewanella yellow enzyme (SYE-4) as novel enoate reductase, Tetrahedron 68 (2012) 7619–7623.

[106] D. Mangan, I. Miskelly, T.S. Moody, A three-enzyme system involving an ene-reductase for generating valuable chiral building blocks, Adv. Synth. Catal. 354 (2012) 2185–2190.

[107] J. Pietruszka, M. Schoelzel, Ene reductase-catalysed synthesis of (*R*)-profen derivatives, Adv. Synth. Catal. 354 (2012) 751–756.

[108] C.K. Winkler, D. Clay, S. Davies, P. O'Neill, P. McDaid, S. Debarge, J. Steflik, M. Karmilowicz, J.W. Wong, K. Faber, Chemoenzymatic asymmetric synthesis of pregabalin precursors via asymmetric bioreduction of β-cyanoacrylate esters using ene-reductases, J. Org. Chem. 78 (2013) 1525–1533.

[109] K. Durchschein, S. Wallner, P. Macheroux, K. Zangger, W.M.F. Fabian, K. Faber, Unusual C=C bond isomerization of an α,β-unsaturated γ-butyrolactone catalysed by flavoproteins from the old yellow enzyme family, ChemBioChem. 13 (2012) 2346–2351.

[110] C. Stueckler, T.C. Reiter, N. Baudendistel, K. Faber, Nicotinamide-independent asymmetric bioreduction of C=C-bonds via disproportionation of enones catalyzed by enoate reductases, Tetrahedron 66 (2010) 663–667.

[111] K. Durchschein, W.M.F. Fabian, P. Macheroux, K. Zangger, G. Trimmel, K. Faber, Reductive biotransformation of nitroalkenes via nitroso-intermediates to oxazetes catalyzed by xenobiotic reductase A (XenA), Org. Biomol. Chem. 9 (2011) 3364–3369.

[112] M. Schittmayer, A. Glieder, M.K. Uhl, A. Winkler, S. Zach, J.H. Schrittwieser, W. Kroutil, P. Macheroux, K. Gruber, S. Kambourakis, J.D. Rozzell, M. Winkler, Old yellow enzyme-catalyzed dehydrogenation of saturated ketones, Adv. Synth. Catal. 353 (2011) 268–274.

[113] K. Durchschein, M. Hall, K. Faber, Unusual reactions mediated by FMN-dependent ene- and nitro-reductases, Green Chem. 7 (2013) 1764–1772.

[114] G. de Gonzalo, M.D. Mihovilovic, M.W. Fraaije, Recent developments in the application of Baeyer–Villiger monooxygenases as biocatalysts, ChemBioChem. 11 (2010) 2208–2231.

[115] V. Alphand, R. Wohlgemuth, Applications of Baeyer–Villiger monooxygenases in organic synthesis, Curr. Org. Chem. 14 (2010) 1928–1965.

[116] H. Leisch, K. Morley, P.C.K. Lau, Baeyer–Villiger monooxygenases: more than just green chemistry, Chem. Rev. 111 (2011) 4165–4222.

[117] N.A. Donoghue, D.B. Norris, P.W. Trudgill, The purification and properties of cyclohexanone oxygenase from *Nocardia globerula* CL1 and Acinetobacter NCIB 9871, Eur. J. Biochem. 63 (1976) 175–192.

[118] C.C. Ryerson, D.P. Ballou, C. Walsh, Mechanistic studies on cyclohexanone oxygenase, Biochemistry 21 (1982) 2644–2655.

[119] B.J. Yachnin, T. Sprules, M.B. McEvoy, P.C.K. Lau, A.M. Berghuis, The substrate-bound crystal structure of a Baeyer–Villiger monooxygenase exhibits a criegee-like conformation, J. Am. Chem. Soc. 134 (2012) 7788–7795.

[120] E. Malito, A. Alfieri, M.W. Fraaije, A. Mattevi, Crystal structure of a Baeyer–Villiger monooxygenase, Proc. Natl. Acad. Sci. USA 101 (2004) 13157–13162.

[121] R. Baron, C. Riley, P. Chenprakhon, K. Thotsaporn, R.T. Winter, A. Alfieri, F. Forneris, W.J.H. van Berkeig, P. Chaiyen, M.W. Fraaije, A. Mattevi, J.A. Mc Cammon, Multiple pathways

guide oxygen diffusion into flavoenzyme active sites, Proc. Natl. Acad. Sci. USA 106 (2009) 10603–10608.

[122] P. Chaiyen, M.W. Fraaije, A. Mattevi, The enigmatic reaction of flavins with oxygen, Trends Biochem. Sci. 37 (2012) 373–380.

[123] S. Franceschini, H.L. van Beek, A. Pennetta, C. Martinoli, M.W. Fraaije, A. Mattevi, Exploring the structural basis of substrate preferences in Baeyer–Villiger monooxygenases, J. Biol. Chem. 287 (2012) 22626–22634.

[124] R. Orru, H.M. Dudek, C. Martinoli, P.D.E. Torres, A. Royant, M. Weik, M.W. Fraaije, A. Mattevi, Snapshots of enzymatic Baeyer–Villiger catalysis: oxygen activation and intermediate stabilization, J. Biol. Chem. 286 (2011) 29284–29291.

[125] C. Martinoli, H.M. Dudek, R. Orru, D.E. Edmondson, M.W. Fraaije, A. Mattevi, Beyond the protein matrix: probing cofactor variants in a Baeyer–Villiger oxygenation reaction, ACS Catal. 3 (2013) 3058–3062.

[126] C.V.F. Baldwin, R. Wohlgemuth, J.M. Woodley, The first 200-L scale asymmetric Baeyer-Villiger oxidation using a whole-cell biocatalyst, Org. Process Res. Dev. 12 (2008) 660–665.

[127] R. Snajdrova, G. Grogan, M.D. Mihovilovic, Resolution of fused bicyclic ketones by a recombinant biocatalyst expressing the Baeyer-Villiger monooxygenase gene Rv3049c from Mycobacterium tuberculosis H37Rv, Bioorg. Med. Chem. Lett. 16 (2006) 4813–4817.

[128] M.D. Mihovilovic, P. Kapitan, P. Kapitanova, Regiodivergent Baeyer–Villiger oxidation of fused ketones by recombinant whole-cell biocatalysts, ChemSusChem. 1 (2008) 143–148.

[129] M.J. Fink, T.C. Fischer, F. Rudroff, H. Dudek, M.W. Fraaije, M.D. Mihovilovic, Extensive substrate profiling of cyclopentadecanone monooxygenase as Baeyer–Villiger biocatalyst reveals novel regiodivergent oxidations, J. Mol. Catal. B Enzym. 73 (2011) 9–16.

[130] Y.K. Bong, M.D. Clay, S.J. Collier, B. Mijts, M. Vogel, X. Zhang, J. Zhu, J. Nazor, D. Smith, S. Song, Engineered cylohexanone monooxygenases for synthesis of prazole compounds. USA WO2011071982 A2, 2011.

[131] H.L. van Beek, H.J. Wijma, L. Fromont, D.B. Janssen, M.W. Fraaije, Stabilization of cyclohexanone monooxygenase by a computationally designed disulfide bond spanning only one residue, FEBS Open. Bio. 4 (2014) 168–174.

[132] M. Bucko, A. Schenkmayerova, P. Gemeiner, A. Vikartovska, M.D. Mihovilovic, I. Lacik, Continuous testing system for Baeyer-Villiger biooxidation using recombinant Escherichia coli expressing cyclohexanone monooxygenase encapsulated in polyelectrolyte complex capsules, Enzym. Microb. Technol. 49 (2011) 284–288.

[133] A. Schenkmayerova, M. Bucko, P. Gemeiner, J. Katrlik, Microbial monooxygenase amperometric biosensor for monitoring of Baeyer-Villiger biotransformation, Biosens. Bioelectron. 50 (2013) 235–238.

[134] M.D. Mihovilovic, F. Rudroff, B. Grötzl, P. Kapitan, R. Snajdrova, J. Rydz, R. Mach, Family Clustering of Baeyer–Villiger Monooxygenases Based on Protein Sequence and Stereopreference, Angew. Chem. Int. Ed. 44 (2005) 3609–3613.

[135] D.A. Bianchi, R. Moran-Ramallal, N. Iqbal, F. Rudroff, M.D. Mihovilovic, Enantiocomplementary access to carba-analogs of C-nucleoside derivatives by recombinant Baeyer-Villiger monooxygenases, Bioorg. Med. Chem. Lett. 9 (2013) 2718–2720.

[136] J. Rehdorf, M.D. Mihovilovic, U.T. Bornscheuer, Exploiting the regioselectivity of Baeyer–Villiger monooxygenases for the formation of β-amino acids and β-amino alcohols, Angew. Chem. Int. Ed. 49 (2010) 4506–4508.

[137] J. Rehdorf, M.D. Mihovilovic, M.W. Fraaije, U.T. Bornscheuer, Enzymatic synthesis of enantiomerically pure β-amino ketones, β-amino esters, and β-amino alcohols with Baeyer–Villiger monooxygenases, Chem. Eur. J. 16 (2010) 9525–9535.

[138] Z.-G. Zhang, G.-D. Roiban, J.P. Acevedo, I. Polyak, M.T. Reetz, A new type of stereoselectivity in Baeyer–Villiger reactions: access to E- and Z-olefins, Adv. Synth. Catal. 355 (2013) 99–106.

[139] N. Berezina, V. Alphand, R. Furstoss, Microbiological transformations. Part 51: The first example of a dynamic kinetic resolution process applied to a microbiological Baeyer-Villiger oxidation, Tetrahedron Asymmetry 13 (2002) 1953–1955.

[140] M.C. Gutierrez, R. Furstoss, V. Alphand, Microbiological transformations 60. Enantioconvergent Baeyer–Villiger oxidation via a combined whole cells and ionic exchange resin-catalysed dynamic kinetic resolution process, Adv. Synth. Catal. 347 (2005) 1051–1059.

[141] C. Rodriguez, G. de Gonzalo, A. Rioz-Martinez, D.E.T. Pazmino, M.W. Fraaije, V. Gotor, BVMO-catalysed dynamic kinetic resolution of racemic benzyl ketones in the presence of anion exchange resins, Org. Biomol. Chem. 8 (2010) 1121–1125.

[142] A. Rioz-Martinez, A. Cuetos, C. Rodriguez, G. de Gonzalo, I. Lavandera, M.W. Fraaije, V. Gotor, Dynamic kinetic resolution of alpha-substituted beta-ketoesters catalyzed by Baeyer–Villiger monooxygenases: access to enantiopure alpha-hydroxy esters, Angew. Chem.-Int. Ed. 50 (2011) 8387–8390.

[143] M. Kadow, K. Loschinski, S. Sass, M. Schmidt, U.T. Bornscheuer, Completing the series of BVMOs involved in camphor metabolism of *Pseudomonas putida* NCIMB 10007 by identification of the two missing genes, their functional expression in E. coli, and biochemical characterization, Appl. Microbiol. Biotechnol. 96 (2012) 419–429.

[144] A. Riebel, G. de Gonzalo, M.W. Fraaije, Expanding the biocatalytic toolbox of flavoprotein monooxygenases from *Rhodococcus jostii* RHA1, J. Mol. Catal B. Enzym. 88 (2013) 20–25.

[145] A. Riebel, M.J. Fink, M.D. Mihovilovic, M.W. Fraaije, Type II flavin-containing monooxygenases: a new class of biocatalysts that harbors Baeyer–Villiger monooxygenases with a relaxed coenzyme specificity, ChemCatChem. 6 (2014) 1112–1117.

[146] M.J. Fink, D.V. Rial, P. Kapitanova, A. Lengar, J. Rehdorf, Q. Cheng, F. Rudroff, M.D. Mihovilovic, Quantitative comparison of chiral catalysts selectivity and performance: a generic concept illustrated with cyclododecanone monooxygenase as Baeyer–Villiger biocatalyst, Adv. Synth. Catal. 354 (2012) 3491–3500.

[147] M.J. Fink, M. Schoen, F. Rudroff, M. Schnuerch, M.D. Mihovilovic, Single operation stereoselective synthesis of aerangis lactones: combining continuous flow hydrogenation and biocatalysts in a chemoenzymatic sequence, ChemCatChem. 5 (2013) 724–727.

[148] N. Oberleitner, C. Peters, J. Muschiol, M. Kadow, S. Sass, T. Bayer, P. Schaaf, N. Iqbal, F. Rudroff, M.D. Mihovilovic, U.T. Bornscheuer, An enzymatic toolbox for cascade reactions: a showcase for an *in vivo* redox sequence in asymmetric synthesis, Chemcatchem. 5 (2013) 3524–3528.

[149] J.W. Song, E.Y. Jeon, D.H. Song, H.Y. Jang, U.T. Bornscheuer, D.K. Oh, J.B. Park, Multistep enzymatic synthesis of long-chain alpha, omega-dicarboxylic and omega-hydroxycarboxylic acids from renewable fatty acids and plant oils, Angew. Chem.-Int. Ed. 52 (2013) 2534–2537.

[150] J. Liu, Z. Li, Cascade biotransformations via enantioselective reduction, oxidation, and hydrolysis: preparation of (*R*)-delta-lactones from 2-alkylidenecyclopentanones, ACS Catal. 3 (2013) 908–911.

[151] S.E. Lewis, Applications of biocatalytic arene ipso, *ortho, cis*-dihydroxylation in synthesis, Chem. Commun. (Cambridge, UK) 50 (2014) 2821–2830.

[152] D.J.Y.D. Bon, B. Lee, M.G. Banwell, I.A. Cade, Biocatalytically-derived cis-1,2-dihydrocatechols Enantiopure and versatile building blocks for chemical synthesis, Chimica Oggi-Chemistry Today 30 (2012) 22–27.

[153] T. Hudlicky, Recent chemoenzymatic total syntheses of natural and unnatural products: codeine, balanol, pancratistatin, and oseltamivir, Pure Appl. Chem. 82 (2010) 1785–1796.

[154] T. Hudlicky, J.W. Reed, Celebrating 20 years of SYNLETT-special account on the merits of biocatalysis and the impact of arene *cis*-dihydrodiols on enantioselective synthesis, Synlett. (2009) 685–703.

[155] R. Friemann, K. Lee, E.N. Brown, D.T. Gibson, H. Eklund, S. Ramaswamy, Structures of the multicomponent Rieske non-heme iron toluene 2,3-dioxygenase enzyme system, Acta Crystallogr. Sect. D-Biological. Crystallogr. 65 (2009) 24–33.

[156] D.R. Boyd, N.D. Sharma, M.V. Hand, M.R. Groocock, N.A. Kerley, H. Dalton, J. Chima, G.N. Sheldrake, Stereodirecting substituent effects during enzyme-catalyzed synthesis of *cis*-dihydrodiol metabolites of 1,4-disubstituted benzene substrates, J. Chem. Soc.-Chem. Commun. (1993) 974–976.

[157] D.R. Boyd, T.D.H. Bugg, Arene cis-dihydrodiol formation: from biology to application, Org. Biomol. Chem. 4 (2006) 181–192.

[158] R.A. Johnson, Microbial arene oxidations, Org. React. (Hoboken, NJ, US) **63**, (2004) 117–264.

[159] M.G. Charest, D.R. Siegel, A.G. Myers, Synthesis of (-)-tetracycline, J. Am. Chem. Soc. 127 (2005) 8292–8293.

[160] M.G. Charest, C.D. Lerner, J.D. Brubaker, D.R. Siege, A.G. Myers, A convergent enantioselective route to structurally diverse 6-deoxytetracycline antibiotics, Science 308 (2005) 395–398.

[161] M.G. Banwell, A.J. Edwards, G.J. Harfoot, K.A. Jolliffe, M.D. McLeod, K.J. McRae, S.G. Stewart, M. Vogtle, Chemoenzymatic methods for the enantioselective preparation of sesquiterpenoid natural products from aromatic precursors, Pure Appl. Chem. 75 (2003) 223–229.

[162] B. Sullivan, I. Carrera, M. Drouin, T. Hudlicky, Symmetry-based design for the chemoenzymatic synthesis of oseltamivir (tamiflu) from ethyl benzoate, Angew. Chem. Int. Ed. 48 (2009) 4229–4231.

[163] T. Hudlicky, J.D. Price, F. Rulin, T. Tsunoda, Efficient, Enantiodivergent Synthesis of (+)-Pinitol and (-)-Pinitol, J. Am. Chem. Soc. 112 (1990) 9439–9440.

[164] L. Werner, A. Machara, B. Sullivan, I. Carrera, M. Moser, D.R. Adams, T. Hudlicky, J. Andraos, Several generations of chemoenzymatic synthesis of oseltamivir (tamiflu): evolution of strategy, quest for a process-quality synthesis, and evaluation of efficiency metrics, J. Org. Chem. 76 (2011) 10050–10067.

[165] B. Buhler, B. Witholt, B. Hauer, A. Schmid, Characterization and application of xylene monooxygenase for multistep biocatalysis, Appl. Environ. Microbiol. 68 (2002) 560–568.

[166] S. Yildirim, J. Zezula, T. Hudlicky, B. Witholt, A. Schmid, Asymmetric dihydroxylation of cinnamonitrile to trans-3-[(5S,6R)-5,6-dihydroxycyclohexa-1,3-dienyl]-acrylonitrile using chlorobenzene dioxygenase in *Escherichia coli* (pTEZ30), Adv. Synth. Catal. 346 (2004) 933–942.

CHAPTER 10

Enzymatic C—C Bond Formation

Pere Clapés
Department Química Biológica y Modelización Molecular, Instituto de Química Avanzada de Cataluña, IQAC-CSIC, Barcelona, Spain

Stereoselective carbon–carbon bond formation is a pivotal process in the asymmetric construction of the skeletal frameworks of complex molecules from smaller and simpler starting materials [1]. Catalytic C—C coupling is particularly valuable in asymmetric synthesis because of its potential for stereodivergent product generation [2], by which multiple stereoisomeric products can be derived from common synthetic building blocks [3,4]. Complementary to conventional chemical catalysis using chiral Lewis acids or organocatalysis, biocatalytic conversions by means of carboligation enzymes offer a unique tool to perform such preparative transformations in a sustainable, environmentally benign fashion [5–8]. Carboligation enzymes are finding increasing acceptance as chiral catalysts for the asymmetric C—C bond formation because of their high selectivities and catalytic efficiencies [9–17]. Using enzyme catalysis, molecular complexity can be rapidly built up under mild conditions, without a need for protecting sensitive or reactive functional groups, with high chemical efficiency and often with uncompromised stereochemical fidelity. Among the enzymes widely used for C—C bond formation, hydroxynitrile lyases, thiamine diphosphate (ThDP)-dependent carboligases, and aldolases have been the most significant. Many hydroxynitrile lyase enzymes are commercially available. However, with the exception of a transketolase and a couple of aldolases (FruA and RibA/DERA), most other enzymes catalyzing the C—C bond formation reaction are not yet available commercially at the present time. This will change as more chemists employ carboligases in synthesis that will increase demand for these enzymes.

1 HYDROXYNITRILE LYASES

Cyanohydrins are important intermediates in the production of useful pharmaceuticals, agrochemicals, and other biologically active compounds [16,18]. Hydroxynitrile lyases (HnLs, also known as oxynitrilases) are a class of carboligase enzymes that catalyze the stereoselective, reversible addition of HCN to

Organic Synthesis Using Biocatalysis. http://dx.doi.org/10.1016/B978-0-12-411518-7.00010-X

aldehydes or ketones furnishing cyanohydrins [16,19]. Since one stereogenic center is formed, it is possible to obtain two enantiomers from a common prochiral acceptor electrophile by selecting the appropriate pair of HnLs. However, under predominantly aqueous conditions, the chemical addition of HCN to the carbonyl group prevailed over the enzymatic catalysis resulting in the formation of racemic mixtures. The synthetic interest in this class of enzymes started growing with the discovery that the spontaneous chemical addition of HCN to aldehydes is more or less suppressed in water immiscible organic solvents. Hence, the biocatalytic addition of HCN to the carbonyl group in organic media allows straightforward access to optically active (*R*)- as well as (*S*)-cyanohydrins [16]. In organic media, it is advantageous to work with the enzyme deposited onto a solid support, for example, Celite or cellulose. This facilitates mass transfer and avoids protein aggregation due to the organic media [20–22]. Aqueous–organic two phase systems can also be used as the reaction medium. HCN can be generated *in situ* from sodium cyanide and acetic acid or by transhydrocyanation with acetone cyanohydrin [11]. *Safety:* Hydrogen cyanide is an extremely toxic (inhalation, ingestion, or skin contact may prove fatal) colorless liquid/gas, and should only be used in a properly vented fume hood or glove box with appropriate caution. Work should not be carried out alone.

HnLs are naturally widespread: ~3000 plant species and a small number of other organisms are known to possess these enzymes [19,23]. At present, the HnLs from 11 cyanogenic plants have been purified and characterized [24,25]. The most significant from the synthetic point of view are: (*S*)-*Me*HnL from *Manihot esculenta* (EC 4.1.2.37), (*S*)-*Hb*HnL from *Hevea brasiliensis* (EC 4.1.2.39), (*S*)-*Sb*HnL from *Sorghum bicolor* (EC 4.1.2.11) and (*R*)-*Pa*HnL from *Prunus amygdalus* (commonly known as bitter almond; EC 4.1.2.10) [10,11,25–31].

1.1 HnL-Catalyzed Addition of HCN to Aldehydes

Some successful examples of HnL-catalyzed preparations of (*R*)- and (*S*)-cyanohydrins from HCN and aldehydes are presented in Tables 10.1 and 10.2 using HnLs from different sources.

(*R*)-*Pa*HnL has been extensively investigated for preparative applications using a broad spectrum of aldehyde acceptors and it yields predominantly the (*R*)-configured adducts (Table 10.1). It is possible to use a crude extract from almond meal instead of the purified enzyme with similar efficacy [10]. Among organic solvents, diisopropyl ether (iPr$_2$O) was one of the best for obtaining high enantiomeric excess (ee) values (Table 10.1). Use of a purified enzyme adsorbed onto microcrystalline cellulose (Avicel®) or Celite [20] in nonaqueous systems, for example, iPr$_2$O with 2% v/v water added, almost completely suppressed the spontaneous (nonenzyme-catalyzed) HCN addition, which significantly improved the chiral purities of the adducts [29]. Similarly, utilizing unpurified almond meal in organic solvents with small amounts of aqueous phase (4% v/v)) provided cyanohydrins with ees of up to 99% [16,27,38,39].

Table 10.1 (*R*)-Hydroxynitrile Lyase from *P. amygdalus* (*Pa*HnL) Catalyzed the Addition Reaction of HCN to Aldehydes

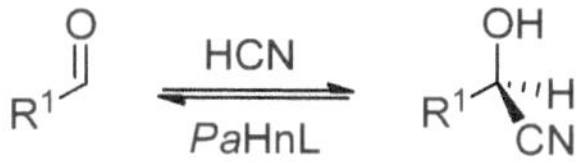

R^1	Yield (%) (Time, h)	ee (%)	Solvent	References
$CH_2CH_2CH_3$	99 (16)	98	iPr_2O^a	[29]
$CH_2CH_2CH_3$	100 (19)	95	iPr_2O^b	[28]
$CH_2(CH_2)_2CH_3$	100 (13)	97	iPr_2O^b	[28]
$CH_2(CH_2)_3CH_3$	100 (18)	94	iPr_2O^b	[28]
$CH_2(CH_2)_5CH_3$	98 (28)	87	iPr_2O^b	[28]
$CH(CH_3)_2$	99 (15)	83	iPr_2O^b	[28]
$CH_2CH(CH_3)_2$	100 (15)	94	iPr_2O^b	[28]
$C(CH_3)_3$	84(4.5)	83	iPr_2O^a	[28]
$CH{=}CHCH_3$	94(5.5)	95	iPr_2O^a	[29]
$CH{=}CHCH_3$	100 (41)	99	$EtOAc^c$	[27]
CH_2SCH_3	98(16)	96	iPr_2O^a	[29]
$C(CH_3)_2CH_2CH_2OH$	84(1.8)	89	iPr_2O^a	[11]
Ph	96(3)	>99	iPr_2O^a	[11]
CH_2CH_2Ph	94(45)	90	iPr_2O^a	[11]
m-Ph(OPh)	99(192)	98	$EtOAc^a$	[11]
p-Ph(OMe)	47 (89)	99	$EtOAc^c$	[27]
2-furyl	96(4)	99	iPr_2O^a	[11]
2-(5-CH_3)furyl	70 (17)	99	$EtOAc^d$	[27]
3-thienyl	95(6)	>99	iPr_2O^a	[29]
3-pyrydyl	97(3)	82	iPr_2O^a	[11]

[a]*Enzyme adsorbed on crystalline cellulose (Avicel), addition of HCN.*
[b]*Two phase system, using almond meal, acetone cyanohydrin as HCN source.*
[c]*Almond meal and HCN generated with sodium cyanide and acetic acid. Reaction conducted at 4°C.*
[d]*Almond meal and HCN generated with sodium cyanide and acetic acid.*

Concerning the complementary (*S*)-HnLs (Table 10.2), (*S*)-*Sb*HnL catalyzes the addition of HCN to aromatic and heteroaromatic aldehydes exclusively and does not tolerate aliphatic aldehydes [10,31,35,39]. Optimal reaction conditions were similar to those found for (*R*)-*Pa*HnL. Moreover, the stereoselectivity in organic solvents is high and also comparable with that obtained with (*R*)-*Pa*HnL [10,11,16]. However, to date, only a few literature data are reported on the synthetic applications of (*S*)-*Sb*HnL because of the limited acceptor substrate spectrum [16].

Table 10.2 (*S*)-Hydroxynitrile Lyase from *M. esculenta* (EC 4.1.2.37) ((*S*)-*Me*HnL), *H. brasiliensis* (EC 4.1.2.39) ((*S*)-*Hb*HnL) and, *S. bicolor* (EC 4.1.2.11) ((*S*)-*Sb*HnL) Catalyzed the Addition Reaction of HCN to Aldehydes [32]

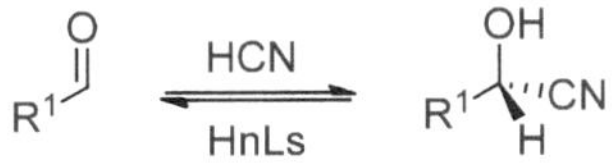

R^1	HnLS	Yield % (Time, h)	ee (%)	Solvent	References
$CH_2CH_2CH_3$	*Hb*HnL	51 (0.5)	75	*tert*BuOMe[a]	[33]
$CH_2CH_2CH_3$	*Me*HnL	36 (0.5)	69	*i*Pr_2O[b]	[34]
$CH_2(CH_2)_2CH_3$	*Me*HnL	58 (0.5)	80	*i*Pr_2O[b]	[34]
$CH_2(CH_2)_3CH_3$	*Me*HnL	39 (2)	92	*i*Pr_2O[b]	[34]
$CH_2CH(CH_3)_2$	*Me*HnL	91 (6.5)	95	*i*Pr_2O[b]	[34]
$CH_2CH(CH_3)_2$	*Hb*HnL	38 (–[d])	88	*tert*BuOMe[a]	[33]
$C(CH_3)_3$	*Me*HnL	80 (8.8)	94	*i*Pr_2O[b]	[34]
$C(CH_3)_3$	*Hb*HnL	49 (–[d])	78	*tert*BuOMe[a]	[33]
$CH{=}CHCH_3$	*Me*HnL	100 (1)	92	*i*Pr_2O[b]	[34]
Ph	*Sb*HnL	91(3)	97	*i*Pr_2O[c]	[35]
Ph	*Me*HnL	100 (7)	98	*i*Pr_2O[b]	[34]
Ph	*Hb*HnL	67 (–[d])	99	*tert*BuOMe[a]	[36]
CH_2CH_2Ph	*Hb*HnL	88 (–[d])	93	*tert*BuOMe[a]	[36]
m-Ph(OPh)	*Sb*HnL	93(144)	96	*i*Pr_2O[c]	[35]
m-Ph(OPh)	*Hb*HnL	87(144)	99	*tert*BuOMe[a]	[36]
p-Ph(OMe)	*Sb*HnL	93 (20)	89	*i*Pr_2O[c]	[35]
p-Ph(OMe)	*Hb*HnL	49 (–[d])	95	*tert*BuOMe[a]	[36]
2-furyl	*Sb*HnL	80 (9)	80	*i*Pr_2O[c]	[37]
2-furyl	*Hb*HnL	55 (–[d])	98	*tert*BuOMe[a]	[36]
3-thienyl	*Sb*HnL	95 (27)	98	*i*Pr_2O[c]	[37]
3-thienyl	*Me*HnL	98 (4)	98	*i*Pr_2O[b]	[34]
3-thienyl	*Hb*HnL	49 (–[d])	99	*tert*BuOMe[a]	[36]

[a] *Aqueous–organic (tert-butyl methyl ether) two phase system. Recombinant soluble HbHNL. HCN generated from KCN and citric acid [38].*
[b] *Enzyme adsorbed on nitrocellulose. HCN added to the reaction medium.*
[c] *Enzyme adsorbed on crystalline cellulose (Avicel). HCN added to the reaction medium.*
[d] Reaction time not reported.

(*S*)-*Me*HnL and (*S*)-*Hb*HnL (Table 10.2) are highly homologous proteins and of paramount interest because they possess a broad acceptor aldehyde selectivity similar to that for (*R*)-*Pa*HnL [10,11,16,32]. The enantioselectivity of (*S*)-*Me*HnL adsorbed on nitrocellulose in *i*Pr_2O is generally high, whereas the same enzyme adsorbed on cellulose gave only moderate to low stereoselectivity [34,40]. The reactions with (*S*)-*Hb*HnL are fully enantioselective and are usually conducted in buffer-organic two-phase systems consisting of a concentrated aqueous enzyme solution and an immiscible organic solvent, for example, *tert*-butyl methyl ether [36]. Moreover, (*S*)-*Hb*HnL overexpressed in *Pichia pastoris* can be obtained in sufficient amounts to be used for preparative purposes [23,33]. Groundbreaking results with this HnL were obtained with the synthesis of the (*S*)-cyanohydrin of 3-phenoxybenzaldehyde, a precursor for some important synthetic pyrethroids [41].

Example: Synthesis of Optically Active Cyanohydrins [28] – General Methodology

Caution: Hydrogen cyanide is an extremely toxic (inhalation, ingestion, or skin contact may prove fatal) colorless liquid/gas, and should only be used in a properly vented fume hood or glove box with appropriate caution. Work should not be carried out alone.

The general procedure for preparing optically active cyanohydrins is as follows: defatted ground almond meal (0.2 g) was incubated with citrate buffer (0.02 M, pH 5.5) (500 μL) in a 10 mL reaction vessel. After 15 min, *i*Pr_2O (sufficient to bring the final volume to 5 mL) was added. The freshly distilled substrate aldehyde (1.0 mmol) was added and the reaction was started by adding acetone cyanohydrin (1.5 equiv.). The mixture was agitated at 420 rpm on a rotary shaker at room temperature. When commercial mandelonitrile lyase was used as a catalyst, the enzyme deposited onto Celite (100 mg of immobilized preparation containing 1.6 mg of mandelonitrile lyase) was employed in the place of almond meal. The crude reaction mixtures were normally extracted with an organic solvent, for example, chloroform. The combined extracts were dried over Na_2SO_4, concentrated and further purified, if necessary, by column chromatography.

1.2 HnL-Catalyzed Addition of HCN to Ketones

Most of the aforementioned hydroxynitrile lyases are able to catalyze the addition of HCN to ketones to furnish (*R*)- or (*S*)-ketone cyanohydrins (Table 10.3). Using *Pa*HnL as catalyst, alkyl methyl ketones gave moderate yields and high optical purities, whereas with alkyl ethyl ketones the yields and optical purities were lower [39,42]. The substrate scope has been extended to methyl phenyl ketones, cyclic, bicyclic, heterocyclic, as well as silicon-containing compounds [16]. Concerning (*S*)-ketone cyanohydrins, excellent results were achieved with aliphatic and aromatic ketones, for example, acetophenone cyanohydrin, using (*S*)-*Hb*HnL (40% conversion, 99% ee., Table 10.3) or (*S*)-*Me*HnL (87% conversion, 98% ee

Table 10.3 Addition of HCN to Ketones Catalyzed by Hydroxynitrile Lyase from *P. amygdalus* ((*R*)-*Pa*HnL) and *H. brasiliensis* (EC 4.1.2.39) ((*S*)-*Hb*HnL)

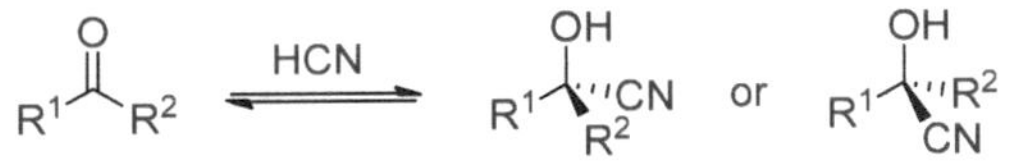

R^1	R^2	HnLS	Yield % (Time h)	ee (%)	Solvent	References
Me	$CH_2CH_2CH_3$	*Pa*HnL	70 (43)	97 (*R*)	*i*Pr_2O[a]	[42]
Me	$CH_2(CH_2)_2CH_3$	*Pa*HnL	90 (42)	98 (*R*)	*i*Pr_2O[a]	[42]
Me	$CH_2(CH_2)_3CH_3$	*Pa*HnL	88 (72)	98 (*R*)	*i*Pr_2O[a,b]	[42]
Me	$CH_2CH(CH_3)_2$	*Pa*HnL	57 (40)	98 (*R*)	*i*Pr_2O[a]	[42]
Et	$CH_2CH_2CH_3$	*Pa*HnL	33 (72)	85 (*R*)	*i*Pr_2O[a,b]	[42]
Et	$CH_2(CH_2)_2CH_3$	*Pa*HnL	21 (72)	90 (*R*)	*i*Pr_2O[a,b]	[42]
Me	$C(CH_3)_3$	*Pa*HnL	73 (24)	99 (*R*)	*i*Pr_2O[c]	[39]
Me	$CH(CH_3)_2$	*Hb*HnL	99 (0.25)	98 (*S*)	*tert*BuOMe[d]	[36]
Me	$CH_2CH(CH_3)_2$	*Hb*HnL	86 (0.25)	99 (*S*)	*tert*BuOMe[d]	[36]
Me	$CH_2(CH_2)_2CH_3$	*Hb*HnL	59 (0.25)	99 (*S*)	*tert*BuOMe[d]	[36]
Me	Ph	*Hb*HnL	40 (0.25)	99 (*S*)	*tert*BuOMe[d]	[36]
Me	CH_2Ph	*Hb*HnL	74 (0.25)	95 (*S*)	Aqueous Buffer[e]	[33]
Me	$CH_2Ph(m\text{-}CF_3)$	*Me*HnL	67 (17)	97 (*S*)	*i*Pr_2O[f]	[47]
Me	$CH_2Ph(p\text{-}Br)$	*Me*HnL	77 (17)	90 (*S*)	*i*Pr_2O[f]	[47]

[a]*Enzyme adsorbed on crystalline cellulose (Avicel), addition of HCN. Reaction conducted at 0°C.*
[b]*Reaction conducted at 20°C.*
[c]*Almond meal in tartarate buffer 7% (v/v), 0.1 M, pH 5.4 at 25°C.*
[d]*Aqueous-organic (tert-butyl methyl ether) two phase system. Recombinant soluble HbHnL. HCN generated from KCN and citric acid.*
[e]*Citrate (pH 3.75 to 4.0).*
[f]*Citrate (0.1 M, pH 4.5)-diisopropyl ether two-phase system. Reactions conducted at 5°C.*

[43]). Hetrocyclic ketone derivatives, for example, tetrahydrofuran-3-one and tetrahydro-2*H*-3-pyranone, were also accepted by (*R*)-*Pa*HnL and (*S*)-*Hb*HnL yielding moderate ees (up to 81 %) [44]. A plethora of examples can be found in the literature and reviews [11,16,19,23,24,29,30,45,46].

2 CARBON–CARBON BOND FORMATION USING THIAMIN DIPHOSPHATE (THDP)-DEPENDENT CARBOLIGASES

Thiamin diphosphate (ThDP)-dependent carboligases are a class of enzymes involved in a variety of metabolic pathways where they catalyze a broad range of bond cleavage and bond formation reactions between carbon and hydrogen,

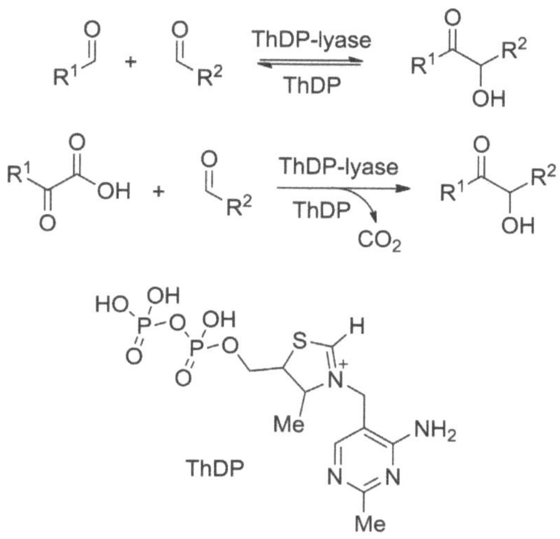

SCHEME 10.1
ThDP-lyases catalyze the umpolung formation of carbon–carbon bonds of aldehydes (acyloin condensation) to afford chiral 2-hydroxy ketones (acyloins) using an aldehyde or 2-ketoacid as donor substrate.

oxygen, sulfur, or nitrogen, as well as between carbon atoms [6,14,48]. This section deals with ThDP-lyases that catalyze the stereoselective umpolung formation of carbon–carbon bonds between aldehydes (e.g., acyloin reactions) to afford chiral 2-hydroxy ketones (acyloins) (Scheme 10.1) [14,48,49]. This will include transketolases that catalyze the transfer of acyl anion equivalents to aldehydes.

The first step in the catalytic cycle is nucleophilic addition of the anion of the cofactor thiamine diphosphate (ThDP) to the aldehyde. Deprotonation yields an enamine carbanion that serves as the nucleophile for the subsequent C—C bond-forming step. Alternatively, for the decarboxylase type ThDP-lyases, the carbonyl group of a 2-ketoacid donor substrate reacts with the anion of ThDP to yield the corresponding 2-hydroxy acid adduct. After CO_2 is released, the enamine nucleophile is formed. Subsequently, the carbonyl acceptor reacts with the enamine to form the 2-hydroxy ketone. The stereoselectivity is governed by the enzyme that specifically discriminates between the two possible enantiotopic faces of the acceptor [14,48,49].

In enzymatic cross-acyloin condensations, four different chiral acyloins can be expected to be formed from the cross- and homo-acyloin condensations (Scheme 10.2). Product distribution (chemoselectivity) will depend on the properties of the ThDP-dependent carboligase, that is, the binding sequence of donor and acceptor to the active site, the substrates, and their stoichiometric ratio [14].

Studies on the C—C bond formation stereoselectivity by ThDP-dependent enzymes reveals that the (*R*)-enantiomer is by far the major product in most cases

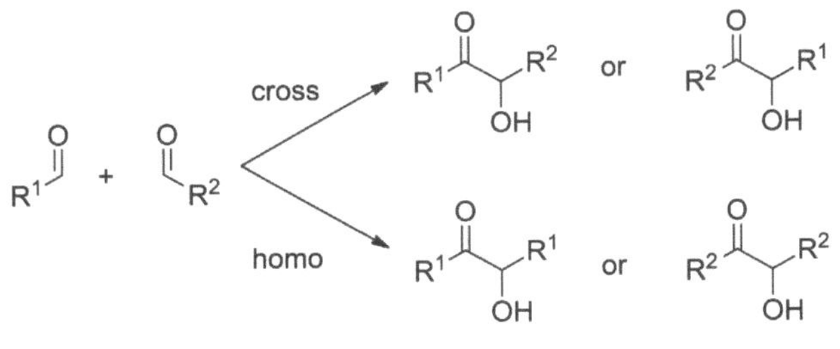

SCHEME 10.2
Chemoselectivity of acyloin condensation catalyzed by ThDP-dependent carbolygases.

[14]. An exception is the benzoylformate decarboxylase (BFD) from *Pseudomonas putida* that can catalyze the acyloin condensation, furnishing (*S*)-configured 2-hydroxy ketone [50].

ThDP-dependent carboligases can catalyze acyloin condensations of aliphatic or aromatic donor aldehydes to aliphatic or aromatic acceptors, thus allowing the production of many useful building blocks (Tables 10.4–10.7). These are the most common reactions catalyzed by ThDP-dependent carboligases such as the pyruvate decarboxylase (PDC) from *Saccharomyces cerevisiae* (*Sc*PDC), *Zymomonas mobilis* (*Zm*PDC), *Acetobacter pasteurianus* (*Ap*PDC), benzoylformate decarboxylase (BFD) from *P. putida* (*Pp*BFD), and benzaldehyde lyase (BAL) from *Psedomonas. fluorescens BioVar I*, which have been recognized as powerful biocatalysts [6,14,48].

PDC enzymes usually prefer small aliphatic aldehydes or 2-ketoacid compounds as donors and aliphatic or aromatic aldehydes as acceptors, furnishing the corresponding acetoin and phenylacetyl carbinol derivatives [14]. A number of PDC variants have been obtained by structure-guided directed mutagenesis with improved activity and stereoselectivity and the results have been summarized in previous reviews [6,14,48,49,56,61,62].

Benzoylformate decarboxylase (BFD) from *P. putida* (Table 10.4, entries 1–5) catalyzes the acyloin condensation of benzaldehyde or benzoyl formate derivatives to acetaldehyde, furnishing (*S*)-configured acyloins. BDF showed a preference for aromatic aldehydes as donors and aliphatic aldehydes as acceptors, although with lower reaction rates than BAL [14,55]. The *meta*- and *para*-substituted benzaldehyde derivatives as well as heteroaromatic, olefinic, and cyclic aliphatic aldehydes are transformed with good to excellent conversions and enantioselectivities [14,55,56,63,64]

Benzaldehyde lyase (BAL) from *Pseudomonas fluorescens* (Table 10.4, entries 6–15) is one of the most efficient catalysts for the homo- and cross-carboligation reaction of aromatic and aliphatic aldehydes, because of its broad substrate range and its high (*R*)-stereoselectivity [14,57,58,65]. The ability of BAL to catalyze the cleavage of benzoins can also be instrumental in the resolution of racemic mixtures of these compounds [65].

Table 10.4 ThDP-Dependent Lyases-Catalyzed Acyloin Condensations of Aliphatic or Aromatic Donor Aldehydes to Aliphatic or Aromatic Acceptors

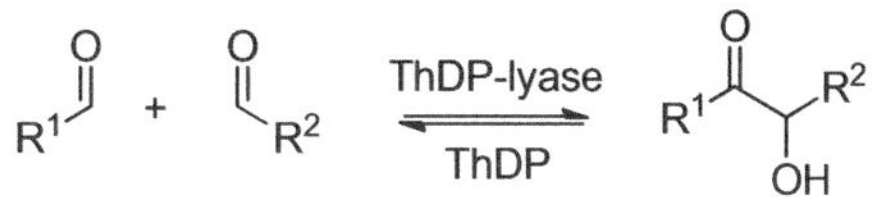

Entry	R¹	R²	Product	Enzyme	Yield (%)	ee (%)	Conditions	References
1	Ph	Me	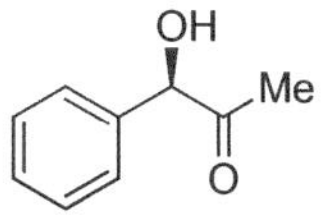	*Cu*PDC Wole cells	96	98	Buffer[a]	[54]
2	Ph(*m*-OMe)	Me	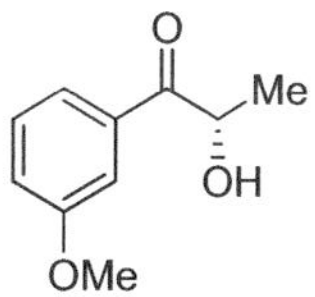	*Pp*BFD	97	96	Buffer[b]	[55]
3	Ph(*m*-O*i*Pr)	Me	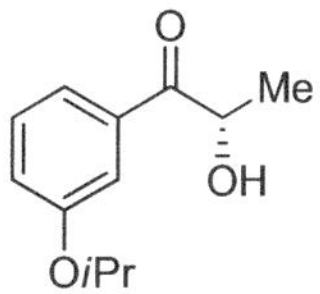	*Pp*BFD	91	>99	Buffer[b]	[55]
4	Ph	Ph	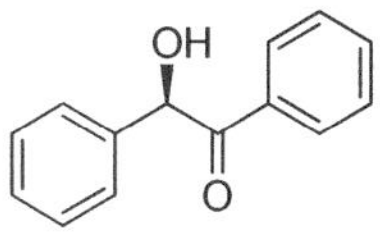	*Pp*BFD	70	>99	DMSO[c] 20%	[56]

(Continued)

Table 10.4 ThDP-Dependent Lyases-Catalyzed Acyloin Condensations of Aliphatic or Aromatic Donor Aldehydes to Aliphatic or Aromatic Acceptors *(cont.)*

Entry	R¹	R²	Product	Enzyme	Yield (%)	ee (%)	Conditions	References
5	Ph(*p*-Me)	Ph(*p*-Me)	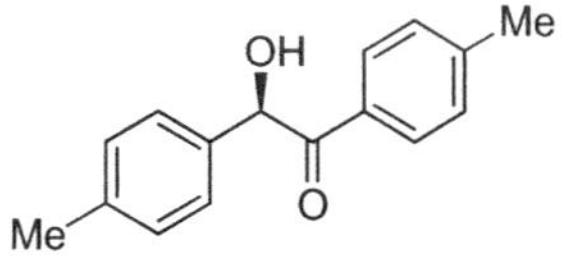	*Pp*BFD	69	>99	DMSO[c] 20%	[56]
6	Ph	Ph	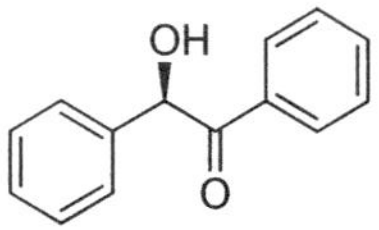	BAL	94	>98	DMSO[c] 20%	[57]
7	2-furyl	2-furyl	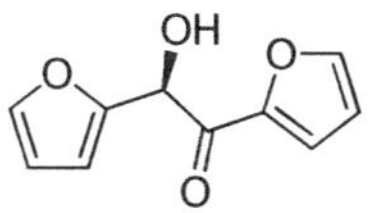	BAL	90	>95	DMSO[c] 20%	[57]
8	Ph(*p*-OH)	Ph(*p*-OH)	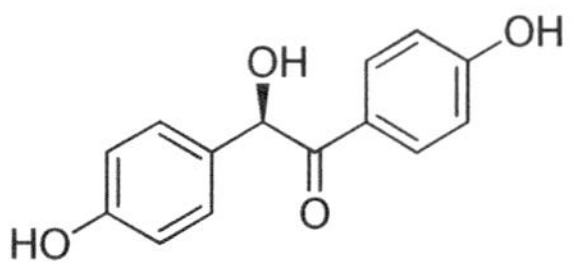	BAL	94	>95	DMSO[c] 20%	[57]
9	Ph	$CH(CH_3)_2$	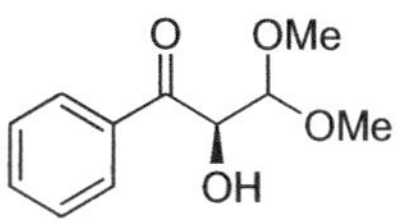	BAL	91	>98	DMSO[c] 20%	[57]
10	Ph(*o*-F,*p*-F)	$CH(CH_3)_2$	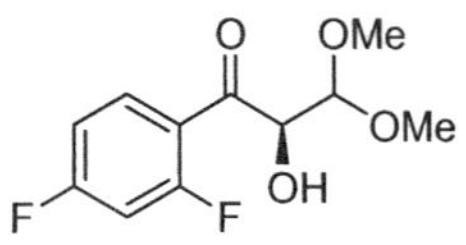	BAL	87	89	DMSO[c] 20%	[57]

11	CH_2OCH_2Ph	CH_2OCH_2Ph	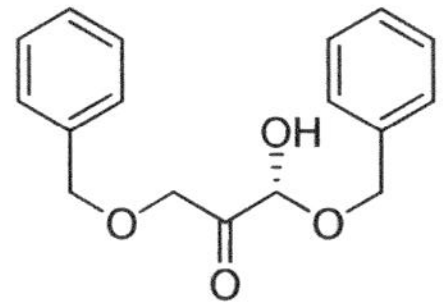	BAL	91	95	DIPE[d]	[58]
12	$CH_2CH(CH_3)_2$	$CH_2CH(CH_3)_2$	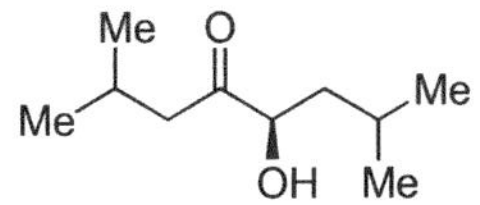	BAL	>90	89	DMSO[c] 20%	[59]
13	CH=CHPh	Me	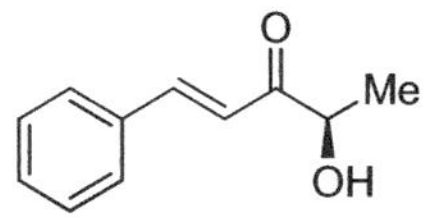	BAL	80	80	Buffer[b]	[60]
14	C(Me)=CHPh	Me	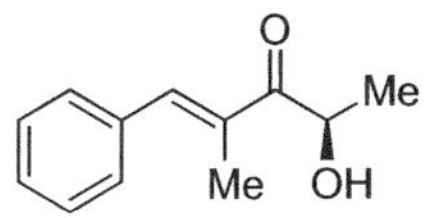	BAL	75	>96	Buffer[b]	[60]
15	$CH=CHCH_2CH_2CH_3$	Me	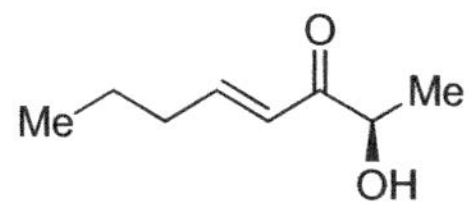	BAL	80	87	Buffer[b]	[60]

[a]Buffer MOPS (2.5 M, pH 6.5), the pH rose to pH 7.2–7.6 during biotransformation.
[b]Potassium phosphate buffer (50 mM, pH 7.0)
[c]DMSO and potassium phosphate buffer (50 mM, pH 7.0)
[d]Diisopropylether (DIPE) (25 vol%)-MOPS buffer (50mM, pH 7)(75 vol%) two-phase system.

Example: Synthesis of (*R*)-2-Hydroxy-3, 3-dimethoxy-1-phenylpropan-1-one, (Table 10.4, entry 9; R^1 = Ph, R^2 = $CH(OMe)_2$) [57].

Benzaldehyde (2 mmol; other aromatic aldehydes can also be used) was dissolved in potassium phosphate buffer (80 mL, 50 mM, pH 7.0), containing $MgSO_4$ (2.5 mM) and ThDP (0.15 mM)) containing DMSO (20% v/v) and dimethoxyacetaldehyde (2 mmol). After the addition of BAL (20 U, prepared as indicated in Ref. [57], not commercially available), the reaction was allowed to stand at 25°C. The reaction was monitored by HPLC with a chiral column. After 24 h, additional BAL (20–40 U) and dimethoxyacetaldehyde (4 mmol) were added. After 48 h (checked by TLC), the reaction mixture was filtered and extracted with dichloromethane (3 × 50 mL). After drying the collected organic phase over $MgSO_4$, removal of the solvent under reduced pressure gave the crude product, which was then purified by flash column chromatography to give the desired compounds. (*R*)-2-hydroxy-3,3-dimethoxy-1-phenylpropan-1-one was obtained in 91% yield and ee >98%(Table 10.4, entry 9).

The ThDP-dependent 2-succinyl-5-enolpyruvyl-6-hydroxy-3-cyclohexene-1-carboxylate synthase (MenD) from *Escherichia coli* K12, uses ketoacids as donor substrates. MenD was found to catalyze the acyloin condensation of α-ketoglutarate to a broad range of aldehydes to furnish (*R*)-configured α-hydroxyketones with high stereoselectivities (Table 10.5) [14]. Pyruvate and oxalacetate are also accepted as donors, but at a lower rate [53].

The flavoenzyme from *Yersinia pseudotuberculosis* (YerD) catalyzes *in vivo* the acyloin condensation of pyruvate to the ketone, cytidine-5′-diphosphate-3,6-di(deoxy)-4-keto-D-glucose [52]. Investigations on its synthetic capabilities revealed that this enzyme can catalyze aldehyde–ketone cross-acyloin condensations, furnishing tertiary alcohols (Table 10.6) [52]. Small cyclic aliphatic ketone derivatives, for example, cyclohexanone and methylated cyclohexanones, tetrahydro-2*H*-pyran-3-one (Table 10.6), open-chain ketones containing an ether moiety, diketones, and ketones containing an α- or β-ketoesters were all successfully converted [52].

The ThDP-dependent enzyme from *Serratia marcescens* (PigD), is an interesting carboligase since it has the ability to catalyze umpolung, Stetter type 1,4-additions to α,β-unsaturated ketones using pyruvate as donor substrate (Table 10.7) [51]. Aliphatic, aromatic as well as heterocyclic α,β-unsaturated ketones selectively furnish the 1,4-adducts in the presence of pyruvate with PigD catalysis. Pyruvate is the preferred donor substrate, but 2-oxobutanoate can also be employed depending on the acceptor [51].

2.1 Transketolases

In vivo, transketolases (TK) reversibly transfer a terminal hydroxyacetyl unit among phosphorylated aldoses and ketoses as acceptor and donor substrates, respectively, in the oxidative pentose phosphate pathway. The synthetic value

Table 10.5 ThDP-Dependent 2-Succinyl-5-enolpyruvyl-6-hydroxy-3-cyclohexene-1-carboxylate synthase (MenD) from *E. coli* K12 Catalyzed Acyloin Condensations Using α-Ketoglutarate and Pyruvate Donors, Respectively

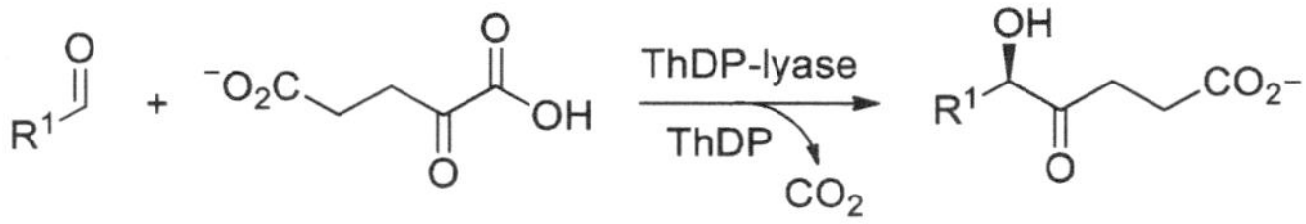

R¹	Product	Enzyme	Yield (%)	ee (%)	Conditions	References
Ph	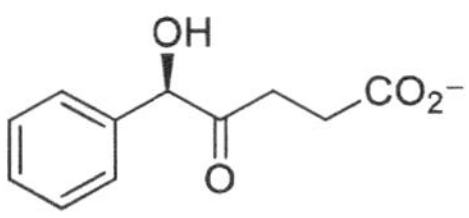	MenD[a]	65	94	DMSO[b] 20%	[53]
Ph(*o*-F)	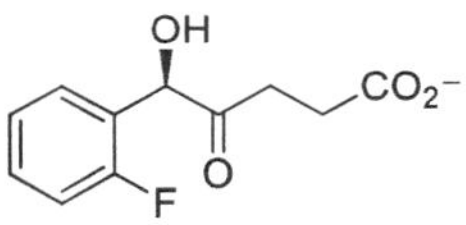	MenD[a]	87	95	DMSO[b] 20%	[53]
Ph(*m*-I)	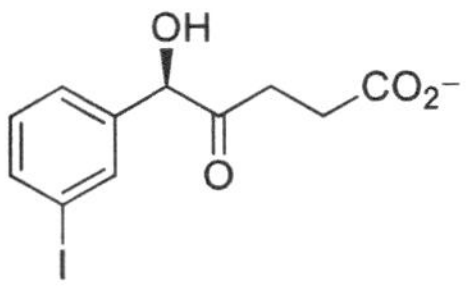	MenD	97	96	DMSO[b] 20%	[53]

[a]*MenD (2-succinyl-5-enolpyruvyl-6-hydroxy-3-cyclohexene-1-carboxylate synthase from E. coli K12.*
[b]*DMSO and potassium phosphate buffer (50 mM, pH 7.0)*

of this enzyme derives from the fact that the natural ketose donor components can be effectively replaced by hydroxypyruvate [66,67]. As mentioned before, the reactive nucleophile is formed by a spontaneous decarboxylation that renders the overall preparative addition to an aldehyde substrate essentially irreversible. TK adds the hydroxyacetyl moiety stereospecifically to the *re*-face of the acceptor aldehyde leading to products of (3*S*)-configuration (Table 10.8). Furthermore, TK kinetically discriminates efficiently racemic α-hydroxyaldehyde acceptors with very high (*R*)-enantiopreference. Thereby, enantiomerically and diastereomerically homogeneous ketose-type structures having a (3*S*,4*R*) or D-*threo* configurations are produced from racemic starting aldehydes. Although 2-hydroxyaldehydes are the preferred substrates, differently substituted, nonphosphorylated alkyl, allyl, aromatic, or heterocyclic moieties have been demonstrated to be converted, albeit at somewhat reduced reaction rates, and many corresponding products have been isolated on a preparative scale [68].

Table 10.6 ThDP-Dependent Flavoenzyme from *Y. pseudotuberculosis* (YerD) Catalyzes Aldehyde–Ketone Cross-Acyloin Condensation Furnishing Tertiary Alcohols [52]

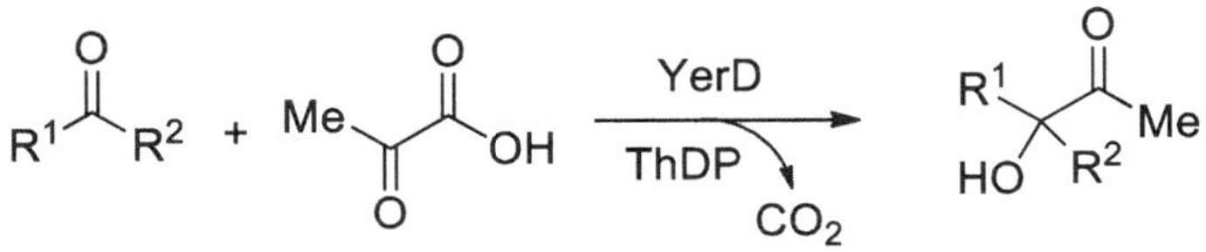

Ketone	Product	Yield[a] (Conv.) (%)	ee (%)
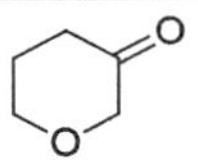	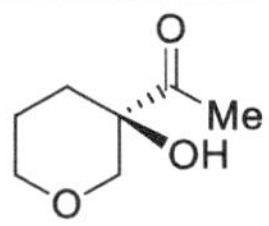	34 (97)	84
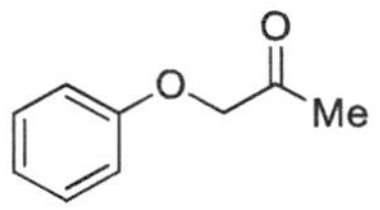	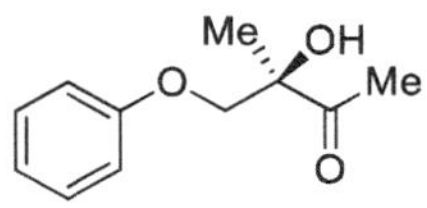	24 (48)	91
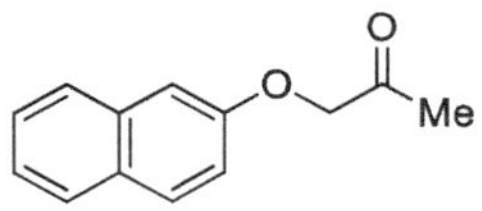	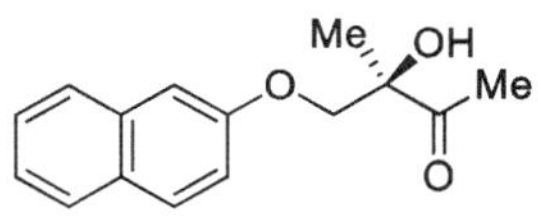	14 (20)	96
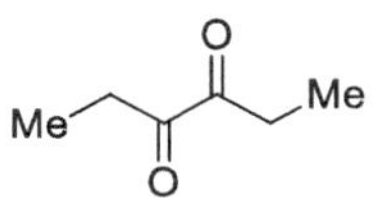	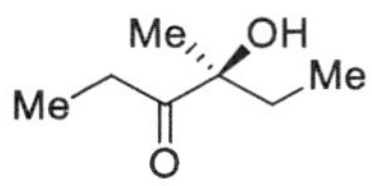	34 (58)	84

[a]Reactions conducted in MTBE:water two-phase system at 30°C. The purification procedures are not optimized. This can explain the differences between the yield, as isolated yield and the conversion, which means the substrate conversion during the reaction measured normally by HPLC.

Example: Preparation of 5-O-Benzyl D-threo-pentulose (Table 10.8, entry 10) [69]

(±)-O-Benzylglyceraldehyde (2.5 g, 13.8 mmol) was dissolved/suspended in water (300 mL) after which the pH of the solution was adjusted to pH 7.0, by careful titration with aq. NaOH (5 M), followed by the addition of thiamine pyrophosphate (150 mg, 0.326 mmol), and magnesium chloride (13 mg, 78 mmol). Readjustment of the pH to 7.0 by addition of aq. NaOH (0. 1 M) was followed by the addition of transketolase (600 U, the *E. coli* enzyme is commercially available from Sigma-Aldrich) and bovine serum albumin (50 mg). The reaction was initiated by the addition of potassium hydroxypyruvate (0.98 g, 6.9 mmol) and the pH was maintained at 7.0 by autotitration with aq. HCl (I M). After 20 h, terahydrofuran (20 mL) was added to aid solubility of the remaining

Table 10.7 ThDP-Dependent Enzyme from *S. marcescens* (PigD) Catalyze Acyloin Condensation of Pyruvate to α,β-Unsaturated Ketones in Stetter Type 1,4-Additions [51]

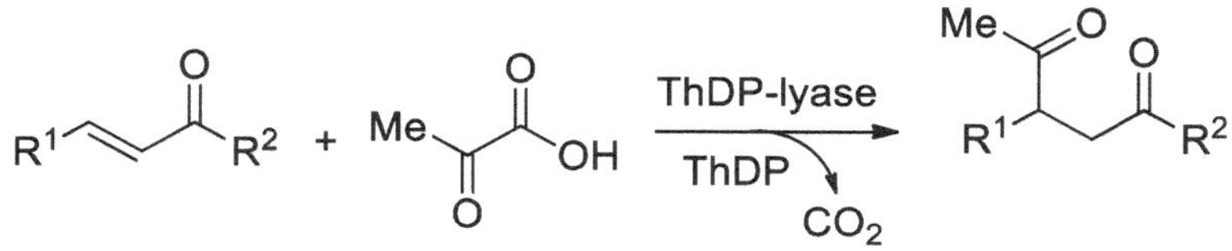

R¹	R²	Product	Yield[a]	ee (%)
$CH_2(CH_2)_4CH_3$	Me	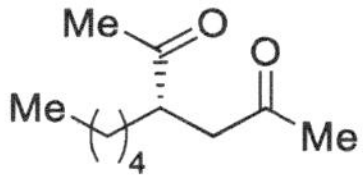	38	>99
$CH_2(CH_2)_5CH_3$	Me	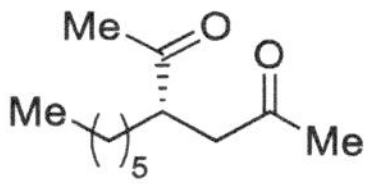	30	>99
Ph(*o*-OH)	Ph	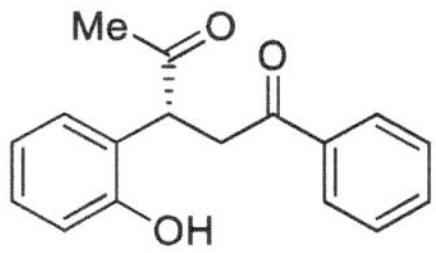	39	>94

[a]*Reactions conducted in DMSO:potassium phosphate buffer (84 mM, pH 7.0) at 30°C.*

substrate and hence ensure complete reaction. The reaction was terminated after 24 h by which time uptake of aq. HCl had ceased. Addition of flash silica (~4 g) followed by removal of the water/THF under reduced pressure and flash chromatography (ethyl acetate/petroleum ether (2:3) then ethyl acetate/methanol (9:1) afforded the title product (1.3 g, 5.3 mmol, 76%).

3 ENZYMATIC ALDOL ADDITIONS

Aldolases are a class of lyases that catalyze the reversible, stereoselective addition of an aldol donor component (nucleophile) to an acceptor component (electrophile) [9,75–81]. Typical products are 3-hydroxy ketones, a structural element that is frequently incorporated in the framework of complex natural products. Mechanistically, aldolases promote the abstraction of the aldol donor α-proton, thereby generating a carbon nucleophile bound at the active site. In type I aldolases, an enamine is produced by covalent binding to a conserved lysine residue while in the type II aldolases, an enediol is formed by chelating coordination to an essential transition metal cation (mostly Zn^{2+}), which is acting as a Lewis acid promotor. Aldolases are often highly selective for the donor substrate, tolerating

Table 10.8 Transketolase from Different Sources Catalyzed Aldol Transfer of Hydroxyacetyl Moiety from Hydroxypyruvate to Aldehydes

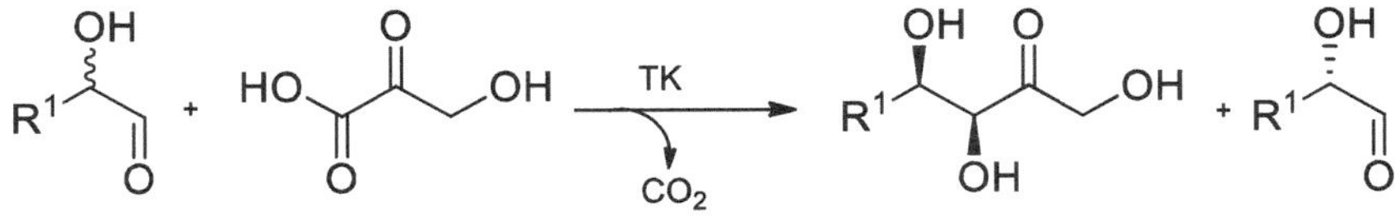

Entry	R1	TK source	Yield[a] (%)	ee[c] of recovered aldehyde	References
1	H	*E.coli*	74	–[d]	[69]
2	CH_2OH	Spinach leaves	48[b]	nd	[70,71]
3	Me	*S. cerevisiae*	88[b]	>95	[68]
4	CH_2CH_3	*S. cerevisiae*	80–90[b]	>95	[68,72,73]
5	CH_2OCH_3	*S. cerevisiae*	72–80[b]	>95	[68,72]
6	$CH_2CH=CH_2$	*S. cerevisiae*	90[b]	95	[68]
7	$CH_2SCH_2CH_3$	*S. cerevisiae*	74[b]	99	[72]
8	CH_2F	*S. cerevisiae*	79[b]	96	[72]
9	CH_2CN	*S. cerevisiae*	82[b]	97	[72]
10	CH_2OCH_2Ph	*E. coli*	75–80[b]	98	[69,74]

[a]*Racemic aldehyde was used unless noted otherwise.*
[b]*Yield based on the R acceptor fraction.*
[c]*The* syn:anti *ratio of the adduct was >99:1 in all the selected examples.*
[d]*Not determined.*

only small structural, isosteric modifications. According to this donor specificity, the enzymes are further categorized as dihydroxyacetone phosphate (DHAP), pyruvate, glycine, and acetaldehyde dependent aldolases (Figure 10.1).

An exception is the dihydroxyacetone (DHA) utilizing aldolases such as the D-fructose-6-phosphate aldolase that also accepts hydroxyacetone, hydroxybutanone, and glycoaldehyde. Such broad donor tolerance is almost unique among aldolases [8,82]. Aldolases can accept a broad structural variety of aldol acceptors, and this is what makes them highly important for synthetic applications.

From a stereochemical point of view, the approach of the aldol acceptor to the enzyme–nucleophile complex is stereospecific, following an overall retention mechanism. On the other hand, the facial differentiation of the acceptor aldehyde carbonyl is usually high but, in some instances, depends on the substrate structure, and the stereochemical outcome does not always follow that of the natural acceptor. Although enzymatic aldol addition is a reversible reaction, for most applications, the C—C bond formation is favored for thermodynamic reasons because the primary aldol adducts can often form more stable cyclic ketal

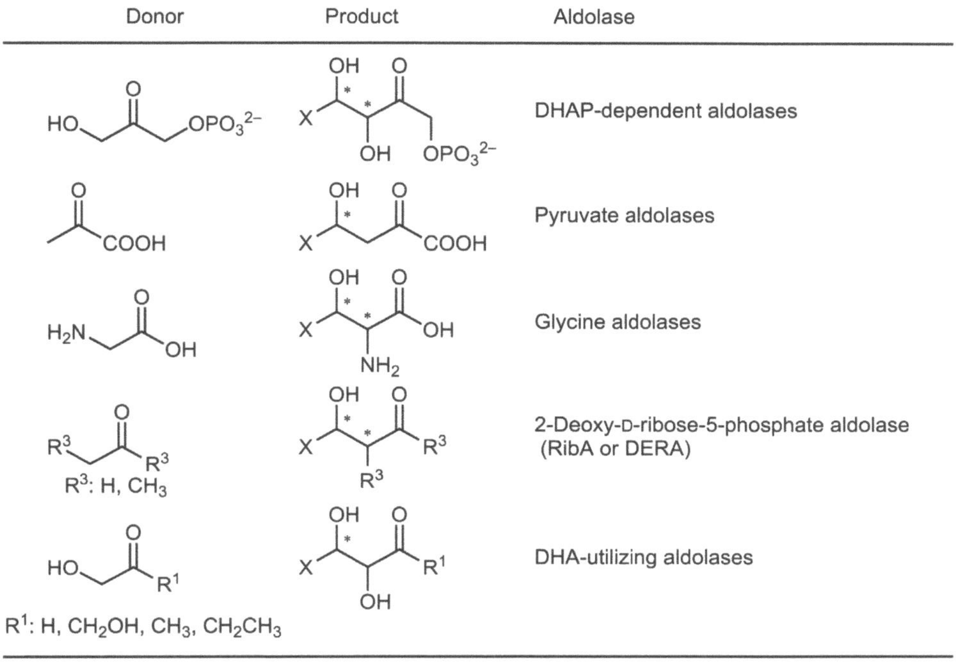

FIGURE 10.1
Classification of aldolases by their selectivity for the donor substrate.

isomers [5,77,83]. In other instances, the product fraction at equilibrium can be increased by driving the reaction with an excess of either donor or acceptor concentration. Selection has to be made according to the relative cost of donor or acceptor or by their inhibitory properties against the aldolase activity.

3.1 Aldol Addition of Dihydroxyacetone Phosphate to Aldehydes

Aldol addition of DHAP to aldehydes is catalyzed by DHAP-dependent aldolases. Two stereogenic centers are formed and therefore four possible stereoisomers can be obtained. Although nature has evolved a set of four distinct stereocomplementary types (Scheme 10.3), so far, only three of the known DHAP-dependent aldolases, namely the D-fructose-1,6-bisphosphate aldolase (FruA), L-rhamnulose-1-phosphate aldolase (RhuA), and L-fuculose-1-phosphate aldolase (FucA), have found broad synthetic applicability due to their high stereoselectivity and broad acceptor tolerance [5,77]. DHAP-dependent aldolases are highly selective for the nucleophilic substrate DHAP, tolerating only few isosteric modifications [84–88].

DHAP is a chemically unstable compound, particularly under alkaline conditions, and decomposes into inorganic phosphate and methyl glyoxal, both of which inhibit the aldolase. Therefore, an economical access to this compound

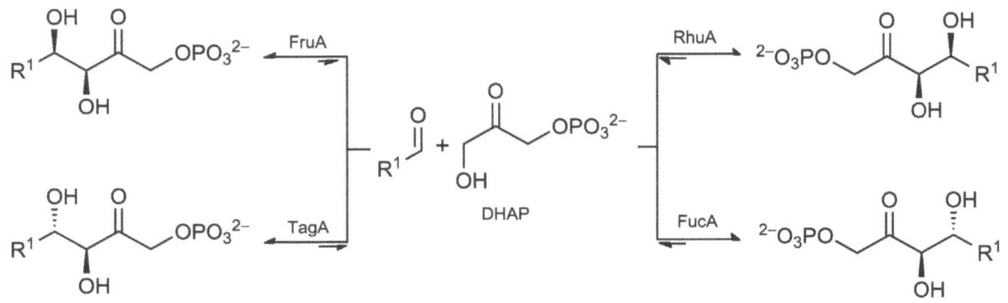

SCHEME 10.3
Stereocomplementary set of DHAP-dependent aldolases. D-Fructose-1,6-bisphosphate aldolase (FruA), L-rhamnulose-1-phosphate aldolase (RhuA), L-fuculose-1-phosphate aldolase (FucA), D-tagatose-1,6-bisphosphate aldolase (TagA)

is essential for the preparative exploitation of these enzymes. DHAP can be prepared chemically, enzymatically, or by a combination of both techniques (for an extensive review of methods to prepare DHAP, see Ref. [89]). Inexpensive and straightforward DHA donor can also be used in the presence of high concentrations of arsenate salts [90–92]. This is possible by virtue of the reversible formation of arsenate esters that mimic phosphorylated substrates [93]. However, the method is problematic, due to the high toxicity of arsenate salts and the low practical reaction rates achieved. *In situ* formation of a DHA-borate ester that is also able to mimic DHAP can also be used, and the toxicity of borate is much lower than that of arsenate. However, the method is so far restricted for the aldol additions catalyzed by RhuA aldolase [94,95].

For any one of the DHAP-dependent aldolases, the absolute configuration at the newly created stereocenter (C_3, the one arising from DHAP addition) is invariably conserved upon reaction with any electrophile due to mechanistic constraints [77]. The stereoselectivity at C_4 (the one originating from the addition to the aldehyde carbonyl) is high in many instances, but in some cases its configuration depends on the acceptor structure and may be reversed relative to that obtained with the natural acceptor substrate [77].

Concerning the reaction systems, hydrophobic aldehydes need cosolvents to enhance their solubility and up to 20% (v/v) dimethylformamide (DMF) or dimethylsulfoxide (DMSO) is usually tolerated by the enzyme without compromising its activity. Emulsions are alternative systems for hydrophobic acceptors, but substrate partitioning between phases may cause rate limitations, particularly for substrates with high K_M values [96]. Examples of aldol additions catalyzed by DHAP-dependent aldolases are summarized in Tables 10.9 and 10.10 and Schemes 10.4 and 10.5. A plethora of examples are extensively described in the literature and summarized in several recent reviews [5,77,82].

Nitroaldehydes are a particular case of nitrogen containing aldehydes because of the high proton acidity next to the nitro group. Thus, upon addition of DHAP, they

Table 10.9 DHAP-Dependent Aldolases Catalyzed Aldol Addition of Dihydroxyacetone Phosphate to Aldehydes

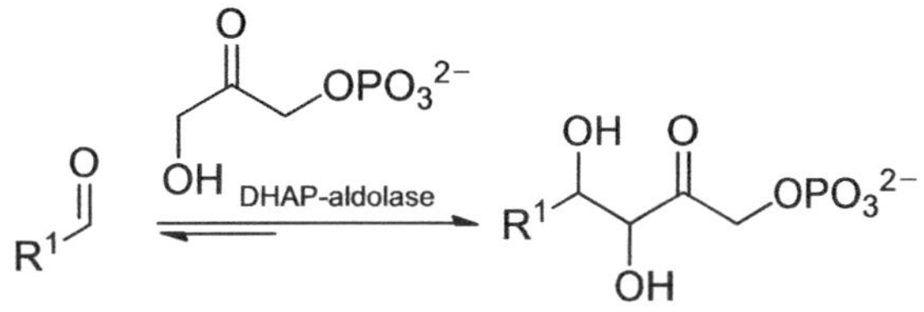

Entry	R¹	Product	Enzyme	Yield %	*syn/anti de*	References
1	$CH(CH_3)_2$	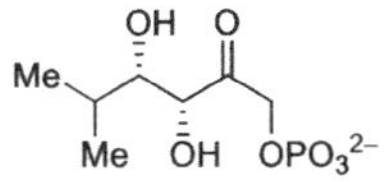	RhuA[a]	88	97:3	[103]
2	(*S*)-$CH(OH)CH_2Cl$	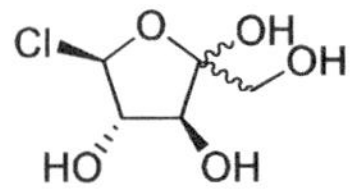	FruA[b]	93[c]	>98:2	[104]
3	(*S*)-$CH(OH)CH_2OH$	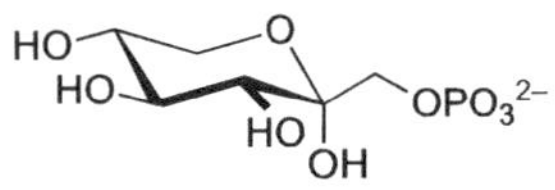	RhuA[a]	84	>97:3	[103]
4	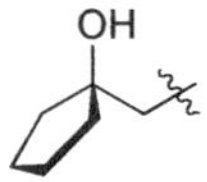	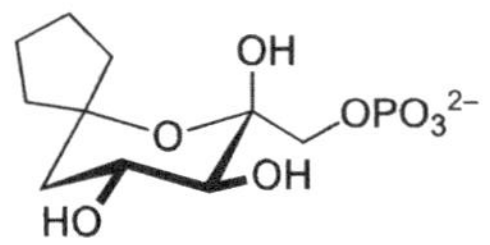	FruA[b]	86	>98:2	[105]
5	(*S*)-$CH(OH)CH_2SH$	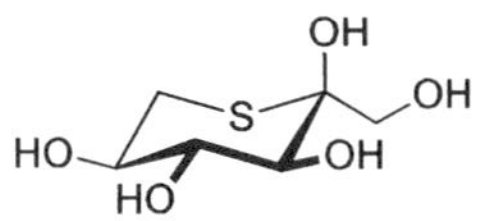	FucA[a]	78[c]	>98:2	[106]

[a]From *E. coli*.
[b]From rabbit muscle (RAMA).
[c]Yield of unphosphorylated material after treatment with acid phosphatase.

Table 10.10 DHAP-Dependent Aldolases Catalyzed Aldol Addition of Dihydroxyacetone Phosphate to Cbz (or similar) Derivatives of Amino or Azido Aldehydes

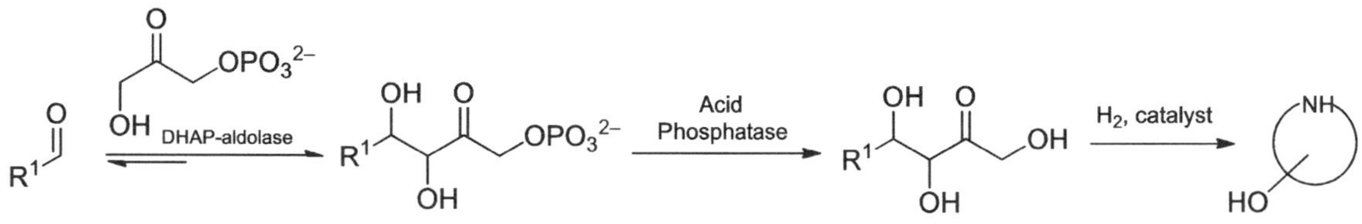

Entry	R¹	Product	Enzyme[a]	Conversion (%) to Aldol adduct	Overall yield of iminocyclitol (%)	References
1	CH_2NHCbz	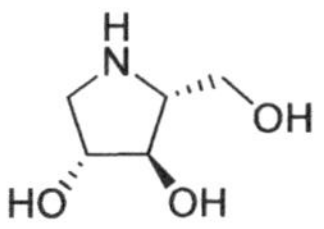	FruA	83	50	[97]
2	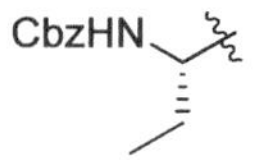	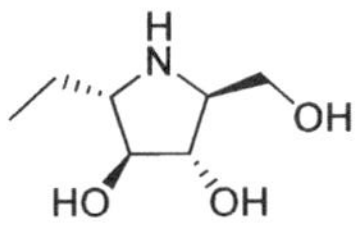	RhuA	77	63	[98]
3	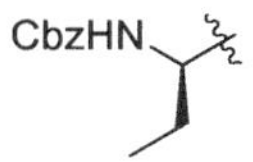	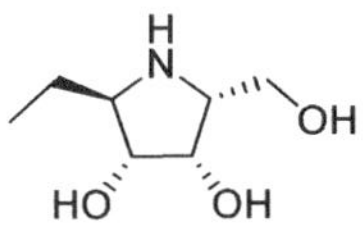	FucA	62	20	[99]
4	(*S*)-$CH(OH)CH_2N_3$	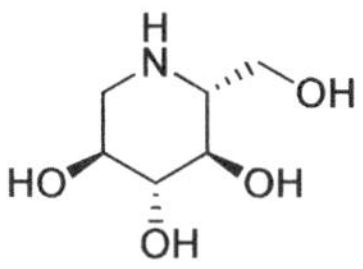	FruA	nd	64	[99]

5	(*R*)-CH(OH)CH_2N_3	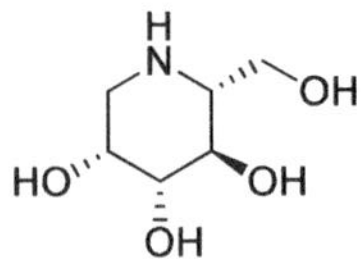	FruA	nd	80	[99]
6	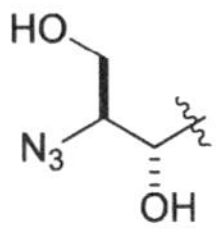	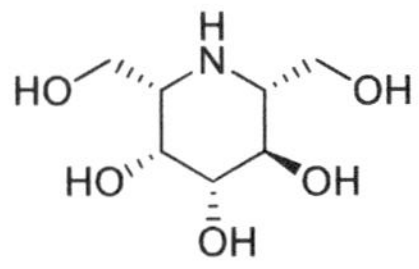	FruA	68	60	[100]
7	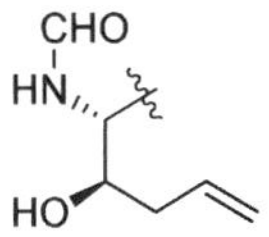	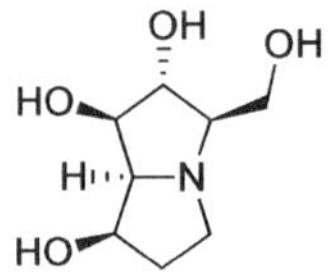	FruA	25[b]	17	[101]
8	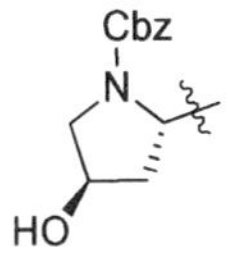	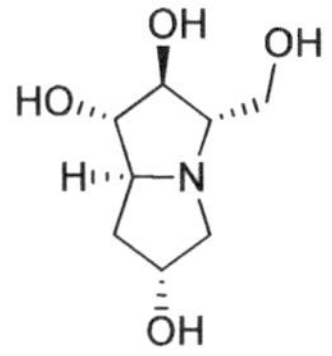	FucA F131A	49[b]	17	[102]

nd, not determined.

[a]*FruA from rabbit muscle (RAMA); RhuA and FucA from E. coli. FucA F131A substitution of the phenylalanine131 by alanine in FucA wild-type.*

[b]*Yield of unphosphorylated material after treatment with acid phosphatase.*

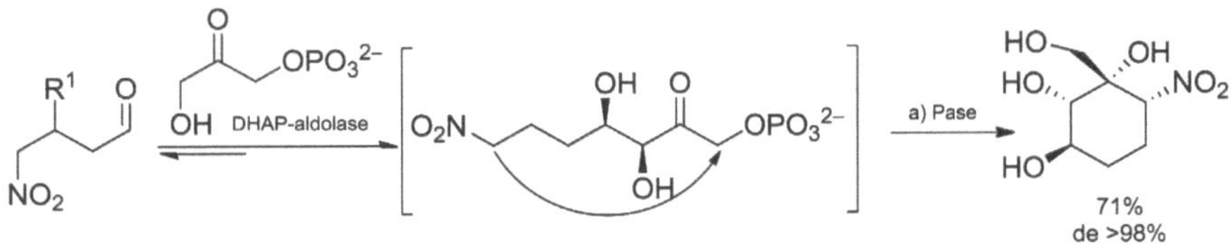

SCHEME 10.4
Aldol addition of DHAP to nitroaldehydes catalyzed by FruA from rabbit muscle (RAMA), followed by spontaneous nitroaldol cyclization (Henry reaction).

undergo an intramolecular chemical cyclization step (Henry reaction), making use of the installed electrophilic carbonyl unit (Scheme 10.4). This twofold C—C bond forming reaction cascade delivers nitrocyclitiols that after reduction, furnished aminocyclitol analogues of interest as inhibitors of intestinal glycosidases [107–109].

Long-chain dialdehydes can be also used as substrates and can react independently at both electrophilic sites, undergoing a bidirectional chain elongation by twofold enzymic aldolization, without the need for isolation of intermediates [110]. If hydroxylation is suitably placed in the substrate, this furnishes C-glycosidically linked saccharide units that can be considered as disaccharide mimics (Scheme 10.5) [111].

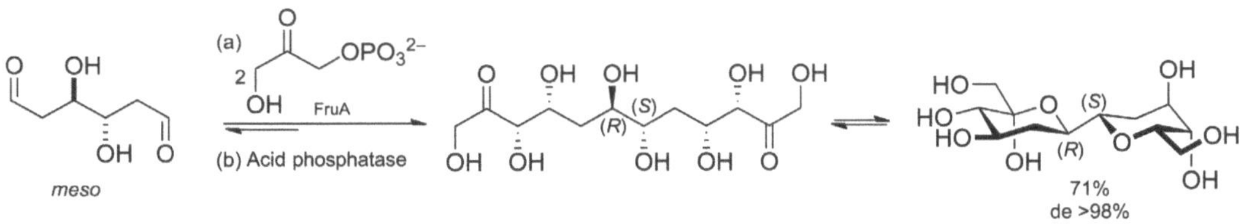

SCHEME 10.5
Bidirectional chain elongation by two fold tandem enzymic aldol addition of DHAP to a *meso*-dihydroxydialdehyde catalyzed by FruA from rabbit muscle (RAMA).

Example: Preparation of D-Sorbose 1-phosphate (Table 10.9, entry 3) [103]

An aqueous solution of D-glyceraldehyde (25 mmol in 300 mL) was combined with a solution of DHAP (20 mmol in 200 mL) freshly adjusted to pH 6.8. The mixture was incubated with RhuA (100 U, prepared as indicated in Ref. [103]; not commercially available) at room temperature for 24 h with monitoring of conversion by thin-layer chromatography (2-propanol/sat. ammonia/water 3:2:1) and enzymic assay for DHAP[i].

[i] Assay conditions for DHAP determination: in a 1 mL spectrophotometric cuvette, NADH (10 μL, 20 mM solution) and DHAP solution of unknown concentration (max ≈15 mM) (10 μL) are dissolved in Tris 50 mM pH 7.8 buffer (975 mL). Then, the absorbance at 340 nm (A1) is measured. After that, a mixture of glycerol phosphate dehydrogenase/triosephosphate isomerase (15 μL Sigma-Aldrich) is added. After a couple of minutes, the absorbance was measured again (A2). The DHAP concentration will be [DHAP] = ((A1−A2)/6.22)×100×(DHAP dilution factor)

The product was isolated by passage through an ion-exchange column (Dowex AGI-X8, HCO_3^- form, 100 mL) and consecutive elution with 200 mM triethylammonium bicarbonate solution. After ion exchange with Dowex AG50W-X8 (H^+) and neutralization with cyclohexylamine, the product was crystallized from aqueous ethanol (90%) to afford colorless crystals of D-sorbose 1-phosphate as *bis*-(cyclohexylammonium) salt; yield: 7.7 g (84%).

Example: Preparation of 1-Deoxynojirimycin (Table 10.10, entry 3) [99,112]

(S)-3-Azido-2-hydroxypropanal (1.4 g, 12 mmol), D-fructose-1,6-bisphophate (1.2 g, 3.0 mmol), FruA from *E. coli* (16 units, FruA from rabbit muscle (RAMA) is commercially available from Sigma-Aldrich), TPI (500 U, Sigma-Aldrich), and $ZnCl_2$ (0.3 mM) in H_2O (50 mL) at pH 6.5 was stirred slowly for 48 h. To the solution was added $BaCl_2.2H_2O$ (3.0 g, 12 mmol) and acetone (~100 mL). The mixture was placed in a freezer at −20°C overnight. The precipitated product was recovered, and treated with Dowex 50 (H^+) in 100 mL water to remove Ba^{+2}. The solution was then adjusted to pH 4.8 and acid phosphatase (150 U, Sigma Aldrich) was added and incubated at 38°C overnight. Ames test for phosphate indicated 100 % hydrolysis of the phosphate ester [113]. The solution was readjusted back to pH 7.0 and lyophilized to give a yellowish product, which was treated with methanol (2 × 50 mL) and filtered to remove insoluble material. Evaporation of the methanol gave a product, which showed a single spot on TLC (R_f = 0.71, EtOAc/MeOH/H_2O 12:6:2). The product was passed through a Dowex 50W (Ba^{+2}) column, using EtOH/H_2O (1:1) as mobile phase. The fractions with product were combined and the ethanol was removed under reduced pressure. To the remaining solution was added one-half its volume of methanol and the solution was treated with H_2 (40 psi) in the presence of Pd/C (10% Pd, 300 mg) for 10 h. The catalyst was removed by filtration and the methanol was removed under vacuum. The remaining solution was concentrated to 4 mL, and chromatographed on a Dowex 50 (Ba^{2+}) column (3.0 × 97 cm) with water as the mobile phase. The fractions (284–310 mL) were collected and lyophilized to yield 0.85 g of *1-deoxynojirimycin* (Table 10.10, entry 3 (80% yield)); $[\alpha]^{22}_D = -29°$ (c = 0.2, MeOH).

3.2 Aldol Addition of 1-Hydroxy-2-alkanones to Aldehydes

The Class I D-fructose-6-phosphate aldolase (FSA) [114] and transaldolase B mutant Phe 178 Tyr (TalB F178Y) [115] both from *E. coli*, catalyze the reversible aldol addition of dihydroxyacetone (DHA) to D-glyceraldehyde 3-phosphate. Thus, they use unphosphorylated DHA and analogs (see further), which greatly simplifies synthetic applications by avoiding the numerous extra efforts in handling the phosphate group. The stereoselectivity and range of accessible aldol products are similar to those discussed for FruA-type reactions. From a preparative point of view, in addition to the DHA nucleophile, FSA also has the ability to accept structurally related DHA analogs such as hydroxyacetone (HA), hydroxybutanone (HB), and glycolaldehyde (GO) as aldol donor components. Most importantly, FSA and TalB F178Y maintain full stereocontrol (*syn*-configuration, (3*S*,4*R*)) over the asymmetric centers at the newly formed C—C bond, which makes available a range of compounds that are inaccessible from FruA catalysis. Concerning the aldol additions of GO, in the presence of other aldehydes, GO invariably acts as a donor. In the competition between self-, that is, furnishing D-(−)-threose, and cross-addition of GO, the extent of the desired cross-aldol reaction depends on the substrate quality of the acceptor aldehyde. It can be further increased by maintaining a low GO concentration during the reaction, for example, by slowly adding this compound with a syringe pump [116]. Similar to other enzymatic aldol additions, the FSA and TalB F178Y-catalyzed reactions are conveniently carried out at ambient conditions in aqueous buffer, or with up to 20% v/v of dimethylformamide as cosolvent to enhance the solubility of hydrophobic aldehydes. Examples of aldol addition reactions of DHA and analogs to aldehydes are summarized in Tables 10.11 and 10.12.

Example: Preparation of 1-Deoxy-D-xylulose (Table 10.11, entry 2) [118]

Glycoladehyde (3.0 g, 50 mmol) and hydroxyacetone (4.63 g, 62.49 mmol) were dissolved in $NaHCO_3$ buffer (50 mM adjusted to pH 7.8) (100 mL). To this solution, FSA aldolase crude powder (0.312 g, 75 U, prepared as indicated in Ref. [118]; not commercially available) dissolved in $NaHCO_3$ buffer (50 mM adjusted to pH 7.8) (50 mL) was added (total final volume 150 mL). After 24 h, no glycoladehyde was detected by TLC ($CHCl_3$/MeOH, 5:1) and the reaction was stopped by adding MeOH (100 mL). The mixture was filtered through activated charcoal and the filtrate adjusted to pH 5 and lyophilized. Purification on silica gel eluted with $CHCl_3$/MeOH from 1:0 to 17:3 yielded in three runs 3.35 g (51%) of 1-deoxy-D-xylulose as a white solid; $[\alpha]^{22}_D = +54.1°$ ($c = 1.11$ in H_2O). The ^1H and ^{13}C NMR data showed it to be a mixture of two cyclic, that is, α-furanose and β-furanose, and one acyclic forms.

Example: Preparation of (2R,3R,4R)-2-Hydroxymethylpiperidine-3,4-diol (D-fagomine (Table 10.12, entry 4) [120]

Step 1 Aldol condensation: Dihydroxyacetone (DHA, 2.1 g, 22.9 mmol) and FSA aldolase powder (2.09 g, 3445 U, prepared as indicated in Ref. [120]; not commercially available) were dissolved in 50 mM boric/borate buffer, pH 7.0 (155 mL) and cooled to 4°C. To this mixture, *N*-Cbz-aminoaldehyde (4.71 g, 22.9 mmol) dissolved in DMF (40 mL) at 4°C was added. The reaction mixture was placed in a reciprocal shaker (120 rpm) at 4°C. After 24 h, MeOH (200 mL) was added to the mixture to stop the reaction and centrifuged (1000 rpm) at 10°C for 40 min. The supernatant was filtered on a 0.45 μm filter and purified by preparative HPLC to give benzyl ((3*R*,4*S*)-3,4,6-trihydroxy-5-oxohexyl)carbamate (4.7 g, 69 %) as a white solid. Step 2 Preparation of D-fagomine: Pd/C (100 mg) was added to a solution of benzyl ((3*R*,4*S*)-3,4,6-trihydroxy-5-oxohexyl)carbamate (373 mg, 1.26 mmol) in H_2O/EtOH 9:1 (50 mL). The reaction mixture was shaken under H_2 (50 psi) overnight at rt. After removal of the catalyst by filtration through

Table 10.11 Aldol Addition of Dihydroxyacetone, Hydroxyacetone, Hydroxybutanone, and Glycolaldehyde to Standard Aldehydes Catalyzed by FSA Wild-Type and Variants and $TalB^{F178Y}$

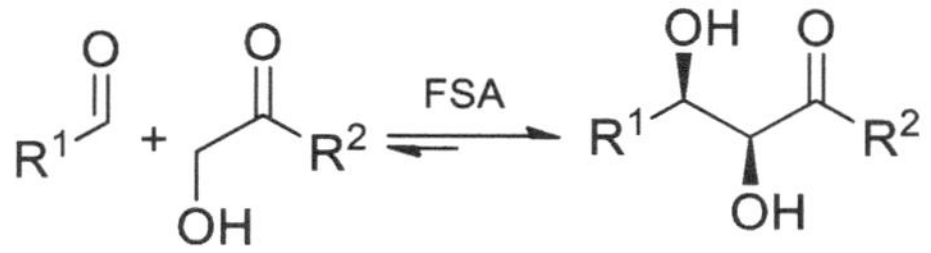

R^2: CH_2CH_3, CH_2OH, CH_3, H

Entry	R^1	Product	Biocatalyst	Yield[a] (%)	Ref
1	CH_2OH	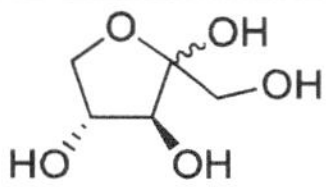	FSA^{A129S}	28	[121]
2	CH_2OH	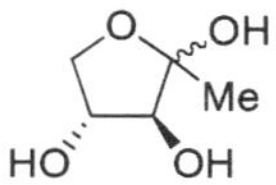	FSA	51	[118]
3	CH_2CH_2OH	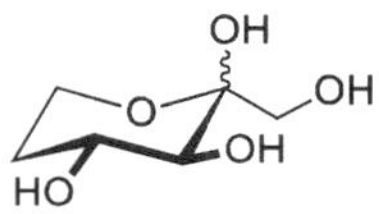	$TalB^{F178Y}$	>75%	[122]
4	CH_2CH_2OH	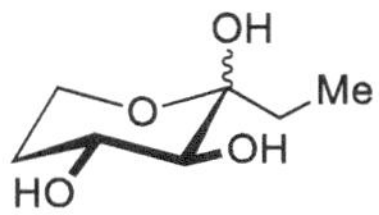	FSA	>75%	[122]
5	(*S*)-CH(OH)CH_2OH	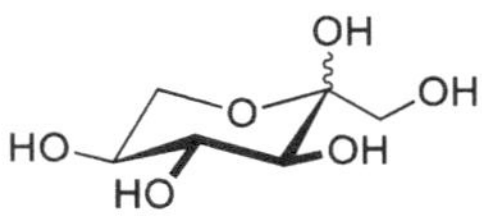	$TalB^{F178Y}$	>75%	[122]
6	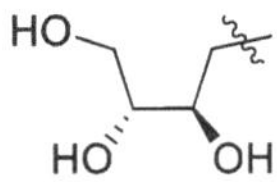	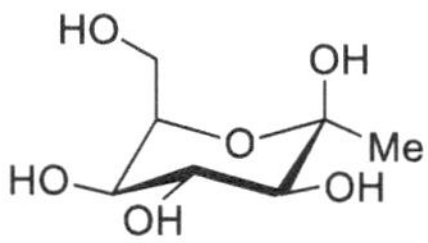	FSA	58	[118]
7	CH_2OH	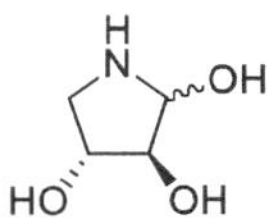	FSA	67	[116]
8	(*S*)-CH(OH)CH_3	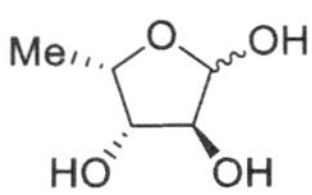	FSA	50	[116]

[a]*Under the detection limits of high-field NMR spectroscopy, no diastereomers other than those with the 3S,4R configuration were detected (i.e., de >97%). FSA^{A129S}, FSA^{A129S} mutant and $TalB^{F178Y}$, Talb B F178Y mutant.*

Table 10.12 Aldol Addition of Dihydroxyacetone, Hydroxyacetone, Hydroxybutanone, and Glycolaldehyde to Nitrogen-Containing Standard Aldehydes Catalyzed by FSA Wild-Type and Variants

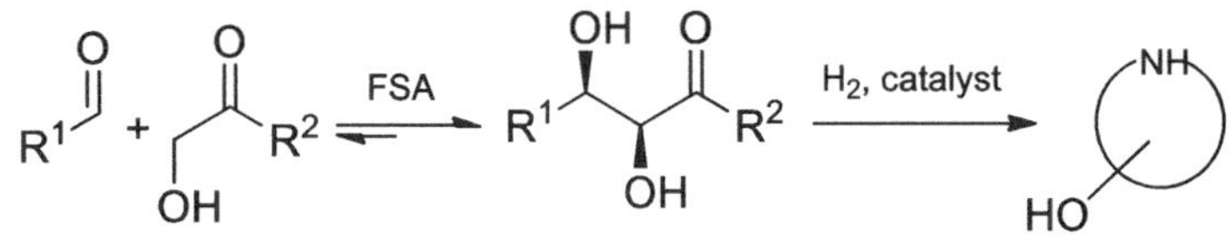

R^2: CH_2CH_3, CH_2OH, CH_3, H

Entry	R^1	Product	Biocatalyst	Yield (aldol adduct, %), iminocyclitol %	de (%)	References
1	CH_2N_3	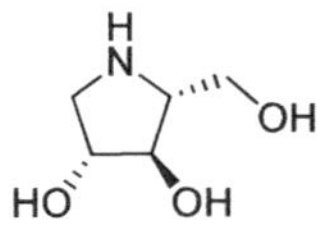	FSA	(ni)[a], 73[b]	>90[e]	[117]
2	CH_2NHCbz	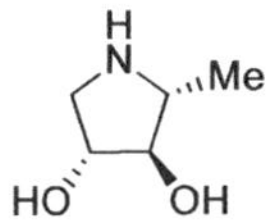	FSA	(75), 56	>97[f]	[118]
3	(*R*)-CH(Me)NHCbz	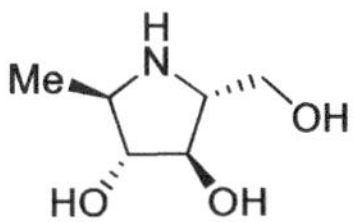	$FSA^{A129S/A165G}$	(82), 80	>97[f]	[119]
4	CH_2CH_2NHCbz	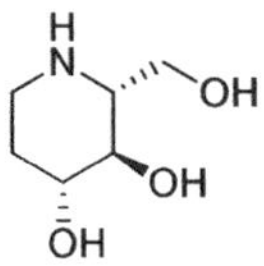	FSA	(69), 51	86	[120]

5	CH_2CH_2NHCbz	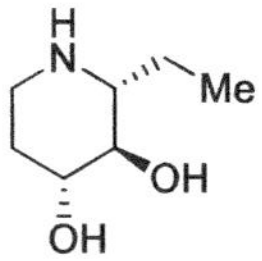	FSA	(nd)[c], 78[d]	>90[e]	[117]
6	(*R*)-$CH(OH)CH_2NHCbz$	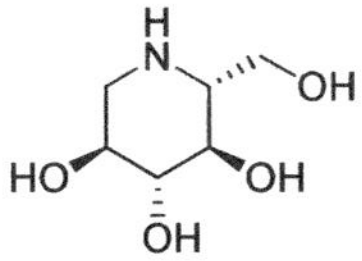	FSA	(ni)[a], 49	>97[f]	[118]
7	(*R*)-$CH(OH)CH_2CH_2NO_2$	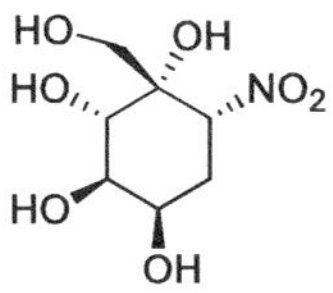	FSA^{A129S}	52	>97[f]	[121]
8	CH_2CH_2NHCbz	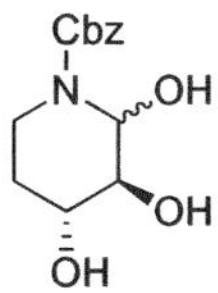	FSA	50[g]	>97[f]	[116]

[a]*not isolated.*
[b]*Iminosugar obtained in one step without purification of aldol adduct.*
[c]*not determined.*
[d]*Iminosugar obtained in two steps.*
[e]*crude product mixtures by NMR showed small quantities of diastereomeric products, typically in the range of 3–5%.*
[f]*Under the detection limits of high-field NMR spectroscopy, no diastereomers other than those with the 3S,4R configuration were detected.*
[g]*The best performance was obtained when glycolaldehyde was added slowly by means of a syringe pump, then the rate of cross-aldol addition was greatly enhanced.*

neutralized and deactivated alumina, the solvent was evaporated under reduced pressure and then lyophilized to give D-fagomine (164 mg, 51 % global yield) as a brown solid; $[\alpha]_D^{22} = +20.4°$ (c = 1.0 in H_2O).

Example: Preparation of (2S,3R)-5-(N-Benzyloxycarbonyl) amino-2,3-dihydroxypentanal (Table 10.12, entry 8) [116]

3-(*N*-benzyloxycarbonyl)aminopropanal (414 mg, 2.0 mmol) was dissolved in DMF (4 mL) and 125 mM triethanolamine pH 7.5 buffer (16 mL) was added. To the milky mixture thus obtained, FSA powder (90 mg, 22.5 U, prepared as indicated in Ref. [116]; not commercially available) was added and the reaction mixture was placed in an orbital shaker (500 rpm) at 25°C. A 4:1 water:DMF solution of glycolaldehyde (10 mL, 132 mg, 2.2 mmol) was added over 48 h using a syringe pump. Additional enzyme (45 mg) was added after 24 h of reaction. After 72 h, the reaction mixture was loaded onto a chromatography column (10x140 mm) filled with Amberlite® XAD16 and equilibrated with water. The product was retained and washed with water (4 column volumes) to eliminate the excess of reactant and enzyme. Finally, the product was eluted with EtOH/H_2O 3:2 and the EtOH evaporated under vacuum. The crude mixture was loaded onto a semi-preparative column XTerra® (19 × 250 mm). Products were eluted using water-CH_3CN gradient from 0 % to 52 % CH_3CN in 30 min. The flow rate was 10 mL/min and the products were detected at 215 nm. Analysis of the fractions was accomplished under gradient conditions (10–70 % of solvent B in 30 min) on the analytical HPLC XBridge® C18, 5 μm, 4.6 × 250 mm column (waters). Pure fractions were pooled and lyophilized to give the title compound (200 mg, 50 %) that was characterized as a mixture of two conformers.

3.3 Aldol Addition of Pyruvate to Aldehydes

Pyruvate aldolases operate in sugar acid metabolism where they catalyze the reversible addition of pyruvate as the common aldol component to aldehydes. Many of them are of type I, but there also exist examples of type II, although with limited synthetic applications so far [123,124]. The most important pyruvate aldolase is the *N*-acetylneuraminic acid aldolase (NeuA, EC 4.1.3.3), also known as sialic acid aldolase, which has a preference for *N*-acetyl-D-mannosamine or configurationally related aldohexose sugars and derivatives as electrophiles. Other important pyruvate aldolases from synthetic point of view are 2-keto-3-deoxy-D-*manno*-octosonate aldolase (KdoA, EC 4.1.2.23), which prefers D-arabinose as the acceptor, and two functionally related enzymes, 2-keto-3-deoxy-6-phospho-D-galactonate aldolases (KDPGal aldolase or GalA; EC 4.1.2.21), and 2-keto-3-deoxy-6-phospho-D-gluconate aldolase (KDPGlc aldolase or GlcA; EC 4.1.2.14) that accept small aldehyde electrophiles but with a preference for D-glyceraldehyde 3-phosphate (Scheme 10.6). All of them create a single stereocenter upon aldol addition to a specific group of acceptor aldehydes (Scheme 10.6). In these reactions, the equilibrium is less favorable in the direction of synthesis that usually requires an excess of pyruvate to achieve a

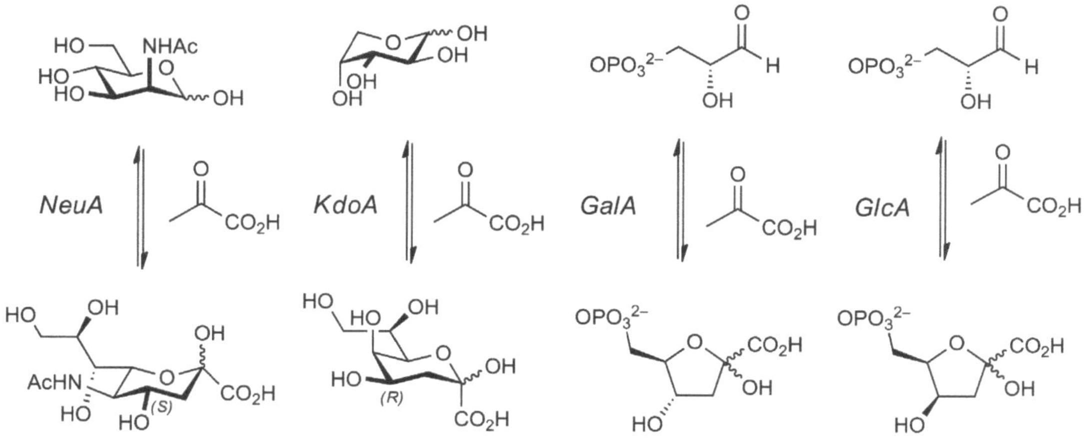

SCHEME 10.6
Stereocomplementary sets of pyruvate dependent aldolases. *N*-acetylneuraminic acid aldolase (NeuA; EC 4.1.3.3), 2-keto-3-deoxy-manno-octosonate (KDO) aldolase (KdoA; EC 4.1.2.23), 2-keto-3-deoxy-6-phospho-D-gluconate aldolases (GlcA; EC 4.1.2.14)) and 2-keto-3-deoxy-6-phospho-D-galactonate aldolases (GalA; EC4.1.2.21).

preparative useful conversion. Product isolation from reaction mixtures is facilitated by enzymatic decomposition of excessive pyruvate by pyruvate decarboxylase (PDC) to yield volatile CO_2 and acetaldehyde.

3.3.1 N-ACETYLNEURAMINIC ACID ALDOLASE (NEUA)

The NeuA enzyme promotes a reversible *si*-face attack of pyruvate on the aldehyde carbonyl group of *N*-acetyl-D-mannosamine to furnish sialic acid with a new stereogenic center of (4*S*) absolute configuration (Scheme 10.6) [77]. Structural modifications at the amide function are tolerated, for example, acyl variations or replacement by a hydroxyl group, but the enzyme prefers aldehydes equal to or larger than aldopentoses (Table 10.13). In most cases, the enzyme is highly stereoselective and yields the (4*S*)-configuration [77,125]. On the other hand, some carbohydrates are converted with random or even inverse stereoselectivity for the C_4 configuration (Table 10.13, entries 8 and 9) [126–130]. The critical determinant appears to be the C_3-configuration in the electrophilic substrate. This means that the stereochemical outcomes of such aldol reactions are (at least partially) determined by the substrate [129,130]. A three-point binding model has been proposed that takes into account the conformational substrate preference to predict the direction of synthesis [77].

NeuA can tolerate disaccharides carrying a reducing D-Man or D-ManNAc, even when containing a sterically more demanding 1,4-glycosidic linkage (Table 10.14) [137,138].

Table 10.13 ***N*-Acetylneuraminic Acid Aldolase Catalyzed Stereoselective Aldol Addition of Pyruvate to *N*-Acetylmannosamine and Analogs**

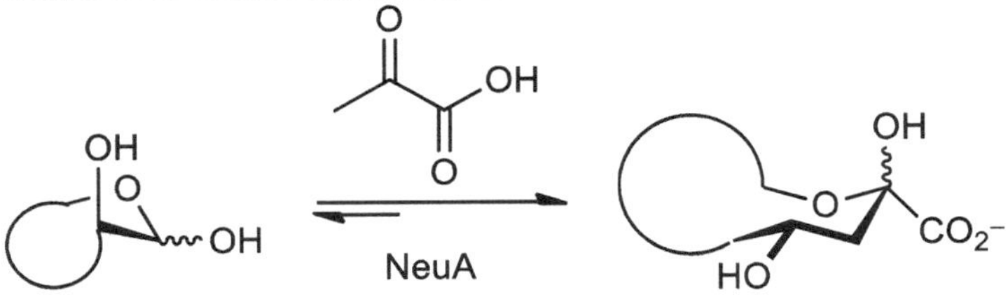

Entry	Substrate	Product	Yield %	4*S*/4*R*	References
1		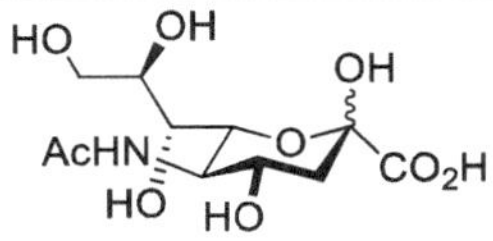	85	>99:1	[131,132]
2		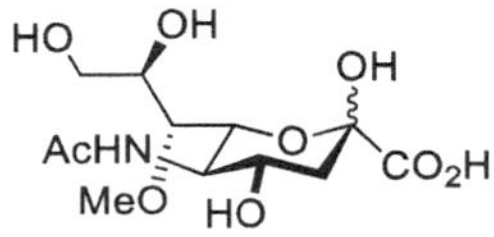	70	>99:1	[131]
3		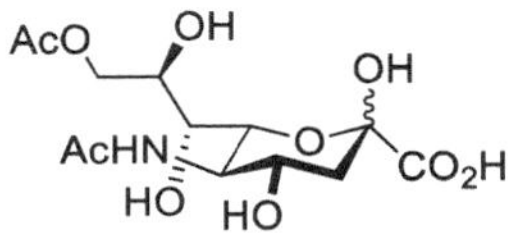	84	>99:1	[125]
4	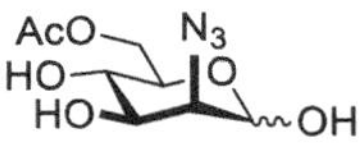	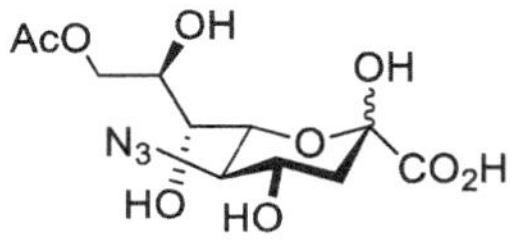	80	>99:1	[133,134]
5	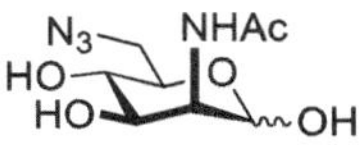	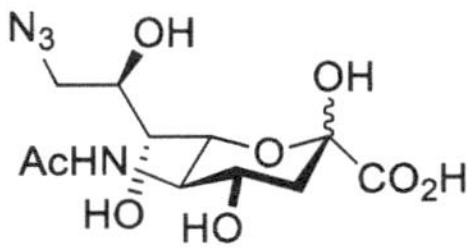	84	>99:1	[135]
6	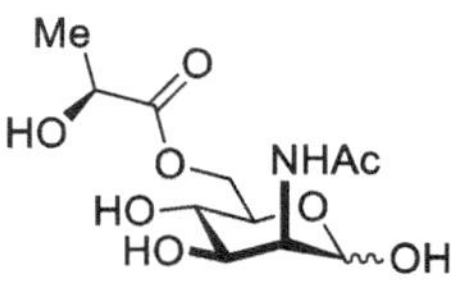	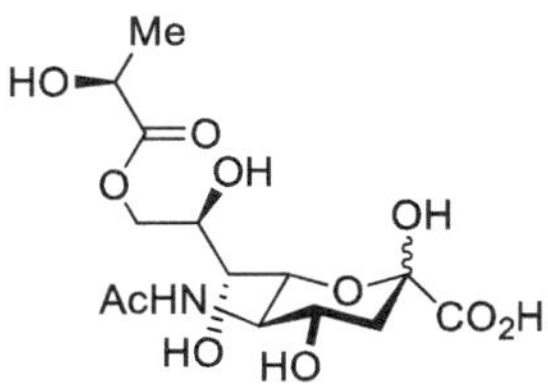	53	>99:1	[135]
7		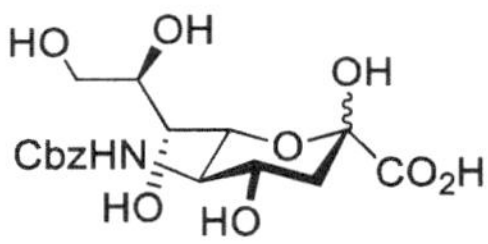	75	>99:1	[134,136]

Table 10.13 *N*-Acetylneuraminic Acid Aldolase Catalyzed Stereoselective Aldol Addition of Pyruvate to *N*-Acetylmannosamine and Analogs *(cont.)*

Entry	Substrate	Product	Yield %	4*S*/4*R*	References
8	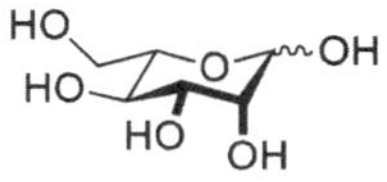	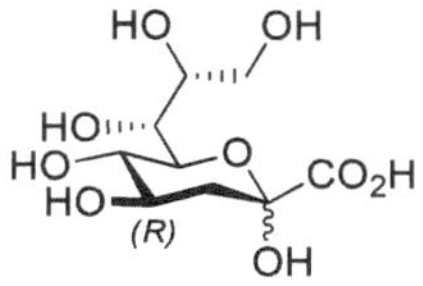	>80	1:99	[127,128]
9	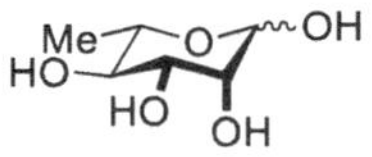	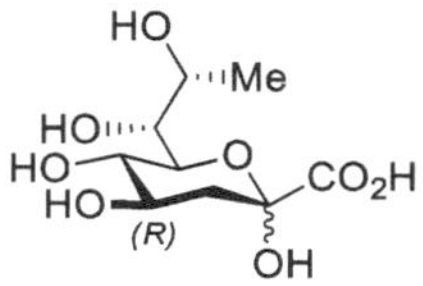	>80	1:99	[127,128]

Table 10.14 *N*-Acetylneuraminic Acid Aldolase Catalyzed Stereoselective Aldol Addition of Pyruvate to Disaccharides

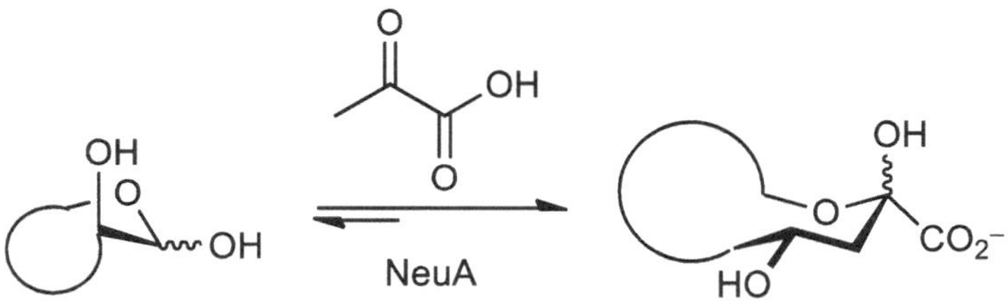

Substrate	Product	Yield %	References
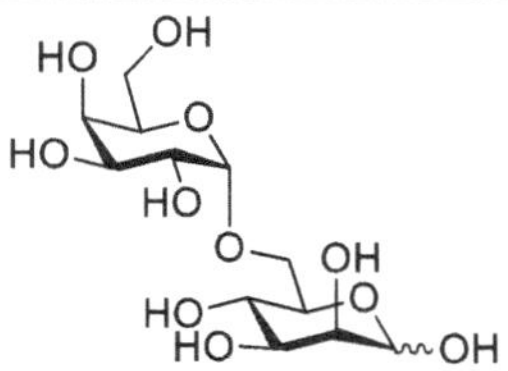	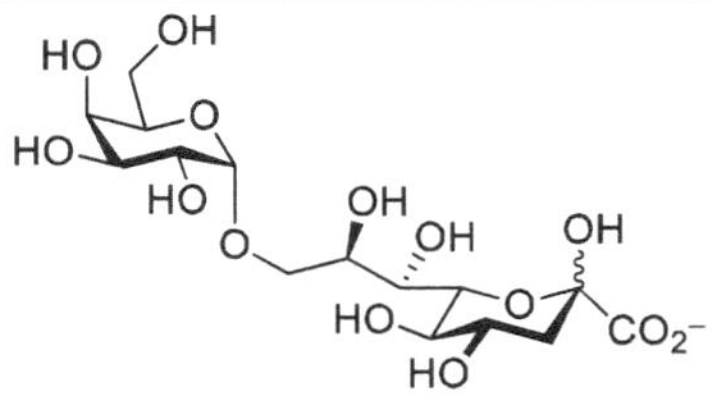	81	[137]
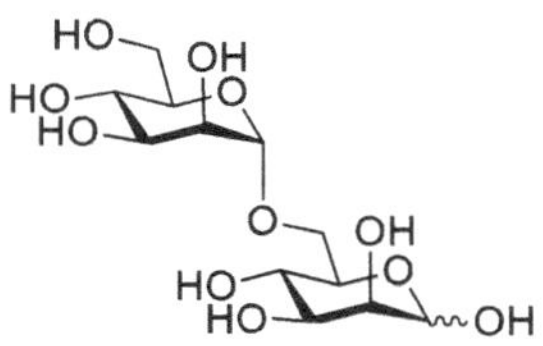	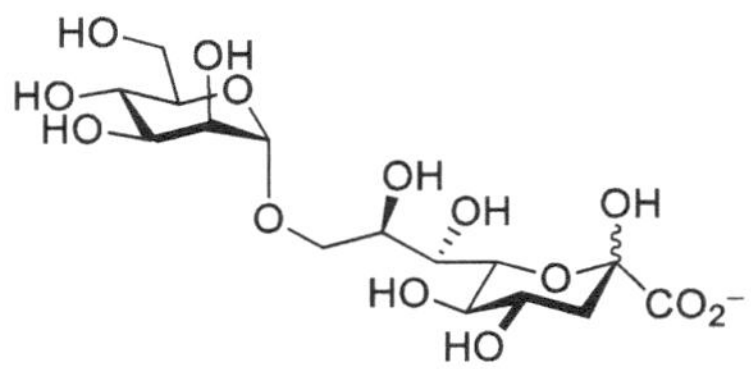	78	[137]

(Continued)

Table 10.14 *N*-Acetylneuraminic Acid Aldolase Catalyzed Stereoselective Aldol Addition of Pyruvate to Disaccharides *(cont.)*

Substrate	Product	Yield %	References
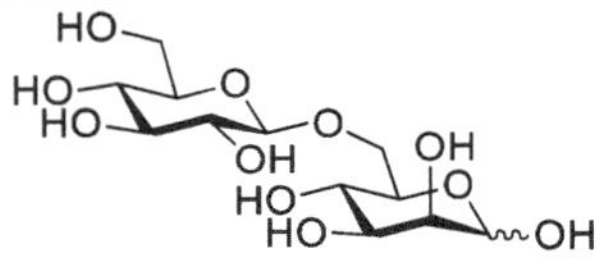	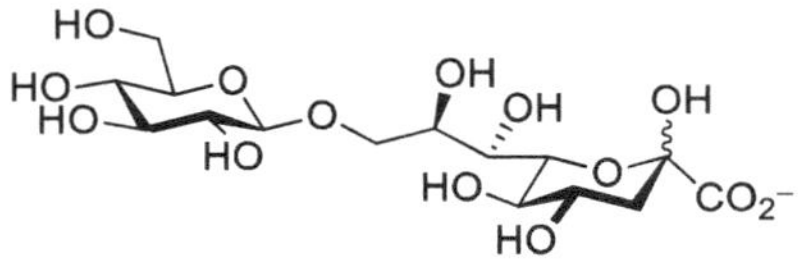	83	[137]

Example: Preparation of 9-Azido-9-Deoxy-*N*-Acetylneuraminic Acid (Table 10.13, Entry 5) [135]

Pyruvate (0.5 M), 6-azido-6-deoxy-*N*-acetylmannosamine (50 mg) and dithiothreitol (10 mM) were dissolved in 0.1 M potassium phosphate buffer (pH 7.5) (10 mL) containing NeuA aldolase (3 U, prepared as indicated in Ref. [135]; not commercially available). The starting material was consumed in 14 h. The solution was lyophilized and the mixture was purified with Bio-Gel P-2 gel filtration (3 × 90 cm) chromatography, eluted with water at a flow rate of 6 mL/40 min, at 4°C. The fractions containing the product were pooled and freeze-dried to give the title compound in 84% yield.

3.3.1.1 General Procedure for the Synthesis of Compounds of Table 10.14

The disaccharide (50–100 mg, 1 equiv.) and pyruvate (5 equiv. as sodium salt) were dissolved in H_2O (5 mL). *tris*(Hydroxymethyl)aminomethane (Tris)/HCl buffer (1 mL, 1 M, pH 7.5) was added, followed by addition of the NeuA from *E. coli* (5 mg, prepared as indicated in the references, not commercially available). The reaction solution was brought to 10 mL in volume by adding H_2O. The reaction mixture was incubated at 37 °C with agitation at 140 rpm. The reaction was monitored by thin-layer chromatographic analysis with developing solvent (EtOAc:MeOH: H_2O:HOAc 5:3:2:0.1, by volume) and stained with *p*-anisaldehyde sugar stain. When the reaction was completed, EtOH/H_2O 95:5 (10 mL) was added to the reaction mixture. The precipitates were separated by centrifugation and discarded. The supernatant was concentrated by rotary evaporation and the product was purified on a Bio-Gel P-2 gel filtration column.

3.3.2 *2-KETO-3-DEOXY-MANNO-OCTOSONATE (KDO) ALDOLASE (KDOA)*

The KdoA enzyme is involved in the catabolism of the eight-carbon sugar, 2-Keto-3-deoxy-manno-octosonate, by reversibly degrading it to D-arabinose and pyruvate [139]. KdoA has broad substrate specificity for aldehydes while pyruvate was found to be irreplaceable (Table 10.15) [139,140]. The stereochemical course of the aldol addition is invariable, always involving *re*-face attack on the aldehyde carbonyl. On the basis of the results published so far, it may be concluded that a (3*R*)-configuration of the acceptor is necessary (but not sufficient), and that stereochemical requirements at C_2 are less stringent [140].

Table 10.15 Aldol Addition of Pyruvate to D-Arabinose and Analogs Catalyzed by 2-Keto-3-deoxy-manno-octosonate (KDO) aldolase (KdoA).

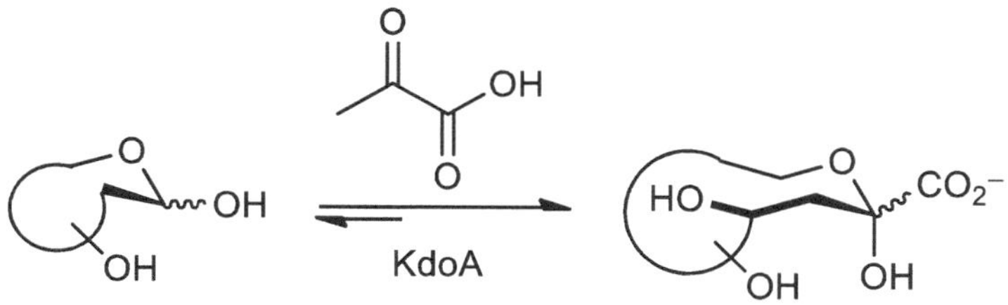

Entry	Substrate	Product	Yield %	Ref
1	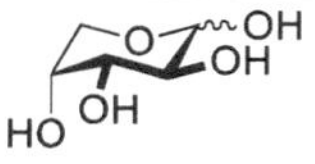	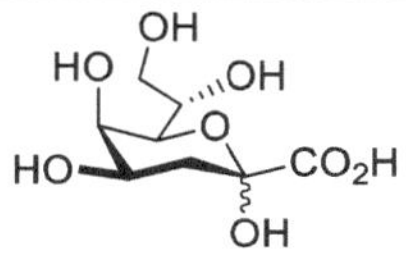	67 (37)[a]	[140]
2	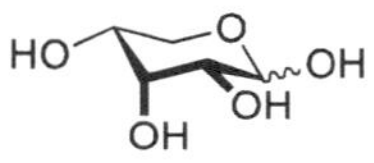	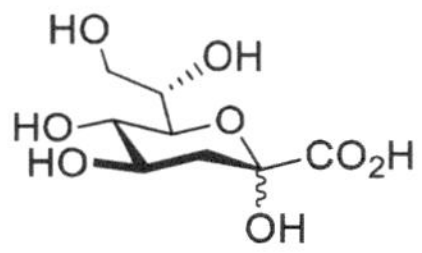	57	[140]
3	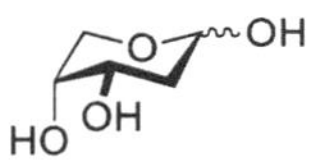	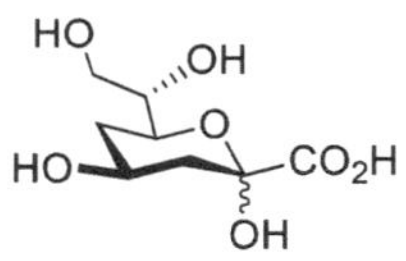	47	[140]
4	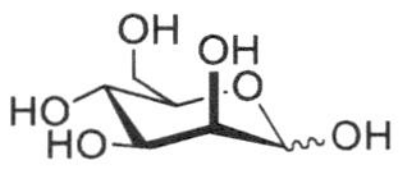	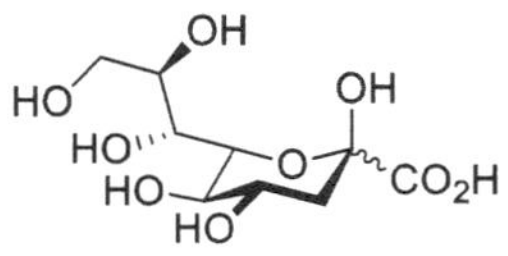	61	[140]
5	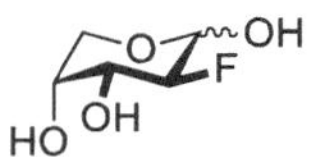	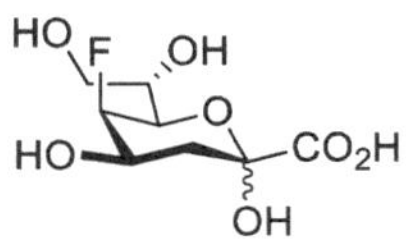	33	[140]

[a] *Recrystallized yield.*

Example: Preparation 3,5-dideoxy-5-fluoro-α-D-manno-2-octulosonate (Table 10.15, entry 5) [140]

2-Deoxy-2-fluoro-D-arabinose (340 mg, 2.25 mmol), pyruvate (2.074 g, 28.9 mmol, as sodium salt), dithiothreitol (1.7 mg), NaN_3, (2.3 mg), and phosphate buffer (pH 7.5, 50 mM, 1.12 mL) were added to the KdoA solution (3.0 mL,

24 U, prepared as indicated in Ref. [140]; not commercially available). After the pH was adjusted to 7.5, the volume was made up to 10.0 mL. The mixture was stirred under N_2 at room temperature for 7 days. The pH was lowered to 2.5 by addition of Dowex 50W-X8 (H^+ form), and the mixture was kept at 0°C for 1 h. The precipitate was removed by centrifugation and the excess of pyruvate was removed as follows. The mixture was diluted to 80 mL, and the pH was adjusted to 6.5 by the addition of aqueous ammonia (2 N). Pyruvate decarboxylase (0.2 mL, 12.5 U) was then added, and the mixture was stirred at room temperature with bubbling of N_2 (1.5 L/min). The pH was monitored and occasionally adjusted between 6.0 and 6.5, by addition of Dowex 50W-X8 (H^+ form). The decarboxylase was periodically added to the mixture (0.2 mL each) at an interval of 30 min, to avoid inactivation, which is caused by the rapid formation of acetaldehyde. The total amount of the enzyme was 3.2 mL (200 units). The reaction mixture was further stirred overnight. Then the mixture was centrifuged, and the supernatant was diluted to 100 mL and applied to a column of Dowex 1-X8 (20–50 mesh, HCO_3^- form, bed volume of 100 mL). The pH of the eluent and washings was readjusted to 5.5, and the resulting solution was again applied to the same column to ensure the adsorption of the desired product. After the column was washed with water, the desired product was eluted with a linear gradient of 0 to 0.3 M NH_4HCO_3. The product was further purified on a BioGel P-2 column (bed volume of 20 mL) to give the title compound (192 mg, 33%).

3.3.3 2-KETO-3-DEOXY-6-PHOSPHO-D-GLUCONATE (GLCA) AND 2-KETO-3-DEOXY-6-PHOSPHO-D-GALACTONATE ALDOLASES (GALA)

GlcA and GalA are pyruvate-dependent aldolases that catalyze reversible formation of 2-keto-3-deoxy-6-phospho-D-gluconate and 2-keto-3-deoxy-6-phospho-D-galactonate, respectively, from pyruvate and D-glyceraldehyde 3-phosphate with complementary stereochemistry (Scheme 10.6). GlcA enzymes can be found in different microorganisms and offer a rather broad substrate tolerance for polar, short-chain aldehydes, albeit with rather low reaction rates (Table 10.16) [141–145]. Simple aliphatic or aromatic aldehydes are not converted. GlcA mutants were identified with up to 2000-fold improved selectivity for unnatural substrates and 40-fold improved catalytic efficiency [146]. Similarly, GalA from *E. coli* [147] has been engineered for improved acceptance of both nonphosphorylated substrates and L-configured aldehydes to facilitate the development of enzymatic syntheses [148,149]. Furthermore, macrophomate synthase (MPS) from *Macrophoma commelinae* can be used to access the same type of reaction products but using oxaloacetate as donor to form the pyruvate enolate intermediate and liberating CO_2. Such a strategy is reminiscent of using ThDP decarboxylases for acyloin condensations (Table 10.16) [150,151].

Example: Preparation of Sodium 5,6-O-isopropylidene-3-deoxy-D-erythro-2-hexulosonate (Table 10.16, entry 6) [153,154]

D-Glyceraldehyde-acetonide (1 g), pyruvate (1.2 g, as sodium salt) and GlcA from *Sulfolobus solfataricus* (10 mg, 51.4 U/mg prepared as indicated in Refs [153] and [154]; not commercially available) were dissolved in 25 mM sodium phosphate, pH

Table 10.16 Addition of Pyruvate to Aldehydes Catalyzed by 2-Keto-3-deoxy-6-phospho-D-gluconate (GlcA), 2-Keto-3-deoxy-6-phospho-D-galactonate aldolases (GalA), and Macrophomate Synthase (MPS)

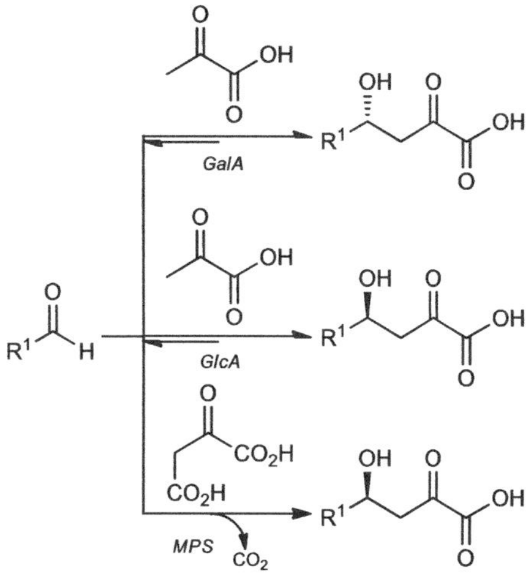

Entry	R¹	Product	Aldolase	Yield %	4S:4R ratio	References
1	COOH	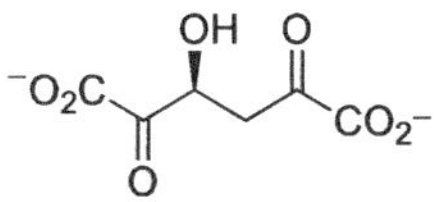	GlcA *E. coli*	70	95:5	[143]
2	(*R*)-CH(OH)CH_2OH	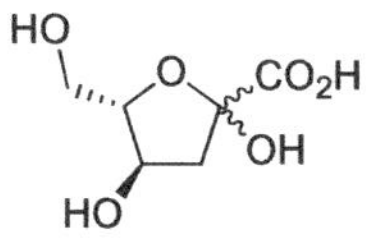	GlcA *P. putida*	24–31	>97:3	[141,145]
3	(*R*)-CH(OH)CH_3	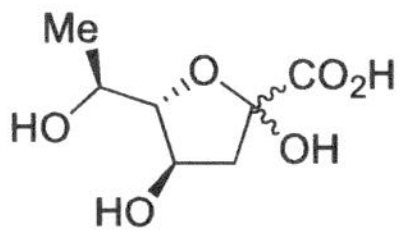	GlcA *P. putida*	30	>97:3	[141,145]

(Continued)

Table 10.16 Addition of Pyruvate to Aldehydes Catalyzed by 2-Keto-3-deoxy-6-phospho-D-gluconate (GlcA), 2-Keto-3-deoxy-6-phospho-D-galactonate aldolases (GalA), and Macrophomate Synthase (MPS) *(cont.)*

Entry	R[1]	Product	Aldolase	Yield %	4*S*:4*R* ratio	References
4	2-pyridyl	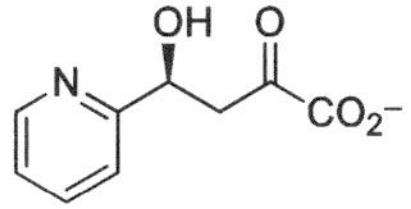	GlcA *P. putida* or *E. coli*	69	>99:1	[144,152]
5	2-pyridyl	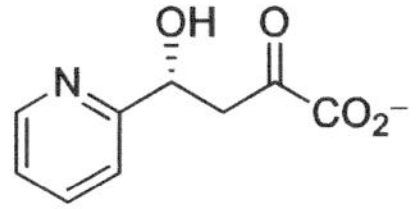	GalA *E. coli*	89–90[a]	1:99	[147,152]
6	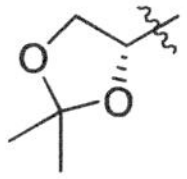	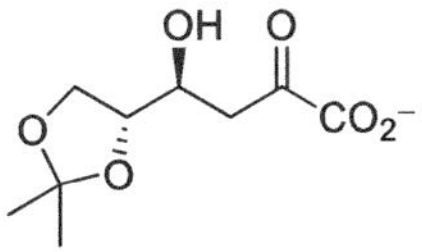	GlcA *S. solfataricus*	61	>92:6	[153,154]
7	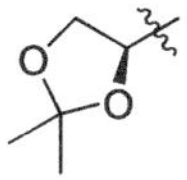	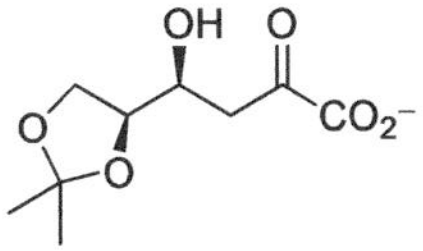	GlcA *S. solfataricus*	42	>94:6	[153]
8	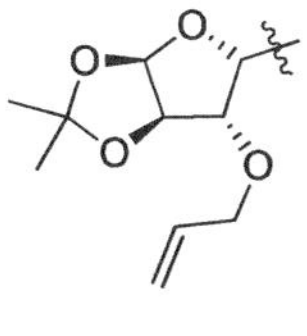	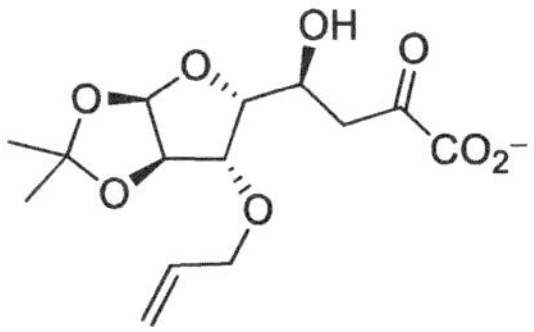	MPS	82	>95:5	[151]
9	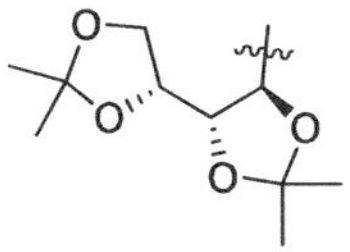	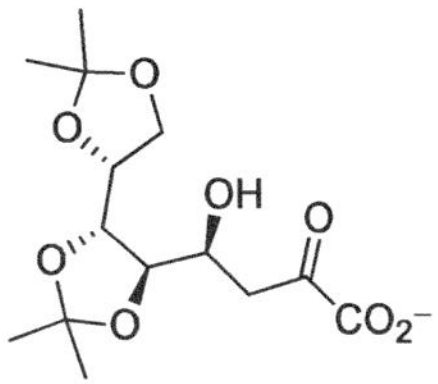	MPS	65	>95:5	[151]

[a]*Estimated by the authors after HPLC purification; S. solfataricus, P. putida.*

7.0 (50 mL). This solution was heated and shaken at 50°C for 3 h. The reaction mixture was then filtered and lyophilized. The residue was dissolved in methanol, filtered, and dried under vacuum to afford the title compound in >92% de as determined by ^1H NMR. The crude product was then dissolved in boiling absolute ethanol (~100) mL and allowed to crystallize over 20 h at 4°C to afford a white crystalline powder that was washed with ice-cold absolute ethanol and dried to give the title compound (1.1 g, 61% yield and >>86 % de), $[\alpha]^{25}_D = +6.0°$ (c = 3, H_2O).

3.4 Aldol Addition of Glycine to Aldehydes

Threonine aldolases (ThrA; EC4.1.2.5) and serine hydroxymethyltransferases (SHMT; EC 2.1.2.1) are pyridoxal-5′-phosphate (PLP) dependent aldolases that catalyze the aldol addition of glycine to aldehydes [155–158]. Since two new stereogenic centers are formed, four possible stereoisomers can be formally obtained. However, contrary to the case of DHAP-dependent aldolases, the four set of stereocomplementary enzymes have not been found in nature yet (Scheme 10.7) [157,159,160].

All enzymes exquisitely control the stereochemistry at the stereogenic center generated from the nucleophile (donor): (2*S*) or (2*R*) for the L- or D-threonine type aldolases, respectively. With respect to the β-carbon, threonine aldolases are usually nonstereoselective giving, in many instances, diastereomeric mixtures. Hence, L-threonine aldolases can be further divided into three sub-types: L-threonine aldolases (LThrA), which preferentially cleave L-threonine (L-Thr); L-*allo*-threonine aldolases (L-*allo*ThrA), which cleave L-*allo*-threonine (L-*allo*-Thr); and L-low specificity threonine aldolases, which accept both L-Thr and L-*allo*-Thr as substrates. Regarding the D-series, only D-low specificity threonine aldolases (DTA) have so far been found in nature [157,161]. In aldol additions catalyzed by ThrA, the diastereochemical outcome crucially depends upon the kinetic selectivity of the enzyme. Because of the equilibrium nature of these reactions, long reaction times usually strongly compromise the stereoselectivity at the

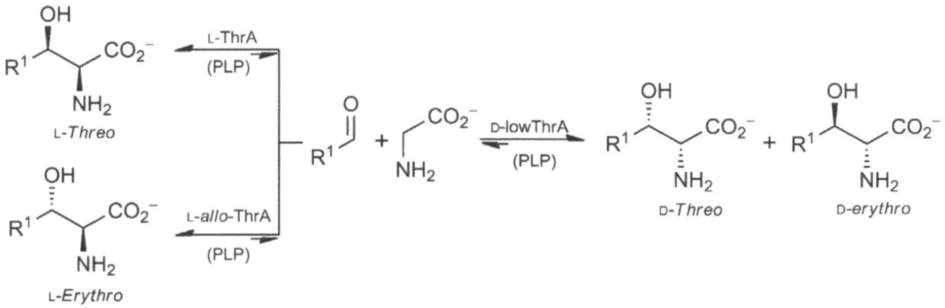

SCHEME 10.7
Aldol addition of glycine to aldehydes catalyzed by threonine aldolases. PLP, Pyridoxal-5'-phosphate.

β-carbon, but, on the other hand, this can be instrumental for obtaining stereocomplementary products using low-specificity glycine aldolases [162,163]. Unless the aldol adduct produced can form a stable ring structure, these reactions are usually characterized by an unfavorable equilibrium constant in direction of synthesis. Therefore, for preparative reactions, a large excess of glycine is often used to increase the yields. In some instances, short reaction times or low reaction temperature, for example, 4 °C instead of 25 °C, can improve the stereoselectivity under kinetic control at the expense of the reaction yields [158]. The kinetic diastereoselectivity of ThrA enzymes can be also exploited in a racemate resolution approach via the selective cleavage of the unwanted component from chemically prepared racemic material. Examples of aldol additions of glycine to aldehydes catalyzed by glycine aldolases are summarized in Table 10.17.

Example: Synthesis of (2S,3R,4R)-2-Amino-4-(benzyloxycarbonylamino)-3-hydroxypentanoic acid ammonium salt (Table 10.17, entry 10) [158]

(*R*)-Benzyl 1-methyl-2-oxoethylcarbamate (423 mg, 2 mmol) was dissolved in DMF (6.8 mL). Then, glycine solution (423 mg, 2 mmol in 27.2, mL H_2O) adjusted to pH 6.5 was added while stirring at 4°C with a vortex mixer. This was followed by the addition of a solution of pyridoxal-5-phosphate (190 mL, 0.3 mmol). The reaction was started by adding recombinant SHMT from *Streptococcus thermophilus* (20 U, prepared as indicated in Ref. [158]; not commercially available). The reaction mixture was placed on a horizontal shaking bath (100 rpm) at 4°C. The reactions were followed by HPLC until the product reached a maximum level. The enzymatic reactions were stopped by adding MeOH/HOAc 39:1 (v/v) and the crude was filtered through Celite/activated charcoal (95:5) and the cake was washed with MeOH. The combined filtrate was evaporated under vacuum to remove MeOH. The residue thus obtained was dissolved in H_2O and the unreacted aldehyde and other hydrophobic impurities were extracted with EtOAc (3 × 150 mL). The aqueous layer containing the product was freeze dried and then purified by semipreparative RP-HPLC on a Perkin–Elmer 250 × 25 mm column, filled with C18, 5 μm-type stationary phase, and eluted with a gradient of 0.1% (v/v) TFA in CH_3CN (8 to 48% v/v in 30 min). The flow rate was 10 mL/min, and the products were detected at 215 nm. The pure fractions were pooled and lyophilized providing the title compound, 170 mg (30%), 99% pure, as determined by HPLC; $[\alpha]^{20}_D = -14.8°$ (*c* = 0.49 DMSO).

3.5 Self- and Cross-Aldol Reactions Catalyzed by 2-Deoxy-D-ribose 5-phosphate aldolase

2-Deoxy-D-ribose 5-phosphate aldolase (RibA, or "DERA" EC 4.1.2.4.) uses acetaldehyde as the nucleophile to catalyze enzymatic self- and cross-aldol reactions [165,166]. *In vivo*, RibA catalyzes the reversible addition of acetaldehyde to D-glyceraldehyde 3-phosphate furnishing 2-deoxy-D-ribose 5-phosphate (Scheme 10.8).

Table 10.17 Aldol Addition of Glycine to Aldehydes Mediated by Threonine Aldolases

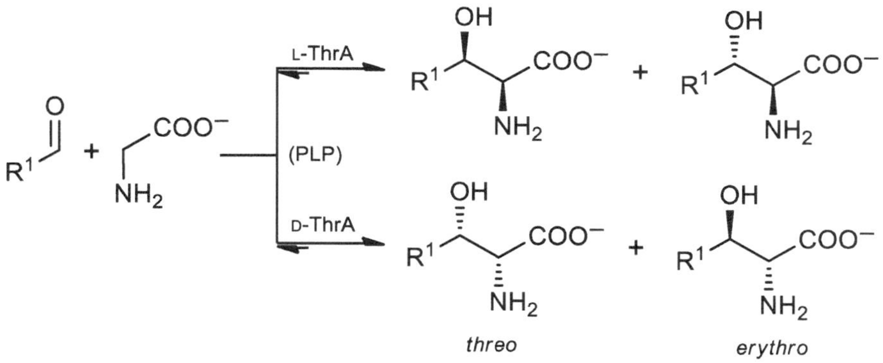

Entry	R^1	Biocatalyst	Yield	D:L	*threo:erythro*	References
1	Me	L-ThrA *E. coli*	34	>1:99	1:99	[163]
2	Me	D-ThrA *X. oryzae*	63	>99:1	50:50[a]	[163]
3	$CH(CH_3)_2$	L-ThrA *E. coli*	17	>1:99	2:94	[163]
4	$CH(CH_3)_2$	D-ThrA *X. oryzae*	49	>99:1	93:7	[163]
5	CH_2OCH_2Ph	L-ThrA *E. coli*	36	>1:99	6:94	[163]
6	CH_2OCH_2Ph	D-ThrA *X. oryzae*	80	>99:1	70:30	[163]
7	CH_2OCH_2Ph	SHMT *S. thermofilus*	40	>1:99	3:97	[158]
8	$CH_2O(CH_2)_3OCH_2Ph$	L-ThrA *C. humicola*	45	>1:99	8:92	[162]
9	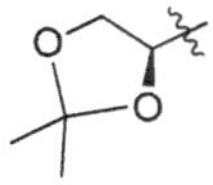	L-ThrA *E. coli*	35	>1:99	4:96	[163]
10	(*R*)-CH(Me)NHCbz	SHMT *S. thermofilus*	30	>1:99	0:100	[158]
11	(*R*)-CH(Me)NHCbz	L-ThrA *E. coli*	20	>1:99	84:16	[158]
12	(*S*)-CH(Me)NHCbz	SHMT *S. thermofilus*	30	>1:99	>2:98	[158]
13	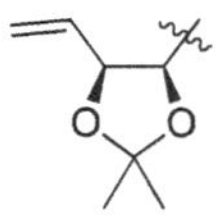	L-ThrA *C. humicola*	52	>1:99	>2:98	[164]

PLP, Pyridoxal-5'-phosphate; [Gly],concentration of glycine; [Ald], concentration of aldehyde.
[a] *Separation of both diastereoisomers by chemical phosphorylation and subsequent dephosphorylation with acid phosphatase.*

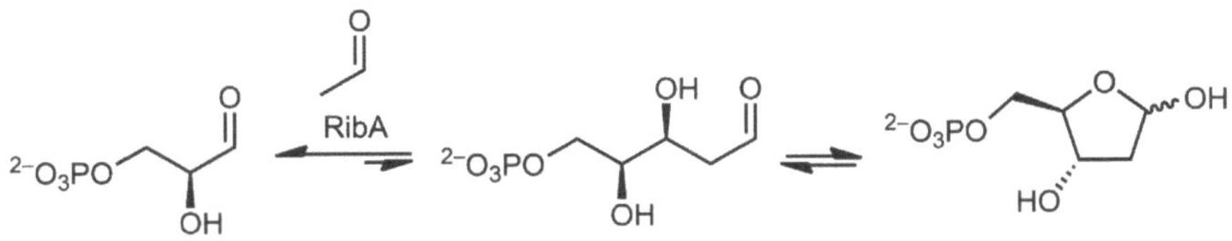

SCHEME 10.8
RibA-catalyzed the reversible aldol addition of acetaldehyde to D-glyceraldehyde-3-phosphate to furnish 2-deoxy-D-ribose 5-phosphate.

RibA also tolerates many other aldehyde acceptors, and can even utilize other donor nucleophiles, for example, propionaldehyde, acetone, and fluoroacetone, but at much lower rates, causing limitations at preparative scale (Table 10.18) [165–167]. For RibA from *E. coli*, nonphosphorylated acceptors show activities as low as 0.03–0.4% of that of the natural substrate. Therefore, the yields in some instances are low (<20%) and require large amounts of enzyme catalyst. On the other hand, 2-hydroxyaldehydes are relatively good acceptors, among which the D-isomers are better substrates than the L-isomers (Table 10.18). RibA-catalyzed reactions provide access to β-hydroxy aldehydes and ketones with the stereoselective creation of one stereogenic center.

Table 10.18 RibA-Catalyzed Aldol Addition of Acetone, Fluoroacetone, and Acetaldehyde Donors to Various Aldehydes

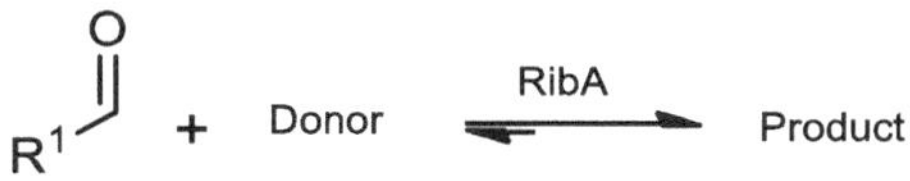

Entry	R^1	Donor	Product	Yield %	References
1	(*R*)-CH(OH) $CH_2OPO_3^{2-}$	CH_3CHO	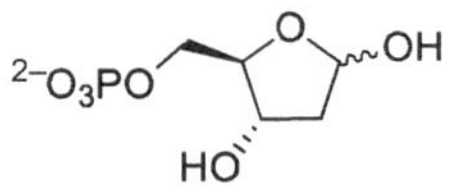	86	[165]
2	$CH(CH_3)_2$	CH_3COCH_3	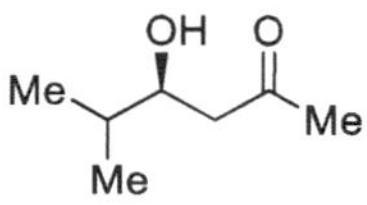	40	[165]
3	$CH(CH_3)_2$	CH_3-$COCH_2F$	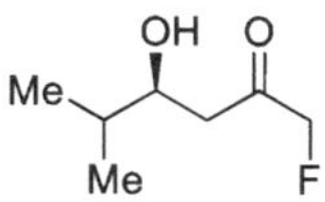	40	[165]
4	(*R*)-CH(OH)CH_2OH	CH_3CHO	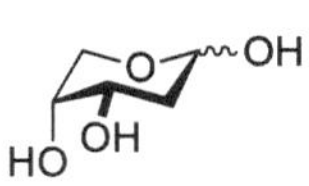	65	[166]

Table 10.18 RibA-Catalyzed Aldol Addition of Acetone, Fluoroacetone, and Acetaldehyde Donors to Various Aldehydes *(cont.)*

Entry	R^1	Donor	Product	Yield %	References
5	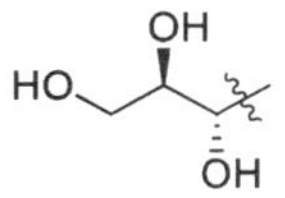	CH_3CHO	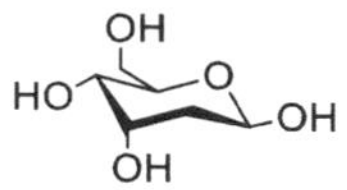	62	[166]
6	(*R*)-$CH(OH)CH_2N_3$	CH_3COCH_3	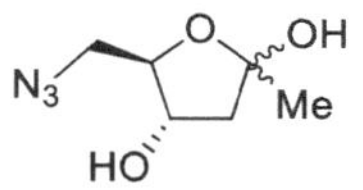	66	[166]
7	(*R*)-$CH(OH)CH_2N_3$	CH_3CO-CH_2F	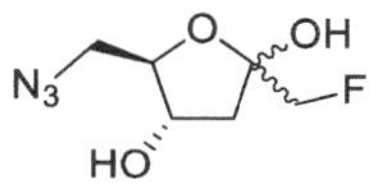	70	[166]
8	(*R*)-$CH(OH)CH_2N_3$	CH_3CHO	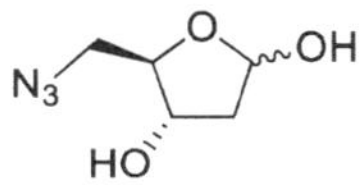	76	[166]
9	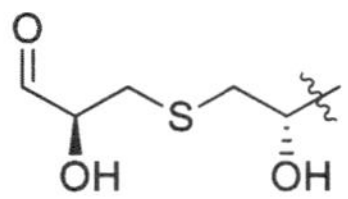	CH_3CHO	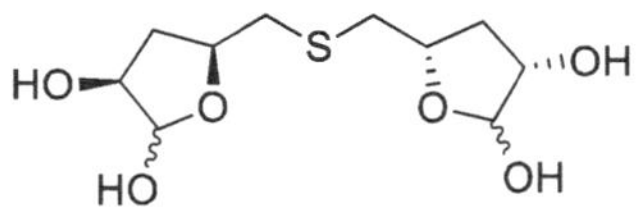	58[a]	[167]
10	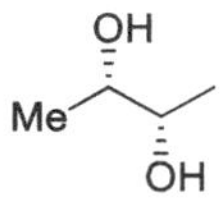	CH_3CHO	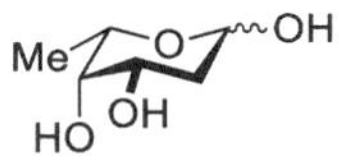	51	[167]
11	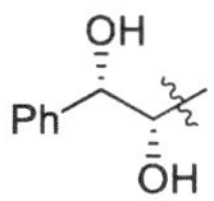	CH_3CHO	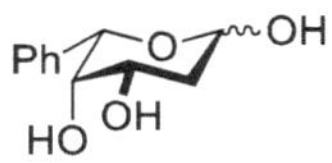	46	[167]

[a]*The substrate contains two aldehyde functions and a two-fold enzymatic tandem aldol addition reaction took place*

Example: Synthesis of 5-Azido-2,5-dideoxy-D-erythro-pentofuranose (Table 10.18, entry 8)) [166]

Acetaldehyde (132 mg, 3 mmol) and D-(*R*)-3-azido-2-hydroxypropanal (115 mg, 1 mmol) were dissolved in 0.1 M triethanolamine buffer, pH 7.3 (10 mL) containing EDTA (1 mM). To this solution, RibA (400 units, available from Sigma-Aldrich) was added. The resulting solution was stirred in the dark for 2 days under N_2. The reaction was quenched by adding 2 volumes of acetone. The mixture was then incubated in ice for 20 min and centrifuged to remove the precipitated enzyme. After removal of the solvent under reduced pressure, the residue was purified by silica gel chromatography (ethyl acetate/hexane (2:1), R_f 0.12) to give the title compound (0.76 mmol, 76% yield).

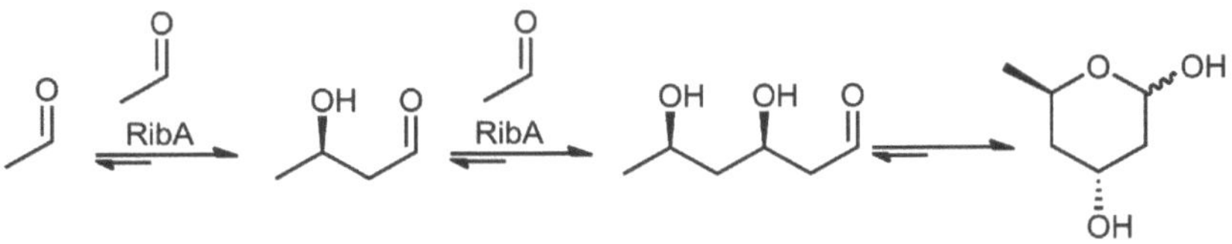

SCHEME 10.9
Two sequential aldol additions of acetaldehyde to acetaldehyde catalyzed by RibA.

3.5.1 SEQUENTIAL TWO-STEP ALDOL ADDITIONS OF ACETALDEHYDE TO ALDEHYDES

The most attractive feature of this enzyme from synthetic viewpoint is that the initial aldol adducts are aldehydes that can in turn act as acceptor in a second aldolase-catalyzed reaction. Therefore, as the FSA with glycolaldehyde, the first aldol addition furnishes another aldehyde that can be used by RibA or by other aldolases as acceptor in consecutive cascade aldol reactions (e.g., see Scheme 10.9) [5,80,167–171].

The key to a successful two-step aldol strategy is that the first aldol reaction product must not form a stable cyclic hemiacetal (Table 10.19) [167,168]. The starting aldehyde acceptor must therefore be carefully selected or strategically protected to prevent the reaction stopping after the first addition step. Higher order sequences are usually impossible after two successive additions due to

Table 10.19 RibA-Catalyzed One Pot Sequential Two-Aldol Addition of Acetaldehyde

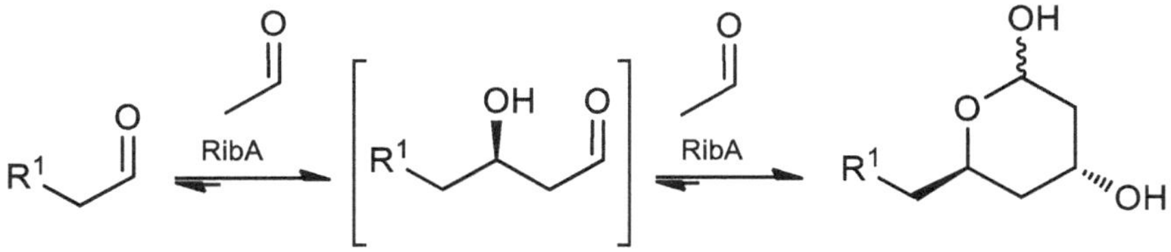

Entry	R^1	Reaction time	Yield (%)	References
1	OMe	6 days	65	[167]
2	CH_2COOH	6 days	80	[167]
3	Cl	6 days	70	[168]
4	Cl	5 h[a]	86	[172]
5	H	2 weeks	64	[170]

[a]*With a RibA F200I variant [172].*

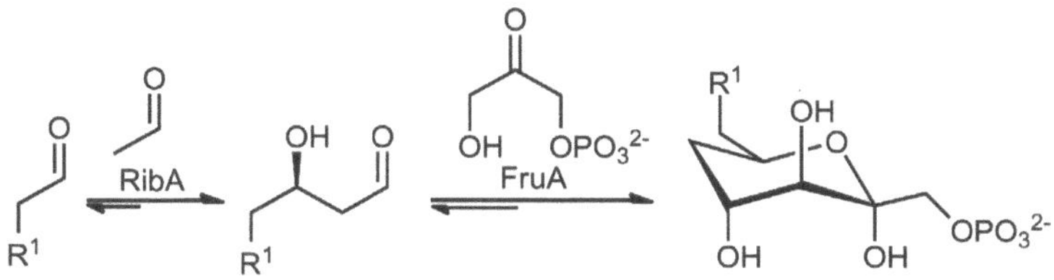

R^1: OMe; N$_3$; OMOM; Cl; OH

SCHEME 10.10
One-pot carboligation reaction of RibA catalyzed aldol addition of acetaldehyde to an acetaldehyde derivative and FruA-catalyzed aldol addition of DHAP to the aldol aldehyde generated *in situ*.

the intramolecular pyranose cyclization involving the first installed aldol hydroxyl group and the final aldehyde acceptor carbonyl. The cascade aldol reactions normally require extended reaction times and high substrate and enzyme concentrations on preparative scales. Because wild-type RibA is inactivated by high aldehyde concentrations, commercial applications awaited the development of an engineered enzyme mutants that can be used in the industrial scale synthesis of the statin precursor (3*R*,5*S*)-6-chloro-2,4,6-trideoxyhexapyranose from chloroacetaldehyde/acetaldehyde (Table 10.19, entry 4) [172].

Consecutive aldol additions by combination of RibA and other aldolases such as FruA (Scheme 10.10) or NeuA (Scheme 10.11) have also been studied for the synthesis of nonnatural sugars [169,170]. In all these reactions, the main driving force is the formation of a stable hemiketal of the target product. In the combination of RibA and FruA from rabbit muscle (RAMA), the aldehyde formed after the first RibA catalyzed aldol addition reaction underwent a second

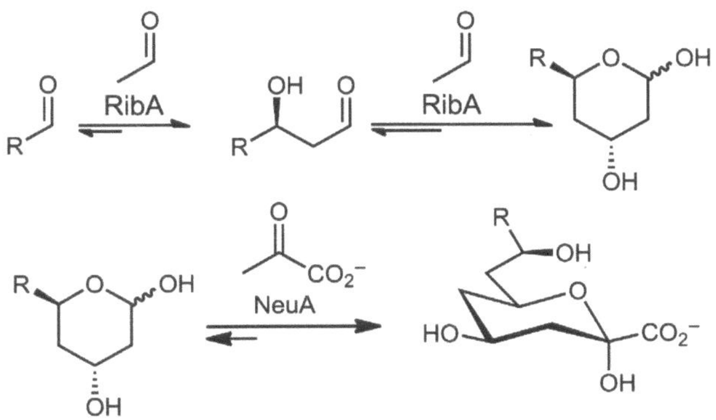

R^1: CH$_3$, CH$_2$OMe

SCHEME 10.11
Two-step carboligation reaction of RibA catalyzed aldol addition of acetaldehyde to an acetaldehyde derivative and NeuA catalyzed aldol addition of pyruvate to the aldol from RibA catalysis.

aldol addition of DHAP catalyzed by FruA to yield 5-deoxy ketose derivatives (Scheme 10.10). Initial substrate concentrations were of paramount importance. An excess of acetaldehyde gives products derived from RibA-catalyzed two-sequential aldol reactions. However, sufficient amount of acceptor aldehyde and acetaldehyde is necessary to get a high concentration of the acceptor for second FruA-catalyzed reaction, particularly when dealing with acceptors with high Km values [169].

The combination of RibA and NeuA was successfully applied in two consecutive steps (Scheme 10.11) [170]. First RibA catalyzed two consecutive aldol additions furnished exclusively the trideoxyhexoses. The pure trideoxyhexoses were then incubated with pyruvate and NeuAc aldolase to give the target products (Scheme 10.11).

Example: Preparation of 2,4,6,7-tetradeoxy-D-erythro-octapyranuronic acid (Table 10.19, entry 2) [167]

Acetaldehyde (264 mg, 6 mmol) and succinic semialdehyde (204 mg, 2 mmol) were dissolved in water (20 mL) and adjusted to pH 7.5. To this solution, RibA (1000 U available from Sigma-Aldrich) was added. After stirring in the dark for 4 days under N_2 at room temperature, the reaction was quenched by addition of acetone (40 mL), then cooled to 0°C for 20 min, and centrifuged to remove the precipitated protein. After removal of the solvent under reduced pressure, the residue was purified by flash chromatography (silica, EtOAc/MeOH, 6:l to 1:1) to give the title compound (305 mg, 80%) as a mixture of anomers (α/β 1:5).

References

[1] D. Enders, K.-E. Jaeger (Eds.), Asymmetric synthesis with chemical and biological methods, Wiley-VCH Verlag GmbH & Co KGaA, Weinheim, 2007.

[2] B.-C. Hong, R.Y. Nimje, Catalytic C—C bond formation in natural products synthesis. Highlights from the years 2000–2005, Curr. Org. Chem. 10 (2006) 2191–2225.

[3] S. Bhowmick, K.C. Bhowmick, Catalytic asymmetric carbon–carbon bond-forming reactions in aqueous media, Tetrahedron Asymmetry 22 (2011) 1945–1979.

[4] L. Yang, H. Huang, Asymmetric catalytic carbon-carbon coupling reactions via C—H bond activation, Catal. Sci. Technol. 2 (2012) 1099–1112.

[5] P. Clapés, W.-D. Fessner, Enzymatic direct aldol additions, in: G.A. Molander (Ed.), Science of Synthesis. Stereoselective Synthesis 2. Stereoselective Reactions of Carbonyl and Imino Groups, Georg Thieme Verlag KG, Suttgart (Germany), 2011, pp. 677–734.

[6] M. Müller, G.A. Sprenger, M. Pohl, CC bond formation using ThDP-dependent lyases, Curr. Opin. Chem. Biol. 17 (2013) 261–270.

[7] M. Müller, Recent developments in enzymatic asymmetric C—C bond formation, Adv. Synth. Catal. 354 (2012) 3161–3174.

[8] P. Clapés, X. Garrabou, Current trends in asymmetric synthesis with aldolases, Adv. Synth. Catal. 353 (2011) 2263–2283.

[9] W.-D. Fessner, Enzymatic asymmetric synthesis using aldolases, in: R.N. Patel (Ed.), Stereoselective Biocatalalysis, Marcel Dekker, Inc, New York, 2000, pp. 239–265.

[10] J. Brussee, A. Van Der Gen, Biocatalysis in the enantioselective formation of chiral cyanohydrins, valuable building blocks in organic synthesis, in: R.N. Patel (Ed.), Stereoselective Biocatalysis, Marcel Dekker, Inc, New York, USA, 2000, pp. 289–320.

[11] F. Effenberger, Hydroxynitrile lyases in stereoselective synthesis, in: R.N. Patel (Ed.), Stereoselective Biocatalysis, Marcel Dekker, Inc, New York, USA, 2000, pp. 321–342.

[12] O.P. Ward, A. Singh, Enzymatic asymmetric synthesis by decarboxylases, Curr. Opin. Biotechnol. 11 (2000) 520–526.

[13] O.P. Ward, M.V. Baev, Decarboxylases in stereoselective catalysis, in: N.P. Ramesh (Ed.), Stereoselective Biocatalysis, Marcel Dekker, Inc, New York, USA, 2000, pp. pp267–287.

[14] M. Pohl, C. Dresen, M. Beigi, M. Müller, Acyloin and benzoin condensations, in: K. Drauz, H. Groeger, O. May (Eds.), Enzyme Catalysis in Organic Synthesis, third ed., Wiley-VCH Verlag GmbH & Co. KGaA, Weinheim, Germany, 2011, pp. 919–945.

[15] L. Hilterhaus, A. Liese, Industrial application and processes using carbon-carbon lyases, in: K. Drauz, H. Groger, O. May (Eds.), Enzyme Catalysis in Organic Synthesis (3rd), Wiley-VCH Verlag GmbH & Co. KGaA, Weinheim, Germany, 2011, pp. 991–1000.

[16] M. Gruber-Khadjawi, M.H. Fechter, H. Griengl, Cleavage and formation of cyanohydrins, in: K. Drauz, H. Groeger, O. May (Eds.), Enzyme Catalysis in Organic Synthesis, third ed., Wiley-VCH Verlag GmbH & Co. KGaA, Weinheim, Germany, 2011, pp. 947–990.

[17] W.-D. Fessner, Aldol reactions, in: K. Drauz, H. Groger, O. May (Eds.), Enzyme Catalysis in Organic Synthesis (3rd), Wiley-VCH Verlag GmbH & Co. KGaA, Weinheim, Germany, 2011, pp. 857–917.

[18] R.J.H. Gregory, Cyanohydrins in nature and the laboratory: biology, preparations, and synthetic applications, Chem. Rev. 99 (1999) 3649–3682.

[19] T. Purkarthofer, W. Skranc, C. Schuster, H. Griengl, Potential and capabilities of hydroxynitrile lyases as biocatalysts in the chemical industry, Appl. Microbiol. Biotechnol. 76 (2007) 309–320.

[20] E. Wehtje, P. Adlercreutz, B. Mattiasson, Formation of C—C bonds by mandelonitrile lyase in organic solvents, Biotechnol. Bioeng. 36 (1990) 39–46.

[21] M. Persson, D. Costes, E. Wehtje, P. Adlercreutz, Effects of solvent, water activity and temperature on lipase and hydroxynitrile lyase enantioselectivity, Enzyme Microb. Technol. 30 (2002) 916–923.

[22] D. Costes, E. Wehtje, P. Adlercreutz, Hydroxynitrile lyase-catalyzed synthesis of cyanohydrins in organic solvents: Parameters influencing activity and enantiospecificity, Enzym. Microb. Technol. 25 (1999) 384–391.

[23] H. Griengl, H. Schwab, M. Fechter, The synthesis of chiral cyanohydrins by oxynitrilases, Trends Biotech 18 (2000) 252–256.

[24] M. Schmidt, H. Griengl, Oxynitrilases: from cyanogenesis to asymmetric synthesis, in: W.-D. Fessner, A. Archelas, D.C. Demirjian, R. Furstoss, H. Griengl, K.E. Jaeger, E. Morís-Varas, R. Öhrlein, M.T. Reetz, J.L. Reymond, M. Schmidt, S. Servi, P.C. Shah, W. Tischer, F. Wedekind (Eds.), Biocatalysis – From Discovery to Application, Springer Berlin Heidelberg, Berlin, Germany, 1999, pp. 193–226.

[25] M. Dadashipour, Y. Asano, Hydroxynitrile lyases: insights into biochemistry, discovery, and engineering, ACS Catal. 1 (2011) 1121–1149.

[26] F. Effenberger, T. Ziegler, S. Förster, Enzyme-catalyzed cyanohydrin synthesis in organic solvents, Angew. Chem. Int. 26 (1987) 458–460.

[27] P. Zandbergen, J. Van Der Linden, J. Brussee, A. Van Der Gen, Synthesis of optically active cyanohydrins using almond meal, Synth. Commun. 21 (1991) 1387–1391.

[28] T.T. Huuhtanen, L.T. Kanerva, Enzyme-catalyzed synthesis of optically active aliphatic cyanohydrins, Tetrahedron Asymmetry 3 (1992) 1223–1226.

[29] F. Effenberger, Synthesis and reactions of optically active cyanohydrins, Angew. Chem. Int. 33 (1994) 1555–1564.

[30] F. Effenberger, S. Forster, H. Wajant, Hydroxynitrile lyases in stereoselective catalysis, Curr. Opin. Biotechnol. 11 (2000) 532–539.

[31] U. Niedermeyer, M.-R. Kula, Enzyme-catalyzed synthesis of (*S*)-cyanohydrins, Angew. Chem. Int. 29 (1990) 386–387.

[32] M. Schmidt, S. Hervé, N. Klempier, H. Griengl, Preparation of optically active cyanohydrins using the (*S*)-hydroxynitrile lyase from *Hevea brasiliensis*, Tetrahedron 52 (1996) 7833–7840.

[33] H. Griengl, A.V. Hickel, D. Johnson, M. Schmidt, C. Kratky, H. Schwab, Enzymatic cleavage and formation of cyanohydrins: a reaction of biological and synthetic relevance, Chem. Commun. (1997) 1933–1940.

[34] S. Förster, J. Roos, F. Effenberger, H. Wajant, A. Sprauer, The first recombinant hydroxynitrile lyase and its application in the synthesis of (*S*)-cyanohydrins, Angew. Chem. Int. 35 (1996) 437–439.

[35] F. Effenberger, B. Hörsch, S. Förster, T. Ziegler, Enzyme-catalyzed synthesis of (*S*)-cyanohydrins and subsequent hydrolysis to (*S*)-α-hydroxy-carboxylic acids, Tetrahedron Lett. 31 (1990) 1249–1252.

[36] H. Griengl, N. Klempier, P. Pöchlauer, M. Schmidt, N. Shi, A.A. Zabelinskaja-Mackova, Enzyme catalyzed formation of (*S*)-cyanohydrins derived from aldehydes and ketones in a biphasic solvent system, Tetrahedron 54 (1998) 14477–14486.

[37] F. Effenberger, J. Eichhorn, Stereoselective synthesis of thienyl and furyl analogues of ephedrine, Tetrahedron Asymmetry 8 (1997) 469–476.

[38] E. Kiljunen, L.T. Kanerva, Novel (*R*)-oxynitrilase sources for the synthesis of (*R*)-cyanohydrins in diisopropyl ether, Tetrahedron Asymmetry 8 (1997) 1225–1234.

[39] E. Kiljunen, L.T. Kanerva, Approach to(*R*)- and (*S*)-ketone cyanohydrins using almond and apple meal as the source of (*R*)-oxynitrilase, Tetrahedron Asymmetry 8 (1997) 1551–1557.

[40] H. Bühler, F. Effenberger, S. Förster, J. Roos, H. Wajant, Substrate specificity of mutants of the hydroxynitrile lyase from *Manihot esculenta*, Chembiochem 4 (2003) 211–216.

[41] M. Elliott, A.W. Farnham, N.F. Janes, P.H. Needham, D.A. Pulman, Synthetic insecticide with a new order of activity, Nature 248 (1974) 710–711.

[42] F. Effenberger, S. Heid, (*R*)-Oxynitrilase catalyzed synthesis of (*R*)-ketone cyanohydrins, Tetrahedron Asymmetry 6 (1995) 2945–2952.

[43] H. Bühler, A. Bayer, F. Effenberger, A convenient synthesis of optically active 5, 5-disubstituted 4-amino- and 4-hydroxy-2(5H)-furanones from (S)-ketone cyanohydrins, Chem. Eur. J. 6 (2000) 2564–2571.

[44] M. Avi, M.H. Fechter, K. Gruber, F. Belaj, P. Pöchlauer, H. Griengl, Hydroxynitrile lyase catalyzed synthesis of heterocyclic (*R*)- and (*S*)-cyanohydrins, Tetrahedron 60 (2004) 10411–10418.

[45] J. Holt, U. Hanefeld, Enantioselective enzyme-catalyzed synthesis of cyanohydrins, Curr. Org. Synth. 6 (2009) 15–37.

[46] J.N. Andexer, J.V. Langermann, U. Kragl, M. Pohl, How to overcome limitations in biotechnological processes – examples from hydroxynitrile lyase applications, Trends Biotechnol. 27 (2009) 599–607.

[47] C. Roberge, F. Fleitz, D. Pollard, P. Devine, Synthesis of optically active cyanohydrins from aromatic ketones: evidence of an increased substrate range and inverted stereoselectivity for the hydroxynitrile lyase from *Linum usitatissimum*, Tetrahedron Asymmetry 18 (2007) 208–214.

[48] M. Müller, D. Gocke, M. Pohl, Thiamin diphosphate in biological chemistry: exploitation of diverse thiamin diphosphate-dependent enzymes for asymmetric chemoenzymatic synthesis, Febs J. 276 (2009) 2894–2904.

[49] M. Pohl, B. Lingen, M. Müller, Thiamin-diphosphate-dependent enzymes: new aspects of asymmetric C—C bond formation, Chem. Eur. J. 8 (2002) 5288–5295.

[50] M. Knoll, M. Müller, J. Pleiss, M. Pohl, Factors mediating activity, selectivity, and substrate specificity for the thiamin diphosphate-dependent enzymes benzaldehyde lyase and benzoylformate decarboxylase, Chembiochem 7 (2006) 1928–1934.

[51] C. Dresen, M. Richter, M. Pohl, S. Lüdeke, M. Müller, The enzymatic asymmetric conjugate umpolung reaction, Angew. Chem. Int. 49 (2010) 6600–6603.

[52] P. Lehwald, R. Michael, R. Caroline, L. Hung-Wen, M. Michael, Enantioselective intermolecular aldehyde–ketone cross-coupling through an enzymatic carboligation reaction, Angew. Chem. Int. 49 (2010) 2389–2392.

[53] A. Kurutsch, M. Richter, V. Brecht, G.A. Sprenger, M. Müller, MenD as a versatile catalyst for asymmetric synthesis, J. Mol. Catal. B Enzym. 61 (2009) 56–66.

[54] G. Satianegara, M. Breuer, B. Hauer, P. Rogers, B. Rosche, Enzymatic (*R*)-phenylacetylcarbinol production in a benzaldehyde emulsion system with *Candida utilis* cells, Appl. Microbiol. Biotechnol. 70 (2006) 170–175.

[55] H. Iding, T. Dünnwald, L. Greiner, A. Liese, M. Müller, P. Siegert, J. Grötzinger, A.S. Demir, M. Pohl, Benzoylformate decarboxylase from *Pseudomonas putida* as stable catalyst for the synthesis of chiral 2-hydroxy Ketones, Chem. Eur. J. 6 (2000) 1483–1495.

[56] A.S. Demir, T. Dünnwald, H. Iding, M. Pohl, M. Müller, Asymmetric benzoin reaction catalyzed by benzoylformate decarboxylase, Tetrahedron Asymmetry 10 (1999) 4769–4774.

[57] A.S. Demir, Ö. Şeşenoglu, P. Dünkelmann, M. Müller, Benzaldehyde lyase-catalyzed enantioselective carboligation of aromatic aldehydes with mono- and dimethoxy acetaldehyde, Org. Lett. 5 (2003) 2047–2050.

[58] P. Ayhan, I. Simsek, B. Cifci, A.S. Demir, Benzaldehyde lyase catalyzed enantioselective self and cross condensation reactions of acetaldehyde derivatives, Org. Biomol. Chem. 9 (2011) 2602–2605.

[59] P.D. De Maria, M. Pohl, D. Gocke, H. Groger, H. Trauthwein, T. Stillger, L. Walter, M. Muller, Asymmetric synthesis of aliphatic 2-hydroxy ketones by enzymatic carboligation of aldehydes, Eur. J. Org. Chem. (2007) 2940–2944.

[60] A. Cosp, C. Dresen, M. Pohl, L. Walter, C. Röhr, M. Müller, α,β-Unsaturated aldehydes as substrates for asymmetric C—C bond forming reactions with thiamin diphosphate (ThDP)-dependent enzymes, Adv. Synth. Catal 350 (2008) 759–771.

[61] M. Pohl, G.A. Sprenger, M. Müller, A new perspective on thiamine catalysis, Curr. Opin. Biotechnol. 15 (2004) 335–342.

[62] Z. Guo, A. Goswami, V.B. Nanduri, R.N. Patel, Asymmetric acyloin condensation catalyzed by phenylpyruvate decarboxylase. Part 2. Substrate specificity and purification of the enzyme, Tetrahedron Asymmetry 12 (2001) 571–577.

[63] T. Dünnwald, A.S. Demir, P. Siegert, M. Pohl, M. Müller, Enantioselective synthesis of (S)-2-hydroxypropanone derivatives by benzoylformate decarboxylase catalyzed C–C bond formation, Eur. J. Org. Chem. 2000 (2000) 2161–2170.

[64] T. Dünnwald, M. Müller, Stereoselective formation of *bis*(α-hydroxy ketones) via enzymatic carboligation, J. Org. Chem. 65 (2000) 8608–8612.

[65] P. Dünkelmann, D. Kolter-Jung, A. Nitsche, A.S. Demir, P. Siegert, B. Lingen, M. Baumann, M. Pohl, M. Müller, Development of a donor–acceptor concept for enzymatic cross-coupling reactions of aldehydes: the first asymmetric cross-benzoin condensation, J. Am. Chem. Soc. 124 (2002) 12084–12085.

[66] N.J. Turner, Applications of transketolases in organic synthesis, Curr. Opin. Biotechnol. 11 (2000) 527–531.

[67] S. Takayama, G.J. Mcgarvey, C.-H. Wong, Microbial aldolases and transketolases: new biocatalytic approaches to simple and complex sugars, Annu. Rev. Microbiol. 51 (1997) 285–310.

[68] Y. Kobori, D.C. Myles, G.M. Whitesides, Substrate specificity and carbohydrate synthesis using transketolase, J. Org. Chem. 57 (1992) 5899–5907.

[69] K.G. Morris, M.E.B. Smith, N.J. Turner, M.D. Lilly, R.K. Mitra, J.M. Woodley, Transketolase from *Escherichia coli*: a practical procedure for using the biocatalyst for asymmetric carbon–carbon bond synthesis, Tetrahedron Asymmetry 7 (1996) 2185–2188.

[70] J. Bolte, C. Demuynck, H. Samaki, Utilization of enzymes in organic chemistry: transketolase catalyzed synthesis of ketoses, Tetrahedron Lett. 28 (1987) 5525–5528.

[71] C. Demuynck, J. Bolte, L. Hecquet, H. Samaki, Enzymes as reagents in organic chemistry: transketolase-catalyzed synthesis of D-[1,2-$^{13}C_2$]xylulose, Carbohydr. Res. 206 (1990) 79–85.

[72] F. Effenberger, V. Null, T. Ziegler, Preparation of optically pure L-2-hydroxyaldehydes with yeast transketolase, Tetrahedron Lett. 33 (1992) 5157–5160.

[73] D.C. Myles, P.J. Andrulis, G.M. Whitesides, A transketolase-based synthesis of (+)-exo-brevicomin, Tetrahedron Lett. 32 (1991) 4835–4838.

[74] A.J. Humphrey, S.F. Parsons, M.E.B. Smith, N.J. Turner, Synthesis of a novel *N*-hydroxypyrrolidine using enzyme catalyzed asymmetric carbon-carbon bond synthesis, Tetrahedron Lett. 41 (2000) 4481–4485.

[75] C.-H. Wong, Engineering enzymes for chemoenzymatic synthesis. Part I: practical routes to aza-sugars and complex carbohydrates, Trends Biotechnol. 10 (1992) 337–341.

[76] H.J.M. Gijsen, L. Qiao, W. Fitz, C.-H. Wong, Recent advances in the chemoenzymic synthesis of carbohydrates and carbohydrate mimetics, Chem. Rev. 96 (1996) 443–473.

[77] W.-D. Fessner, C. Walter, Enzymic C—C bond formation in asymmetric synthesis, Top. Curr. Chem. 184 (1997) 97–194.

[78] W.-D. Fessner, Enzyme mediated C—C bond formation, Curr. Opin. Chem. Biol. 2 (1998) 85–97.

[79] W.-D. Fessner, V. Helaine, Biocatalytic synthesis of hydroxylated natural products using aldolases and related enzymes, Curr. Opin. Biotechnol. 12 (2001) 574–586.

[80] W.-D. Fessner, Biocatalytic C—C bond formation in asymmetric synthesis, in: D. Enders, K.-E. Jaeger (Eds.), Asymmetric Synthesis with chemical and biological methods, Wiley-VCH Verlag GmbH & Co KGaA, Weinheim, 2007, pp. pp351–375.

[81] P. Clapés, W.-D. Fessner, G.A. Sprenger, A.K. Samland, Recent progress in stereoselective synthesis with aldolases, Curr. Opin. Chem. Biol. 14 (2010) 154–167.

[82] P. Clapés, J. Joglar, Enzyme-Catalyzed Aldol Additions, in: R. Mahrwald (Ed.), Modern Methods in Stereoselective Aldol Reactions, John Wiley & Sons Ltd, 2013, pp. 475–528.

[83] R.N. Goldberg, Y.B. Tewari, Thermodynamics of enzyme-catalyzed reactions: part 4. Lyases, J. Phys. Chem. Ref. Data 24 (1995) 1669–1698.

[84] H.L. Arth, W.-D. Fessner, Practical synthesis of 4-hydroxy-3-oxobutylphosphonic acid and its evaluation as a bio-isosteric substrate of DHAP aldolase, Carbohydr. Res. 305 (1998) 313–321.

[85] M. Schuster, W.F. He, S. Blechert, Chemical-enzymatic synthesis of azasugar phosphonic acids as glycosyl phosphate surrogates, Tetrahedron Lett. 42 (2001) 2289–2291.

[86] W.-D. Fessner, G. Sinerius, Enzymes in organic synthesis. 7. Synthesis of dihydroxyacetone phosphate (and isosteric analogs) by enzymic oxidation: sugars from glycerol, Angew. Chem. Int. 33 (1994) 209–212.

[87] R. Duncan, D.G. Drueckhammer, A pseudo-isomerization route to aldose sugars using aldolase catalysis, J. Org. Chem. 60 (1995) 7394–7395.

[88] R. Duncan, D.G. Drueckhammer, Preparation of deoxy sugars via aldolase-catalyzed synthesis of 1-deoxy-1-thioketoses, J. Org. Chem. 61 (1996) 438–439.

[89] M. Schümperli, R. Pellaux, S. Panke, Chemical and enzymatic routes to dihydroxyacetone phosphate, Appl. Microbiol. Biotechnol. 75 (2007) 33–45.

[90] J.R. Durrwachter, D.G. Drueckhammer, K. Nozaki, H.M. Sweers, C.-H. Wong, Enzymic aldol condensation/isomerization as a route to unusual sugar derivatives, J. Am. Chem. Soc. 108 (1986) 7812–7818.

[91] D.G. Drueckhammer, J.R. Durrwachter, R.L. Pederson, D.C. Crans, L. Daniels, C.-H. Wong, Reversible and *in situ* formation of organic arsenates and vanadates as organic phosphate mimics in enzymatic reactions: mechanistic investigation of aldol reactions and synthetic applications, J. Org. Chem. 54 (1989) 70–77.

[92] R. Schoevaart, F. Vanrantwijk, R.A. Sheldon, Facile enzymatic aldol reactions with dihydroxyacetone in the presence of arsenate, J. Org. Chem. 66 (2001) 4559–4562.

[93] R. Lagunas, A. Sols, Arsenate induced activity of certain enzymes on their dephosphorylated subtrates, FEBS Lett. 1 (1968) 32–34.

[94] M. Sugiyama, Z.Y. Hong, L.J. Whalen, W.A. Greenberg, C.-H. Wong, Borate as a phosphate ester mimic in aldolase-catalyzed reactions: practical synthesis of L-fructose and L-iminocyclitols, Adv. Synth. Catal. 348 (2006) 2555–2559.

[95] X. Garrabou, J. Calveras, J. Joglar, T. Parella, J. Bujons, P. Clapés, Highly efficient aldol additions of DHA and DHAP to N-Cbz-amino aldehydes catalyzed by L-rhamnulose-1-phosphate and L-fuculose-1-phosphate aldolases in aqueous borate buffer, Org. Biomol. Chem. 9 (2011) 8430–8436.

[96] L. Espelt, P. Clapés, J. Esquena, A. Manich, C. Solans, Enzymatic carbon–carbon bond formation in water-in-oil highly concentrated (gel emulsions), Langmuir 19 (2003) 1337–1346.

[97] L. Espelt, T. Parella, J. Bujons, C. Solans, J. Joglar, A. Delgado, P. Clapés, Stereoselective aldol additions catalyzed by dihydroxyacetone phosphate dependent aldolases in emulsion systems: preparation and structural characterization of linear and cyclic aminopolyols from aminoaldehydes, Chem. Eur. J. 9 (2003) 4887–4899.

[98] J. Calveras, M. Egido-Gabás, L. Gómez, J. Casas, T. Parella, J. Joglar, J. Bujons, P. Clapés, Dihydroxyacetone phosphate aldolase-catalyzed synthesis of structurally diverse polyhydroxylated pyrrolidine derivatives and evaluation of their glycosidase inhibitory properties, Chem. Eur. J. 15 (2009) 7310–7328.

[99] C.H. Von Der Osten, A.J. Sinskey, C.F. Barbas III, R.L. Pederson, Y.F. Wang, C.-H. Wong, Use of a recombinant bacterial fructose-1,6-diphosphate aldolase in aldol reactions: preparative syntheses of 1-deoxynojirimycin, 1-deoxymannojirimycin, 1,4-dideoxy-1,4-imino-D-arabinitol, and fagomine, J. Am. Chem. Soc. 111 (1989) 3924–3927.

[100] I. Henderson, K. Laslo, C.-H. Wong, Chemoenzymic synthesis of homoazasugars, Tetrahedron Lett. 35 (1994) 359–362.

[101] A. Romero, C.-H. Wong, Chemo-enzymatic total synthesis of 3-epiaustraline, australine, and 7-epialexine, J. Org. Chem. 65 (2000) 8264–8268.

[102] X. Garrabou, L. Gomez, J. Joglar, S. Gil, T. Parella, J. Bujons, P. Clapés, Structure-guided minimalist redesign of L-fuculose-1-phosphate aldolase active site. Expedient synthesis of novel polyhydroxylated pyrrolizidines and their inhibitory properties against glycosidases and intestinal disaccharidases, Chem. Eur. J. 16 (2010) 10691–10706.

[103] W.-D. Fessner, G. Sinerius, A. Schneider, M. Dreyer, G.E. Schulz, J. Badia, J. Aguilar, Enzymes in organic synthesis. Part 1. Diastereoselective, enzymatic aldol addition with L-rhamnulose- and L-fuculose-1-phosphate aldolases from *E. coli*, Angew. Chem. Int. 30 (1991) 555–558.

[104] P. Page, C. Blonski, J. Perie, An improved chemical and enzymic synthesis of new fructose derivatives for import studies by the glucose transporter in parasites, Tetrahedron 52 (1996) 1557–1572.

[105] W.-D. Fessner, C. Walter, Enzymes in organic synthesis. 3. Artificial metabolism applied to the one-pot asymmetric synthesis of branched saccharides, Angew. Chem. Int. 31 (1992) 614–616.

[106] W.-C. Chou, L. Chen, J.-M. Fang, C.-H. Wong, A new route to deoxythio sugars based on aldolases, J. Am. Chem. Soc. 116 (1994) 6191–6194.

[107] L. El Blidi, M. Ahbala, J. Bolte, M. Lemaire, Straightforward chemo-enzymatic synthesis of new aminocyclitols, analogues of valiolamine and their evaluation as glycosidase inhibitors, Tetrahedron Asymmetry 17 (2006) 2684–2688.

[108] L. El Blidi, D. Crestia, E. Gallienne, C. Demuynck, J. Bolte, M. Lemaire, A straightforward synthesis of an aminocyclitol based on an enzymatic aldol reaction and a highly stereoselective intramolecular Henry reaction, Tetrahedron Asymmetry 15 (2004) 2951–2954.

[109] L. El Blidi, Z. Assaf, F.C. Bres, H. Veschambre, V. Thery, J. Bolte, M. Lemaire, Fructose-1,6-bisphosphate aldolase-mediated synthesis of aminocyclitols (analogues of valiolamine) and their evaluation as glycosidase inhibitors, Chem. Catal. Chem. 1 (2009) 463–471.

[110] M. Petersen, M.T. Zannetti, W.-D. Fessner, Tandem asymmetric C—C bond formations by enzyme catalysis, Top. Curr. Chem. 186 (1997) 87–117.

[111] O. Eyrisch, W.-D. Fessner, Enzymes in organic synthesis. 10. Disaccharide mimetics by enzymic tandem aldol additions. Angew, Chem. Int. 34 (1995) 1639–1641.

[112] R.L. Pederson, M.J. Kim, C.-H. Wong, A combined chemical and enzymic procedure for the synthesis of 1-deoxynojirimycin and 1-deoxymannojirimycin, Tetrahedron Lett. 29 (1988) 4645–4648.

[113] H.U. Bergmeyer, E. Bernt, Fructose-1, 6-diphosphate Aldolase UV Assay, Manuel Method, in: H.U. Bergmeyer (Ed.), Methods of Enzymatic Analysis, Second Edition, Academic Press, New York, USA, 1974, pp. 1100–1105.

[114] M. Schürmann, G.A. Sprenger, Fructose-6-phosphate aldolase is a novel class I aldolase from *Escherichia coli* and is related to a novel group of bacterial transaldolases, J. Biol. Chem. 276 (2001) 11055–11061.

[115] S. Schneider, T. Sandalova, G. Schneider, G.A. Sprenger, A.K. Samland, Replacement of a phenylalanine by a tyrosine in the active site confers fructose 6-phosphate aldolase activity to the transaldolase of *Escherichia coli* and human origin, J. Biol. Chem. 283 (2008) 30064–30072.

[116] X. Garrabou, J.A. Castillo, C. Guérard-Hélaine, T. Parella, J. Joglar, M. Lemaire, P. Clapés, Asymmetric self- and cross-aldol reaction of glycolaldehyde catalyzed by D-fructose-6-phosphate aldolase, Angew. Chem. Int. 48 (2009) 5521–5525.

[117] M. Sugiyama, Z. Hong, P.H. Liang, S.M. Dean, L.J. Whalen, W.A. Greenberg, C.-H. Wong, D-Fructose-6-phosphate aldolase-catalyzed-one-pot synthesis of iminocyclitols, J. Am. Chem. Soc. 129 (2007) 14811–14817.

[118] A.L. Concia, C. Lozano, J.A. Castillo, T. Parella, J. Joglar, P. Clapés, D-Fructose-6-phosphate aldolase in organic synthesis: cascade chemical-enzymatic preparation of sugar-related polyhydroxylated compounds, Chem. Eur. J. 15 (2009) 3808–3816.

[119] M. Gutierrez, T. Parella, J. Joglar, J. Bujons, P. Clapés, Structure-guided redesign of D-fructose-6-phosphate aldolase from *E. coli*: remarkable activity and selectivity towards acceptor substrates by two-point mutation, Chem. Commun. 47 (2011) 5762–5764.

[120] J.A. Castillo, J. Calveras, J. Casas, M. Mitjans, M.P. Vinardell, T. Parella, T. Inoue, G.A. Sprenger, J. Joglar, P. Clapés, Fructose-6-phosphate aldolase in organic synthesis: preparation of D-fagomine, *N*-alkylated derivatives, and preliminary biological assays, Org. Lett. 8 (2006) 6067–6070.

[121] J.A. Castillo, C. Guérard-Hélaine, M. Gutiérrez, X. Garrabou, M. Sancelme, M. Schürmann, T. Inoue, V. Hélaine, F. Charmantray, T. Gefflaut, L. Hecquet, J. Joglar, P. Clapés, G.A. Sprenger, M. Lemaire, A mutant D-fructose-6-phosphate aldolase (Ala129Ser) with improved affinity towards dihydroxyacetone for the synthesis of polyhydroxylated compounds, Adv. Synth. Catal. 352 (2010) 1039–1046.

[122] M. Rale, S. Schneider, G.A. Sprenger, A.K. Samland, W.-D. Fessner, Broadening deoxysugar glycodiversity: natural and engineered transaldolases unlock a complementary substrate space, Chem. Eur. J. 17 (2011) 2623–2632.

[123] W. Wang, S.Y.K. Seah, Purification and biochemical characterization of a pyruvate-specific class II aldolase, HpaI. Biochemistry 44 (2005) 9447–9455.

[124] H. Ling, G. Wang, Y. Tian, G. Liu, H. Tan, SanM catalyzes the formation of 4-pyridyl-2-oxo-4-hydroxyisovalerate in nikkomycin biosynthesis by interacting with SanN, Biochem. Biophys. Res. Commun. 361 (2007) 196–201.

[125] M.J. Kim, W.J. Hennen, H.M. Sweers, C.-H. Wong, Enzymes in carbohydrate synthesis: *N*-acetylneuraminic acid aldolase catalyzed reactions and preparation of *N*-acetyl-2-deoxy-D-neuraminic acid derivatives, J. Am. Chem. Soc. 110 (1988) 6481–6486.

[126] C. Augé, C. Gautheron, S. David, A. Malleron, B. Cavayé, B. Bouxom, Sialyl aldolase in organic synthesis: from the trout egg acid 3-deoxy-D-glycero-D-galacto-2-nonulosonic acid (KDN) to branched-chain higher ketoses as possible new chirons, Tetrahedron 46 (1990) 201–214.

[127] C. Gautheron-Le Narvor, Y. Ichikawa, C.-H. Wong, A complete change of stereoselectivity in sialic acid aldolase reactions: a novel synthetic route to the KDO type of nine-carbon L-sugars, J. Am. Chem. Soc. 113 (1991) 7816–7818.

[128] C.H. Lin, T. Sugai, R.L. Halcomb, Y. Ichikawa, C.-H. Wong, Unusual stereoselectivity in sialic acid aldolase-catalyzed aldol condensations: synthesis of both enantiomers of high-carbon monosaccharides, J. Am. Chem. Soc. 114 (1992) 10138–10145.

[129] U. Kragl, A. Goedde, C. Wandrey, N. Lubin, C. Auge, New synthetic applications of sialic acid aldolase, a useful catalyst for KDO synthesis. Relation between substrate conformation and enzyme stereoselectivity, J. Chem. Soc. Perkin Trans. 1 (1994) 119–124.

[130] W. Fitz, J.-R. Schwark, C.-H. Wong, Aldotetroses and C(3)-modified aldohexoses as substrates for *N*-acetylneuraminic acid aldolase: a model for the explanation of the normal and the inversed stereoselectivity, J. Org. Chem. 60 (1995) 3663–3670.

[131] C. Auge, S. David, C. Gautheron, A. Malleron, B. Cavaye, Preparation of six naturally occurring sialic acids with immobilized acylneuraminate pyruvate lyase, New J. Chem. 12 (1988) 733–744.

[132] U. Kragl, D. Gygax, O. Ghisalba, C. Wandrey, Enzymatic, two-step synthesis of *N*-acetylneuraminic acid in a enzyme-membrane reactor, Angew. Chem. 103 (1991) 854–855.

[133] C. Augé, B. Bouxom, B. Cavayé, C. Gautheron, Scope and limitations of the aldol condensation catalyzed by immobilized acylneuraminate pyruvate lyase, Tetrahedron Lett. 30 (1989) 2217–2220.

[134] M.A. Sparks, K.W. Williams, C. Lukacs, A. Schrell, G. Priebe, A. Spaltenstein, G.M. Whitesides, Synthesis of potential inhibitors of hemagglutination by influenza virus: chemoenzymic preparation of N-5 analogs of *N*-acetylneuraminic acid, Tetrahedron 49 (1993) 1–12.

[135] J.L.C. Liu, G.J. Shen, Y. Ichikawa, J.F. Rutan, G. Zapata, W.F. Vann, C.-H. Wong, Overproduction of CMP-sialic acid synthetase for organic synthesis, J. Am. Chem. Soc. 114 (1992) 3901–3910.

[136] P. Zhou, H.M. Salleh, J.F. Honek, Facile chemoenzymic synthesis of 3-(hydroxymethyl)-6-epicastanospermine, J. Org. Chem. 58 (1993) 264–266.

[137] S. Huang, H. Yu, X. Chen, Disaccharides as sialic acid aldolase substrates: synthesis of disaccharides containing a sialic acid at the reducing end, Angew. Chem. Int. 46 (2007) 2249–2253.

[138] H. Yu, S. Huang, H. Chokhawala, M. Sun, H. Zheng, X. Chen, Highly efficient chemoenzymatic synthesis of naturally occurring and non-natural α-2,6-linked sialosides: a *P. damsela* α-2,6-sialyltransferase with extremely flexible donor-substrate specificity, Angew. Chem. Int. 45 (2006) 3938–3944.

[139] M.A. Ghalambor, E.C. Heath, The biosynthesis of cell wall lipopolysaccharide in *Escherichia coli*. V. Purification and properties of 3-deoxy-D-manno-octulosonate aldolase, J. Biol. Chem. 241 (1966) 3222–3227.

[140] T. Sugai, G.J. Shen, Y. Ichikawa, C.-H. Wong, Synthesis of 3-deoxy-D-manno-2-octulosonic acid (KDO) and its analogs based on KDO aldolase-catalyzed reactions, J. Am. Chem. Soc. 115 (1993) 413–421.

[141] M.C. Shelton, I.C. Cotterill, S.T.A. Novak, R.M. Poonawala, S. Sudarshan, E.J. Toone, 2-Keto-3-deoxy-6-phosphogluconate aldolases as catalysts for stereocontrolled carbon–carbon bond formation, J. Am. Chem. Soc. 118 (1996) 2117–2125.

[142] N.C. Floyd, M.H. Liebster, N.J. Turner, A simple strategy for obtaining both enantiomers from an aldolase reaction: preparation of L- and D-4-hydroxy-2-ketoglutarate, J. Chem. Soc. Perkin Trans 1 (1992) 1085–1086.

[143] I.C. Cotterill, M.C. Shelton, D.E.W. Machemer, D.P. Henderson, E.J. Toone, Effect of phosphorylation on the reaction rate of unnatural electrophiles with 2-keto-3-deoxy-6-phosphogluconate aldolase, J. Chem. Soc. Perkin Trans. 1 (*1998*) 1335–1342.

[144] D.P. Henderson, I.C. Cotterill, M.C. Shelton, E.J. Toone, 2-Keto-3-deoxy-6-phosphogalactonate aldolase as a catalyst for stereocontrolled carbon–carbon bond formation, J. Org. Chem. 63 (1998) 906–907.

[145] S.T. Allen, G.R. Heintzelman, E.J. Toone, Pyruvate aldolases as reagents for stereospecific aldol condensation, J. Org. Chem. 57 (1992) 426–427.

[146] M. Cheriyan, E.J. Toone, C.A. Fierke, Mutagenesis of the phosphate-binding pocket of KDPG aldolase enhances selectivity for hydrophobic substrates, Protein Sci. 16 (2007) 2368–2377.

[147] M.J. Walters, V. Srikannathasan, A.R. Mcewan, J.H. Naismith, C.A. Fierke, E.J. Toone, Characterization and crystal structure of *Escherichia coli* KDPGal aldolase, Bioorg. Med. Chem. 16 (2008) 710–720.

[148] S. Fong, T.D. Machajewski, C.C. Mak, C.-H. Wong, Directed evolution of D-2-keto-3-deoxy-6-phosphogluconate aldolase to new variants for the efficient synthesis of D- and L-sugars, Chem. Biol. 7 (2000) 873–883.

[149] N. Wymer, L.V. Buchanan, D. Henderson, N. Mehta, C.H. Botting, L. Pocivavsek, C.A. Fierke, E.J. Toone, J.H. Naismith, Directed evolution of a new catalytic site in 2-keto-3-deoxy-6-phosphogluconate aldolase from *Escherichia coli*, Structure 9 (2001) 1–9.

[150] J.M. Serafimov, D. Gillingham, S. Kuster, D. Hilvert, The putative diels-alderase macrophomate synthase is an efficient aldolase, J. Am. Chem. Soc. 130 (2008) 7798–7799.

[151] D.G. Gillingham, P. Stallforth, A. Adibekian, P.H. Seeberger, D. Hilvert, Chemoenzymatic synthesis of differentially protected 3-deoxysugars, Nat. Chem. 2 (2010) 102–105.

[152] M.J. Walters, E.J. Toone, Pyruvate aldolases in chiral carbon-carbon bond formation, Nat. Protoc. 2 (2007) 1825–1830.

[153] H.J. Lamble, M.J. Danson, D.W. Hough, S.D. Bull, Engineering stereocontrol into an aldolase-catalyzed reaction, Chem. Commun. Jan (2005) 124–126.

[154] H.J. Lamble, N.I. Heyer, S.D. Bull, D.W. Hough, M.J. Danson, Metabolic pathway promiscuity in the archaeon *Sulfolobus solfataricus* revealed by studies on glucose dehydrogenase and 2-keto-3-deoxygluconate aldolase, J. Biol. Chem. 278 (2003) 34066–34072.

[155] A. Saeed, D.W. Young, Synthesis of L-β-hydroxy amino acids using serine hydroxymethyltransferase, Tetrahedron 48 (1992) 2507–2514.

[156] H. Ogawa, T. Gomi, M. Fujioka, Serine hydroxymethyltransferase and threonine aldolase: are they identical? *Int*, J. Biochem. Cell Biol. 32 (2000) 289–301.

[157] K. Fesko, C. Reisinger, J. Steinreiber, H. Weber, M. Schürmann, H. Griengl, Four types of threonine aldolases: similarities and differences in kinetics/thermodynamics, J. Mol. Catal. B Enzym. 52–53 (2008) 19–26.

[158] M.L. Gutierrez, X. Garrabou, E. Agosta, S. Servi, T. Parella, J. Joglar, P. Clapés, Serine hydroxymethyl transferase from *Streptococcus thermophilus* and L-threonine aldolase from *Escherichia coli* as stereocomplementary biocatalysts for the synthesis of β-hydroxy-β,ω-diamino acid derivatives, Chem. Eur. J. 14 (2008) 4647–4656.

[159] J.Q. Liu, T. Dairi, N. Itoh, M. Kataoka, S. Shimizu, H. Yamada, Diversity of microbial threonine aldolases and their application, J. Mol. Catal. B Enzym. 10 (2000) 107–115.

[160] J. Steinreiber, K. Fesko, C. Reisinger, M. Schürmann, F. Van Assema, M. Wolberg, D. Mink, H. Griengl, Threonine aldolases – an emerging tool for organic synthesis, Tetrahedron 63 (2007) 918–926.

[161] N. Dückers, K. Baer, S. Simon, H. Gröger, W. Hummel, Threonine aldolases – screening, properties and applications in the synthesis of non-proteinogenic β-hydroxy-α-amino acids, Appl. Microbiol. Biotechnol. 88 (2010) 409–424.

[162] V.P. Vassilev, T. Uchiyama, T. Kajimoto, C.-H. Wong, L-Threonine aldolase in organic synthesis: preparation of novel β-hydroxy-α-amino acids, Tetrahedron Lett. 36 (1995) 4081–4084.

[163] T. Kimura, V.P. Vassilev, G.J. Shen, C.-H. Wong, Enzymatic synthesis of β-hydroxy-α-amino acids based on recombinant D- and L-threonine aldolases, J. Am. Chem. Soc. 119 (1997) 11734–11742.

[164] T. Nishiyama, S.S. Mohile, T. Kajimoto, M. Node, Synthesis of thymine polyoxin C by using L-threonine aldolase-catalyzed aldol reaction, Heterocycles 71 (2007) 1397–1405.

[165] C.F. Barbas Iii, Y.F. Wang, C.-H. Wong, Deoxyribose-5-phosphate aldolase as a synthetic catalyst, J. Am. Chem. Soc. 112 (1990) 2013–2014.

[166] L. Chen, D.P. Dumas, C.-H. Wong, Deoxyribose 5-phosphate aldolase as a catalyst in asymmetric aldol condensation, J. Am. Chem. Soc. 114 (1992) 741–748.

[167] C.-H. Wong, E. Garcia-Junceda, L. Chen, O. Blanco, H.J.M. Gijsen, D.H. Steensma, Recombinant 2-deoxyribose-5-phosphate aldolase in organic synthesis: use of sequential two-substrate and three-substrate aldol reactions, J. Am. Chem. Soc. 117 (1995) 3333–3339.

[168] H.J.M. Gijsen, C.-H. Wong, Unprecedented asymmetric aldol reactions with three aldehyde substrates catalyzed by 2-deoxyribose-5-phosphate aldolase, J. Am. Chem. Soc. 116 (1994) 8422–8423.

[169] H.J.M. Gijsen, C.-H. Wong, Sequential one-pot aldol reactions catalyzed by 2-deoxyribose-5-phosphate aldolase and fructose-1,6-diphosphate aldolase, J. Am. Chem. Soc 117 (1995) 2947–2948.

[170] H.J.M. Gijsen, C.-H. Wong, Sequential three- and four-substrate aldol reactions catalyzed by aldolases, J. Am. Chem. Soc. 117 (1995) 7585–7591.

[171] W.-D. Fessner, Enzyme-catalyzed aldol additions, in: M. Rainer (Ed.), Modern Aldol Reactions. Enolates, Organocatalysis, Biocatalysis and Natural Product Synthesis, Wiley-VCH Verlag GmbH & Co. KgaA, Weinheim, 2004, pp. 201–272.

[172] S. Jennewein, M. Schürmann, M. Wolberg, I. Hilker, R. Luiten, M. Wubbolts, D. Mink, Directed evolution of an industrial biocatalyst: 2-deoxy-D-ribose 5-phosphate aldolase, Biotechnol. J. 1 (2006) 537–548.

CHAPTER 11

Applications of Biocatalysis for Pharmaceuticals and Chemicals

Ramesh N. Patel
SLRP Associates, Consultation in Biotechnology, Bridgewater, NJ, USA

1 INTRODUCTION

In the development of drugs, the synthesis of single enantiomers of intermediates has become increasingly important in the pharmaceutical industry [1]. The advantages of biocatalysis over chemical synthesis are that enzyme-catalyzed reactions are often highly enantioselective and regioselective, and can be carried out under mild conditions. Such high selectivity is extremely desirable in pharmaceutical synthesis since it offers several advantages such as minimizing side reactions and not requiring protection and deprotection steps. Technological advances have established biocatalysis as an environment friendly and economical alternative to traditional chemical synthesis in the pharmaceutical realm. Directed evolution of biocatalysts under process conditions can lead to increased enzyme activity, selectivity, and stability of biocatalysts [2–11].

A number of review articles [12–28] have described the use of enzymes in organic synthesis. This chapter provides many examples of the industrial use of enzymes for the synthesis of pharmaceuticals and of the synthesis of basic and specialty chemicals. Because hydrolytic reactions have been most commonly employed in pharmaceutical synthesis, they are considered first.

2 ENZYMATIC ENANTIOSELECTIVE HYDROLYSIS AND DESYMMETRIZATION

2.1 Preparation of Ethyl (S)-3-Cyano-5-Methylhexanoate for Pregabalin (Lyrica®) Synthesis

(*S*)-(+)-3-aminomethyl-5-methylhexanoic acid (Pregabalin) **1** (Figure 11.1) is a lipophilic GABA (γ-aminobutyric acid) analog developed for the treatment of several central nervous system disorders including epilepsy, neuropathic pain, anxiety, and social phobia [29,30]. Recent studies have shown that pregabalin

Organic Synthesis Using Biocatalysis. http://dx.doi.org/10.1016/B978-0-12-411518-7.00011-1

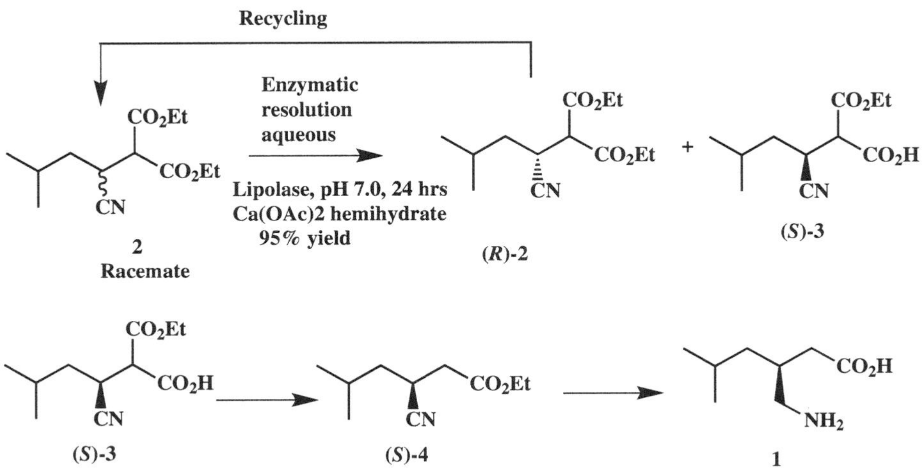

FIGURE 11.1
Pregabalin: enzymatic preparation of ethyl (*S*)-3-cyano-5-methylhexanoate.

is effective at treating chronic pain in disorders such as fibromyalgia and spinal cord injury. In June 2007, pregabalin became the first medication approved by the US Food and Drug Administration specifically for the treatment of fibromyalgia.

During the initial process development for pregabalin, several routes were examined in detail [31–33]. The first generation manufacturing process was selected on the basis of overall yield, cost, and high process throughput, and executed as a racemic synthesis followed by resolution with (*S*)-(+)-mandelic acid. Although this route was cost effective, there were possibilities for improvement by employing an early – as opposed to a late-stage – resolution as well as a recycling step to avoid discarding the undesired enantiomer. The use of asymmetric hydrogenation to reduce 3-cyano-5-methylhex-3-enoic acid [33,34] was a potentially elegant manufacturing process, but it was surpassed in terms of cost effectiveness and environmental performance by an enzymatic route [34].

The enzymatic route was developed from an existing, relatively inexpensive, racemic precursor **2** and used a hydrolase to generate enantiopure precursor (*S*)-**3**. This strategy also required a procedure to recycle unused *R*-**2a** enantiomer as well as chemistry to convert (*S*)-**3** to Pregabalin (Figure 11.1). From a screen of commercially available hydrolases, Lipolase was selected for process development on the basis of its high enantioselectivity (*E* value >200) and activity for the hydrolysis of **2**. Lipolase is commercially produced by Novozymes using a submerged fermentation of a genetically modified *Aspergillus oryzae* strain. Unfortunately, at high substrate input (>1 M) the enzyme was inactivated by the reaction product. Adding divalent ions such as calcium and zinc to the reaction significantly suppressed the inhibition, possibly by forming a complex that remained suspended in the emulsion. The effect of calcium acetate in the reaction medium at higher concentrations of **2** (3 M or 765 g/L) gave conversion values ranging from 42% to 48% after 24 h [34].

Under optimized conditions, the enzymatic process was tested for robustness in multiple runs at 10 kg scale. Three pilot runs at 900 kg (1600 L reactor) scale as well as manufacturing trials at 3.5*t* (8000 L reactor) demonstrated the consistently high performance and scalability of this enzymatic reaction. A heat-promoted decarboxylation of (*S*)-**3** efficiently generated ethyl ester (*S*)-**4**, a known precursor of Pregabalin **1** [34].

This new route dramatically improved process efficiency compared to the first-generation route by setting the stereocenter early in the synthesis and enabling the facile racemization and reuse of (*R*)-**2a**. The chemoenzymatic process also reduced organic solvent usage resulting in a mostly aqueous process. Compared to the first-generation manufacturing process, the new process resulted in higher yields of pregabalin and reductions of waste streams.

Example 1

A suitable reactor was charged a premixed solution of deionized water (2.1 L) and $Ca(OAc)_2$ hemihydrate (282.5 g). Lipolase commercial solution (800 g, 32 g actual protein) was then charged, and the resulting solution was stirred at 400 rpm. Cyano di-ester **2** (10.0 kg) was charged as a liquid in one portion, and an autotitrator maintained the pH at 7.0 using 30% NaOH. After 24 h, the reaction was completed with a reaction yield of 48%. The product was isolated from the reaction mixture in >95% recovery, and >95% purity [34].

To recycle *R*-**2** a suitable vessel was charged with 1 kg of (*R*)-cyanodiester **2** as a toluene solution, followed by 1.42 kg of 21 wt% sodium ethoxide solution in ethanol (1.1 equiv). The mixture was then heated to 80°C for 8–16 h. After completion of the reaction, the mixture was cooled to 4°C and neutralized by adding 0.27 L of glacial acetic acid. The slurry was filtered, and the filtrate was evaporated to afford racemic diester **2**. The material was then distilled to afford *rac*-**2** in more than 95% isolated yield.

2.2 Preparation of C-13 Taxane Side Chain

Among the antimitotic agents, paclitaxel taxol® **5** (Figure 11.2) a complex, polycyclic deterrence, exhibits a unique mode of action on microtubule proteins responsible for the formation of the spindle during cell division as the drug inhibits microtubule depolymerization. Various types of cancers have been treated with paclitaxel including ovarian and metastatic breast cancers. For preparing paclitaxel by a semisynthetic process [35–38], the side-chain, (*R*)-**6**, is required. The enantioselective enzymatic hydrolysis of racemic acetate *cis*-3-(acetyloxy)-4-phenyl-2-azetidinone **7** to the corresponding (*S*)-alcohol **6** and the unreacted desired (*R*)-acetate **7** was demonstrated [39] using lipase PS-30 from *Pseudomonas cepacia* (Amano International Enzyme Company) and BMS lipase (extracellular lipase derived from the fermentation of *Pseudomonas* sp. SC 13856). Reaction yields of >48% (theoretical maximum yield 50%) with enantiomeric excess (ee) of >99.5% were obtained for the (*R*)-acetate.

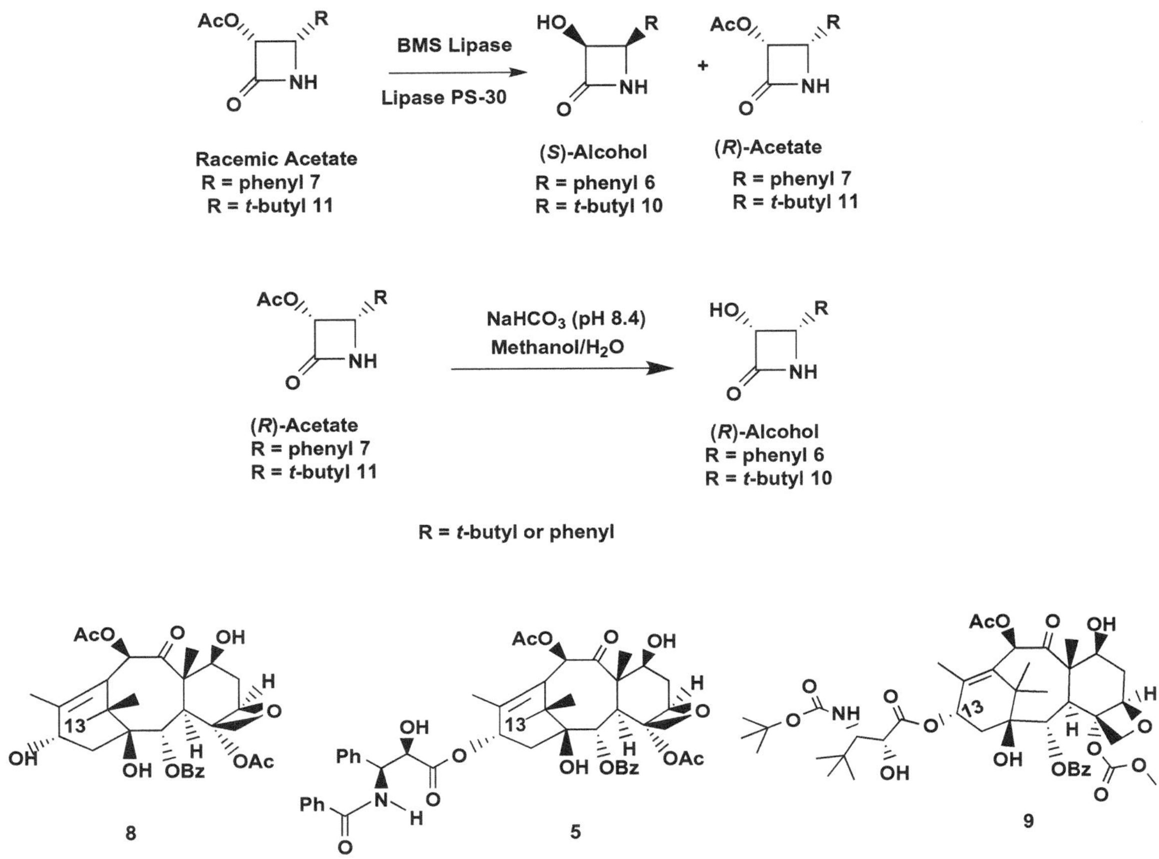

FIGURE 11.2
Paclitaxel: enzymatic preparation of C-13 taxane side chain.

BMS lipase and lipase PS-30 were immobilized on Accurel polypropylene (PP), and the immobilized lipases were reused (10 cycles) without loss of enzyme activity, productivity, or optical purity of the product (*R*)-**7**. The enzymatic process was scaled up to 250 L (2.5 Kg substrate input) using immobilized lipase PS-30. From the reaction batch, (*R*)-7 was isolated in 45 M% yield (theoretical maximum yield 50%) and 99.5% ee. The (*R*)-acetate **7** was chemically converted to (*R*)-alcohol **6**. The C-13 paclitaxel side-chain synthon (*R*)-**6** produced could be coupled to baccatin III **8** after protection and deprotection to prepare Paclitaxel **5** [39].

Due to the poor solubility of Paclitaxel, various groups have developed taxane analogs with greater aqueous solubility. Taxane **9** (Figure 11.2) is a water soluble taxane derivative which when given orally was as effective as i.v. Paclitaxel in five tumor models [40]. The chiral intermediate (3*R*-*cis*)-3-acetyloxy-4-(1,1-dimethylethyl)-2-azetidinone, (*R*)-**10** was prepared for the semisynthesis of the new orally active taxane **9** [41]. The enantioselective enzymatic hydrolysis of *cis*-3-acetyloxy-4-(1,1-dimethylethyl)-2-azetidinone **11** to the corresponding undesired (*S*)-alcohol **10** and unreacted desired (*R*)-acetate **11** was carried out using the same immobilized lipase PS-30 (Amano International Enzyme Co.). Reaction yields of more than 48% (theoretical maximum yield 50%) with (ee) greater than 99% were obtained for the (*R*)-acetate **11**. Acetoxy β-lactam (*R*)-**11** was hydrolysed to hydroxy β-lactam (*R*)-**10** for use in the semisynthesis of **9** [41].

Preparative scale hydrolysis of racemic acetate **11** was carried out using immobilized lipase PS-30. The reaction mixture in 7.5 L of 10 mM potassium phosphate buffer (pH 7.0) contained 229 g of substrate and 57 g of the immobilized lipase on Accurell polypropylene. The reaction was carried out at 40°C, pH 7.0, and 200 rpm in a 10 L reactor for 390 min. At the end of the reaction, the immobilized enzyme was recovered and reused. The product (*R*)-**10** was isolated from the reaction mixture as a white solid in 87% recovery; >99.9% ee. [41].

2.3 Atorvastatin (Lipitor): Enzymatic Desymmetrization of 3-Hydroxyglutaronitrile

Today, HMG-CoA reductase inhibitors (collectively known as statins) have worldwide sales of approximately $20 billion, led by Pfizer's Lipitor **12** (Atorvastatin). Atorvastatin and AstraZeneca's Crestor **13** (Rosuvastatin) (Figure 11.3)

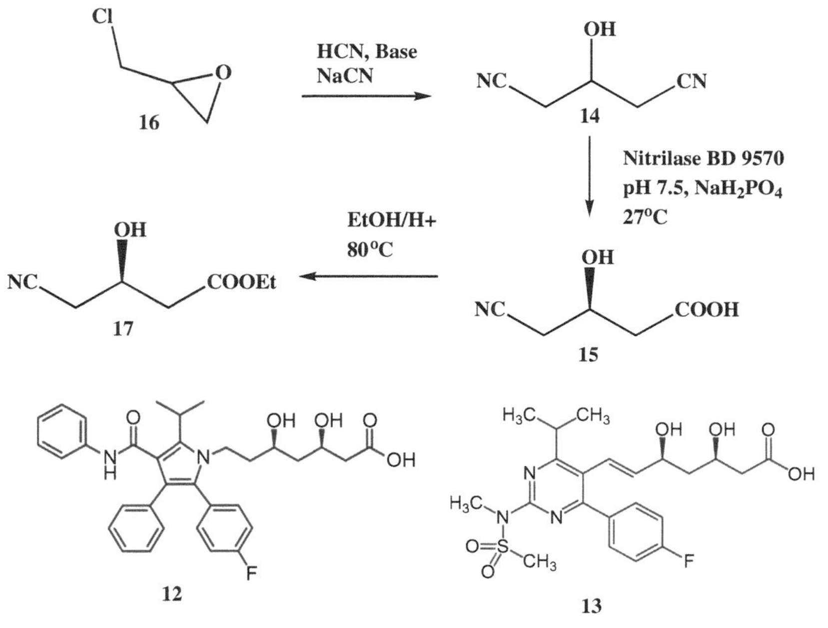

FIGURE 11.3
Atorvastatin: enzymatic desymmetrization of 3-hydroxyglutaronitrile.

are completely synthetic molecules, unlike the earlier generation of statins that were fungal metabolites or semisynthetic derivatives thereof. The synthetic statins share the chiral 3, 5-dihydroxy acid side chain also found in the natural products, which is essential for activity and represents the greatest challenge for preparation of these drugs [42].

Gene site saturation mutagenesis (GSSM) technology was applied for the directed evolution of a nitrilase by Diversa. By screening genomic libraries prepared from environmental samples collected around the globe, they discovered over 200 unique nitrilases that allowed mild and selective hydrolysis of organonitriles to carboxylic acid derivatives [43]. The nitrilase effectively catalyzes the desymmetrization of the prochiral substrate 3-hydroxyglutaronitrile **14** to afford (*R*)-4-cyano-3-hydroxybutyric acid **15**, a precursor to Lipitor. Through GSSM, a mutagenesis technique that effects the combinatorial saturation of each amino acid in the protein to each of the other 19 proteinogenic amino acids, combined with a novel high-throughput mass spectroscopy assay [43–46], a number of improved variants were identified. Transformation and expression of this collection of genes in *Escherichia coli*, furnished a comprehensive library of single-site enzyme mutants. The nitrilase GSSM library accessed every single amino acid variant and contained 10,528 genetic variants. It was screened in an attempt to identify mutant enzymes with improved enantioselectivity. A number of improved variants were identified, the best of which is the Ala190His mutant that yields product with 98.5% ee at 3 M substrate loading and a volumetric productivity of 619 g/L/d [43]. Using the best mutant nitrilase enzyme, an efficient process for synthesizing the Lipitor intermediate (*R*)-**15** was developed [43].

Subsequently, Dowpharma and Chirotech Technology Ltd developed a three-stage process that started by reacting low-cost epichlorohydrin **16** with cyanide to give 3-hydroxyglutaronitrile (3-HGN) **14**. The second stage utilizes a nitrilase-catalyzed desymmetrisation of 3-HGN **14**. The nitrilase reaction was optimized to work at 3 M (330 g/L) substrate concentration, pH 7.5, 27 °C. Under these conditions, with an enzyme loading of 6 wt%, 100% conversion, and 99% ee product **15** was obtained in 16 h. This material was then esterified to give the target compound, ethyl (*R*)-4-cyano-3-hydroxybutyrate **17** [47]. A summary of experimental procedure is described further.

Example 2

A nitrilase enzyme (BD9570, 0.1 unit/mg, Diversa) was used to desymmetrize prochiral substrate **14** to afford (*R*)-**15**. In this process, an aqueous solution of 100 mM NaH_2PO_4 at pH 7.5 (510 mL) was added to lyophilized nitrilase enzyme powder (15.15 g). The mixture was stirred at 27°C (to rehydrate the lyophilized enzyme powder) for 40 min. 3-Hydroxyglutaronitrile (252.5 g) was then charged over 10 min. The mixture was stirred at 27°C for 16 h and then cooled to 2°C prior to acidification to give pH 2. Celite was charged (25 g) and the slurry filtered. The filtrate was extracted with methyl ethyl ketone. The combined methyl ethyl ketone extracts were evaporated *in vacuo* (15 mbar, 40°C) to yield the product **15** as a brown liquid (240.8 g, 81% yield; ee 98.8%).

For conversion of (*R*)-4-cyano-3-hydroxybutyrate **15** to ethyl (*R*)-4-cyano-3-hydroxybutyrate **17, 15** (239.5 g) was dissolved in absolute ethanol (480 mL). Concentrated sulphuric acid (3.3 mL) was charged and the mixture was heated at reflux (80°C) for 2.25 h, after which time the conversion was complete. Ethanol was evaporated *in vacuo* to yield the crude product **17** (290 g, yield 99%; ee 98.7%) which was purified by two passes through a wiped film evaporator. The final product **17** was obtained as a colorless liquid (181.5 g, 62% yield) in 97% purity and 98.8% ee. [47].

2.4 Melatonin Receptor Agonist: Enantioselective Enzymatic Hydrolysis of Racemic 1-{2′,3′-Dihydro benzo[b]furan-4′-yl}-1,2-oxirane

Epoxides are versatile intermediates in organic synthesis and are easily prepared from various starting materials. The inherent polarity and strain of the three-membered ring structure make them susceptible to reacting with a large number of reagents like electrophiles, nucleophiles, acids, bases, reducing agents, and some oxidizing agents [48]. The most common currently available chemical methods for preparing chiral epoxides are asymmetric alkene oxidations as developed by Sharpless, Jacobsen, and Shi [49–51].

Epoxide hydrolases (EH) are promising biocatalysts for the preparation of chiral epoxides and vicinal diols and can be effectively used in the resolution of racemic epoxides. There are number of reviews published on the synthetic application of epoxide hydrolase and preparation of chiral epoxides [52–57] and additional information can be found elsewhere in this book.

The (*S*)-epoxide **18a** (Figure 11.4) is a key intermediate in the synthesis of a melatonin receptor agonist **19** [58]. The enantiospecific hydrolysis of the racemic 1-{2′, 3′-dihydro benzo[b]furan-4′-yl}-1, 2-oxirane **18** to the corresponding (*R*)-diol **20**, and unreacted *S*-epoxide **18a** has been demonstrated [59]. Two *Aspergillus niger* strains (SC 16310, SC 16311) and *Rhodotorula glutinis* SC 16293

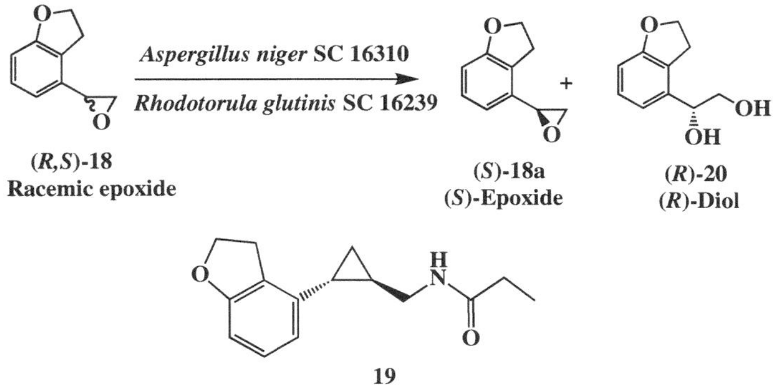

FIGURE 11.4
Melatonin receptor agonist: enantioselective enzymatic hydrolysis of racemic 1-{2′,3′-dihydro benzo[b]furan-4′-yl}-1,2-oxirane.

selectively hydrolyzed the (*R*)-epoxide, leaving behind the (*S*)-epoxide **18a** in >95% ee and 45% yield (theoretical maximum yield is 50%). Several solvents at 10% v/v were evaluated in an attempt to improve the optical purity and yield. A number of solvents were found to have significant effects on both the extent of hydrolysis and the optical purity of unreacted (*S*)-epoxide **18a** for the hydrolysis catalyzed by *R. glutinis* SC 16293. Most solvents gave a lower ee product and slower reaction rate than that of reactions without any solvent supplement; however, MTBE gave a reaction yield of 45% and an ee of 99.9% for unreacted (*S*)-epoxide **18a** with an increase in the *E* value of enzyme from 14 in buffer to 67 in buffer-MTBE mixture.

Example 3

Hydrolysis of racemic epoxide **18** was carried out using cells of *R. glutinis*. The cells (150 g) were suspended in 500 mL of 100 mM phosphate buffer pH 8.0 in a jacketed three-necked flask fitted with an air driven stirrer, condenser and stopper. A solution of 2.5 g racemic epoxide **18**, in 50 mL of methyl *tert*butyl ether was added to the stirred cell suspension. After 5.5 h at 28°C, the reaction was stopped. NaCl (180 g) was added and the mixture was extracted twice with an equal volume of ethyl acetate. The combined ethyl acetate layers were dried over anhydrous Na_2SO_4, filtered and the solvent was removed by a rotary evaporator to provide (*S*)-epoxide **18a** in 48% solution yield. A portion (2.2 g) of the reaction product was dissolved in a mixture of heptane:ethyl acetate (1:1) and repeatedly extracted with 0.10 M solution of $Na_2B_4O_7$ in water. The organic layer provided 0.95 g of (*S*)-epoxide **18a** with ee of >99% while the aqueous layer provided 1.02 g of (*R*)-diol **20** with ee of 79% [59].

2.5 Preparation of Chiral Intermediate for NK1/NK2 Dual Antagonists

Tachykinins are a group of biologically active neuropeptide hormones that are widely distributed throughout the nervous system. They are implicated in a variety of biological processes such as pain transmission, inflammation, vasodilatation, and secretion [60].The effect of tachykinins is modulated via the specific G-protein coupled receptors like NK1 and NK2. Thus nonpeptide NK-receptors antagonists are potentially useful in the treatment of variety of chronic diseases including asthma, bronchospasm, arthritis, and migraine [61].

The structure–activity relationship of several non-peptide NK1/NK2 antagonists has led to the discovery of a new class of oxime-based NK1/NK2 dual antagonists [62,63] such as compound **21** (Figure 11.5). The biological activity of **21** resides mainly in the (*R*,*R*)-diastereomer. An enzymatic process for desymmetrizing the prochiral diethyl 3-[3′, 4′-dichlorophenyl] glutarate **22** to the corresponding (*S*)-monoester **23** was developed. A screen of about 200 commercial hydrolases revealed several capable of hydrolyzing **22** to both (*S*)-**23** and (*R*)-**23** with good to excellent enantioselectivity. Out of 11 candidates with pro-*S* selectivity, lipase B from *Candida antarctica*, Chirazyme L-2, was selected for further development [64].

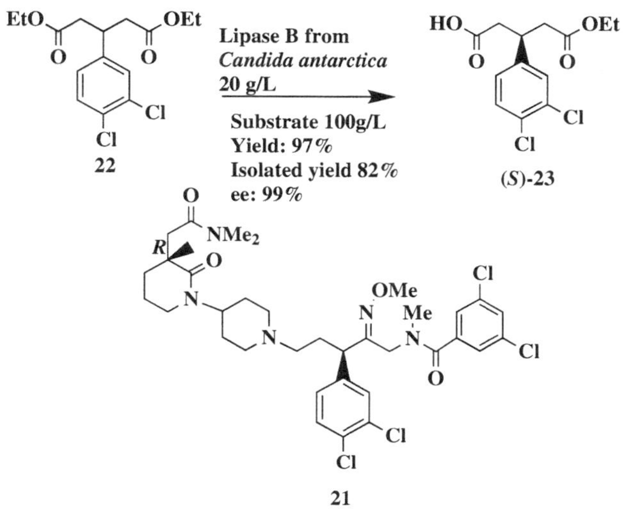

FIGURE 11.5
Enzymatic preparation of chiral intermediate for NK1/NK2 dual antagonists.

The reaction was carried out using immobilized lipase B at 100 g/L substrate input and 20 g/L immobilized enzyme input. A reaction yield of 97% and an ee of >99% were obtained for the desired (*S*)-monoester **23** [64]. The process was scaled up to produce 200 kg of product in 80% overall isolated yield.

Example 4

Diethyl 3-[3′,4′-dichlorophenyl]glutarate **22** (50 kg) was dispersed in 10 mM sodium phosphate buffer pH ~8.0 (450 L) with agitation at 40°C in a 1200-L glass-lined reactor equipped with an in-vessel pH probe. The reaction was initiated by adding 6.25 kg of immobilized Chirazyme L-2 lipase B. The pH was maintained between 7.9 and 8.1 with 0.5 M NaOH. A 93% conversion yield was achieved following 43 h incubation. The hydrolysis was stopped by removing the enzyme by filtration. The enzyme filter cake was washed with 100 mM phosphate buffer pH ~8.0 (~125 L). The filter cake wash and filtrate were combined in an 800 L reactor and the pH was adjusted to ~4.2 using 10% H_2SO_4 followed by isolation of the precipitated product by filtration. The acid precipitate was rinsed *in situ* with water (~100 L) and dried on trays in an air dryer, yielding 40.5 kg product in >99% ee. The dried cakes from two reaction batches (~79 kg total), both made with the fresh supply of the enzyme, were mixed with 6 kg of Supercel and MTBE (320 L). Insoluble material and Supercel were removed by filtration in a sparkler and rinsed with MTBE (110 L). The rinse was combined with the filtrate and concentrated threefold by vacuum evaporation. The MTBE concentrate was mixed with *n*-heptane (640 L) and vacuum evaporated to remove the MTBE. Removal of MTBE resulted in precipitation of the desired (S)-monoester. The precipitated slurry was filtered and dried to **(S)-23,** 75.2 kg (82.1% yield), >99% ee [64].

DNA family shuffling was used to create a chimeric lipase B protein with improved activity towards diethyl 3-[3,4′-dichlorophenyl]glutarate. Three homologous lipases from *C. antarctica* ATCC 32657, *Hyphozyma* sp. CBS 648.91 and *Cryptococcus tsukubaensis* ATCC 24555 were cloned and shuffled to generate a diverse gene library. Using a high-throughput screening assay, a chimeric lipase B protein having 20-fold higher activity towards the substrate was identified [65]. In addition, the stability characteristics of several highly active chimeric proteins were also improved as a result of family shuffling. The half-life at 45 °C and melting point (Tm) of one chimera exceeded those of lipase B from *C. antarctica* ATCC 32657 by 11-fold and 6.4 °C, respectively.

The thermostability of the lipase B from *Candia antarctica* was also improved by directed evolution. Two mutants, 23G5 and 195F1, were generated with over a 20-fold increase in half-life at 70 °C compared with the wild-type lipase B [66].

2.6 Chemokine Receptor Modulators: Enzymatic Desymmetrization of Dimethyl Ester

The chiral monoester, (1*S*,2*R*)-2-(methoxycarbonyl)cyclohex-4-ene-1-carboxylic acid **24a** (Figure 11.6) is a key intermediate for the synthesis of a potential drug candidate **27** for the modulation of chemokine receptor 2 (CCR2) activity useful in treatment of rheumatoid arthritis [67–69]. Both the (1*S*,2*R*)-monoester **24a** and its enantiomer (1*R*,2*S*)-monoester **24b** were obtained by resolving the

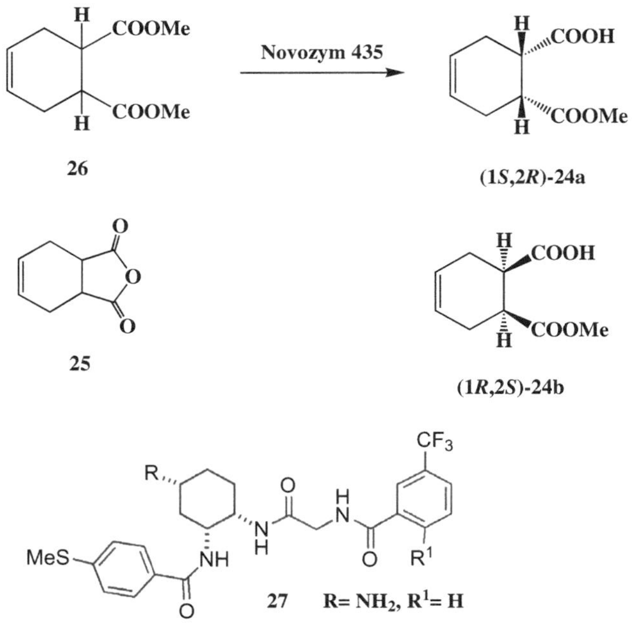

FIGURE 11.6
Chemokine receptor modulators: enzymatic desymmetrization of dimethyl ester.

racemic acid **24** with alkaloids [70,71]. A desymmetrization process has been developed that affords either the (1*S*,2*R*)-monoester **24a** or its enantiomer (1*R*,2*S*)-monoester **24b** from mesoanhydride **25** by alcoholysis using cinchona alkaloids [72]. The synthesis (1*R*,2*S*)-monoester **24b** was demonstrated by porcine liver enzyme-catalyzed desymmetrization of the dimethyl ester **26** [73,74].

At Bristol-Myers Squibb, the quinine-catalyzed alcoholysis of the anhydride **25** initially provided kilogram quantities of the (1*S*,2*R*)-monoester **24a** with 90.8% ee. To improve the enantioselectivity of desired **24a**, alternative enzymatic processes were evaluated [75]. Screening of various enzymes was carried out to prepare (1*S*,2*R*)-monoester **24a** from dimethyl ester **26**. After evaluating yield and optical purity of desired product, reaction rate, and cost of enzyme, the immobilized lipase from *C. antarctica* (Novozym 435) was chosen for further development. After optimizing the reaction parameters such as pH, temperature, substrate and enzyme input, a small scale reaction of 50 mL (57.2 g) of dimethyl ester **26** afforded (1*S*,2*R*)-monoester **24a** in 96% yield and >99.9% ee after 24 h reaction.

Two reparative batches were carried out at 1.73 kg input of dimethyl ester **26** to afford 3.4 kg of 1*S*,2*R*-monoester **24a** [75].

Example 5

A typical reaction used a 30-L jacketed reactor fitted with a pH stat to maintain constant pH by automatic addition of 10 M NaOH. K_2HPO_4 buffer (1 M, 15 L, pH 8.5) was stirred at 40°C in a 30 L reactor, and Novozym 435 (150 g) was added followed by the dimethyl ester **26** (1.73 kg). The mixture was gently stirred for 27 h, keeping the pH constant at 8.5 by addition of NaOH. After 27 h the reaction was completed. The reaction mixture was filtered through a Buchner funnel (no. 1 filter paper). The recovered enzyme was washed with water (2 × 500 mL). The combined aqueous filtrate was washed with MTBE (3 L) to remove any unreacted dimethyl ester **26**. The aqueous layer was acidified to pH 2 with concentrated HCl (2.33 L) and extracted with MTBE (3 × 5 L). The combined MTBE layers were washed with water (2 × 500 mL), evaporated to dryness on a rotary evaporator (106 torr, bath 35°C, 1 h) to provide the desired (1*S*,2*R*)-monoester **24a** as a yellow oil (1.675 kg, and 1.58 kg for two batches) yields (corrected for MTBE and water) 98.1–99.8%, HPLC purity of 98.3–99.2% and ee >99.9% [75].

3 ENZYMATIC ESTERIFICATION, TRANSESTERIFICATION, AND AMIDATION

3.1 Antithrombotic Agent: Enzymatic Preparation of 2-Chloromandelic acid Esters

Clopidogrel bisulfate 28 (Plavix, Figure 11.7) is an orally active inhibitor of platelet aggregation marketed as an antithrombotic agent. It is an adenosine diphosphate (ADP) receptor antagonist indicated for the reduction of atherosclerotic events including myocardial infarction, ischemic stroke, and vascular death in patients with atherosclerosis. Clopidogrel works by helping to prevent

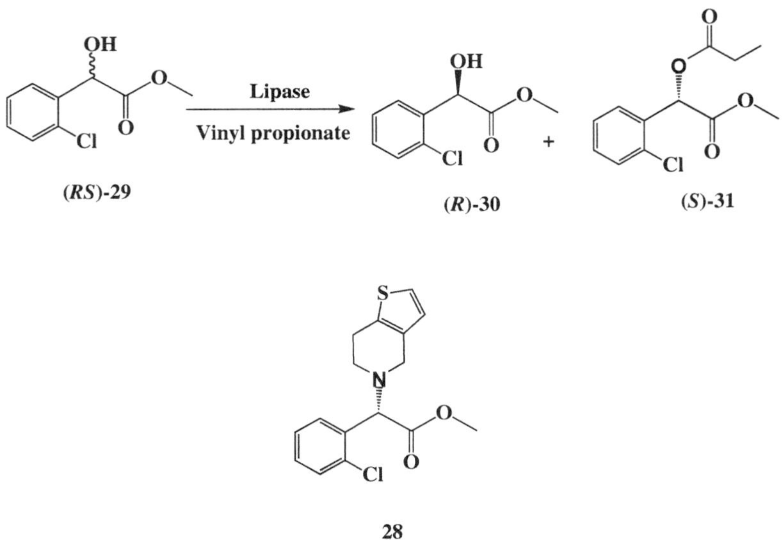

FIGURE 11.7
Antithrombotic agent: enzymatic preparation of 2-chloromandelic acid esters.

harmful blood clots. Recent studies have shown that clopidogrel is more effective in blocking platelet aggregation than aspirin and ticlopidine even at much lower dosage [76–80].

An efficient process for resolving methyl 2-chloromandelate (*R,S*)-**29** was developed using a lipase-mediated transesterification. Among 11 hydrolytic enzymes examined, *C. antarctica* lipase A (CAL-A) showed the highest enantioselectivity and reaction rate toward methyl (*S*)-2-chloromandelate, yielding (*S*)-**31** and methyl (*R*)-2-chloromandelate **30.** Moreover, (*R*)-**30** was obtained in enantiomerically pure form (99% ee) and 41% yield through the lipase-mediated resolution under solvent-free condition [81]. On a preparative scale reaction, immobilized CAL-A (4.8 g) was added to a solution of (*R,S*)-**29** (500 g) and vinyl propionate (500 g) which gave (*R*)-**30** as colorless oil (205 g, 41% yield), 99.3% ee. CAL-A maintained its catalytic activity during 13 cycles of repeated use without significant decrease in enantioselectivity, indicating that the method is economical and easy to scale-up for commercial production of methyl (*R*)-2-chloromandelate [81].

3.2 Enzymatic Preparation of (5S)-4,5-Dihydro-1H-pyrrole-1,5-dicarboxylic acid, 1-(1,1-Dimethylethyl)-5-ethyl Ester

The synthesis of dipeptidyl peptidase-IV (DPP4) inhibitor, Saxagliptin **32** (Figure 11.8) required (5*S*)-5-aminocarbonyl-4,5-dihydro-1H-pyrrole-1-carboxylicacid,1-(1,1-dimethyl ethyl)-ester **33** [82–84]. Direct chemical ammonolyses were hindered by the requirement for aggressive reaction conditions, which resulted in unacceptable levels of amide racemization and side-product

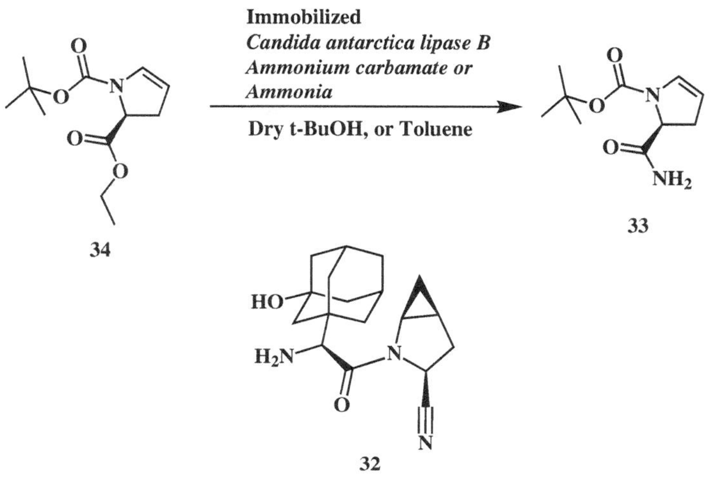

FIGURE 11.8
Saxagliptin: enzymatic preparation of (5*S*)-4,5-dihydro-1H-pyrrole-1,5-dicarboxylic acid, 1-(1,1-dimethylethyl)-5-ethyl ester.

formation. Milder, two-step hydrolysis–condensation protocols using coupling agents such as 4-(4,6-dimethoxy-1,3,5-triazin-2-yl)-4-methylmorpholinium chloride (DMT-MM) were compromised by reduced overall yields [85]. To address this issue, a biocatalytic procedure was developed, based upon the *C. antarctica* lipase B (CALB)-mediated ammonolysis of (5*S*)-4,5-dihydro-1H-pyrrole-1,5-dicarboxylic acid, 1-(1,1-dimethylethyl)-5-ethyl ester **34** with ammonium carbamate to furnish **33** without racemization and with low levels of side-product formation [86].

Experiments utilized process stream ester feed, which consisted of approximately 22% w/v (0.91 M) of the ester in toluene. The low solubility of free ammonia in toluene precluded its use, so solid ammonium carbamate was employed instead. Reactions were performed using a mixture of neat process feed, ammonium carbamate (71 g/L) and biocatalyst (25 g/L) with shaking at 400 rpm and 50 °C. Under these conditions, CALB provided racemization-free amide with yields of 69%, together with 21% of side-products (by HPLC). The inclusion of various additives was investigated to solve potential inhibitory phenomena, shifting the equilibrium towards amide synthesis, and reducing side-product formation. Drying agents such as calcium chloride gave significant improvement (79% amide and 13% side products). Calcium chloride is known to complex alcohols as well as act as a dessicant, and its presumed binding of ethanol released during the course of amide formation may have served to mitigate any deleterious effects of this alcohol on CALB catalysis. A dramatic increase in amide yield (to 84 and 95%) was achieved by including Sodalime and Ascarite, respectively, at 200 g/L in the reaction headspace. Sodalime and Ascarite likely adsorbed and thereby removed the CO_2 by-product from decomposition of ammonium carbamate. A further increase in yield to 98% was attained via the combined use of 100 g/L of calcium chloride and 200 g/L of Ascarite. In the preparative-scale reaction, the process ester feed (Ester **34**, 220 g/L) was reacted with 90 g/L of ammonium carbamate, 33 g/L of CALB, 110 g/L calcium chloride and 216 g/L of Ascarite (in the headspace). Complete conversion of ester was achieved after 3 days at 50°C, with the formation of 96% (182 g/L) of amide **33** and 4% of side products, and after workup 98% potency amide **33** of >99.9% ee was isolated in 81% yield [86].

3.3 Enzymatic Preparation of (*S*)-3-Benzoylthio-2-Methylpropanoic Acid

Captopril (1-[(2*S*)-3-mercapto-2-methylpropionyl]-L-proline) **35** and Zofenopril **36** are angiotensin converting enzyme inhibitors (Figure 11.9). They are used as antihypertensive agents through suppression of the renin-angiotensin-aldosterone system [87–89]. Both prevent the conversion of angiotensin I to angiotensin II by inhibition of ACE. The potency of Captopril and Zofenopril as an inhibitor of ACE depends critically on the configuration of the mercaptoalkanoyl moiety; the compounds with the (*S*)-configuration are about 100 times more active than their corresponding (*R*)-isomers [87–89]. The required 3-mercapto-(2*S*)-methylpropionic acid moiety was prepared from the microbially-derived chiral 3-hydroxy-(2*R*)-methylpropionic acid, which is obtained by the hydroxylation of isobutyric acid [90].

In an alternative approach to prepare the chiral side-chain of Captopril and Zofenopril, the lipase-catalyzed enantioselective esterification of racemic 3-benzoylthio-2-methylpropanoic acid **37** in an organic solvent was demonstrated to yield (*R*)-methyl ester **38** and unreacted acid enriched in the desired (*S*)-**37a** [91]. Using lipase PS-30 with toluene as solvent and methanol as nucleophile, the desired (*S*)-**37a** was obtained in 37% yield (maximum theoretical yield is 50%) with 97% ee. The amounts of water and the concentration of

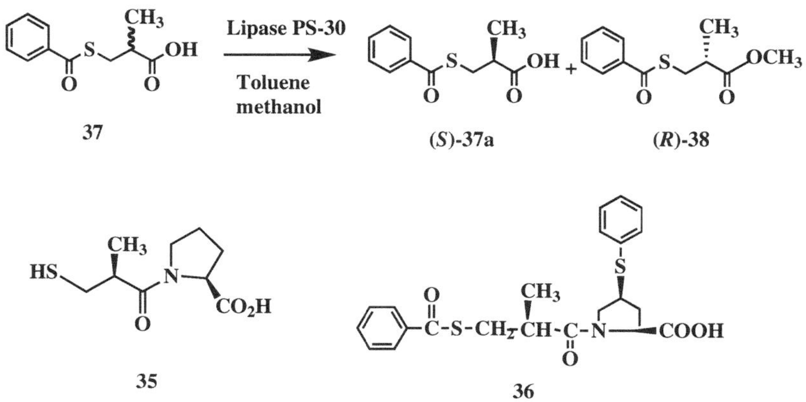

FIGURE 11.9
Zofenopril: enzymatic preparation of (*S*)-3-benzoylthio-2-methylpropanoic acid.

methanol supplied in the reaction mixture were very critical. Water was used at 0.1% concentration in the reaction mixture. Water levels higher than 1% decreased the reaction rate due to the aggregation of enzyme in the organic solvent and mass transfer limitation.

The rate of esterification also decreased as the methanol to substrate ratio was increased from 1:1 to 4:1. Higher methanol concentrations inhibited the esterification reaction by stripping the essential water from the enzyme. Crude lipase PS-30 was immobilized on Accurel polypropylene (PP) with absorption efficiencies of 98.5%. The immobilized lipase efficiently catalyzed the esterification reaction at substrate input of 224 g/L with 0.4 M methanol, 0.1% water, and 120 g/L immobilized enzyme giving a 45% reaction yield of (*S*)-**37a** with 97.7% ee. The immobilized enzyme under identical conditions gave a similar ee and yield of product in twenty three additional reaction cycles without any loss of activity and productivity [91]. (S)-3-benzylthio-2-methylpropanoic acid, (*S*)-**37** is a key chiral intermediate in the synthesis of Zofenopril [92] and Captopril [93].

3.4 Lobucavir: Regioselective Enzymatic Transesterification

Lobucavir **39** (Figure 11.10), a cyclobutyl guanine nucleoside analog, was developed as an antiviral agent for the treatment of herpes virus and hepatitis B [94]. A prodrug **40** in which one of the two hydroxyls is coupled to valine has also been considered for development. Synthesis of the lobucavir L-valine prodrug **40** requires regioselective coupling of one of the two hydroxyl groups of lobucavir **39** with valine. Such regioselective aminoacylation are difficult by chemical methods, but achievable by an enzymatic approach [95]. Enzymatic processes

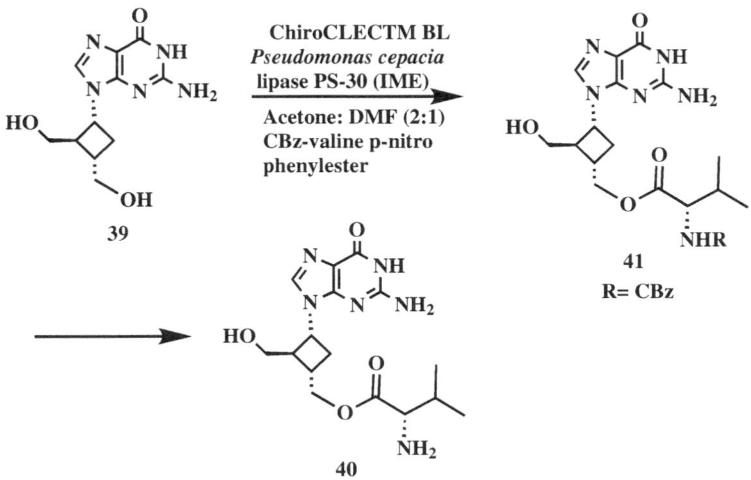

FIGURE 11.10
Lobucavir: regioselective enzymatic transesterification.

were developed for the synthesis of CBz-valine ester of **41**, by transesterification of lobucavir using ChiroCLEC™ BL (61% yield) or more selectively by using lipase from *P. cepacia* (84% yield) [95].

Example 6

Lobucavir 3**9** (7.00 g, 26.39 mmol), *N*-Cbz-L-valine *p*-nitrophenyl ester (35 g) and 10.5 g Chiro-CLEC BL (immobilized enzyme) were added to a jacketed 1 L three-necked round-bottom flask equipped with an air-driven stirrer and a condenser maintained at 5°C by a circulating water bath. Acetone (490 mL) followed by 210 mL DMF were added and the reaction was carried out at 45°C. The recovered immobilized enzyme was re-used. The reaction was scaled up to a 4.5 Kg batch and excess p-nitrophenyl ester of CBz-L-valine was also recovered [95].

4 ENZYMATIC ACYLATION

4.1 Farnesyl Transferase Inhibitor: Enzymatic Resolution of Substituted (6,11-Dihydro-5H-benzo-[5,6]cyclohepta[1,2-β]pyridin-11-yl)piperidines)

An enzymatic process has been developed for preparing the chiral intermediate (+)-**42a** (Figure 11.11) needed for the synthesis of compound SCH66336 **43**, a selective farnesyl transferase inhibitor involved in binding *Ras* which moderates cell proliferation [96]. *Ras* proteins have been observed in 30% of all human

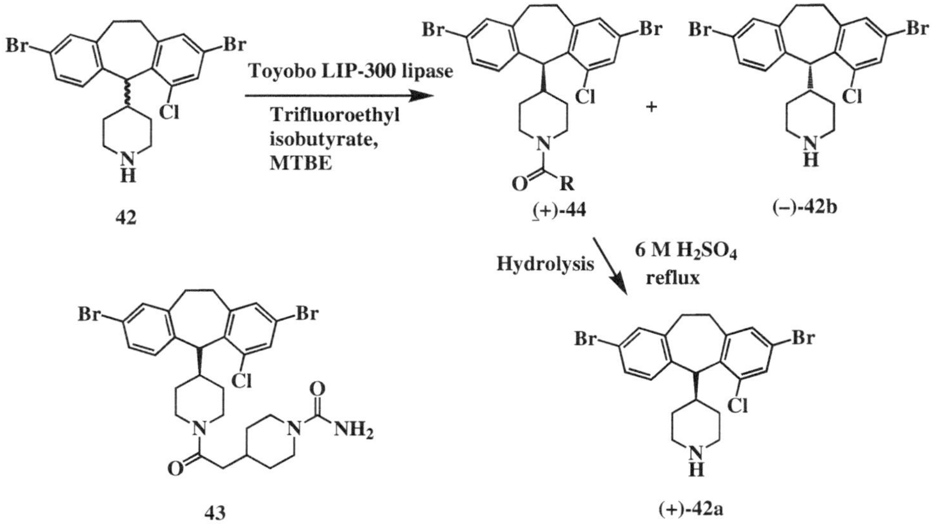

FIGURE 11.11
Farnesyl transferase inhibitor: enzymatic resolution of substituted (6,11-dihydro-5H-benzo-[5,6]cyclohepta[1,2-β]pyridin-11-yl)piperidines.

cancers, including 50% of colon cancers and 95% of pancreatic cancers [97]. Resolution of racemic substituted (6,11-dihydro-5H-benzo-[5,6]cyclohepta[1,2-β]pyridin-11-yl)piperidines **42** was demonstrated to yield the desired (+)-**44** and unreacted (−)-**42b** by an enzymatic acylation process. In the case of racemic **42**, the molecule exists as a pair of enantiomers due to atropisomerism about the exocyclic double bond.

From screen of 233 commercially available enzyme preparations under acylating conditions, Toyobo lipase LIP-300 was selected as the catalyst of choice. Acylation reactions carried out in *t*-butyl methyl ether with trifluoroethyl isobutyrate as acylating agent resulted in isobutyrylation of the (+)-enantiomer, which was easily separated from the unwanted (−)-**42b**. Hydrolysis of the isobutyramide (+)-**44** with 6M sulfuric acid yielded the desired (+)-**42a** in high enantiomeric excess [98].

Example 7

The reaction was carried out using a mixture of racemic **42** (1.01 g) in MTBE (100 mL) which was dried by azeotropic distillation reducing the volume to 50 mL. This solution was transferred to a flask containing Toyobo LIP-300 (2 g) which was held under a stream of N_2 for 2 h. After being shaken for 0.5 h, trifluoroethyl isobutyrate (1.0 mL) was added, and the reaction was shaken at 200 rpm at room temperature for 20 h. A reaction yield of 46% with an ee of 97% was obtained for the *N*-isobutyryl derivative (+)-**44**. The undesired isomer (−)-**42b** was recovered from the resolution step, racemized, and resubjected to enzymatic acylation to increase material throughput. A mixture of (−)-**42b** (100 g, 83.0% ee) and diethyleneglycol dibutyl ether (500 mL) was heated at 210°C under N_2 for 18 h to obtain complete racemization [98].

4.2 Regioselective Enzymatic Acylation of Ribavirin

Ribavirin **45** (Figure 11.12) is an antiviral agent used in combination with α-2β interferon to treat hepatitis C [99,100]. Although this therapy is effective against hepatitis C virus, it has several side effects [101]. To improve the pharmacokinetic profile and reduce side effects, a ribavirin prodrug was considered for development. In a series of preclinical evaluations, the alanine ester of ribavirin **46** showed improved bioavalibility and reduced side effects. The synthesis of **46** required the acylation of unprotected ribavirin. The chemical acylation gave a mixture of mono-, di-, and triacylated products. An enzymatic process was developed for the regioselective acylation of ribavirin **46** with the oxime ester of L-carbobenzyloxy-alanine **47** to give the desired **48** using *C. antarctica* lipase B (in the form of Novozym 435 or Chirazyme L-2). Chemical deprotection of **48** gave **46**. On preparative scale, the coupling of **49** with acetone oxime in the presence of di-*t*-butyl dicarbonate in THF was carried out giving **47** in >96% yield. At the end of the reaction, the reaction mixture was diluted threefold with THF, ribavirin was added, and the acylation was initiated by addition of the Novozym 435. After 24 h at 60°C, the product **46** was isolated in 85% yield [102].

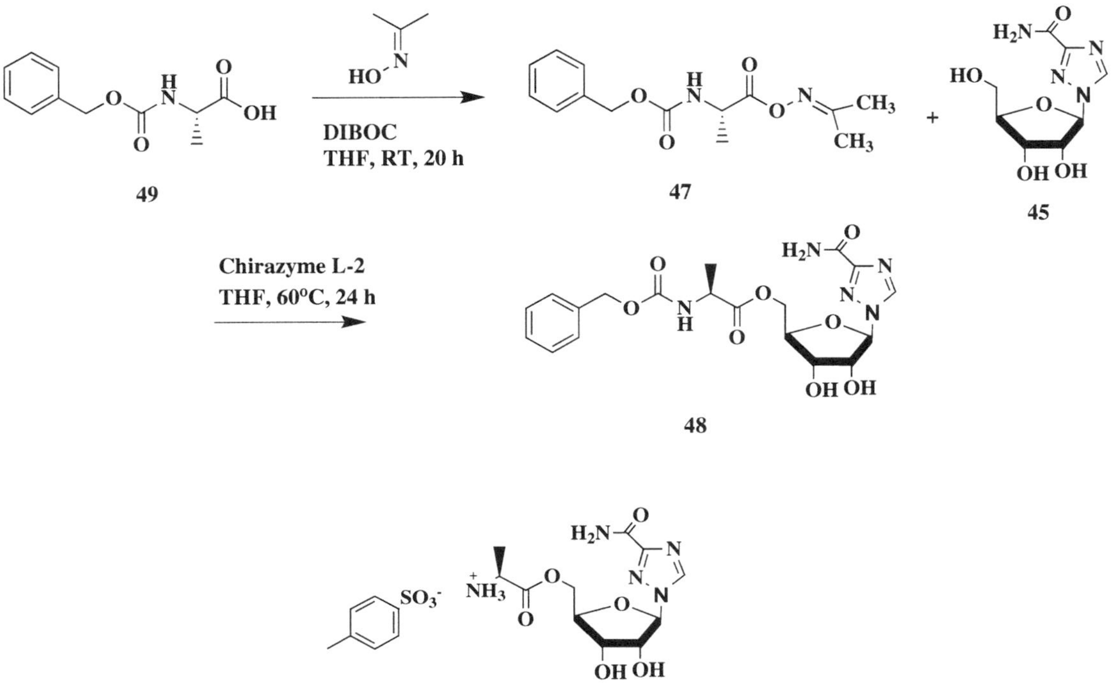

49
47
45
48
46

FIGURE 11.12
Regioselective enzymatic acylation of ribavirin.

Example 8

A pilot scale reaction was carried out with 24.8 kg of CBz-Ala-OH **49**, 8.5 kg of acetone oxime, 28.3 kg of di-*t*-butyl dicarbonate and 6.9 kg of pyridine in THF (260 L) with stirring at 22°C.

After 24 h, 18 kg of ribavirin **45** and 14.4 kg of Chirazyme L-2 and THF (495 L) were added, and the mixture was agitated at 55–60°C for 24 h. At the end of reaction, Celite was added, the reaction mixture was filtered, and the filtrate was concentrated to about 216 L by atmospheric pressure distillation. After cooling to 40–50°C, MTBE (250 L) was slowly added. Following precipitation, the mixture was cooled to 25°C over 1 h, and the precipitated product was filtered. The filter-cake was dried under vacuum at 50°C for 12–18 h. It was then resuspended in 15 volumes of water and stirred at 45°C for about 7 h to remove unreacted ribavirin. The suspension was then filtered and dried in an air-draft oven at about 45°C to give 27.5 kg of product **48** in 83% yield, 99.3% purity, >99.9% de [102].

4.3 Xemilofiban: Enzymatic Preparation of a β-Aminoester

Xemilofiban **50** (Figure 11.13) and related compounds are peptido-mimetics based on the RGD (arginine–glycine–aspartic acid) sequence of fibrinogen

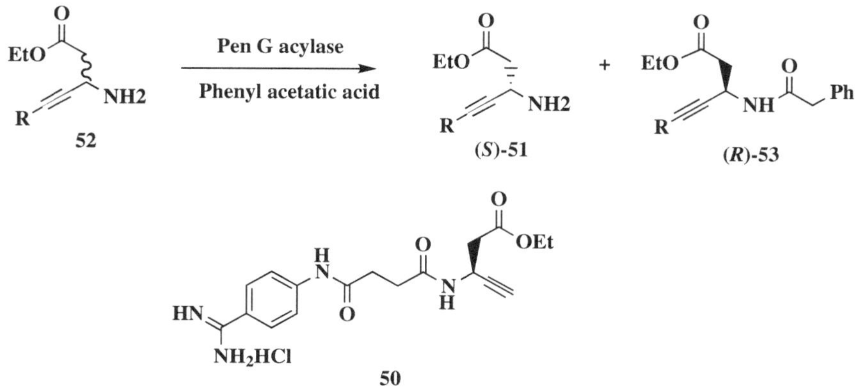

FIGURE 11.13
Xemilofiban: enzymatic preparation of a β-aminoester.

that effectively disrupts the platelet–fibrinogen interaction, leading to inhibition of platelet aggregation and prevention of thrombus formation [103,104].

The optically active β-amino ester (*S*)-**51** has been used in the synthesis of xemilofiban. The resolution of (±)-**52** was efficiently achieved by enzymatic acylation using immobilized penicillin G amidase (PGA-450) and phenylacetic acid as the acyl donor. The reaction was carried out at pH 5.7 and 28 °C to yield the phenylacetamide (*R*)-**53** and the unreacted β–amino ester (*S*)-**51** which were isolated in >99% ee and 98% ee, respectively. This bioconversion was used to resolve 6.1 kg of amine, from which the required (*S*)-**51** was isolated in >96% ee and >43% yield [105].

Example 9

Bioconversions were performed on a 70-L scale in a stainless steel tank equipped with a bottom filter screen. The phenylacetate salt of amine **52** (6.1 kg) was dissolved in 70 L of water at 25°C. The pH was adjusted (as needed) to 5.6 and 525 000 units of PGA-450 was added to initiate reaction. After 22 h, the product was recovered by adding 35 L of chilled (refrigerated, 4°C) MTBE followed by 7.9 L of 2.5 M $KHCO_3/K_2CO_3$ (1.25 M each). The suspension was pumped out of the tank through the filter screen, leaving the wet enzyme in the tank. The layers were separated in a 120 L glass separatory funnel, the aqueous suspension was re-extracted with 21 L of chilled MTBE and the aqueous layer was finally discarded. The combined organic extracts were extracted with 19.8 L of 0.5 N HCl. After separating the layers, the organic layer was re-extracted with 14 L of 10 mM HCl. To the combined aqueous layers 35 L of chilled MTBE was added, pH was adjusted to 7.9 with potassium carbonate/bicarbonate solution, and extracted. After separation, the aqueous layer was again extracted with 17.5 L of MTBE. The MTBE extracts were combined, dried over sodium sulfate, and evaporated using a rotary evaporator. The desired product (S)-**51** was isolated in >96% ee and >43% yield. The recovered enzyme was used for three cycles [105].

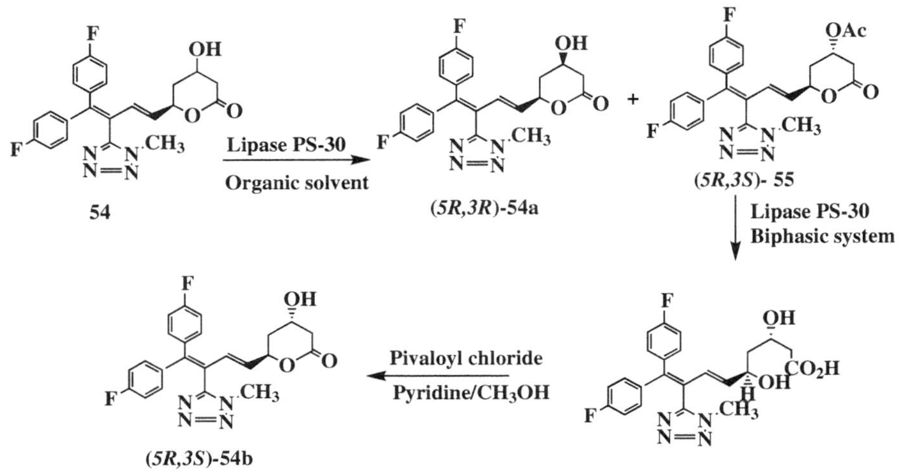

FIGURE 11.14
Anticholesterol drug: enantioselective enzymatic acylation of racemic alcohol 54.

4.4 Anticholesterol Drug: Enantioselective Enzymatic Acylation of Racemic Alcohol 54

The (5*R*,3*R*)-alcohol **54a**, [4-[4α,6ß(E)]]-6-[4, 4-bis[4-fluorophenyl)-3-(1-methyl-1H-tetrazol-5-yl)-1,3-butadienyl]tetrahydro-4-hydroxy-2H-pyren-2-one (Figure 11.14) is a potential new anticholesterol drug which acts by inhibition of HMG CoA reductase [106].

Using an enzymatic diastereoselective acetylation process, the (5*R*,3*R*)-alcohol **54a** was prepared from racemic **54** [107]. Lipase PS-30 efficiently catalyzed the acetylation of racemic **54** (4 g/L) to yield the (5*R*,3*S*)-acetate **55** and the unreacted, desired (5*R*,3*R*)-alcohol **54a**. A reaction yield of 49 M% and an ee of 98.5% was obtained for (5*R*,3*R*)-alcohol **54a** when the reaction was conducted in toluene in the presence of isopropenyl acetate as an acyl donor. In methyl ethyl ketone at 50 g/L substrate concentration, a reaction yield of 46 M% and ee of 96% were obtained for (5*R*,3*R*)-alcohol **54a** [107].

4.4.1 *PREPARATIVE SCALE*

The enzymatic process was scaled-up to a 640 L preparative batch using immobilized lipase PS-30 (lipase PS-30 immobilized on Accurell polypropylene). The substrate input was 4 g/L, with 0.05% water, isopropenyl acetate was the acyl donor, and toluene was the solvent. The reaction was carried out at 37 °C, 150 RPM for 24 h. At the end of the reaction, the enzyme was filtered off with Niagara plate filter and reused for the next cycle. The toluene filtrate was extracted with methanol: water (50:50 v/v). The methanol content of the recovered extract was reduced to 20% and the entire contents were extracted with ethyl acetate. The ethyl acetate layer was collected and concentrated under reduced pressure and the product was crystallized from heptane. The product (5*R*,3*R*)-alcohol **54a** was

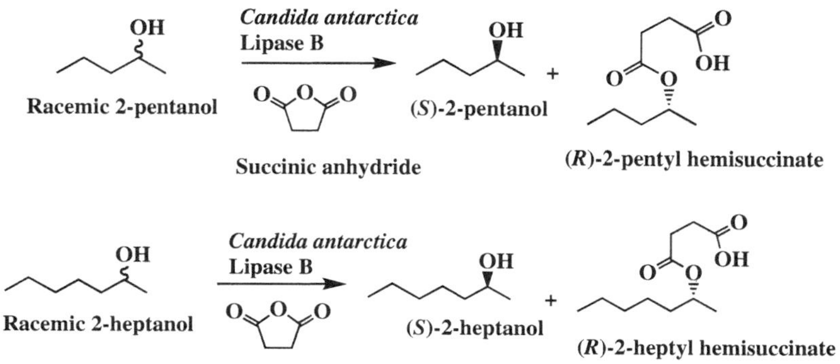

FIGURE 11.15
Anti-Alzheimer drug: enzymatic resolution of secondary alcohols.

isolated in 35 M% overall yield with 98.5% ee and 99.5% chemical purity. The (5*R*,3*S*)-acetate **55** produced by this process was enzymatically hydrolyzed by lipase PS-30 in a biphasic system to prepare the corresponding (5*R*,3*S*)-alcohol **54b** [107].

4.5 Anti-Alzheimer Drug: Enzymatic Resolution of Secondary Alcohols

(*S*)-2-Pentanol (Figure 11.15) is an intermediate in the synthesis of several potential anti-alzheimer's drugs which inhibit β-amyloid peptide release and/or its synthesis [108,109]. The enzymatic resolution of racemic 2-pentanol and 2-heptanol by lipase B from *C. antarctica* has been demonstrated [110].

Commercially available lipases were screened for the enantioselective acetylation of racemic 2-pentanol in an organic solvent (hexane) in the presence of vinyl acetate as an acyl donor. Immobilized *C. antarctica* lipase B gave a reaction yield of 49% (theoretical maximum yield 50%) and 99% ee for (*S*)-2-pentanol at 100 g/L substrate input. Among acylating agents tested, succinic anhydride was found to be of choice due to easy recovery of the (*S*)-2-pentanol at the end of the reaction. Reactions were carried out using racemic 2-pentanol as solvent as well as substrate. Using 0.68 mole equivalent of succinic anhydride and 13 g of lipase B per Kg of racemic 2-pentanol, a reaction yield of 43 M% and an ee of >98% was obtained for (*S*)-2-pentanol, isolated in 38% overall yield. A neat reaction mixture from acylation of racemic 2-pentanol (1.2 L) containing 399 g of (*S*)-2-pentanol was used to develop the recovery process. A portion of the reaction mixture (600 mL) containing (*S*)-2-pentanol (193.3 g) was cooled to −12 to −14 °C and the precipitates were removed by filtration. The clarified crude oil was distilled, with the (*S*)-2-pentanol distilling at 51 to 67 °C under vacuum (20–25 mm Hg) to yield 165 g of product with 98% ee and 99.5% purity.

The resolution of 2-heptanol was also carried out using same immobilized lipase B under similar conditions to give a reaction yield of 44 M% with an ee of >99% of (*S*)-2-heptanol, isolated in 40 M% overall yield [110].

5 KETOREDUCTASES AND CHIRAL ALCOHOLS

5.1 Enzymatic Preparation of (*R*)-3-(4-Fluorophenyl)-2-hydroxy propionic acid

(*R*)-3-(4-fluorophenyl)-2-hydroxy propionic acid **56** (Figure 11.16) is a building block for the synthesis of AG7088, a rhinovirus protease inhibitor **57** [111,112]. A continuous enzymatic process for the efficient synthesis of (*R*)-3-(4-fluorophenyl)-2-hydroxy propionic acid at multikilogram scale with a high space-time yield (560 g/L/day) using a membrane reactor [113] was developed. The product was generated in excellent > 99.9% ee and 70% overall yield. Using this method, 23 kg of (*R*)-**56** was prepared.

The key step in this process was an aqueous enzymatic reduction of keto acid salt **58** to the corresponding (*R*)-hydroxy acid **56** using D-lactate dehydrogenase (D-LDH) and formate dehydrogenase (FDH) to regenerate the NADH cofactor (Figure 11.16).

The process began from 4-fluorobenzaldehyde and hydantoin, which are inexpensive and readily available. Upon condensation and saponification, α-keto acid salt **58** was obtained as a white solid in one step in good yields (77–82%). To scale-up the process, both D-LDH and FDH have to be recycled for the process to be economically feasible. A membrane reactor was used to allow recycling of both D-LDH and FDH. The key part of the reactor was an ultrafiltration membrane unit (10,000 MW cut off in Millipore Pellicon 2 Mini Filter Module), which allows the permeation of small molecules but not macromolecules such as enzymes. In operation, the reactor is initially charged with D-LDH and FDH

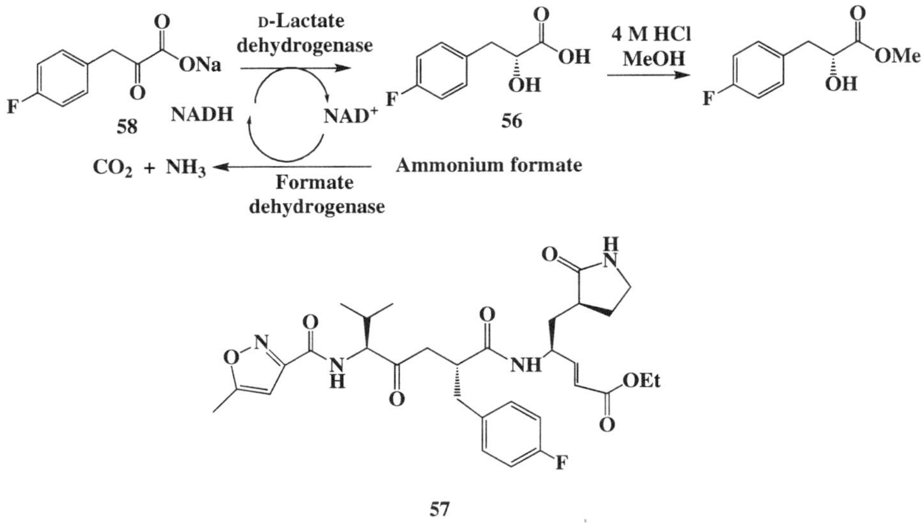

FIGURE 11.16
Rhinovirus protease inhibitor: enzymatic preparation of (*R*)-3-(4-fluorophenyl)-2-hydroxy propionic acid.

before the start of the reaction. An aqueous mixture, which consists of substrate 58, ammonium formate and a catalytic amount of NAD^+, was then continuously fed into the reactor by a peristaltic pump. The substrate solution was mixed with enzymes inside the reactor by a circulation pump. The product was collected continuously as an effluent out of the filtration membrane unit. In this fashion, both enzymes are retained inside the reactor by the membrane leading to a high turnover. D-LDHs from *Leuconostoc mesenteroides* (Sigma-Aldrich) and FDH from *Candida boidinii* (Julich Fine Chemicals) were used [113].

Example 10

The reaction was carried out in a 22 L reactor with EDTA (3.35 g), mercaptoethanol (1.41g), ammonium formate (908 g), and sterile water (18.0 L), which was degassed prior to addition of keto acid salt **58** (800 g). The solution was filtered through a 0.2 μm filter and transferred to a clean 22-L reactor. NAD^+ (23.88 g) was added and the pH adjusted to 6.3 by adding 1 N HCl. This substrate solution was then fed into a membrane reactor with ultrafiltration membrane for enzymatic reduction. The reactor was previously filled with an aqueous mixture of enzymes (D-LDH, 400 units mL-1 with activity 20 units mg-1 and FDH, 20 units mL-1 with activity 76 units mL-1). An appropriate feed rate was used to maintain a conversion of > 90%. The circulation rate was kept between 15 and 30 times that of the feed rate. The aqueous effluent solution thus obtained was adjusted to pH 3.0 with 2 N HCl and extracted with MTBE (5 L). The organic layer was evaporated to obtain 972 g of acid **56** as an off-white solid in a yield of 88%, >90% purity and >99.9% ee. In this process a total of 14.5 kg of **56** was prepared with a productivity of approximately 560 gram per liter per day with good overall 72% yields [113]. To evaluate the optical purity, **56** was converted to methylester by esterification and the ee of methylester was found to be >99.9%.

5.2 Enzymatic Preparation of 6-Hydroxybuspirone

Buspirone (Buspar®, 59, Figure 11.17) is a drug used for the treatment of anxiety and depression, thought to produce its effects by binding to the serotonin 5HT1A receptor [114–116]. Mainly as a result of hydroxylation reactions, it is extensively converted to various metabolites and blood concentrations return to low levels a few hours after dosing [117]. A major metabolite, 6-hydroxybuspirone, produced by the action of liver cytochrome P450 CYP3A4, was present at much higher concentrations in human blood than buspirone itself. For development of 6-hydroxybuspirone as a potential antianxiety drug, preparation and testing of the two enantiomers as well as the racemate was of interest. An enantioselective microbial reduction process was developed for the reduction of 6-oxobuspirone **60** to (*R*)-6-hydroxybuspirone **61a** or (*S*)-6-hydroxybuspitone **61b**. About 150 microbial cultures were screened for the enantioselective reduction of **60**. *Rhizopus stolonifer* SC 13898, *Neurospora crassa* SC 13816, *Mucor racemosus* SC 16198, and *Pseudomonas putida* SC 13817 gave >50% reaction yields and >95% ee of (*S*)-6-hydroxybuspirone **61a**. The yeast strains *Hansenula polymorpha* SC 13845 and *Candida maltosa* SC 16112 gave (*R*)-6-hydroxybuspirone in >60% reaction yield and >97% ee [118]. The NADPH-dependent (*R*)-reductase (RHBR)

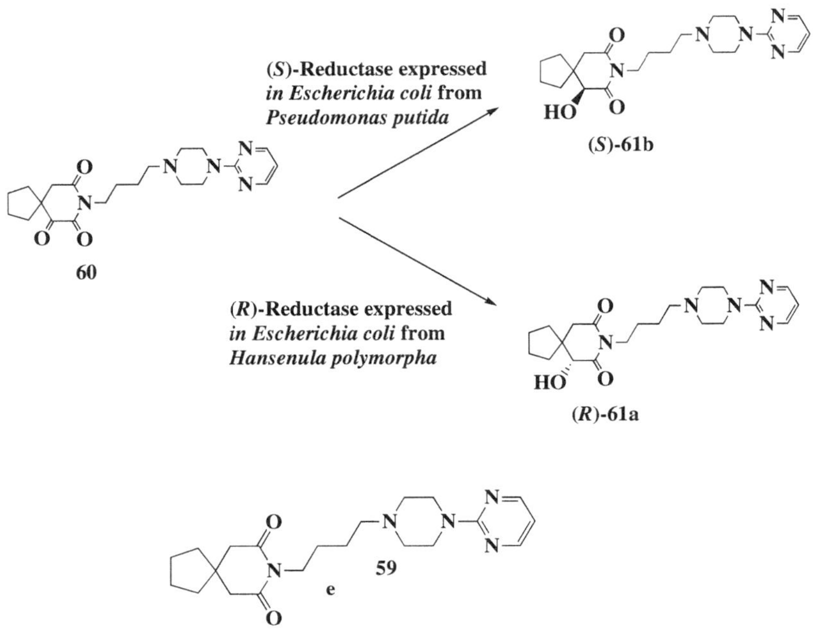

FIGURE 11.17
Enzymatic preparation of 6-hydroxybuspirone.

from *H. polymorpha* SC 13845 was purified to homogeneity, its *N*-terminal and internal amino acid sequences were determined and the corresponding gene was cloned and expressed in *E. coli*. To regenerate the NADPH required for reduction, glucose-6-phosphate dehydrogenase gene from *Saccharomyces cerevisiae* was cloned and coexpressed in the same *E. coli* strain. Recombinant cultures coexpressing (*R*)-reductase (RHBR) and glucose 6-phosphate dehydrogenase catalyzed the reduction of 6-ketobuspirone to (*R*)-6-hydroxybuspirone **61a** in 99% yield and 99.9% ee at 50 g/L substrate input [119].

The NADH-dependent (*S*)-reductase (SHBR) from *P. putida* SC 16269 was also purified to homogeneity, its *N*-terminal and internal amino acid sequences were determined and the corresponding gene was cloned and expressed in *E. coli*. To regenerate the NADH required for reduction, the NAD^+ dependent formate dehydrogenase gene from *Pichia pastoris* was also cloned and co-expressed in the same *E. coli* strain. Recombinant *E. coli* coexpressing (*S*)-reductase and formate dehydrogenase was used to catalyze the reduction of 6-ketobuspirone to (*S*)-6-hydroxybuspirone **61b**, in >98% yield and >99.8% ee at 50 g/L substrate input [119].

5.3 Enzymatic Reduction Process for Synthesis of Intermediate for LTD 4 Antagonists

After the slow reacting substance of anaphylaxis (SRS-A) and its relation to the leukotrienes (LTC4, LTD4, and LTE4) and asthma was discovered, the search for

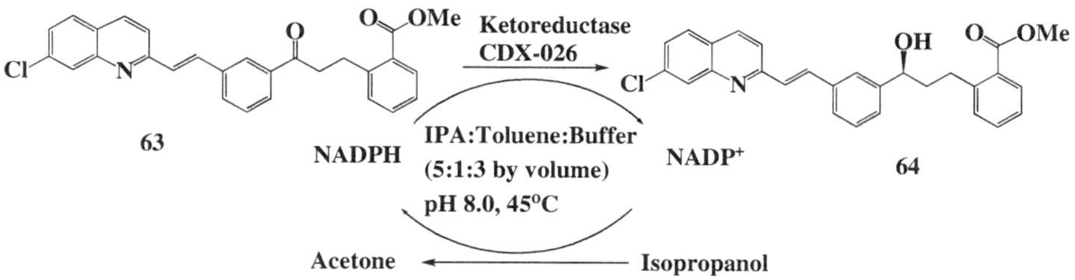

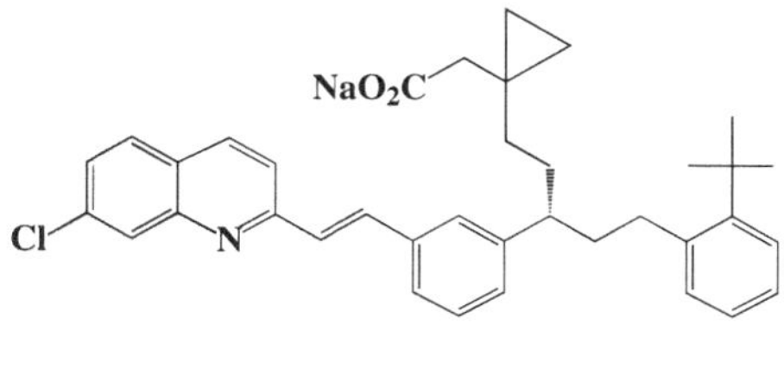

FIGURE 11.18
Enzymatic reduction for synthesis of intermediate for LTD 4 antagonists.

leukotriene antagonists has been intensive as useful for the treatment of asthma. As part of an ongoing program for the development of specific LTD4 antagonists, Merck identified Montelukast **62** (Figure 11.18) [120–122]. The original synthetic route for producing montelukast required the stereoselective reduction of ketone **63** to the (*S*)-alcohol **64** as the key step. Subsequent displacement of the alcohol provides the (*R*)-configured final product [122–124]. The original strategy for reducing **63** to alcohol **64** required stoichiometric amounts of the chiral reducing agent (−)-chlorodiisopino camphenylborane [(−)-DIP-chloride]. Although (−)-DIP-chloride is selective and avoids the side reactions, it is also corrosive and moisture-sensitive, causing burns if it is allowed to contact the skin. The reaction must be carried out at −20 to −25 °C to achieve the best stereoselectivity. The quench and extractive work-up generate large volumes of waste solvent for disposal.

An enzymatic alternative for reducing ketone **63** was developed by Codexis. A ketoreductase was developed by directed evolution using high throughput screens that mimicked the actual process conditions (slurry of the ketone substrate and high isopropanol concentration). Beneficial mutations obtained during each round were recombined and new mutations were introduced, guided by ProSAR. The productivity of the final enzyme was improved 2000-fold and stability was also substantially increased [125,126].

Example 11

The reaction was carried out with a slurry of 2-[3-2[7-chloro-2-quinolinyl]ethenyl]phenyl]-3-oxopropyl]-1-methoxycarbonylbenzene **63** (230 kg) in a mixture of isopropyl alcohol (5 vol), toluene (1 vol), and triethanolamine buffer of pH 8.0 (3 vol). Ketoreductase

CDX-026 (9.2 kg) and the cofactor NADP-Na (0.23 kg) were added to the reaction mixture and the reaction was carried out at 40–45°C under stirring for 40–45 h. At the end of the reaction, saturated NaCl solution (4 vol) and ethyl acetate (10 vol) were added to the reaction mixture. The enzyme was filtered at 60–70°C using a pressure nutsche. The organic phase containing the product was separated. The aqueous layer was extracted with ethyl acetate (2 vol), and the organic phase was combined with the product layer. The combined organic phase was dried and concentrated to give the crude (S)-2-[3-2[7-chloro-2-quinolinyl]ethenyl]phenyl-3-hydroxypropyl]benzoic acid methyl ester **64**. The crude product was purified by recrystallisation from 70% aqueous methanol (14 vol) to get pure product **64** as monohydrate 3 (233 kg), 97.2% yield, >99.9% ee [125,126].

5.4 Enzymatic Synthesis of (*R*)-2-Hydroxy-3, 3-dimethylbutanoic acid for Thrombin Inhibitor

Thrombin is a trypsin-like protease enzyme that plays a critical role in intrinsic and extrinsic blood coagulation. As a result of the enzymatic activation of numerous coagulation factors, thrombin is activated to cleave fibrinogen, producing fibrin, which is directly responsible for blood clotting. An imbalance between these factors and their endogenous activators and inhibitors can give rise to a number of disease states such as myocardial infarction, unstable angina, stroke, ischemia, restenosis following angioplasty, pulmonary embolism, deep vein thrombosis, and arterial thrombosis [127,128]. Consequently, the aggressive search for a potent, selective, and bioavailable thrombin inhibitor is widespread [129]. An intensive effort by Merck has led to the identification of thrombin inhibitor **65** [130]. The synthesis of **65** required a key chiral intermediate (*R*)-hydroxy ester **66**. An enzymatic process was developed for the asymmetric reduction of ketoester **67** to (*R*)-**66** using commercially available ketoreductase from Codexis KRED1001 (Figure 11.19). The cofactor NADPH required for this reaction was regenerated using glucose dehydrogenase. The reaction mixture contained ketoester **67** (45 g/L), NADPH (0.25 g/L), glucose

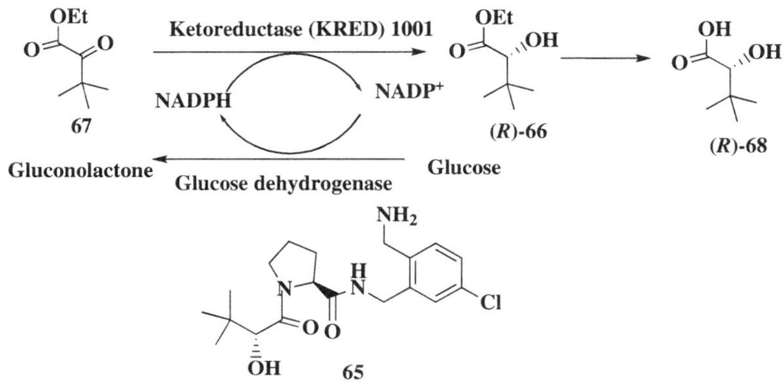

FIGURE 11.19
Thrombin inhibitor: enzymatic synthesis of (*R*)-2-Hydroxy-3,3-dimethylbutanoic acid.

dehydrogenase (1 g/L), KRED1001 (1 g/L), and glucose (70 g/L) in a phosphate buffer at 25 °C with the pH being maintained at 7.0. The conversion was quantitative and occurred with an extremely high degree of stereoselectivity (>500:1; *R*/*S*). The hydroxy ester (*R*)-**66** could be isolated as an oil and then saponified to the corresponding enantiomerically pure hydroxy acid (*R*)-**68** without epimerization. But in a more direct process, a simple pH adjustment (pH >13) of the crude enzymatic reaction mixture (post-reduction) immediately saponified the ethyl ester (*R*)-**66**. After pH adjustment (pH 2), the desired product (*R*)-**68** was extracted into ethyl acetate and the KRED, GDH, NADPH, and glucose remained in the aqueous phase. Removal of the ethyl acetate and crystallization from heptane afforded the enantiomerically pure (*R*)-3,3-dimethyl-2-hydroxybutyric acid (*R*)-**68** in an 82% isolated yield (>99.5% ee). The process was successfully demonstrated at 200 g scale [131].

5.5 Atorvastatin: Enzymatic Preparation of (*R*)-4-cyano-3-hydroxybutyrate

An enzymatic process for the preparation of ethyl (*R*)-4-cyano-3-hydroxybutyric acid **69** (Figure 11.20), a key intermediate for the synthesis of Atorvastatin **12** [132] was developed by Codexis [133]. In this process, first the enzymatic synthesis of ethyl (*S*)-4-chloro-3-hydroxybutyric acid derivative **70** was carried out by ketoreductase-catalyzed conversion of 4-chloro-3-ketobutyric acid derivative **71** [133]. The genes encoding halohydrin dehydrogenase (HHDH) from *Agrobacterium tumefaciens*, ketoreductase from *Candida magnoliae*, glucose

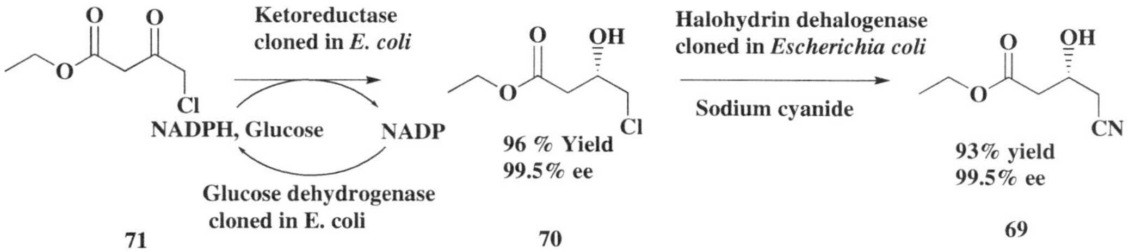

71 70 69

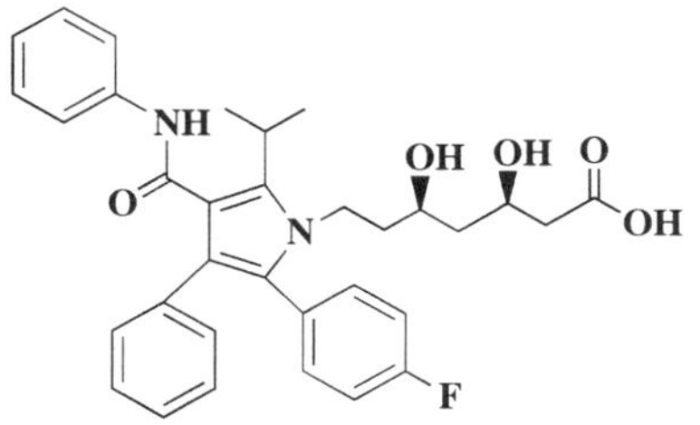

12

FIGURE 11.20
Atorvastatin, rosuvastatin: enzymatic preparation of (*R*)-4-cyano-3-hydroxybutyrate.

dehydrogenase from *Bacillus subtilis* and formate dehydrogenase from *C. boidinii* were separately cloned into *E. coli* BL21. Through several generations of DNA shuffling, GDH activity was improved by a factor of 13 and KRED activity by a factor of 7. The enantioselectivity of the improved KRED remained >99.5%. With the improved enzymes, the reaction was completed in 8 h with substrate input of 160 g/L, KRED loading to 0.57 g/L, and GDH loading to 0.38 g/L.

Similarly after many rounds of DNA shuffling, with screening in the presence of higher concentrations of product, the product inhibition was overcome and the halohydrin dehalogenase activity was increased >2500-fold compared to the wild-type enzyme.

Example 12

To produce ethyl (S)-4-chloro-3-hydroxybutyrate **70**, the reaction mixture contained 570 mL of 100 mM triethanolamine, D-glucose (298 g), KRED (854 mg), GDH (578 mg) and Na-NADP (98 mg) at pH 7.0 and 25°C. The reaction was started by the addition of substrate **71** (240 g) while maintaining the pH at 6.9 with 4 N NaOH feeding. The reaction was completed in 8 h to give 232.61 g (96%) of **71**, as a light yellow-colored liquid. The enantiomeric excess of **70** was >99.5%. For the conversion of ethyl (S)-4-chloro-3-hydroxybutyrate **70** to ethyl (*R*)-4-cyano-3-hydroxybutyrate **69**, a 1 L jacketed three-neck round bottom flask equipped with a rubber septum, a pH electrode connected to a pH stat, and a mechanical stirrer was charged with 400 mL 500 mM NaCN. The whole system was airtight since gaseous HCN is generated. The pH of the solution was 11.2. The vessel was sealed and the pH was adjusted to 7.5 using conc. H_2SO_4 (7 mL). HHDH was charged as an aqueous solution (1.05 g in 20 mL de-ionized water) and the reaction mixture was heated to 40°C. Then, the pH stat control was started and 70 g of **70** was added with a syringe over 10 min. The pH stat maintained the pH at 7.3 with the addition of a 25% NaCN solution containing 0.25% NaOH. The product **69** was recovered 61.6 g (93%) with chemical purity of 99.5% and >99.5% ee after 50 h reaction time [133].

5.6 Cholesterol-Lowering Agents: Enzymatic Preparation of Ethyl (3*S*,5*R*)-dihydroxy-6-(benzyloxy)hexanoate 72

Ethyl (3*R*,5*S*)-dihydroxy-6-(benzyloxy)hexanoate 72 (Figure 11.1) is a key chiral intermediate for the synthesis of Atorvastatin 12, and Rosuvastatin 13, anticholesterol drugs that act by inhibiting HMG CoA reductase [132–134]. The enantioselective reduction of a diketone, ethyl 3,5-dioxo-6-(benzyloxy) hexanoate 73 to ethyl (3*R*,5*S*)-dihydroxy-6-(benzyloxy) hexanoate 72 (Figure 11.21) was demonstrated by cell suspensions of *Acinetobacter calcoaceticus* SC 13876 [135,136]. On reduction of 73 by cell suspensions, the *syn*-4 and *anti*-8 dihydroxy esters were formed in the ratio of about 87:13, 83:17, 76:24 after 24 h at 2, 5 and 10 g/L of substrate input, respectively. There was no significant peak due to a monohydroxy ester. Chiral HPLC determined that the desired *syn*-(3*R*,5*S*)-72 was the major product with 99.4% ee. Almost complete (>95%) conversion of the diketoester 73 to dihydroxy ester 72 in 24 h

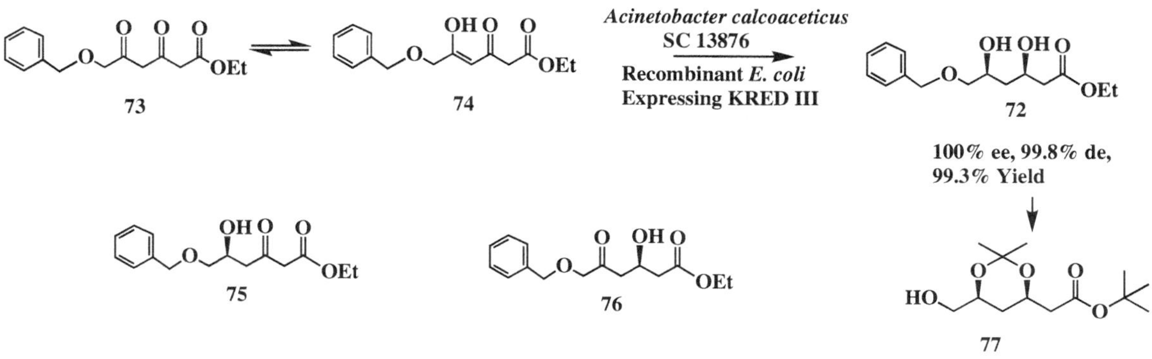

FIGURE 11.21
Enzymatic preparation of (3*S*,5*R*)-dihydroxy-6-(benzyloxy) hexanoic acid, ethyl ester.

was seen up to a substrate concentration of 10 g/L and cell concentration of 200 g/L [135,136].

A mixture of ethyl 3-keto-5-hydroxy **75** (major) and 5-keto-3-hydroxy **76** (minor) was obtained from partial microbial reduction of ketoester **73**. Another mixture of ethyl 3-keto-5-hydroxy **75** (minor) and 5-keto-3-hydroxy **76** (major) was obtained from chemical reduction of ketoester **73**. These two mixtures were subjected to microbial reduction by *Acinetobacter sp* SC13874 cells for 6 h. The results indicated that the second reduction of the monohydroxy compound by SC13874 cells was quite enantioselective. Reduction of the 3-keto-5-hydroxy **75** provided predominantly the (3*R*)-hydroxy, while reduction of the 3-hydroxy-5-keto ester **76** provided predominantly the (5*S*)-hydroxy compound [136].

Three different ketoreductases were purified to homogeneity from cell extracts of *A. calcoaceticus* SC 13876 and their biochemical properties were compared. Reductase I only catalyzes the reduction of ethyl diketoester **73** to its monohydroxy products whereas reductase II catalyzes the formation of dihydroxy products from monohydroxy substrates. A third reductase (III) was identified which catalyzes the reduction of diketoester **73** to *syn*-(3*R*,5*S*)-dihydroxy ester **72** [136]. Reductase III has been cloned and expressed in *E. coli* [137] and reduced the diketoester **73** to *syn*-(3*R*,5*S*)-dihydroxy ester **72** in 99.3% yield, 100% ee, and 99.8% de [137]. Ethyl (3*R*,5*S*)-dihydroxy-6-(benzyloxy) hexanoate **72** is the precursor of Kaneka alcohol **77**, a key intermediate for the synthesis of Arotvastatin **12**, and Rosuvastatin **13**.

5.7 Enzymatic Preparation of (1*S*,2*R*)-[3-chloro-2-hydroxy-1-(phenylmethyl) propyl]-carbamic acid,1,1-dimethyl-ethyl ester

Atazanavir **78** (Figure 11.22) is an acyclic aza-peptidomimetic, a potent HIV protease inhibitor [138–140] approved by the Food and Drug Administration for the treatment of Acquired Immunodeficiency Syndrome (AIDS). An

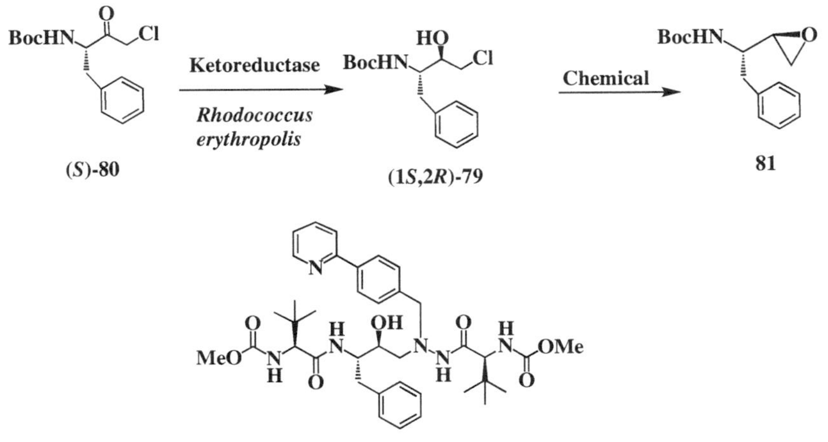

FIGURE 11.22
Atazanavir: enzymatic preparation of (1*S*,2*R*)-[3-chloro-2-hydroxy-1-(phenylmethyl) propyl]-carbamic acid,1,1-dimethyl-ethyl ester.

enzymatic process was developed to prepare (1*S*, 2*R*)-[3-chloro-2-hydroxy-1-(phenylmethyl) propyl]carbamic acid, 1,1-dimethylethyl ester **79**, a key chiral intermediate in the synthesis of atazanavir. The diastereoselective reduction of (1*S*)-[3-chloro-2-oxo-1-(phenylmethyl)propyl] carbamic acid, 1,1-dimethylethyl ester **80** was carried out using *Rhodococcus, Brevibacterium, and Hansenula* strains to provide **79**. Three strains of *Rhodococcus* gave >90% yield with a diastereomeric purity of >98% and an ee of 99.4% [141]. An efficient single-stage fermentation-biotransformation process was developed with cells of *R. erythropolis* SC13845 for the reduction of ketone **80** to **79**. The reaction proceeded in 95% yield with a diasteromeric purity of 98.2% and an ee of 99.4% at a substrate input of 10 g/L. The reduction process was further improved by generating mutants and selecting the best-suited for converting **80** to (1*S*,2*R*)-**79** at substrate input of 60 g/L. Subsequently, (1*S*,2*R*)-**79** was converted to epoxide **81** and used in the synthesis of atazanavir [142]. Chemical reduction of chloroketone **80** using $NaBH_4$ produces the undesired chlorohydrin diastereomer [143].

Recently, Codexis developed a ketoreductase which catalyzed the reduction of **80** to (1*S*,2*R*)-**79**. Wild type recombinant ketoreductase gave only 5% conversion and 80% enantiopurity after 24 h at substrate input of 20 g/L and enzyme input of 5 g/L. The enzyme was evolved based on ProSAR generated library and screening of entire library of mutants. They developed a highly stereoselective and diastereoselective reduction process with isopropanol in water as the solvent and NADP+ as the cofactor by a custom-evolved ketoreductase. The evolved enzyme operated at 200 g/L substrate, 1 g/L enzyme and 0.01 g/L NADP inputs with 99.9% conversion and >99% enantiomeric purity in 24 h reaction time [144].

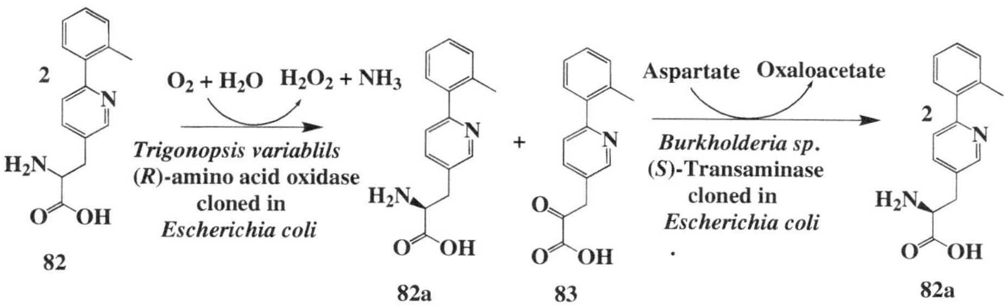

FIGURE 11.23
GLP-1 peptide: enzymatic preparation of (*S*)-amino-3-[3-{6-(2-methylphenyl)} pyridyl]-propionic acid.

6 ENZYMATIC TRANSAMINATION REACTIONS

6.1 Enzymatic Preparation of (*S*)-amino-3-[3-{6-(2-methylphenyl)} pyridyl]-propionic acid

(*S*)-amino-3-[3-{6-(2-methylphenyl)} pyridyl]-propionic acid **82a** (Figure 11.23) is a key intermediate required for synthesis of GLP-1 mimics or GLP-1 receptor modulators. Such receptor modulators are potentially useful for the treatment of type II diabetes treatment [145,146].

(*S*)-Amino-3-[3-{6-(2-methylphenyl)}pyridyl]-propionic acid **82a** was prepared by an enzymatic deracemization process using a combination of two enzymes: (*R*)-amino acid oxidase from *Trigonopsis variabilis* cloned and expressed in *E. coli* and an (*S*)-aminotransferase from *Sporosarcina ureae,* also cloned and expressed in *E. coli* [147]. Racemic amino acid **82** was used as a substrate and (*S*)-aspartate was used as amino donor. An (*S*)-aminotransferase was also purified from a soil organism identified as *Burkholderia sp.* and cloned and expressed in *E. coli* and used in this process [147]. This process was scaled up to 70 L scale.

Example 13

Frozen *rec E. coli* SC16541 cells (12 kg) cells containing (S)-aminotransferase and *rec E. coli* SC16544 cells (2.4 kg) containing (*R*)-amino acid oxidase were suspended in 46 L of 0.1 M phosphate buffer (pH 8.0) containing 5 μM pyridoxal phosphate (PLP) with agitation at 200 rpm. The 24% cell suspension was passed twice through a microfluidizer. Polyethylene amine was added to 0.2% concentration, mixed for 30 min, followed by centrifugation to clarify the extract. To a 100 L vessel containing 29 L of water, stirring at 100 rpm, were added sodium aspartate monohydrate (1.67 kg), ascorbic acid (0.33 kg), racemic amino acid **82** monosulfate monohydrate (1 kg, potency 60.7%), and propylene glycol (3.33 kg). The solution was adjusted to pH 7.0. Subsequently, catalase (0.133 kg), pyridoxal phosphate (167 mg), clarified cell extract (40 L), and SAG-5693 antifoam (133 g) were added to the reactor. The reaction was carried out at 30°C, 250 rpm, pH 7.5 and 80 L/min aeration. The bioconversion was completed after 25 h and product **82a** was isolated from the reaction mixture in 66% yield (580 g) with 99.9% enantiomeric purity.

In an alternate process, the enzymatic dynamic resolution of racemic amino acid **82** was also demonstrated by combining amino acid oxidase with chemical reduction. (*R*)-selective oxidation with Celite-immobilized (*R*)-amino acid oxidase from *T. variabilis* expressed in *E. coli* in combination with chemical imine reduction with borane-ammonia gave a 75% in process yield and 100% ee of (*S*)-amino acid **82a** [147].

In another alternate enzymatic process, racemic amino acid **82** was first treated with (*R*)-amino acid oxidase for 4 h to convert **82** to a mixture of (*S*)-amino acid **82a** and keto acid **83**. Subsequently, an amino acid dehydrogenase, NADP and glucose dehydrogenase were added to the same reaction mixture to convert **83** to (*S*)-amino acid **82a**, providing this product in 54% isolated yield and 99% ee.

6.2 Enzymatic Synthesis of Thiazepine

[4*S*-(4a,7a,10ab)]-1-Octahydro-5-oxo-4-[[(phenylmethoxy)carbonyl]amino]-7H-pyrido-[2,1-β] [1,3]thiazepine-7-carboxylic acid **84** is a key intermediate in the synthesis of Omapatrilat (Vanlev) **85** [148,149]. An enzymatic process (Figure 11.24) was developed that oxidized the ε-amino group of (*S*)-lysine in

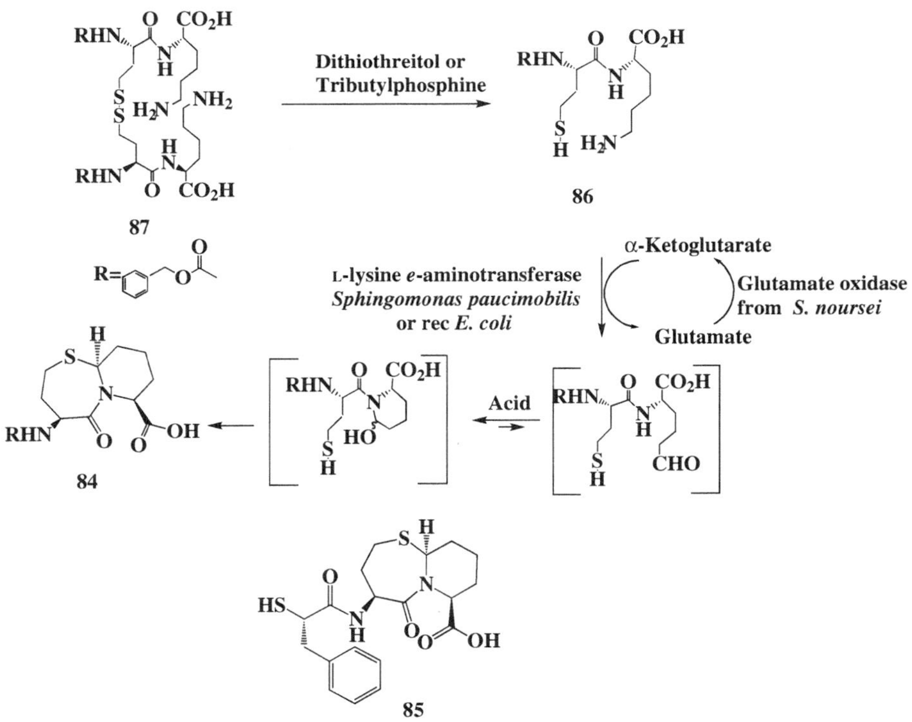

FIGURE 11.24
Vanlev: enzymatic synthesis of thiazepine.

the thiol **86** generated *in situ* from disulfide N[2]-[N[[(phenylmethoxy)carbonyl] L-homocysteinyl] L-lysine)-1,1-disulphide **87** to produce compound **84** using L-lysine ε-aminotransferase [LAT] from *Sphingomonas paucimobilis* SC 16113 [150,151]. This enzyme was cloned and overexpressed in *E. coli* and a process was developed using the recombinant enzyme. The aminotransferase reaction required α-ketoglutarate as the amine acceptor. The glutamate formed during this reaction was recycled back to α-ketoglutarate by glutamate oxidase [GOX] from *Streptomyces noursei* SC 6007.

The biotransformation of compound **87** to compound **84** was carried out using LAT from *E. coli* GI724[pal781-LAT] in the presence of α-ketoglutarate and dithiothreitol or tributylphosphine and glutamate oxidase. Reaction yields of 65–67 M% were obtained. To reduce the cost of producing two enzymes, the transamination reactions were carried out in the absence of GOX and with higher levels of α-ketoglutarate. The reaction yield in the absence of GOX averaged only about 33–35 M%; however, the reaction yield increased to 70 M%, by increasing the α-ketoglutarate level to 40 mg/mL (a 10-fold increase in concentration) and conducting the reaction at 40 °C, equivalent to that in the presence of GOX [150].

Example 14

In a 5-L bioreactor containing 22 g of substrate **87** in 2.5 L of 100 mM phosphate buffer, pH 7.8, 5 mM EDTA was added and the pH was adjusted to 12 by addition of 5 N NaOH while stirring (200 rpm) at room temperature. The pH was maintained at 12 until all solids were dissolved. The pH was then readjusted to 7.8 with slow addition of concentrated phosphoric acid. The temperature of the bioreactor was adjusted to 30°C and 6.75 g of DTT, 19 g of α-ketoglutarate (disodium salt) and 100 mg of pyridoxal phosphate were added. An additional 1.4 L of 100 mM phosphate buffer containing 5 mM EDTA was then added. The reaction was started by addition of 100 mL of *E. coli* GI724[pal781-LAT] cell extract containing 2200 units of enzyme. Glutamate oxidase solution (2000 units) was added at 30 min, 1 h, 1.5 h and 3.5 h. The reaction was carried out at 30°C, pH 7.8, and 200 rpm. The reaction was completed in 4 h and product **84** was isolated in overall 65% yield [150,151].

6.3 CGRP Receptors Antagonist (Antimigraine Drugs): Enzymatic Preparation of (R)-2-amino-3-(7-methyl-1 H-indazol-5-yl)Propanoic Acid

The (*R*)-amino acid, (*R*)-2-amino-3-(7-methyl-1 H-indazol-5-yl)propanoic acid (*R*)-**88a**, (Figure 11.25) is a key intermediate for preparing antagonists of calcitonin gene-related peptide receptors (CGRP receptor) **89** [152] Such antagonists are potentially useful for the treatment of migraine and other maladies [153,154].

(*R*)-Amino acid **88a** was prepared in 68% isolated yield with >99% ee from racemic amino acid **88** using (*S*)-amino acid oxidase from *Proteus mirabilis*

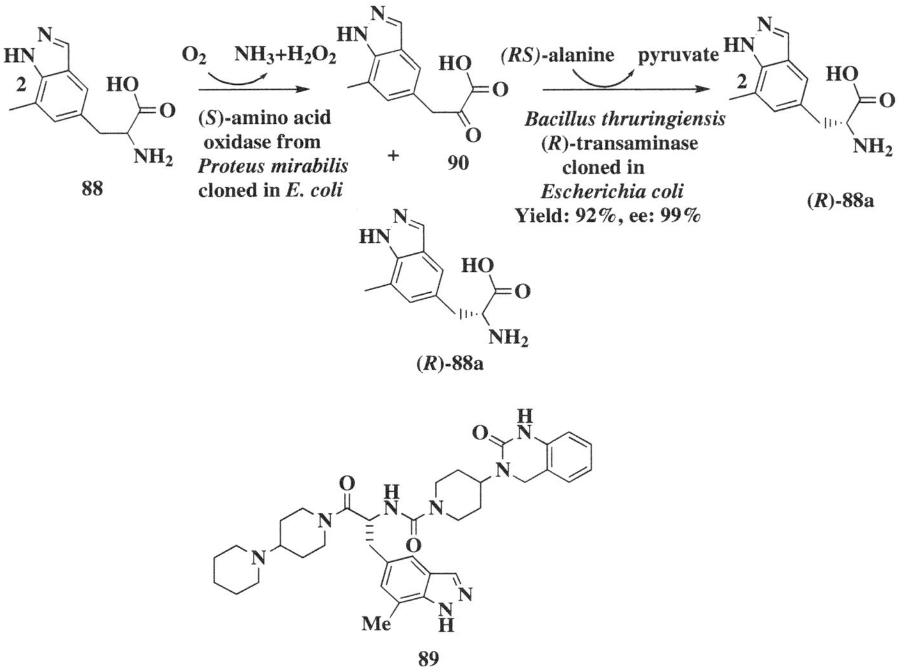

FIGURE 11.25
CGRP receptors antagonist: enzymatic preparation of (*R*)-2-amino-3-(7-methyl-1 H-indazol-5-yl)propanoic acid.

cloned and expressed in *E. coli* in combination with a commercially available (*R*)-transaminase and (*R*)-alanine as the amino donor [155]. Several 1 L batches were run to develop a procedure suitable for scale-up. All contained racemic **88** (20 g), D-alanine (40 g), pyridoxal phosphate monohydrate (2.65 mg,), 100 g of *E. coli* frozen cell paste containing the cloned (*S*)-amino acid deaminase (27 U/g cells wet weight) from *P. mirabilis*, a Biocatalytics (R)-transaminase (200 mg), K_2HPO_4 (6.97 g), and KH_2PO_4 (1.36 g,). The pH was maintained at 7.5. Three batches run under these conditions gave an average 85% solution yield with 96.5% ee. The conversion of 2.9 kg racemic **88** to (*R*)-**88a** was carried out in two batches in a pilot plant. The average yield of the isolated product was 1.79 kg with 98.6% ee [155].

The (*R*)-enantiomer was also prepared in 79% isolated yield with >99% ee from the corresponding keto acid **90** using the (*R*)-transaminase with racemic alanine as the amino donor. An (*R*)-transaminase was identified and purified from a soil organism identified as *Bacillus thuringiensis* and cloned and expressed in *E. coli*. The recombinant (*R*)-transaminase was very effective for preparing (*R*)-**88a** and gave a nearly complete conversion of **90** to (*R*)-**88a** [154].

Example 15

The recombinant (*R*)-transaminase prepared from cell extracts of *rec E. coli* SC16577 (40 g wet cells/200 mL suspension) was suspended in buffer containing 50 mM potassium phosphate pH 7, 1 mM dithiothreitol, and 10 μM pyridoxal phosphate using a microfluidizer at 12000 psi followed by separation of cell debris. The reaction mixture contained 10 mL of this cell extract, keto acid **90** (6.0 g), D,L-alanine (12.0 g), pyridoxal phosphate (2.6 mg), dithiothreitol (15.0 mg) and 0.1 M potassium phosphate buffer pH 7.5 (80 mL). After incubating at 30°C with gentle shaking at 50 rpm for 39 h, the reaction was complete with 95% in-process yield and the product (*R*)-**88a** was isolated in 79% yield (5.4g) with >99% ee [155].

6.4 Sitagliptin: Synthesis of a Chiral Amine

Sitagliptin **91** (sold under the trade name Januvia, Figure 11.26) is an oral antihyperglycemic (anti-diabetic drug) of the dipeptidyl peptidase-4 (DPP-4) inhibitor class. It was developed and is marketed by Merck & Co. This enzyme-inhibiting drug is used either alone or in combination with other oral antihyperglycemic agents (such as metformin or a thiazolidinedione) for treatment of diabetes mellitus type 2.

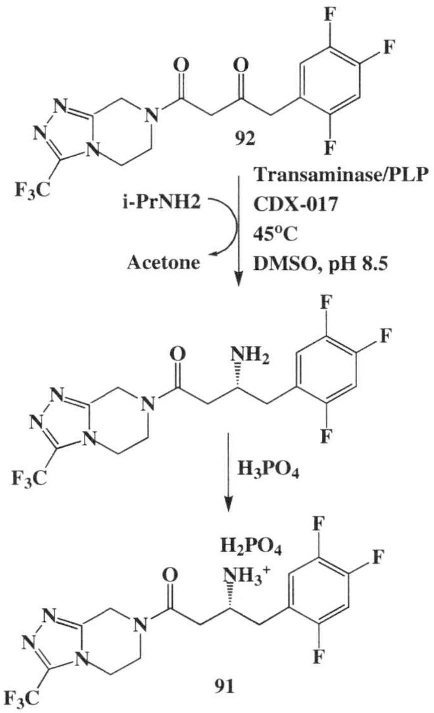

FIGURE 11.26
Sitagliptin: synthesis of Sitagliptin by (*R*)-transaminase.

The original synthesis of sitagliptin [156,157] required the asymmetric hydrogenation of an enamine at high pressure using a rhodium-based chiral catalyst [158]. This reaction suffers from inadequate stereoselectivity and a product stream contaminated with rhodium, necessitating additional purification steps at the expense of yield to upgrade both enantiomeric excess (ee) and chemical purity. By using a transaminase [159–162] scaffold and various protein engineering technologies, an enzymatic process was developed which substantially improved the efficiency of sitagliptin manufacturing. Starting from an enzyme that had the catalytic machinery to perform the desired chemistry but lacked any activity toward the prositagliptin ketone **92**, a substrate walking, modeling, and mutation approach was used to create a transaminase with marginal activity for the synthesis of the chiral amine; this variant was then further engineered *via* directed evolution for practical application in a manufacturing setting. The resulting biocatalyst showed broad applicability toward the synthesis of chiral amines that previously were accessible only *via* enzymatic resolution process [163].

Under optimal conditions [163], the best variant converted 200 g/L prositagliptin ketone **92** (Figure 11.26) to sitagliptin **91** in >99.95% ee by using 6 g/L enzyme in 50% DMSO with a 92% assay yield at the end of reaction. In comparison with the rhodium-catalyzed process, the biocatalytic process provides sitagliptin with a 10–13% increase in overall yield, a 53% increase in productivity, a 19% reduction in total waste, the elimination of all heavy metals, and a reduction in total manufacturing cost; the enzymatic reaction is run in multipurpose vessels, avoiding the need for specialized high-pressure hydrogenation equipment.

6.5 (CRF)-1 Receptor Antagonist: Enzymatic Preparation of (*R*)-Amines

Anxiety and depression are psychiatric disorders that constitute a major health problem throughout world. While numerous marketed treatments exist for both disorders, there remains a continuing need for drugs with increased efficacy and/or reduced side-effect profiles [164,165]. CRF (1) receptor antagonists have been proposed as novel pharmacological treatments for depression, anxiety and stress disorders [164–166].

(*R*)-sec-butylamine **93a** and (*R*)-1-cyclopropylethylamine **94a** (Figure 11.27) are key chiral intermediates for the synthesis of CRF-1 receptor antagonists such as **95** [165,166].The enzymatic resolution process for the preparation of both (*R*)-sec-butylamine and (*R*)-1-cyclopropylethylamine was developed [167]. Screening was carried out to identify strains useful for preparing (*R*)-1-cyclopropylethylamine and (*R*)-*sec*-butylamine from the respective racemic amines with an (*S*)-specific transaminase. Several *Bacillus megaterium* strains as well as several soil isolates were found to have the desired activity. Using an extract of the best strain, *B. megaterium* SC6394, the reaction was shown to be a transamination requiring pyruvate as amino acceptor and pyridoxal

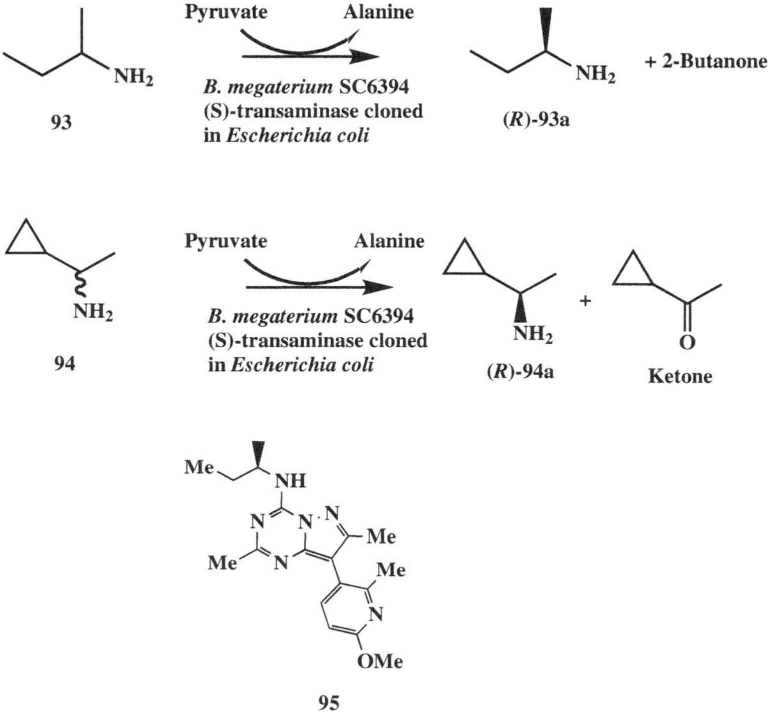

FIGURE 11.27
CRF-1 receptor antagonist: enzymatic preparation of (*R*)-amines.

phosphate as a cofactor. Initial batches of both amines were produced using whole cells of *B. megaterium* SC6394. The transaminase was subsequently purified to homogeneity and the *N*-terminal as well as internal amino acid sequences were determined. These enabled the cloning and overexpression of the transaminase in *E.coli* SC16578. In contrast to the reaction with *B. megaterium* cells, isolated cells (or cell extracts) of recombinant *E. coli* for the resolution process provided several benefits. Neither pH control nor aeration was required and an excess of pyruvate was not consumed by the recombinant cells during preparation of (*R*)-sec-butylamine **93a** and (*R*)-1-cyclopropylethylamine **94a**. The resolution of *sec*-butylamine using whole cells of *E. coli* SC16578 was scaled up to give (*R*)-*sec*-butylamine1/2 H_2SO_4 in 46.6% isolated yield with 99.2% ee. The reaction used 375 g frozen SC16578 *E. coli* cells, racemic sec-butylamine (750 g) and sodium pyruvate (750 g) at pH 8.0 in 15 L potassium phosphate buffer. The reaction was run at 30 °C, 100 rpm and pH 8 and continued until the ee reached >99% (23 h). From the reaction mixture, 585 g of (*R*)-sec-butylamine · 1/2 H_2SO_4 was obtained in 46.6% isolated yield (theoretical maximum yield is 50%) and 99.2% ee. Using the same recombinant (*S*)-transaminase, (*R*)-1-cyclopropylethylamine was obtained in 42% isolated yield (theoretical max. 50%) and 99% ee [167].

7 ENZYMATIC REDUCTIVE AMINATION

The reductive amination of α-keto acids using amino acid dehydrogenases is one of the most useful biocatalytic methods because the enzymes have good stabilities, broad substrate specificities, very high enantioselectivities and can be used at high substrate concentrations since keto acids are highly soluble in aqueous system. A reductive amination process coupled to an enzymatic cofactor regeneration system is the most prominent method for preparing chiral amino acids. For most enzymes, the required cofactor is NADH, although NADPH is required in some cases. Yeast formate dehydrogenase is commonly used for NADH regeneration and glucose dehydrogenase (usually from *Bacillus* species) can regenerate both NADH and NADPH. There are excellent reviews on the amino acid dehydrogenases and examples of their synthetic utilities [168–172].

7.1 Tigemonam: Enzymatic Synthesis of (*S*)-β-Hydroxyvaline

(*S*)-β-hydroxyvaline **96** (Figure 11.28), is a key chiral intermediate required for the total synthesis of the orally active monobactam [173,174], Tigemonam **97**. The synthesis of (*S*)-β-hydroxyvaline **96** from α-keto-β-hydroxyisovalerate **98** by reductive amination using leucine dehydrogenase from *Bacillus sphaericus* ATCC 4525 was demonstrated [175]. The NADH required for this reaction was regenerated by either formate dehydrogenase from *C. boidinii* or glucose dehydrogenase from *B. megaterium*. The required substrate **98** was generated either from α-keto-β-bromoisovalerate or its ethyl esters by hydrolysis with sodium hydroxide *in situ*. In this process, an overall reaction yield of 98% and an ee of 99.8% were obtained for the S-β-hydroxyvaline **96**. Reactions with glucose

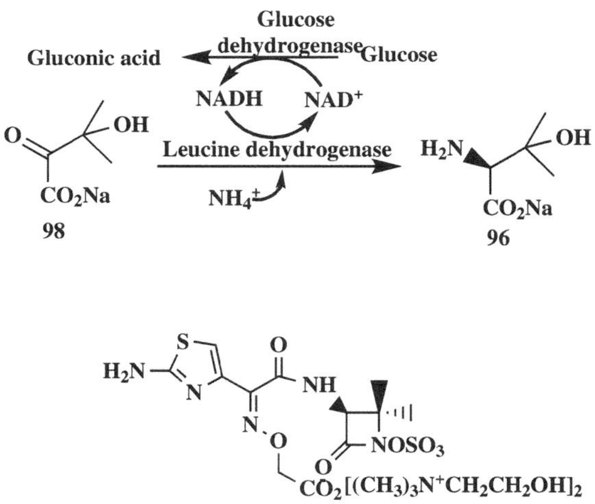

FIGURE 11.28
Tigemonam: enzymatic synthesis of (*S*)-β-hydroxyvaline.

dehydrogenase and leucine dehydrogenase were run at pH 8.5 and 30 °C. The pH was maintained by addition of 3 M NH_4OH and the reactions contained 0.5 M α-keto-β-hydroxyisovalerate, 1 M glucose, 0.2 M NH_4Cl, 0.5 mM NAD^+, 0.01% mercaptoethanol, 44 units leucine dehydrogenase from *B. sphaericus* ATCC 4525, and 29 units glucose dehydrogenase from *B. megaterium*.

7.2 Atazanavir: Enzymatic Synthesis of (*S*)-tertiary-leucine

Atazanavir **78** is an acyclic aza-peptidomimetic and potent HIV protease inhibitor [176,177]. The synthesis of atazanavir required (*S*)-*tertiary* leucine **99** (Figure 11.29). The same building block is also a key chiral amino acid required for a number of other drugs containing peptides such as Atazanavir, Boceprevir and Telaprevir [178,179].

Leucine dehydrogenase and a formate dehydrogenase from *C. boidinii* were used to enzymically reduce 2-ketocarboxylic acids while regenerating the NADH cofactor *in situ* and this constitutes an industrially established method for preparing optically active L-α-amino acids. In particular, this route is suitable for preparing the non-proteinogenic amino acid S-*tert*-leucine, which is produced on the ton scale using this biocatalytic method using membrane reactor. The method is described in detail in the literature [169,180,181]. However, this method required isolated enzymes in purified form.

Soda et al. [182] described the use of a whole-cell catalyst for the enzymatic synthesis of L- and D- amino acids from α-keto acids with *E. coli* cells that overexpress heterologous genes. L-Amino acids were produced with a thermostable (*S*)-amino acid dehydrogenase and formate dehydrogenase (FDH) from α-keto acids and ammonium formate. No exogenous NADH was required as the

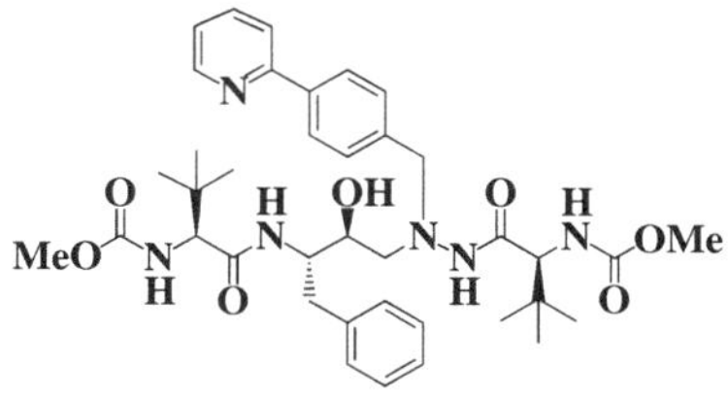

78

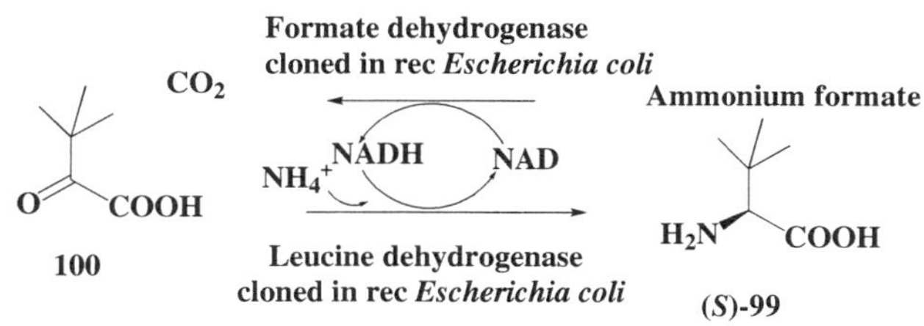

FIGURE 11.29
Atazanavir: enzymatic synthesis of (*S*)-tertiary-leucine.

pre-existing intracellular pool was sufficient. This group constructed plasmids containing – in addition to the FDH gene – genes encoding a variety of amino acid dehydrogenases including those for leucine, alanine and phenylalanine. These recombinant cells were used to produce L-Leucine, L-valine, L-norvaline, L-methionine, L-phenylalanine, and L-tyrosine with high chemical yields (>80%) and high optical purities (up to 100% enantiomeric excess). Stereoselective conversions of various α-keto acids to D-amino acids were also examined with recombinant *E. coli* cells containing a plasmid encoding four heterologous genes: a thermostable D-amino acid aminotransferase, alanine racemase, L-alanine dehydrogenase, and FDH. Optically pure D-enantiomers of glutamate and leucine were obtained.

This publication points out that if preference were to be given to taking advantage of the intracellular pool of NAD^+ in *E.coli*, the final concentration of product would be restricted to about 0.35–0.38 M.

Recently, Groger et al [183] developed a process for preparing enantiomerically enriched (*S*)-α-amino acids or their salts by reacting the corresponding 2-ketocarboxylic acid with an ammonium ion donor in the presence of a whole-cell catalyst which comprises a cloned gene encoding a cofactor-dependent amino acid dehydrogenase and a cloned gene encoding an enzyme which regenerates the cofactor. Substrate addition must be metered such that the stationary concentration of 2-ketocarboxylic acid remains < 500 mM and the external addition of cofactor, based on the total input of substrate, corresponds to < 0.0001 equivalents. In this process, a reaction yield of 84% and ee of 99% was obtained for (*S*)-*tertiary* leucine at 130 g/L substrate **100** input.

7.3 Vanlev: Enzymatic Synthesis of Allysine Ethylene Acetal

(*S*)-2-Amino-5-(1,3-dioxolan-2-yl)-pentanoic acid [(*S*)-allysine ethylene acetal] **101** (Figure 11.30) is one of three building blocks used in an alternative synthesis of Vanlev **85**, an antihypertensive drug which acts by inhibiting angiotensin-converting enzyme (ACE) and neutral endopeptidase (NEP) [184]. The synthesis of **101** was demonstrated by reductive amination of ketoacid acetal **102** using phenylalanine dehydrogenase [PDH] from *Thermoactinomyces intermedius* [185]. The reaction required ammonia and NADH; NAD^+ produced during the reaction was recyled to NADH by the oxidation of formate to CO_2 using formate dehydrogenase [FDH] The *T. intermedius* PDH was cloned and overexpressed in *E.coli* and the recombinant culture was used as a source of PDH. Expression of *T. intermedius* PDH in *P. pastoris*, inducible by methanol, allowed both enzymes to be produced in a single fermentation as methanol-grown cells of *P. pastoris* also highly express a native formate dehydrogenase. A total of 197 kg of (*S*)-**101** was produced in three 1600 L batches using a 5% concentration of substrate **102** with an average yield of 91 M% and ee >98% [185]. (*S*)-allysine ethylene acetal was converted to Vanlev **85** [186].

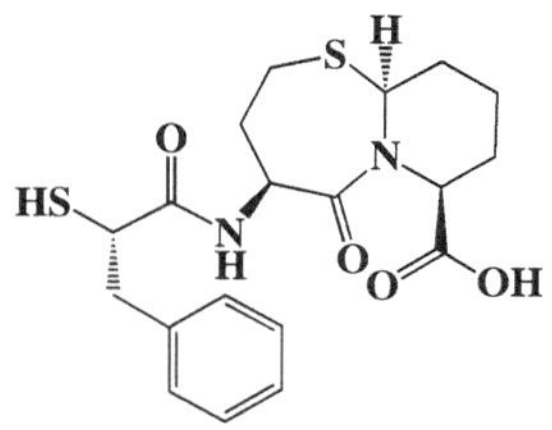

85

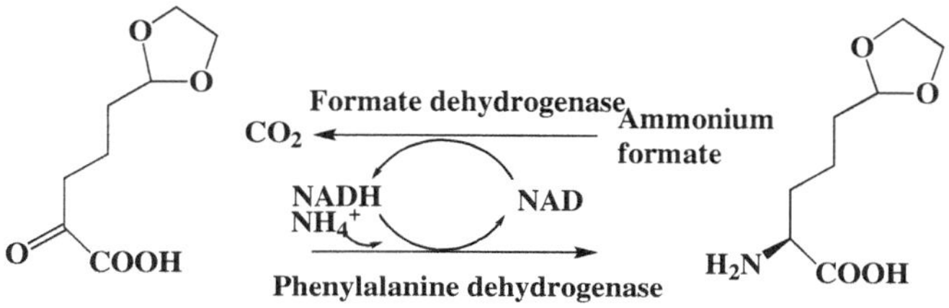

102 (*S*)-101

FIGURE 11.30
Vanlev: enzymatic synthesis of allysine ethylene acetal.

Example 16

In a typical reaction, the lithium salt of **102** (20.0 g) and ammonium formate (7.796 g) were added to a 1 L jacketed reactor and 400 mL of water was added to the reactor and stirred at 350 rpm until the solids were dissolved. The pH was adjusted to 8 by adding concentrated ammonium hydroxide or 99% formic acid as required. Dithiothreitol (65.6 mg) and NAD^+ (282 mg) were added, then the reactor was heated to 40°C and maintained at that temperature. *E. coli BL21(DE3)*(pPDH155K) [SC16144] heat-dried cells (0.5328 g, 666 unit PDH) and *C. boidinii* heat-dried cells (1.954 g, 110 units FDH) were added to the stirred solution and the pH was brought back to 8 by addition of concentrated ammonium hydroxide solution. When the conversion was complete, cells were removed by vacuum filtration after 5% (w/v) Celite admix and precoat. The solution was then passed through a 10 000 MW cutoff membrane to remove large molecular weight solubles such as proteins.

7.4 Saxagliptin: Enzymatic Reductive Amination of 2-(3-hydroxy-1-adamantyl)-2-oxoethanoic acid

Dipeptidyl peptidase 4 (DPP-4) is a ubiquitous proline-specific serine protease responsible for the rapid inactivation of incretins, including glucagon-like peptide 1 (GLP-1) and glucose-dependent insulinotropic peptide. To alleviate the inactivation of GLP-1, inhibitors of DPP-IV are being evaluated for their ability to provide improved control of blood glucose for diabetics [187–189].

Saxagliptin **32** [187–189] (Figure 11.31), a DPP-IV inhibitor developed by Bristol-Myers Squibb, requires (*S*)-*N*-boc-3-hydroxyadamantylglycine **103** as

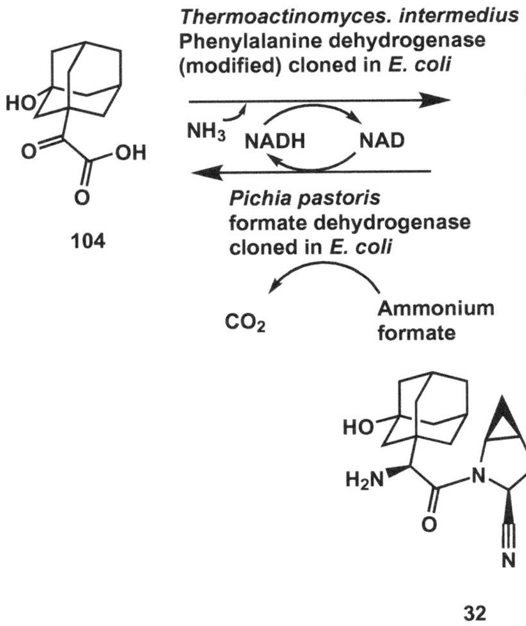

FIGURE 11.31
Saxagliptin: enzymatic reductive amination of 2-(3-hydroxy-1-adamantyl)-2-oxoethanoic acid.

an intermediate. A process for converting keto acid **104** to the corresponding amino acid **105** using (*S*)-amino acid dehydrogenases was developed. A modified form of a recombinant phenylalanine dehydrogenase cloned from *T. intermedius* and expressed in *P. pastoris* or *E. coli* was used for this process. NAD^+ produced during the reaction was recycled to NADH using formate dehydrogenase. The modified phenylalanine dehydrogenase (PDHmod) contained two amino acid changes at the C-terminus along with a 12 amino acid extension at the C-terminus [190].

Production of multi-kg batches was originally carried out with extracts of *P. pastoris* overexpressing the modified phenylalanine dehydrogenase from *T. intermedius* and endogenous formate dehydrogenase. The reductive amination process was further scaled up using a preparation of the two enzymes expressed in single recombinant *E. coli*. The amino acid **105** was directly protected as its Boc derivative without isolation to afford intermediate **103**. Yields before isolation were close to 98% with 100% ee [190].

Reductive amination was also conducted using cell extracts from *E.coli* strain SC16496 expressing both PDHmod and FDH from *P. pastoris*. Cell extracts after polyethyleneimine treatment, clarification and concentration were used to complete the reaction in 30 hrs with >96% yield and >99.9% ee of product **105**. This process has been used to prepare Boc-protected amino acid **103** [190].

Example 17

The reaction was carried out using cell extracts of recombinant *E. coli* (co-expressing modified phenylalanine dehydrogenase and formate dehydrogenase, 266 mL) containing 2456 U phenylalanine dehydrogenase and 8801 U formate dehydrogenase in a 1 L bottle. A solution (266 mL) containing ammonium formate (16.74 g) and keto acid **104** (29.76 g) was brought to pH 8.0 by ammonium hydroxide. NAD^+ (372 mg) and dithiothreitol (81.8 mg,) were added, then the solution was added to the bottle containing the *E. coli* extract. The bottle was maintained at 40°C on a shaker at 400 rpm. After 38 h, the solution contained 31.5 g (100% yield) of amino acid **105**. The product was isolated as Boc-amino acid **103** from the reaction mixture in 88% overall yield with 100% ee.

7.5 Enzymatic Synthesis of (*S*)-Neopentylglycine

The enantioselective synthesis of (*S*)-neopentylglycine **106** (Figure 11.32) was developed by Groeger et al. [191]. Recombinant whole cells containing leucine dehydrogenase and formate dehydrogenase were used in the reductive amination of the corresponding α-keto acid **107**. The desired (*S*)-neopentylglycine was obtained with >95% conversion and a high enantiopurity of >99% ee at substrate concentrations of up to 88 g/L. Spiroheterocyclic compounds (morpholine-4-carboxylic acid amides of heterocyclic cyclohexylalanine and neopentylglycine derivatives and their analogs) are reversible inhibitors of cysteine proteases such as cathepsin S and therefore useful in treating variety of autoimmune diseases [192].

Example 18

The reaction was carried out in a 250 mL flask containing 50 mL of a mixture of 15.0 g of ammonium formate and 26.6 g of the biocatalyst (*E. coli* BW3110/pAM 3.25/pAM10.1) in water. An aqueous solution (12.5 mL) of 4,4-dimethyl-2-oxopentanoic acid sodium salt **107** (0.8 M, pH 7.15) was added to the reaction mixture at 30°C and the pH was adjusted to 7. Additional portions (12.5 mL each) of the aqueous solution of 4,4-dimethyl-2-oxopentanoic acid sodium salt **107** (0.8 M, pH 7.15) were added after 1, 2, 3.5, 5, 6.5, 8, and 9.5 h. The final total volume was 150 mL, and the overall substrate concentration (substrate input) was 0.53 M, corresponding to 88 g/L. After 24 h, a conversion of >95% and an ee of >99% was obtained. The product, L-neopentylglycine **106**, was isolated in 83% overall yield and >99% ee.

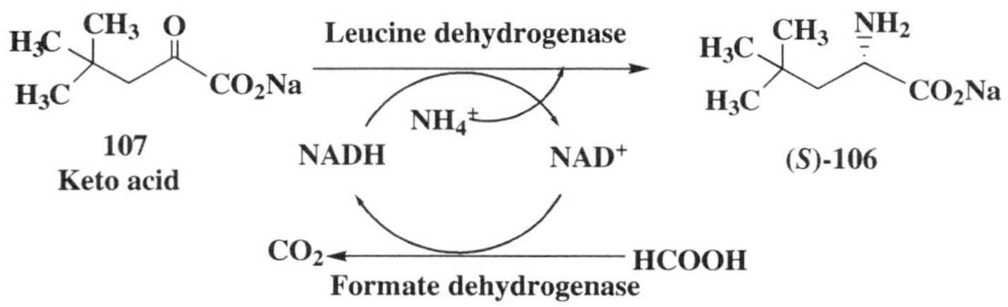

FIGURE 11.32
Enzymatic synthesis of (*S*)-neopentylglycine.

8 ENZYMATIC HYDROXYLATION

8.1 Dioxygenases and Preparation of Dihydrodiols

Hydroxylation reactions are catalyzed by monooxygenase and dioxygenases [193–195]. As detailed elsewhere in this book, toluene dioxygenase (TDO) from *P. putida* F1 was the first characterized aromatic hydrocarbon dioxygenase [193], shown to catalyze *cis*-dihydroxylation of benzene and toluene (Figure 11.33) [194,195]. A great deal of our understanding of the structure, function, and substrate specificity of aromatic hydrocarbon dioxygenases has come from studies of TDO and naphthalene dioxygenase (NDO). To date, more than 100 Rieske nonheme iron oxygenases have been identified based on biological activity or nucleotide sequence identity. More than three hundred arene *cis*-dihydrodiols have been obtained using toluene dioxygenase and these have been extensively used as chiral precursors in organic synthesis [193–200]. Benzene dioxygenase, TDO and NDO also catalyze the dihydroxylation of alkenes. This has been recently developed into a biocatalytic asymmetric dihydroxylation of conjugated mono-and poly-alkenes to obtain enantiopure cyclic *cis*-diols. The diol metabolites were obtained from monosubstituted, *gem*-disubstituted, *cis*-disubstituted, and trisubstituted alkene substrates, using whole cells of *P. putida* strains containing TDO and NDO [193–200]. Dioxygenase selection and alkene type were established as important factors, in the preference for dioxygenase-catalyzed 1,2-dihydroxylation of conjugated alkene or arene groups, and monohydroxylation at benzylic or allylic centers. *cis*-1,2-Diol metabolites from arenes, cyclic alkenes and dienes were generally observed to be enantiopure (> 98%

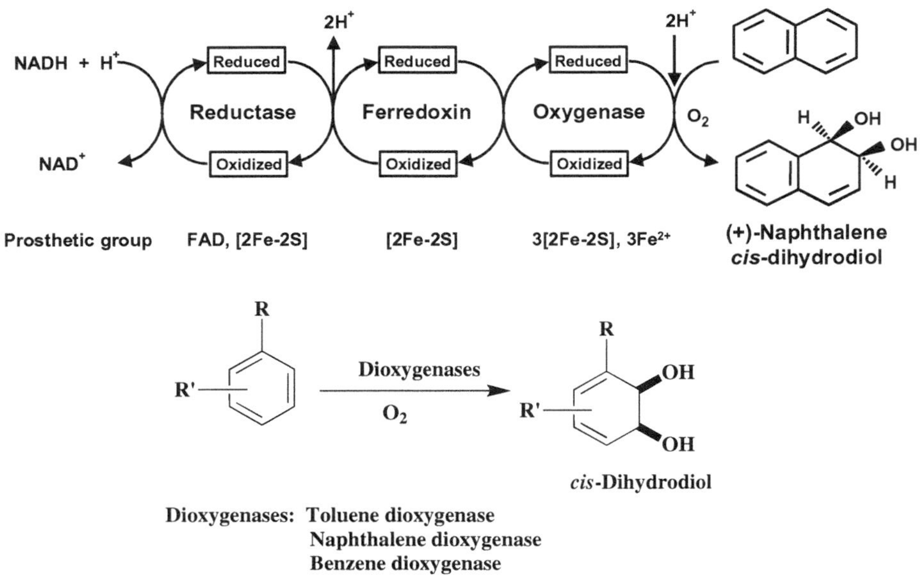

FIGURE 11.33
Dioxygenases and preparation of dihydrodiols.

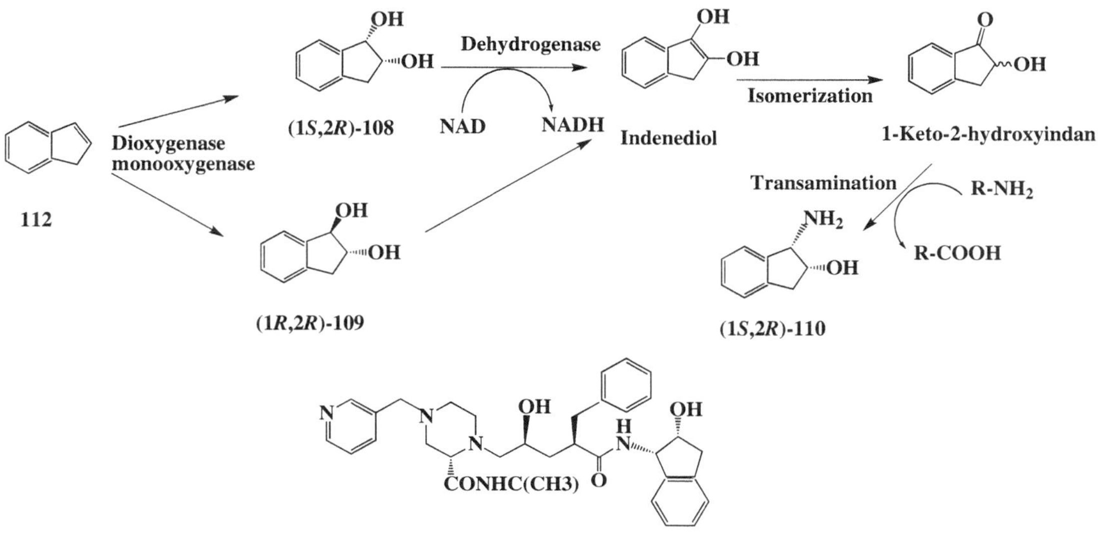

FIGURE 11.34
Enzymatic preparation of *cis*-(1*S*,2*R*)-indandiol or *trans*-(1*R*,2*R*)-indandiol.

ee), while 1,2-diols from acyclic alkenes had lower enantiomeric excess values (<88% ee) [198,200].

Cis-(1*S*,2*R*)-indandiol **108** or *trans*-(1*R*,2*R*)-indandiol **109** (Figure 11.34) are both potential precursors to *cis*-(1*S*,2*R*)-1-aminoindan-2-ol **110**, a key chiral synthon for Crixivan (Indinavir) **111**, an HIV protease inhibitor [201]. Enrichment and isolation of microbial cultures yielded two strains, *Rhodococcus* sp. B 264-1 (MB 5655) and I-24 (MA 7205), capable of oxidizing indene **112** to *cis*-(1*S*,2*R*)-indandiol and *trans*-(1*R*,2*R*)- indandiol, respectively [201]. Isolate MB 5655 was found to have a toluene dioxygenase, while isolate MA 7205 was found to harbor both toluene and naphthalene dioxygenases as well as a naphthalene monooxygenase which catalyzes the aforementioned biotransformation. When scaled up in a 14 L fermentor, MB5655 produced up to 2.0 g/L of *cis*-(1*S*,2*R*)-indandiol **108** with ee >99%. *Rhodococcus* sp MA 7205 cultivated under similar conditions produced up to 1.4 g/L of *trans*-(1*R*,2*R*)-indandiol **109** with an ee of >98%. Process development studies yielded titers >4.0 g/L of *trans*-(1*R*,2*R*)-indandiol [202]. A metabolic engineering approach [203] and directed evolution techniques [204] were evaluated to avoid side reactions, block degradative pathways and enhance the desired reaction (conversion of indene to *cis*-amino indanol **110** or *cis*-indanediol). Multi-parameter flow cytometry was used to assess indene toxicity, and it was shown that concentrations up to 0.25 g/L of indene (0.037 g indene/g dry cell wt.) in batch bioconversions did not influence cell physiology or reaction rate. Using this information, a single phase indene fed-batch bioconversion was carried out. The results indicated that

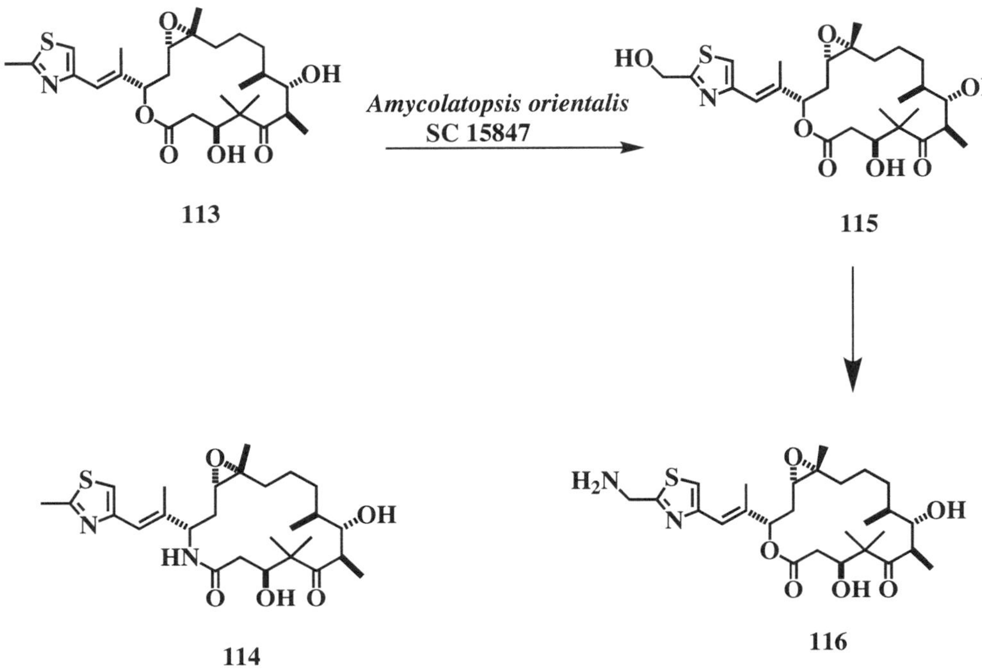

FIGURE 11.35
Epothilones: microbial hydroxylation of Epothiolone B to Epothilone F.

indene supply at a rate of 0.1 g/L/h was feasible without any deleterious effects. *cis*-(1*S*,2*R*)-indandiol **108** production rates were enhanced up to 200 mg/L/h by a combination of suitable indene feeding rates in the stationary phase and by operating with a high biomass concentration [205].

8.2 Epothilones (Anticancer drugs): Microbial Hydroxylation of Epothiolone B to Epothilone F

The clinical success of Paclitaxel has stimulated research into compounds with similar modes of action in order to emulate its antineoplastic efficacy while minimizing its less desirable aspects, which include nonwater solubility, difficult synthesis and emerging resistance. The epothilones are a novel class of microtubule-stabilizing, non-taxane cytotoxic natural products derived from the fermentation of the myxobacterium *Sorangium cellulosum* that trigger apoptosis [206,207]. The natural product epothilone B **113** (Figure 11.35) has demonstrated broad spectrum antitumor activity *in vitro* and *in vivo*, including tumors with Paclitaxel resistance [208]. The role of **113** as a potential paclitaxel successor has initiated interest in its synthesis, resulting in several total syntheses of **113** and various derivatives thereof [209,210]. The epothilone analogs were synthesized in an effort to optimize the water solubility, *in vivo* metabolic stability and antitumor efficacy of this class of antineoplasic agents [211].

Ixabepilone **114**, a novel epothilone derivative was prepared from epothiolne B by Bristol-Myers Squibb Company (BMS), was approved by FDA for the treatment of patients with malignant tumors [212,213].

A microbial hydroxylation process was developed for converting epothilone B **113** to epothilone F **115** by *Amycolatopsis orientalis* SC 15847. Epothilone F is a key intermediate for the synthesis of an anticancer drug **116**. A bioconversion yield of 37–47% was obtained when the process was scaled up to 100–250 L. The epothilone B hydroxylase along with the ferredoxin gene have been cloned and expressed in *Streptomyces rimosus* from *A. orientalis* SC 15847. Mutants and variants thereof this cloned enzyme have been used in the hydroxylation of epothilone B providing 80% yields of epothilone F [214,215].

8.3 Carbovir (Antiviral Drug): Enzymatic Hydroxylation of 2-Cyclopentylbenzoxazole

Hydroxylation and epoxidation reactions catalyzed by monooxygenases are of particular interest in chemical synthesis and have been the subject of numerous engineering studies aimed at understanding their functions and properties as well as making better catalysts [216,217] Cytochrome P450 BM-3 (BM-3; CYP102) from *B. megaterium* hydroxylates long-chain fatty acids at subterminal positions. It is called a 'self-sufficient' P450 because it contains the required diflavin NADPH-P450 reductase on the same polypeptide as the P450 hemoprotein [218] and easily expressed in recombinant *E.coli* [219]. A colorimetric assay for hydroxylation activity has facilitated several recent engineering studies on BM-3, particularly those needing high-throughput activity screens [220,221].

By employing protein engineering and substrate engineering, the biohydroxylation of 2-cyclopentylbenzoxazole **117** (Figure 11.36) to compound **118a and 118b** has been demonstrated by Munzer et al. This is a key intermediate for the synthesis of (−)-carbovir **119**. [222]. Protecting group removal provided end product **120**. In the initial screen, regardless of the microorganisms and fermentation conditions used, only a single diastereoisomer – (*S,S*)-**120b** – could be obtained with synthetically useful diastereoselectivities. Side products were also formed during this transformation [223]. To circumvent these problems, the cytochrome P450 BM-3 hydroxylase from *B. megaterium* was engineered to accept non-natural substrates with enhanced regio- and, enantioselectivities along with higher catalytic rates and total turnover numbers [224,225]. Mutations in the active site, for example, enabled the enzyme to hydroxylate alicyclic, heterocyclic, aromatic and even polyaromatic compounds [226]. Biohydroxylation using the wild type enzyme expressed in *E. coli* gave very low enantioselectivity (1.5%, *S,S*) and high diastereoselectivity (87%) of **120a**. Mutant 139-3 also afforded (*S,S*)-**120a** with high ee (86%) and de (96%). In dramatic contrast, mutant 1-12G was found to produce (*R,R*)-**120a** in high selectivities (89% ee, 94% de) [222,226].

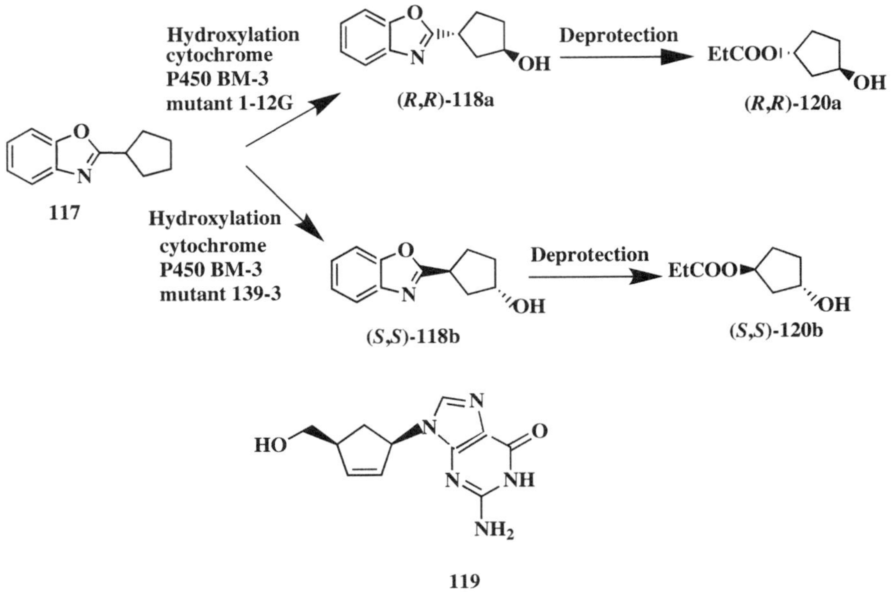

FIGURE 11.36
Carbovir: mnzymatic hydroxylation 2-cyclopentylbenzoxazole.

9 APPLICATION OF BIOCATALYSIS IN BASIC AND SPECIALTY CHEMICALS

9.1 Nicotinamide

Lonza Fine Chemicals manufactures nicotinamide by a process in which 2-methyl-1,5-diaminopentane (a by-product of nylon-6,6 manufacture) is first converted to 3-cyanopyridine **121** by a series of three chemically-catalyzed reactions, then the nitrile is hydrated to nicotinamide **122** (Figure 11.37) using *Rhodococcus rhodochrous* J1 cells immobilized in polyacrylamide particles [227–229]. A continuous feed of 3-cyanopyridine at concentrations of between 10–20 wt% is added in the direction of process flow, with a counter-current feed of biocatalyst in a series of stirred-tank batch bioreactors. The process generates

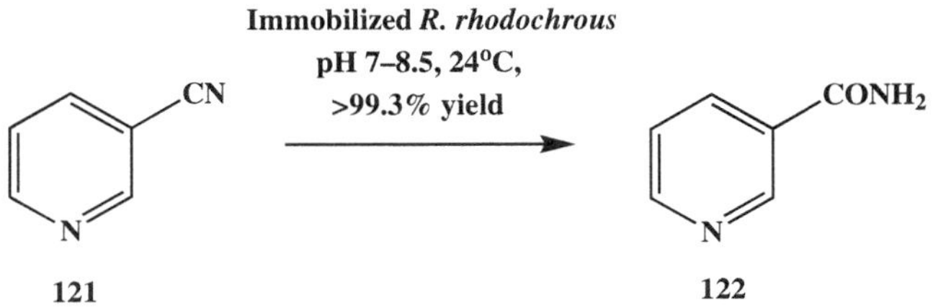

FIGURE 11.37
Enzymatic preparation of nicotinamide.

about 3500 metric tons of nicotinamide per year with a 99.3% selectivity at 100% conversion of 3-cyanopyridine. *R. rhodochrous* J1 has both a high K_M (200 mM) and a low tolerance for 3-cyanopyridine. It also contains a red pigment that requires decolorizing the final product. Recently, Lonza has focused on developing alternate biocatalysts for nicotinamide production, including *Amycolatopsis, Actinomadura*, and *Rhodococcus* [230–232]. *Rhodococcus* sp. FZ4 (DSM 13597) was isolated from soil samples using 3-pyridinaldoxime as inducer of nitrile hydratase (NHase) activity. It is expressed as cobalt-dependent nitrile hydratase with improved properties for production of nicotinamide relative to *R. rhodochrous* J1 .Using unimmobilized FZ4 cells, a total of 1876 g nicotinamide/g FZ4 dry cell wt. was obtained in a single batch reaction to produce approximately 60% (w/v) nicotinamide [230–232].

9.2 Acrylamide

Various nitrile hydratases are continued to be developed and compared to the existing wild-type whole-cell catalysts (*R. rhodochrous* J1) presently used in the commercial production of acrylamide from acrylonitrile. Mitsubishi Rayon has produced mutant enzymes of *R. rhodochrous* J1 with significantly improved thermal stabilities and catalytic activities at 50–70 °C [233,234]. Saint-Etienne, France has licensed and commercialized the manufacture of acrylamide using Mitsubishi Rayon's immobilized *R. rhodochrous* J1 [235] and has independently developed a *Rhodococcus pyridinovorans* whole cell catalyst for this process [236].

An *E. coli* transformant that overexpressed a thermally-stable nitrile hydratase produced by *Comamonas testosteroni* strain 5-MGAM-4D was immobilized in calcium alginate and evaluated for production of acrylamide in batch reactions with catalyst recycle [237,238]. Catalyst productivity decreased with increasing acrylonitrile concentration or reaction temperature, but was relatively insensitive to acrylamide concentration. A total of 975 g acrylamide/g dry cell weight and 89 g acrylamide/L were produced at 5 °C in 206 consecutive batch reactions with catalyst recycle, with initial and final volumetric productivities of 142 and 76 g acrylamide/L/h, respectively. This catalyst productivity was comparable to that of commercial polyacrylamide-immobilized *Rhodococcus* sp. 774 and *P. putida* B23 cells (500 and 850 g acrylamide/g dry cell weight, respectively); however, it is still far less than that of the currently employed polyacrylamide-immobilized *R. rhodochrous* J1 (>7000 g acrylamide/g dcw) [239]. Although the *C. testosteroni* nitrile hydratase had good thermal stability at 35 °C, catalyst productivity at this temperature was significantly less than that at 5 °C. The major determinant of biocatalyst productivity was the concentration of acrylonitrile in the reaction mixture, which was responsible for enzyme inactivation [240,241].

In addition to the commercial production of >30,000 MT per year of acrylamide using *R. rhodochrous* J1, >15,000 MT acrylamide per year are currently produced in the Peoples Republic of China (PRC) using a *Nocardia* sp. microbial nitrile hydratase. In an initial study of the suitability of *Nocardia sp.* 9112–118 for commercial production of acrylamide, the cells were immobilized in alginate

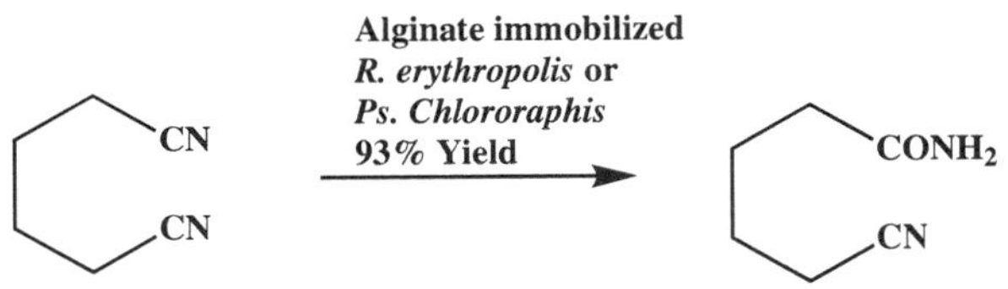

123

124

FIGURE 11.38
Enzymatic preparation of 5-cyanovaleramide.

and used to produce 250 g/L acrylamide at 20°C to 23°C in a 5000 L batch reactor [242]. For the fermentation of *Nocardia* sp. RS, optimization of pH regulation and addition of glucose-Co^{2+} in fed-batch fermentations increased nitrile hydratase activity to 10,195 U/mL of fermentation broth, reportedly among the highest volumetric productivities for all reported fermentations [243]. The acrylamide-tolerance of *Nocardia* sp. RS was improved by feeding acrylonitrile periodically into shake flasks containing growing cells and isolating survivors, a process that yielded the acrylamide-tolerant *Nocardia* sp. strain RS-1 [244,245]. The final acrylamide concentration and percent conversion of acrylonitrile catalyzed by the RS-1 strain was 587 g/L and 99.97%, respectively. This is a 30.6% higher final acrylamide concentration than that produced by the parent RS strain; further optimization of reaction conditions using RS-1 produced a final acrylamide concentration of 641 g/L [244,245].

9.3 5-Cyanovaleramide

A biocatalytic process was commercialized for the production of 5-cyanovaleramide **124** (5-CVAM) from adiponitrile **123** (Figure 11.38). This is a starting material in the manufacture of the herbicide azafenidin. Nitrile hydratase-catalyzed production of 5-CVAM resulted in higher yields of 5-CVAM, higher catalyst productivity, lower byproduct production, and significantly less process waste than alternative chemical methods. The process was first piloted using *R. erythropolis* A4 cells [246,247] immobilized in calcium alginate beads. Reactions were run at 5°C, as enzyme stability decreased markedly above this temperature. Sixty consecutive batch reactions converted a total of 1.1 metric tons of adiponitrile to a 9.0 wt% solution of 5-CVAM in 93% yield, with a catalyst productivity of >1000 kg 5-CVAM/kg dry cell weight. The 5-CVAM product was isolated by distillation of water, dissolution of the resulting solids in boiling methanol, and filtration of insoluble adipamide and salts; the resulting methanolic 5-CVAM solution was used directly in the subsequent azafenidin process step.

Commercial-scale fermentation of *R. erythropolis* A4 required a light-activation step of the microbial NHase [248], which was an additional cost for catalyst manufacture. Further screening of microbial NHases identified *Pseudomonas chlororaphis* B23 as a more effective biocatalyst for 5-CVAM production. *P. chlororaphis* B23 was first isolated and characterized by Yamada and co-workers [249,250], and was later used by the Nitto Chemical Industry Co. for the manufacture of acrylamide from acrylonitrile [251]. In the initial commercial-scale production run, *P. chlororaphis* B23 microbial cells were immobilized in calcium alginate beads, and 58 consecutive batch reactions were run at 5 °C with biocatalyst recycle to convert a total of 12.7 metric tons of adiponitrile to 13.6 metric tons of 5-CVAM (97% conversion, 96% selectivity, 93% yield) as a 19.2 wt% solution in water. The immobilized-cell catalyst was highly regioselective, with less than 5% selectivity for the undesired byproduct adipamide. The catalyst productivity in this first commercial-scale run was 3150 kg 5-CVAM/kg dry cell wt. A total of approximately 150 metric tons of 5-CVAM were prepared by this process for azafenidin manufacture [252,253].

9.4 L-DOPA

Lyases have extensively used in the synthesis of C-N bond [254–257]. Tyrosine phenol lyase (TLP) in *Erwinia herbicola* catalyzed the synthesis of L-DOPA **126** from catechol **125** (Figure 11.39) [258,259]. Ajinomoto Co. Ltd. (Japan) produces L-DOPA by this lyase-biotransformation using suspended *E. herbicola* cells with extremely high TLP activity in a fed batch reactor [258]. Cells are first prepared by cultivation in a medium containing L-tyrosine as an inducer of TLP. The intact cells are then harvested by centrifugation and transferred to the reactor together with the substrate catechol. The one-step biotransformation is more economical than the established chemical route that involves eight chemical steps including an optical resolution [258–260]. L-DOPA is insoluble in the reaction medium, so it precipitates during the reaction from a final concentration of 110 g/L [258,259]. About 250 tons of L-DOPA is supplied per year, and more than half of it is produced via biotransformation.

9.5 L-Phenylalanine

L-Phenylalanine ammonia (PAL) lyase catalyzes the reversible cleavage of ammonia from L-phenylalanine yielding *trans*-cinnamic acid with no need for a cofactor. The reverse reaction (Figure 11.40) is used in industrial biotransformations to produce L-phenylalanine **128** from trans-cinnamic acid **127** [261]. This amino acid is used for the synthesis of the sweetener aspartame and in

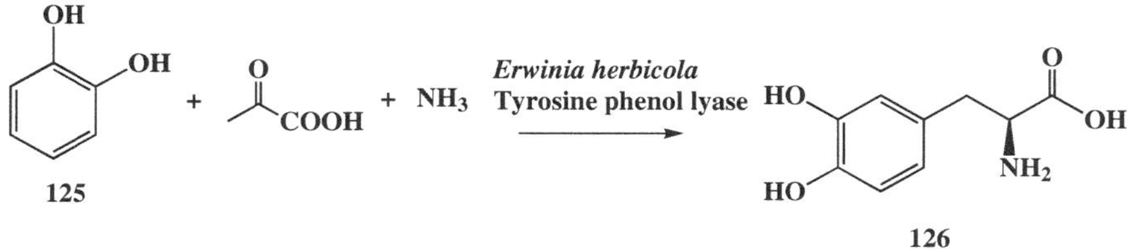

FIGURE 11.39
Enzymatic preparation of L-DOPA.

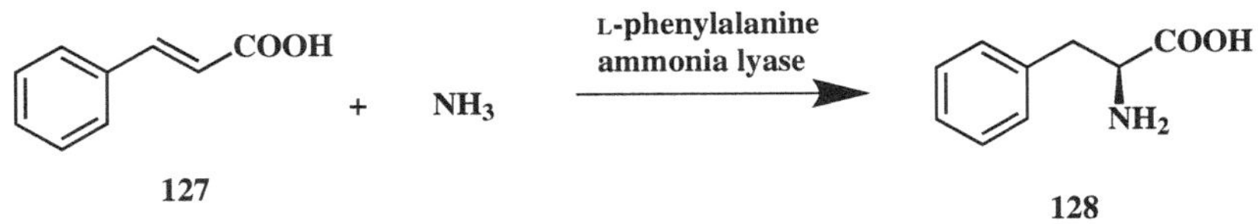

FIGURE 11.40
Enzymatic preparation of L-phenylalanine.

parenteral nutrition. Phenylalanine is also a key chiral building block for the synthesis of the macrolide antibiotic rutamycin B [262].

A fed batch process using suspended *Rhodotorula rubra* cells in an aqueous medium at pH 10.6 was developed [263]. Since PAL from *R. rubra* is oxygen-sensitive, the biotransformation is performed under anaerobic, static condition in fed batch mode with periodic addition of concentrated ammonium cinnamate solution and pH maintenance by CO_2 addition [263]. The fermentative process was improved by starting the process directly from glucose as a substrate and L-phenylalanine is now manufactured by fermentation on a scale of 8–10,000 tons per year [264].

9.6 1,5-Dimethyl-2-piperidone

A commercial process for converting 2-methylglutaronitrile **129** (MGN, a by-product produced during the manufacture of adiponitrile for nylon-6,6) to 1,5-dimethyl-2-piperidone **130** (Figure 11.41) was developed. 1,5-Dimethyl-2-piperidone has desirable properties for electronics, coatings and solvent applications. MGN is first hydrolyzed to the ammonium salt of 4-cyanopentanoic acid (4-CPA) using an immobilized microbial-cell catalyst (*Acidovorax facilis* 72W), which contains a nitrilase [265]. The hydrolysis reaction yields only one of two possible cyanocarboxylic acid ammonium salts with >98% selectivity at 100% conversion [266,267]. Reactions typically produced 170–210 g 4-cyanopentanoic acid per liter, and catalyst productivities of at least 1000 kg 4-cyanopentanoic acid per kg dry cell weight are achieved by recycling the catalyst in consecutive batch reactions. After recovery of the immobilized cells for reuse, the 4-cyanopentanoic acid ammonium salt in the aqueous product

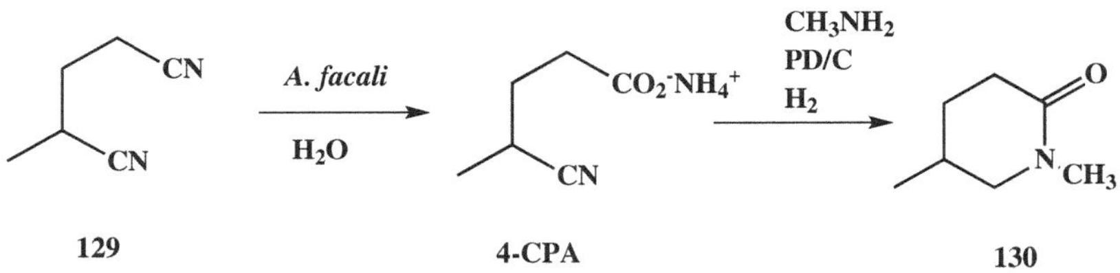

FIGURE 11.41
Enzymatic preparation of 1,5-dimethyl-2-piperidone from 2-methylglutaronitrile.

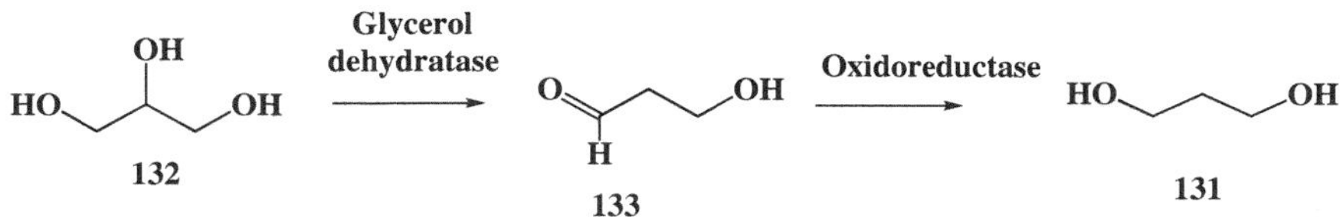

FIGURE 11.42
Enzymatic preparation of 1,3 propanediol from glycerol.

mixture is converted to 1,5-DMPD by low-pressure catalytic hydrogenation in the presence of added methylamine. When compared with an alternate chemical process in which MGN is directly converted *via* hydrogenation to a mixture of 1,3-dimethyl-2-piperidone and 1,5-dimethyl-2-piperidone, the chemoenzymatic process generates less waste and produces a single lactam isomer in higher yield.

9.7 1,3 Propanediol

1,3-Propanediol **131** (1,3-PDO) could potentially be used for many synthetic applications, particularly as a monomer for polycondensations to produce polyesters, polyethers and polyurethanes [268]. 1,3-PDO also has a number of other interesting possibilities in addition to polymer production. It can give improved properties for solvents, adhesives, laminates, resins, detergents and cosmetic [269].

The natural substrate for microbial production of 1,3-PDO **131** is glycerol **132** (Figure 11.42). Glycerol is dehydrated to 3-hydroxypropionaldehyde (3-HPA) by glycerol dehydratase. The product of dehydration reaction, 3-HPA **133**, is reduced to 1,3-PDO by an NAD^+-dependent oxidoreductase [270]. Since 50% of entire cost of microbial production of 1,3-PDO is due to the price of raw materials, raw glycerol from biodiesel production processes was an interesting renewable carbon source for microorganisms that produce 1,3-PDO [271]. Biodiesel production has increased exponentially in past several years and the principal byproduct of its production is glycerol. As the demand and production of biodiesel grows, the quantity of crude glycerol generated will be considerable and its utilization will become an important topic [272].

Glycerol conversion to 1,3-PDO can be carried out by *Clostridia* as well as *Enterobacteriaceae* [268]. Unfortunately, the actual fermentations by the enterobacteria *Klebsiella pneumoniae, Citrobacter freundii* and *Enterobacter agglomerans* also produce a number of other by-products such as ethanol, lactic acid, succinic acid, and 2,3-butanediol. *Clostridium* strains also produce butyric acid and butanol [269]. Three different mixed cultures were tested for production of 1,3-PDO from glycerol. 1,3-PDO was produced as the main product in this mixed culture along with organic acids like acetic and butyric acids as by-products. The yield was in the range of 0.56–0.76 mol 1,3-PDO per mol glycerol consumed depending on the glycerol concentration. A final product concentration as high as 70 g/L was obtained in fed-batch cultivation with a productivity of 2.6 g/L/h [273].

In an effort to improve the industrial production of 1,3-propanediol, Tang et al. [274] engineered a novel polycistronic operon under the control of the temperature-sensitive lambda phage P_LP_R promoter and expressed this construct in *E. coli* K-12 ER2925. The genes for the production of 1,3-PDO in *Clostridium butyricum*, *dhaB1*, and *dhaB2*, which encode the vitamin B_{12}-independent glycerol dehydratase DhaB1 and its activating factor, DhaB2, respectively, were tandemly arrayed along with the *E. coli yqhD* gene, which encodes the 1,3-propanediol oxidoreductase. This set of enzymes can directly convert glycerol to 1,3-PDO. The performance of this recombinant *E. coli* strain was studied in a two-stage fermentation process. During the first stage, a novel high-cell-density fermentation step, there was significant cell growth and the majority of the metabolites produced were organic acids, mainly acetate. During the second stage, glycerol from the fresh medium was rapidly converted to 1,3-PDO following a temperature shift from 30°C to 42°C. The by-products were mainly pyruvate and acetate. During this two-stage process, the overall 1,3-PDO yield and productivity reached 104.4 g/L and 2.61 g/L/h, respectively, and the conversion rate of glycerol to 1,3-PDO reached 90.2% (g/g). The overall fermentation time was 40 h [274].

While outside the scope of this book, it is worth noting that the biological production of 1,3 propanediol from a renewable source such as glucose has been explored for the past several years [275,276]. A recombinant *E. coli* has been constructed that contains two previously known metabolic pathways, one for the conversion of glucose to glycerol and another for conversion of glycerol to 1,3 propanediol [275]. This biocatalyst can now be used to convert glucose to 1,3 propanediol in an economically attractive process.

9.8 2,3-Butanediol

2,3-Butanediol **134** (2,3-BD) is a promising bulk chemical with extensive industrial applications [277,278]. It is also a potentially valuable fuel or a fuel additive. Dehydration of 2,3-BD yields methyl ethyl ketone that is a high quality aviation fuel. [279,280]. There are three isomeric forms of 2,3-BD: the (2*R*,3*R*)- and (2*S*,3*S*)- forms are optically active, while the *meso*-form is optically inactive. Optically pure 2,3-BD can act as an excellent building block in asymmetric synthesis of chiral compounds that contain 2 vicinal stereocenters [281]. Thus, it is desirable to develop a practical technique for producing optically pure 2,3-BD. Native microorganisms generally produce a mixture of two isomers (*meso*- and (2*S*,3*S*)- or *meso*- and (2*R*,3*R*). Metabolic engineering and synthetic biology have been applied to improve this situation [280,281]. (2*S*,3*S*)-BD **134** could be produced from diacetyl (DA) **135** (Figure 11.43) by engineered *E. coli* strains expressing 2,3-butanediol dehydrogenase (2,3-BDH). Glucose was used for the internal cofactor regeneration. The transformation between 2,3-BD and DA was reversible and coupled with the NADH/NAD^+ conversion. By manipulating the

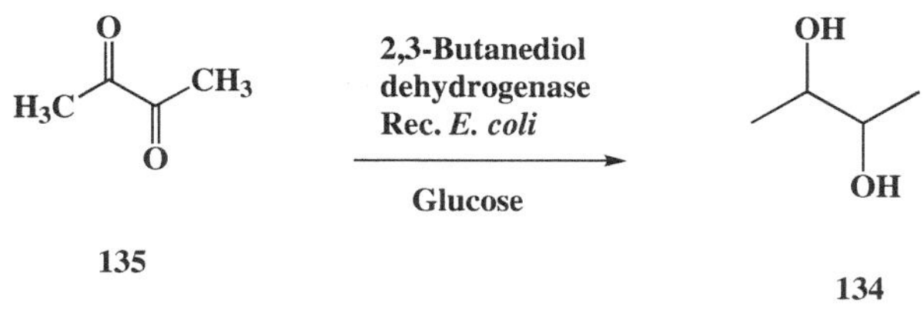

FIGURE 11.43
Enzymatic preparation of 2,3-butanediol from diacetyl.

level of intracellular cofactors, production of (2*S*,3*S*)-BD or DA can be improved [282,283].

Glucose dehydrogenase (GDH) and formate dehydrogenase (FDH) are widely-used for cofactor regeneration and Wang et al. [284] therefore coexpressed FDH or GDH along with 2,3-BDH in *E. coli*. The performance of the two resulting recombinant strains, *E. coli* BL21 (DE3) (pETDuet-*bdhgdh*) and *E. coli* BL21 (DE3) (pETDuet-*bdhfdh*) was evaluated in terms of enzymatic activities, (2*S*,3*S*)-2,3-BD production and intracellular NADH/NAD^+ level. Introduction of FDH resulted in higher (2*S*,3*S*)-BD concentration, productivity and yield from diacetyl, and large increase in the intracellular NADH concentration. In fed-batch bioconversion, the final titer, productivity and yield of (2*S*,3*S*) -BD for diacetyl reached 31.7 g/L and 89.8%, respectively.

9.9 Amino Acids

Degussa has described the use of engineered whole-cell biocatalysts producing enzymes L-carbamoylase, L-hydantoinase and hydantoin racemase to synthesize unnatural L-amino acids [285,286]. They have further extended this dynamic kinetic resolution technology to produce of D-amino acids, by including a D-specific carbamoylase and a D-hydantoinase obtained by directed evolution from the L-specific enzyme [285]. In another known dynamic kinetic resolution process of Degussa, racemic *N*-acetyl amino acids are resolved using an acylase in combination with an *N*-acetyl amino acid racemase. A novel *N*-acetyl amino acid racemase from *A. orientalis*, which is not prone to substrate inhibition, provides a significant improvement over the old process.

A production method for the manufacture of cysteine by fermentation has been developed by Wacker Chemie. Building on this technology, an engineered *E. coli* strain overexpressing the enzyme O-acetylserine sulfhydrylase, which accepts a broad range of substrates, was used to produce a number of unnatural amino acids. This included triazole-1-yl-alanine, (*S*)-hydroxyethyl cysteine and phenylselenocysteine, which were prepared by feeding fermentations with unnatural precursor molecules (Figure 11.44) [287].These novel unnatural amino acids may prove useful for the synthesis of combinatorial (peptide) libraries.

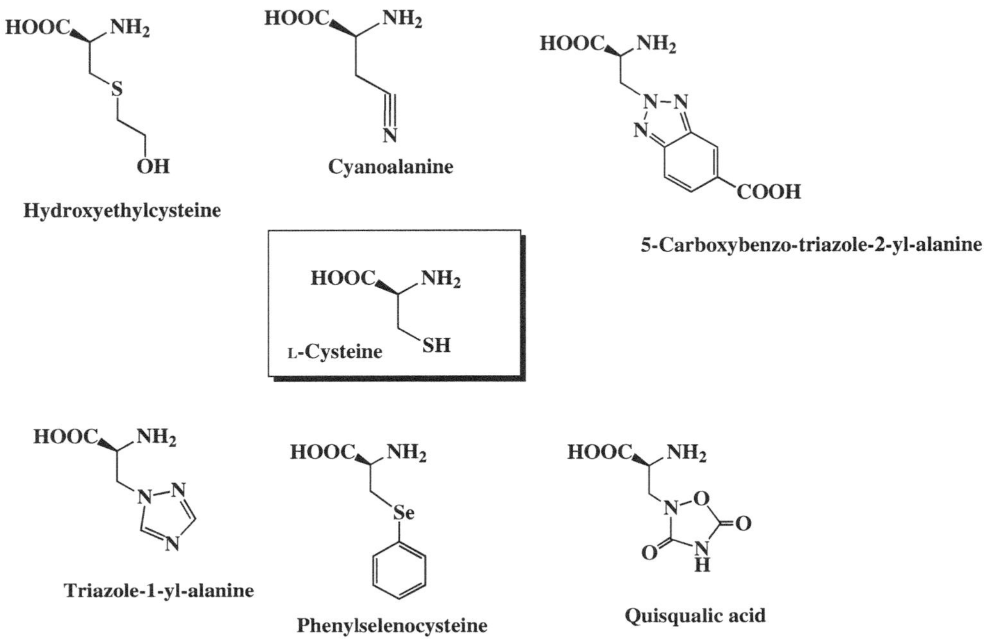

FIGURE 11.44
Enzymatic preparation of unnatural amino acids.

10 CONCLUSIONS

Demand for chiral compounds continues to rise, and high demand from the pharmaceutical industry is fueled by regulations covering chiral active pharmaceutical ingredients (APIs) and the recognition that enantiomers of a chiral compound often have dramatically different biological activities and toxicity. Whereas chiral APIs were usually formulated as racemates in previous times, the preference now is for single enantiomers. Furthermore, the switch from a racemic to a single-enantiomer API is key to the life cycle management, as well as improving the efficacy of drugs and this will contribute to the demand for optically pure compounds. To meet these challenges, the pharmaceutical industry is evaluating a variety of chiral technologies, including separations, salt resolutions, and asymmetric syntheses involving chiral auxiliaries, chemocatalysis, chiral pools as well as biocatalysis to get to single enantiomer API's.

In addition to the selective nature of biocatalysts which is one primary reason for their applications, their impact on "Green chemistry" has been another key driving factor in their applications, especially for large volume basic and specialty chemicals. Often, replacement of an existing chemical process by an alternate biocatalysis-based process results in significant environmental benefits: less waste production and less energy consumption. With the rapid advances in the technologies used to improve biocatalysts and the greater availability of more classes of enzymes, it is hoped that there will be increasing industrial application of biocatalysis in the synthesis of pharmaceuticals and chemicals.

Acknowledgments

The author would like to acknowledge Drs. Ronald Hanson, Animesh Goswami, Amit Banerjee, Venkata Nanduri, Jeffrey Howell, Steven Goldeberg, Robert Johnston, Mary-Jo Donovan, Dana Cazzulino, Thomas Tully, Thomas LaPorte, Larry Parker, John Wasylyk, Michael Montana, Ronald Eiring, Rapheal Ko, Linda Chu, Clyde McNamee, Michael Montana, Paul Cino, Laszlo Szarka, John Scott, Richard Mueller, Robert Waltermire and many chemists and chemical enginerrs at Bristol-Myers Squibb for research collaboration in various projects.

References

[1] Food and Drug Administration, FDA's statement for the development of new stereoisomeric drugs, Chirality 4 (1992) 338–340.

[2] M. Oliver, C.A. Voigt, F.H. Arnold, Enzyme engineering by directed evolution, in:K. Drauz, H. Waldmann (Eds.), Enzyme Catalysis in Organic Synthesis 1 (2002) 95–138 second ed., Wiley-VCH Verlag GmbH, Weinheim, Germany, 2002.

[3] R.J. Kazlauskas, Enhancing catalytic promiscuity for biocatalysis, Curr. Opin. Chem. Biol. 9 (2) (2005) 195–201.

[4] M. Schmidt, M. Bauman, E. Henke, M. Konarzycka-Bessler, U.T. Bornscheuer, Directed evolution of lipases and esterases, Method. Enzymol. (Prot. Eng.) 388 (2004) 199–207.

[5] M.T. Reetz, C. Torre, A. Eipper, R. Lohmer, M. Hermes, B. Brunner, A. Maichele, M. Bocola, M. Arand, A. Cronin, Y. Genzel, A. Archelas, R. Furstoss, Enhancing the enantioselectivity of an epoxide hydrolase by directed evolution, Org. Lett. 6 (2) (2004) 177–180.

[6] C.R. Otey, G. Bandara, J. Lalonde, K. Takahashi, F.H. Arnold, Preparation of human metabolites of propranolol using laboratory-evolved bacterial cytochromes P450, Biotech. Bioeng. 93 (3) (2006) 494–499.

[7] U.T. Bornscheuer, G.W. Huisman, R.J. Kazlauskas, S. Lutz, J.C. Moore, K. Robins, Engineering the third wave of biocatalysis, Nature 485 (2012) 185–194.

[8] G.W. Huisman, J.J. Lalonde, in: R.N. Patel (Ed.), Enzyme Evolution for Chemical Process Applications in Biocatalysis in the Pharmaceutical and Biotechnology Industries, CRC Press, FL, USA, 2007, pp. 717–742.

[9] G.J. Williamsa, A.S. Nelson, A. Berry, Directed evolution of enzymes for biocatalysis and the life sciences, Cell. Mol. Life Sci. 61 (2004) 3034–3046.

[10] H. Zhao, K. Chockalingom, Z. Chen, Directed evolution of enzymes and pathways for industrial biocatalysis, Curr. Opin. Biotechnol. 13 (2) (2002) 104–110.

[11] R. DiCosimo, in: R.N. Patel (Ed.), Nitrilases and nitrile hydratases. in Biocatalysis in the Pharmaceutical and Biotechnology Industries, CRC Press, Florida, 2007, pp. 1–26.

[12] R.N. Patel, Biocatalysis: synthesis of key intermediates for development of pharmaceuticals, ACS Catal. 1 (2011) 1056–1074.

[13] Y. Simeo, W. Kroutil, K. Faber, Biocatalytic deracemization: dynamic resolution, stereoinversion, enantioconvergent processes, and cyclic deracemization, in: R.N. Patel (Ed.), Biocatalysis in the Pharmaceutical and Biotechnology Industries, CRC Press, Florida, 2007, pp. 27–51.

[14] C. Simons, U. Hanefeld, I. Arends, T. Maschmeyer, R. Sheldon, Towards catalytic cascade reactions: asymmetric synthesis using combined chemo-enzymatic catalysts, Top. Catal. 40 (1–4) (2006) 35–44.

[15] N.J. Turner, Enzyme catalyzed deracemization and dynamic kinetic resolution reactions, Curr. Opin. Chem. Biol. 8 (2) (2004) 114–119.

[16] D.E. Robertson, U.T. Bornscheuer, Biocatalysis and biotransformation new technologies, enzymes and challenges, Curr. Opin. Chem. Biol. 9 (2) (2005) 164–165.

[17] T. Ishige, K. Honda, S. Shimizu, Whole organism biocatalysis, Curr. Opin. Chem. Biol. 9 (2) (2005) 174–180.

[18] J.D. Stewart, Green chemical manufacturing with biocatalysis, in: V.H. Grassian (Ed.), Environmental Catalysis, CRC Press, Boca Raton, FL, 2004, pp. 649–665.

[19] D.R. Boyd, T.D.H. Bugg, Arene cis-dihydrodiol formation:from biology to application, Org. Biomol. Chem. 4 (2) (2006) 181–192.

[20] W. Kroutil, H. Mang, K. Edegger, K. Faber, Recent advances in the biocatalytic reduction of ketones and oxidation of sec-alcohols, Curr. Opin. Chem. Biol. 8 (2) (2004) 120–126.

[21] J. Tao, G-Q.Lin, A. Liese, Biocatalysis for the Pharmaceutical Industry: Discovery, Development, and Manufacturing, Wiley-VCH, Verlag & Co. Weinheim, Germany, 2008.

[22] M. Hall, C. Stueckler, H. Ehammer, E. Pointner, G. Oberdorfer, K. Gruber, B. Hauer, R. Stuermer, W. Kroutil, P. Macheroux, K. Faber, Asymmetric bioreduction of C═C bonds using enoate reductases OPR1, OPR3 and YqjM: enzyme-based stereocontrol, Adv. Synth. Catal. 350 (3) (2008) 411–418.

[23] V. Gotor, Pharmaceuticals through enzymatic transesterification and enzymatic aminolysis reactions, Biocat. Biotrans. 18 (2000) 87–103.

[24] R.N. Patel, Synthesis of chiral pharmaceutical intermediates by biocatalysis, Coord. Chem. Rev. 252 (5–7) (2008) 659–701.

[25] A. Goswami, Enzyme catalysis in the synthesis of active pharmaceutical ingredients, in: W.-K. Yeh, H.-C. Yang, J.R. McCarthy (Eds.), Enzyme Technologies: Mutagenomics, Evolution, Biocatalysis, and Biosynthesis, John Wiley and Sons, Hoboken, NJ, 2010, pp. 127–184.

[26] A.S. Wells, G.L. Finch, P.C. Michels, J.W. Wong, Use of enzymes in the manufacture of active pharmaceutical ingredients – a science and safety-based approach to ensure patient safety and drug quality, Org. Process Res. Dev. 16 (2012) 1986–1993.

[27] M.T. Reetz, Biocatalysis in organic chemistry and biotechnology: Past, present, and future, J. Am. Chem. Soc. 135 (2013) 12480–12496.

[28] G.W. Huisman, S.J. Collier, On the development of new biocatalytic processes for practical pharmaceutical synthesis, Curr. Opin. Chem. Biol. 17 (2013) 284–292.

[29] B.A. Lauria-Horner, R.B. Pohl, Pregabal: a new anxiolytic, Expert Opin. Inv. Drug. 12 (4) (2000) 663–672.

[30] I. Selak, Pregabalin (Pfizer), Curr. Opin. Inv. Drug. 2 (6) (2001) 828–834.

[31] M.S. Hoekstra, D.M. Sobieray, M.A. Schwindt, T.A. Mulhern, T.M. Grote, B.K. Huckabee, V.S. Hendrickson, L.C. Franklin, E.J. Granger, G.L. Karrick, Chemical development of CI-1008, Enentiomerically pure anticonvulsant, Org. Process Res. Dev. 1 (1997) 26–38.

[32] M.J. Burk, P.D. de Koning, T.M. Grote, M.S. Hoekstra, G. Hoge, R.A. Jennings, W.S. Kissel, T.V. Le, I.C. Lennon, T.A. Mulhern, J.A. Ramsden, R.A. Wade, An enantioselective synthesis of (S)-(+)-3-aminomethyl-5-methylhexanoic acid via asymmetric hydrogenation, J. Org. Chem. 68 (2003) 5731–5734.

[33] G. Hoge, H.-P. Wu, W.S. Kissel, D.A. Pflum, D.J. Greene, J. Bao, Highly selective asymmetric hydrogenation using three hindered quadrant biphospine rhodium catalyst, J. Am. Chem. Soc. 126 (2004) 5966–5967.

[34] C.A. Martinez, S. Hu, Y. Dumond, J. Tao, P. Kelleher, T.L. Liam, Development of a chemoenzymatic manufacturing process for Pregabalin, Org. Process Res. Dev. 12 (3) (2008) 392–398.

[35] M. Suffness, M.E. Wall, Discovery and development of taxol, in: M. Suffness (Ed.), Taxol: Science and Application, CRC Press, New York, 1995, pp. 3–25.

[36] R. Holton, R. Biediger, P. Joatman, Semisynthesis of taxol and taxotere, in: M. Suffness (Ed.), Taxol: Science and Application, CRC press, New York, 1995, pp. 97–123.

[37] R.N. Patel, Tour de Paclitaxel, Ann. Rev. Microbiol. 98 (1995) 361–395.

[38] E. Baloglu, D.G. Kingston, A new semisynthesis of paclitaxel from baccatin III, J. Nat. Products 62 (7) (1999) 1068–1071.

[39] R.N. Patel, A. Banerjee, R.Y. Ko, J. Howell, W.-S. Li, F. Comezoglu, R. Partyka, L. Szarka, Enzymic preparation of (3R-cis)-3-(acetyloxy)-4-phenyl-2-azetidinone: a taxol side-chain synthon, Biotechnol. Appl. Biochem. 20 (1) (1994) 23–33.

[40] W.C. Rose, B.H. Long, C.R. Fairchild, F.Y. Lee, J.F. Kadow, Preclinical pharmacology of BMS-275183, an orally active taxane, Clin. Cancer Res. 7 (7) (2001) 2016–2021.

[41] R.N. Patel, J.M. Howell, R. Chidambaram, S. Benoit, J. Kant, Enzymatic preparation of (3R)-cis-3-acetyloxy-4-(1,1-dimethylethyl)-2-azetidinone: a side-chain synthon for an orally active taxane, Tetrahedron Asymmetry 14 (23) (2003) 3673–3677.

[42] G. DeSantis, Z. Zhu, W.A. Greenberg, K. Wong, J. Chaplin, S.R. Hanson, B. Farwell, L.W. Nicholson, C.L. Rand, D.P. Weiner, D.E. Robertson, M.J. Burk, An enzyme library approach to biocatalysis: development of nitrilases for enantioselective production of carboxylic acid derivatives, J. Am. Chem. Soc. 124 (2002) 9024–9025.

[43] G. DeSantis, K. Wong, B. Farwell, K. Chapman, Z. Zhu, G. Tomlinson, H. Huang, X. Tan, L. Bibbs, P. Chen, K. Kretz, N.J. Burk, Creation of a productive, highly enantioselective nitrilase through gene site saturation mutagenesis (GSSM), J. Am. Chem. Soc. 125 (38) (2003) 11476–11477.

[44] K.A. Gray, T.H. Richardson, K. Kretz, J.M. Short, F. Bartnek, R. Knowles, L. Kan, P.E. Swanson, D.E. Robertson, Rapid evolution of reversible denaturation and elevated melting temperature in a microbial haloalkane dehalogenase, Adv. Synth. Catal. 343 (2001) 607–617.

[45] M.T. Reetz, M.H. Becker, H.-W. Klein, D. Stockigt, A method for high-throughput screening of enantioselective catalysts, Angew. Chem. Int. Ed. 38 (1999) 1758–1761.

[46] J. Guo, J. Wu, G. Siuzdak, M.G. Finn, Measurement of enantiomeric excess by kinetic resolution and mass spectrometry, Angew. Chem. Int. Ed. 38 (1999) 1755–1758.

[47] S. Bergeron, D.A. Chaplin, J.H. Edwards, B.S.W. Ellis, C.L. Hill, K. Holt-Tiffin, J.R. Knight, T. Mahoney, A.P. Osborne, G. Ruecroft, Nitrilase-catalysed desymmetrisation of 3-hydroxyglutaronitrile: preparation of a statin side-chain intermediate, Org. Process Res. Dev. 10 (2006) 661–665.

[48] J.G. Smith, Synthetically useful reactions of epoxides, Synthesis (1984) 629–656.

[49] R.A. Johnson, K.B. Sharpless, Catalytic asymmetric epoxidation of allylic alcohols, in: Ojima (Ed.), Catalytic Asymmetric Synthesis, 1st ed, VCH, New York, USA, 1993, pp. 103–158.

[50] E.N. Jacobsen, Asymmetric catalytic epoxidation of unfunctionalized olefins, in: Ojima (Ed.), Catalytic Asymmetric Synthesis, 1st ed, VCH, New York, USA, 1995, pp. 159–202.

[51] Z.-X. Wang, Y. Tu, M. Frohn, J.R. Zhang, Y. Shi, An efficient catalytic asymmetric epoxidation method, J. Am. Chem. Soc. 119 (1997) 11224–11235.

[52] W.J. Choi, C.Y. Choi, J.A.M. de Bont, C.A.G.M. Weijers, Continuous production of enantiopure 1,2-epoxyhexane by yeast epoxide hydrolase in a two phase membrane bioreactor, Appl. Microbiol. Biotechnol. 54 (2000) 641–646.

[53] A. Archelas, R. Furstos, Synthetic applications of epoxide hydrolases, Curr. Opin. Chem. Biol. 5 (2001) 112–119.

[54] E.J. de Vries, D. Janssen, Biocatalytic conversion of epoxides, Curr. Opin. Biotechnol. 14 (2003) 414–420.

[55] A. Steinreiber, K. Faber, Microbial epoxide hydrolases for preparative biotransformations, Curr. Opin. Biotechnol. 12 (2001) 552–558.

[56] W.-J. Choi-, C.-Y. Choi, Production of chiral epoxides: epoxide hydrolase catalyzed enantioselective hydrolysis, Biotechnol. Bioprocess Eng. 10 (2005) 167–179.

[57] W.-J. Choi, Biotechnological production of enantiopure epoxides by enzymatic kinetic resolution, Appl. Microbiol. Biotechnol. 84 (2009) 239–247.

[58] J.D. Catt, G. Johnson, D.J Keavy, R.J. Mattson, M.F. Parker, K.S. Takaki, J.P. Yevich, Preparation of benzofuran and dihydrobenzofuran melatonergic agents, US5,856,529, 1999.

[59] A. Goswami, M.J. Totleben, A.K. Singh, R.N. Patel, Stereospecific enzymatic hydrolysis of racemic epoxide: a process for making chiral epoxide, Tetrahedron Asymmetry 10 (16) (1999) 3167–3175.

[60] B. Veronesi, J.D. Carter, R.B. Devlin, S.A. Simon, M. Oortgiesen, Neuropeptides and capsaicin stimulate the release of inflammatory cytokines in a human bronchial epithelial cell line, Neuropeptides (Edinburgh) 33 (6) (1999) 447–456.

[61] L. Yuan, E. Burcher, B.S. Nail, Characterization of tachykinin receptors mediating bronchomotor and vasodepressor responses to neuropeptide γ and substance P in the anesthetized rabbit, Pulm Pharmacol Ther. 11 (1) (1998) 31–39.

[62] P.C. Ting, J.F. Lee, J.C. Anthes, N.-Y. Shih, J.J. Piwinski, Synthesis and NK1/NK2 receptor activity of substituted 4(Z)-(methoxyimino)pentyl-1-piperazines, Bioorg. Med. Chem. Lett. 10 (20) (2000) 2333–2335.

[63] G.A. Reichard, Z.T. Ball, R. Aslanian, J.C. Anthes, N.-Y. Shih, J.J. Piwinski, The design and synthesis of novel NK1/NK2 dual antagonists, Bioorg. Med. Chem. Lett.V 10 (20) (2000) 2329–2332.

[64] M.J. Homan, R. Vail, B. Morgan, V. Sabesan, C. Levy, D.R. Dodds, A. Zaks, Enzymatic hydrolysis of a prochiral 3-substituted glutarate ester, an intermediate in the synthesis of an NK1/NK2 dual antagonist, Adv. Synth. Catal. 343 (6–7) (2001) 744–749.

[65] W.-C. Suen, N. Zhang, L. Xiao, V. Madison, A. Zaks, Improved activity and thermostability of *Candida antarctica* lipase B by DNA family shuffling, Protein Eng. Des. Sel. 17 (2) (2004) 133–140.

[66] N. Zhang, W.-C. Suen, W. Windsor, L. Xiao, V. Madison, A. Zaks, Improving tolerance of *Candida antarctica* lipase B towards irreversible thermal inactivation through directed evolution, Protein Eng. 16 (8) (2003) 599–605.

[67] P.H. Carter, R.J. Cherney, V.W. Rosso, J. Li, *N*-((1r, 2s, 5*r*)-5-(*tert*-butylamino)-2-((*S*)-3-(7-*tert*-butylpyrazolo[1,5-*a*][1,3,5]triazin-4-ylamino)-2-oxopyrrolidin-1yl) cyclohexyl) acetamide, a dual modulator of chemokine receptor activity, crystalline forms and processes, PCT Int. Appl., WO/2005/021500 A1, 2005.

[68] R.J. Cherney, J.B. Brogan, R. Mo, Y.C. Lo, G. Yang, P.B. Miller, P.A. Scherle, B.F. Molino, P.H. Carter, C.P. Decicco, Discovery of trisubstituted cyclohexanes as potent chemokine receptor 2 (CCR2) antagonists, Bioorg. Med. Chem. Lett. 19 (2009) 597–601.

[69] P. Carter, Progress in the discovery of CC chemokine receptor 2 antagonists, 2009–2012, Expert Opin. Ther. Pat. 23 (5) (2013) 549–568.

[70] Y. Chen, S. Tian, L. Deng, A highly enantioselective catalytic desymmetrization of cyclic anhydrides with modified cinchona alkaloids, J. Am. Chem. Soc. 122 (39) (2000) 9542–9543.

[71] C. Bolm, I. Schiffers, C.L. Dinter, A. Gerlach, Practical and highly enantioselective ring opening of cyclic meso-anhydrides mediated by cinchona alkaloids, J. Org. Chem. 65 (21) (2000) 6984–6991.

[72] W.-M. Dai, K.K.Y. Yeung, C.W. Chow, I.D. Williams, Study on enantiomerically pure 2-substituted *N,N*-dialkyl-1-naphthamides: resolution, absolute stereochemistry, and application to desymmetrization of cyclic meso anhydrides, Tetrahedron Asymmetry 12 (11) (2001) 1603–1613.

[73] M. Kurihara, K. Kamiyama, S. Kobayashi, M. Ohno, Diversified synthetic approaches to the carbapenem antibiotics based on symmetrization-asymmetrization concept, Tetrahedron Lett. 26 (47) (1984) 5831–5834.

[74] S. Kobayashi, K. Kamiyama, T. Iimori, M. Ohno, Creation of novel chiral synthons with enzymes and applications to natural product synthesis. 15. Efficient introduction of chiral centers into cyclohexane ring, Tetrahedron Lett. 25 (24) (1984) 2557–2560.

[75] A. Goswami, T.P. Kissick, Enzymatic desymmetrization of dimethyl cylcohex-4- ene-cis-1,2-dicarboxylate to (1S,2R)-2-(methoxycarbonyl)cyclohex-4-ene-1-carboxylic acid, Org. Process Res. Dev. 13 (3) (2009) 483–488.

[76] P.A. Gurbel, C.M. O'Connor, C.C. Cummings, V.L. Serebruany, Clopidogrel: the future choice for preventing platelet activation during cotonary setting, Pharmacol. Res. 40 (2) (1999) 107–111.

[77] A.J. Coukell, A. Markham, Clopidogrel, Drugs 54 (1997) 745–751.

[78] S. Yende, R.G. Wunderink, Effect of clopidogrel on bleeding after coronary artery bypass surgery, Crit. Care Med. 29 (12) (2001) 2271–2275.

[79] L. Wang, Y. Tang, Y. Chen, F. Tian, F, Method for the preparation of clopidogrel and its analogues of methyl-tetrahydrothieno [3, 2-C] pyridine acetate, US Patent 7,932,391, 2011.

[80] L. Wang, J. Shen, Y. Tang, Y. Chen, W. Wen Wang, Z. Cai, Z. Du, Synthetic improvements in the preparation of clopidogrel, Org. Process Res. Dev. 11 (3) (2007) 487–489.

[81] K.-N. Uhm, S.-J. Lee, H.-K. Kim, H.-Y. Kang, Y.J. Lee, Enantioselective resolution of methyl 2-chloromandelate by *Candida antarctica* lipase A in a solvent-free transesterification reaction, J. Mol. Catal. Enzym. 45 (2007) 34–38.

[82] E.M. Sinclair, D.J. Drucker, Glucagon-like peptide 1 receptor agonists and dipeptidyl peptidase IV inhibitors: new therapeutic agents for the treatment of type 2 diabetes, Curr. Opin. Endocrinol. Diabetes 12 (2005) 146–151.

[83] A. Thornberry, A.E. Weber, Discovery of Januvia (Sitagliptin), a selective dipeptidyl peptidase IV inhibitor for the treatment of type 2 diabestes, Curr. Top. Med. Chem. 7 (6) (2007) 557–561.

[84] D.J. Augeri, J.A. Robl, D.A. Betebenner, D.R. Magnin, A. Khanna, J.G. Robertson, A. Wang, L.M. Simpkins, P. Taunk, Q. Huang, S.-P. Han, B. Abboa-Offei, M. Cap, L. Xin, L. Tao, E. Tozzo, G.E. Welzel, D.M. Egan, J. Marcinkeviciene, S.Y. Chan, S.A. Biller, M.S. Kirby, R.A. Parker, L.G. Hamann, Discovery and preclinical profile of saxagliptin (BMS-477118): A highly potent, long-acting, orally active dipeptidyl peptidase IV inhibitor for the treatment of type 2 diabetes, J. Med. Chem. 48 (2005) 5025–5037.

[85] M. Kunishima, C. Kawachi, K. Hioki, S. Terao, S. Tani, Formation of carboxamides by direct condensation of carboxylic acids and amines in alcohols using a new alcohol- and water-soluble condensing agent: DMT-MM, Tetrahedron 57 (8) (2001) 1551–1557.

[86] I. Gill, R.N. Patel, Biocatalytic ammonolysis of (5S)-4,5-dihydro-1H- pyrrole-1,5-dicarboxylic acid, 1-(1,1-dimethylethyl)-5-ethyl ester: Preparation of an intermediate to the dipeptidyl peptidase IV inhibitor Saxagliptin, Bioorg. Med. Chem. Lett. 16 (3) (2006) 705–709.

[87] M. Ondetti, D. Cushman, Inhibitors of angiotensin-converting enzyme for treatment of hypertension, J. Med. Chem. 24 (1981) 355–361.

[88] M. Ondetti, B. Rubin, D. Cushman, Design of potent competitive inhibitors of angiotensin-converting enzyme. Carboxyalkanoyl and mercaptoalkanoyl amino acids, Science 1977 (1977) 441–444.

[89] D. Cushman, M. Ondetti, Inhibitors of angiotensin-converting enzyme, in: G.P. Ellis, G.B. West (Eds.), Progress in Medicinal Chemistry, 17, Elsevier/North Holland Biomedical Press, Wiley Online library, 1980, pp. 40–48.

[90] C. Goodhue, J. Schaeffer, Preparation of L-β-hydroxyisobutyric acid by bacterial oxidation of isobutyric acid, Biotechnol. Bioeng. 13 (1971) 203–214.

[91] R. Patel, J. Howell, A. Banerjee, K. Fortney, L. Szarka, Stereoselective esterification of 3-benzylthio-2-methylpropanoic acid, Appl. Microbiol. Biotechnol. 36 (1991) 29–34.

[92] J. Moniot, Preparation of [N-(2-mercaptomethyl)propionyl]-Lprolines, US Patent Application: CN 88-100862, 1988.

[93] M. Ondetti, A. Miguel, J. Krapcho, Mercaptoacyl derivatives of substituted pralines, US Patent 4316906, 1982.

[94] C. Ireland, P.A. Leeson, J. Castaner, Lobucavir: antiviral, Drugs Future 22 (4) (1997) 359–370.

[95] R.L. Hanson, Z. Shi, D.B. Brzozowski, A. Banerjee, T.P. Kissick, J. Singh, A.J. Pullockaran, J.T. North, J. Fan, J. Howell, S.C. Durand, M.A. Montana, D.R. Kronenthal, R.H. Mueller, R.N. Patel, Regioselective enzymatic aminoacylation of lobucavir to give an intermediate for lobucavir prodrug. Bioorg. Med. Chem. 8 (12) (2000) 2681–2687.

[96] F.G. Njoroge, A.G. Taveras, J. Kelly, S. Remiszewski, A.K. Mallams, R. Wolin, A. Afonso, A.B. Cooper, D.F. Rane, Y.-T. Liu, J. Wong, B. Vibulbhan, P. Pinto, J. Deskus, C.S. Alvarez, J. del Rosario, M. Connolly, J. Wang, J. Desai, R.R. Rossman, W.R. Bishop, R. Patton, L. Wang, P. Kirschmeier, M.S. Bryant, A.A. Nomeir, C.-C. Lin, M. Liu, A.T. McPhail, R.J. Dol, V.M. Girijavallabhan, A.K. Ganguly, (+)-4-[2-[4-(8-Chloro-3,10-dibromo-6,11-dihydro-5H-benzo[5,6] cyclohepta[1,2- b]- pyridin-11(R)-yl)-1-piperidinyl]-2-oxo-ethyl]-1-piperidinecarboxamide (SCH- 66336): A very potent farnesyl protein transferase inhibitor as a novel antitumor agent, J. Med. Chem. 41 (1998) 4890–4902.

[97] M. Barbacid, Ras Genes, Ann. Rev. Biochem. 56 (1987) 779–827.

[98] B. Morgan, A. Zaks, D.R. Dodds, J. Liu, S. Megati, F.G. Njoroge, V.M. Girijavallabhan, Enzymatic kinetic resolution of piperidine atropisomers: Synthesis of a key intermediate of the farnesyl protein transferase inhibitor SCH66336, J. Org. Chem. 65 (18) (2000) 5451–5459.

[99] S. Pianko, J.G. McHutchison, Treatment of hepatitis C with interferon and ribavirin, J. Gastroent. Hepatol. 15 (6) (2000) 581–586.

[100] P. Ferenci, H. Brunner, K. Nachbaur, C. Datz, M. Gschwantler, H. Hofer, R. Stauber, F. Hackl, W. Jessner, M. Rosenbeiger, M.-S. Petra, K. Hegenbarth, A. Gangl, W. Vogel, Combination of interferon indication therapy and ribavirin in chronic hepatitis C, Hepatology 34 (5) (2001) 1006–1011.

[101] H.L. Bonkovsky, D. Stefancyk, K. McNeal, B.F. Banner, Q. Liu, G.M. Zucker, J. Israel, J. Stagias, J. Colker, Comparative effects of different doses of ribavirin plus interferon-(2b for therapy of chronic hepatitis C: results of a controlled, randomized trial, Dig. Dis. Sci. 46 (10) (2001) 2051–2059.

[102] M. Tamarez, B. Morgan, G.S.K. Wong, W. Tong, F. Bennett, R. Lovey, J.L. McCormick, A. Zaks, Pilot-scale lipase-catalyzed regioselective acylation of ribavirin in anhydrous media in the synthesis of a novel prodrug intermediate, Org. Process Res. Dev. 7 (6) (2003) 951–953.

[103] J. Cossy, A. Schmitt, C. Cinquin, D. Buisson, D. Belotti, A very short, efficient and inexpensive synthesis of the prodrug from SC-54701 A platelet aggregation inhibitor, Bioorg. Med. Chem. Lett. 7 (13) (1997) 1699–1700.

[104] T. Yamanaka, M. Ohkubo, F. Takahashi, M. Kato, An efficient synthesis of the orally-active GpIIb/IIIa antagonist FR184764, Tetrahedron Lett. 45 (13) (2004) 2843–2845.

[105] B.H. Landis, P.B. Mullins, K.E. Mullins, P.T. Wang, Kinetic resolution of (-amino esters by acylation using immobilized penicillin amidohydrolase, Org. Process Res. Dev. 6 (4) (2002) 539–546.

[106] S.Y. Sit, R.A. Parker, I. Motoc, W. Han, N. Balasubramanian, J.D. Catt, P.J. Brown, W.E. Harte, M.D. Thompson, J.J. Wright, Synthesis, biological profile, and quantitative structure-activity relationship of a series of novel 3-hydroxy-3- methylglutaryl coenzyme A reductase inhibitors, J. Med. Chem. 33 (11) (1990) 2982–2999.

[107] R.N. Patel, C.M. McNamee, L.J. Szarka, Enantioselective enzymatic acetylation of racemic [4-[4α,6β (*E*)]]-6-[4,4-bis(4-fluorophenyl)-3-(1-methyl-1H- tetrazol-5-yl)-1,3-butadienyl]-tetrahydro-4-hydroxy-2H-pyran-2-one, Appl. Microbiol. Biotechnol. 38 (1992) 56–60.

[108] D. Schenk, D. Games, P. Seubert, Potential treatment opportunities for Alzheimer's disease through inhibition of secretases and A[beta] immunization, J. Mol. Neurosci. 17 (2) (2001) 259–267.

[109] J.E. Audia, E. James, T.C. Britton, C. Thomas, J.J. Droste, J. James, B.K. Folmer, K. Beverly, G.W. Huffman, V. John, L.H. Latimer, T.E. Mabry, J.S. Nissen, Methods and compounds for inhibiting β-amyloid peptide release and/or synthesis, PCT Int. Appl. (1998) 487 WO 9822494 A2.

[110] R.N. Patel, A. Banerjee, V.B. Nanduri, A. Goswami, F.T. Comezoglu, Enzymatic resolution of racemic secondary alcohols by lipase B from *Candida antarctica*, J. Am. Oil Chem. Soc. 77 (10) (2000) 1015–1019.

[111] L.S. Zalman, M.A. Brothers, P.S. Dragovich, R. Zhou, T.J. Prins, S.T. Worland, A.K. Patick, Inhibition of human rhinovirus-induced cytokine production by AG7088, a human rhinovirus 3C protease inhibitor Antimicrob, Agents Chemother. 44 (2000) 1236–1241.

[112] P.S. Dragovich, T.J. Prins, R. Zhou, S.E. Webber, J.T. Marakovits, S.A. Fuhrman, A.K. Patick, D.A. Matthews, C.A. Lee, C.E. Ford, B.J. Burke, P.A. Rejto, T.F. Hendrickson, T. Tuntland, E.L. Brown, J.W. Meador III, R.A. Ferre, J.E.V. Harr, M.B. Kosa, S.T. Worland, Structure-based design, synthesis, and biological evaluation of irreversible human rhinovirus 3C protease inhibitors. 4. incorporation of P_1 lactam moieties as l-glutamine replacements, J. Med. Chem. 42 (1999) 1213–1224.

[113] J. Tao, K. McGee, Development of a Continuous Enzymatic process for the preparation of (*R*)-3-(4-fluorophenyl)-2-hydroxy propionic acid, Org. Process Res. Dev. 6 (2002) 520–524.

[114] H. Jajoo, R. Mayol, J. LaBudde, I. Blair, Metabolism of the antianxiety drug buspirone in human subjects, Drug Metab. Dispos. 17 (6) (1989) 634–640.

[115] R. Mayol, Buspirone metabolite for the alleviation of anxiety US Ser. No. 484,161, US 6,150,365, 2000.

[116] J. Yevich, J. New, W. Lobeck, P. Dextraze, E. Bernstein, D. Taylor, F. Yocca, M. Eison, D. Temple Jr., Synthesis and biological characterization of α-(4- fluorophenyl)-4-(5-fluoro-2-pyrimidinyl)-1-piperazinebutanol and analogs as potential atypical antipsychotic agents, J. Med. Chem. 35 (24) (1992) 4516–4525.

[117] J. Yevich, R. Mayol, J. Li, F. Yocca, (*S*)-6-Hydroxy-buspirone for treatment of anxiety, depression and related disorders, US 2003,022,899.

[118] R. Patel, L. Chu, V. Nanduri, L. Jianqing, A. Kotnis, W. Parker, M. Liu, R. Mueller, Enantioselective microbial reduction of 6-oxo-8-[4-[4-(2-pyrimidinyl)-1-piperazinyl]butyl]-8-azaspiro[4.5]decane-7,9-dione, Tetrahedron Asymmetry 16 (16) (2005) 2778–2783.

[119] S. Goldberg, V. Nanduri, L. Chu, R. Johnston, R. Patel, Enantioselective microbial reduction of 6-oxo-8-[4-[4-(2-pyrimidinyl)-1-piperazinyl]butyl]-8-azaspiro[4.5]decane-7,9-dione: cloning and expression of reductases, Enzym. Microb. Technol. 39 (7) (2006) 1441–1450.

[120] A.O. King, E.G. Corley, R.K. Anderson, R.D. Larsen, T.R. Verhoeven, P.J. Reider, Y.B. Xiang, M. Belley, Y. Leblanc, M. Labelle, P. Prasit, R.J. Zamboni, An efficient synthesis of LTD4 antagonist L-699,392, J. Org. Chem. 58 (1993) 3731–3735.

[121] I. Shinkai, A.O. King, R.D. Larsen, Practical asymmetric synthesis of LTD4 antagonist, Pure Appl. Chem. 66 (7) (1994) 1551–1556.

[122] M. Zhao, A.O. King, R.D. Larsen, T.R. Verhoeven, P.J. Reider, A convenient and economical method for the preparation of DIP-Chloride™ and its application in the asymmetric reduction of aralkyl ketones, Tetrahedron Lett. 36 (1997) 2641–2644.

[123] A. Avdagic, J. Jerkovic, D. Pavlicic, Process for the preparation of montelukast PCT Patent Application WO 2008/009970, 2008.

[124] A. Shafiee, H. Motamedi, A. King, Purification, characterization and immobilization of an NADPH-dependent enzyme involved in the chiral specific reduction of the keto ester M,

an intermediate in the synthesis of an anti-asthma drug, Montelukast, from *Microbacterium campoquemadoensis* (MB5614), Appl. Mivrobiol. Biotechnol. 49 (1998) 709–717.

[125] J. Liang, B. Borup, V. Mitchell, E. Mundorff, J. Lalonde, G.W. Huisman,Ketoreductase polypeptides and uses thereof. Application # patents/app/20090155863 AC12P1712FI, USPC Class: 43512, 2009.

[126] J. Liang, J. Lalonde, B. Borup, V. Mitchell, E. Mundorff, N. Trinh, D.A. Kochrekar, C.R. Nair, G.G. Pai, Development of a biocatalytic process as an alternative to the (−)-DIP-Cl-mediated asymmetric reduction of a key intermediate of montelukast, Org. Process Res. Dev. 14 (2010) 193–198.

[127] J.P. Vacca, New advances in the discovery of thrombin and factor Xa inhibitors, Curr. Opin. Chem. Biol. 4 (2000) 394–400.

[128] T.D. Gladwell, Bivalirud: A direct thrombin inhibitor, Clinical Therapeutics 24 (2002) 38–58.

[129] J.M. Fevig, R.R. Wexler, Anticoagulants: thrombin and factor Xa inhibitors, Ann. Reports Med. Chem. 34 (1999) 81–100.

[130] P.D. Williams, C. Coburn, C. Burgey, M.M. Morrissette, Preparation of triazolopyrimidines as thrombin inhibitors, PCT Int. Appl. (2002) WO2002064211 A1.

[131] T.D. Nelson, C.R. LeBlond, D.E. Frantz, L. Matty, J.V. Mitten, D.G. Weaver, J.C. Moore, J.M. Kim, R. Boyd, P.-Y. Kim, K. Gbewonyo, M. Brower, M. Sturr, K. McLaughlin, D.R. McMasters, M.H. Kress, J.M. McNamara, U.H. Dolling, Stereoselective synthesis of a potent thrombin inhibitor by a novel P2-P3 lactone ring opening, J. Org. Chem. 69 (2004) 3620–3627.

[132] B.D. Roth, The discovery and development of atorvastatin, a potent novel hypolipidemic agent, Progress Med. Chem. 40 (2002) 1–22.

[133] S.K. Ma, J. Gruber, C. Davis, L. Newman, D. Gray, A. Wang, J. Grate, G.W. Huisman, R.A. Sheldon, A green-by-design biocatalytic process for atorvastatin intermediate, Green Chem. 12 (2012) 81–86.

[134] N. Balasubramanian, P.J. Brown, J.D. Catt, W.T. Han, R.A. Parker, S.Y. Sit, J.J. Wright, A potent, tissue-selective, synthetic inhibitor of HMG-CoA reductase inhibitor, in: D.T. Witiak, H.A.I. Newman, D.R. Feller (Eds.), Antilipidemic Drugs: Medicinal, Chemical and Biochemical Aspects, Elsevier Science, New York, 1991.

[135] R.N. Patel, A. Banerjee, C.G. McNamee, i D.B. Brzozowsk, R.L. Hanson, L.J. Szarka, Enantioselective microbial reduction of 3,5-dioxo-6-(benzyloxy) hexanoic acid, ethyl ester, Enzyme Microb. Technol. 15 (12) (1993) 1014–1021.

[136] Z. Guo, Y. Chen, A. Goswami, R.L. Hanson, R.N. Patel, Synthesis of ethyl and t-butyl (3R,5S)-dihydroxy-6-benzyloxyhexanoates via diastereo- and enantioselective microbial reduction, Tetrahedron Asymmetry 17 (1) (2006) 1589–1602.

[137] S. Goldberg, Z. Guo, S. Chen, A. Goswami, R.N. Patel, Synthesis of ethyl- (3*R*,5*S*)-dihydroxy-6-benzyloxyhexanoates via diastereo-and enantioselective microbial reduction: Cloning and expression of ketoreductase IIIfrom *Acinetobacter* sp. SC 13874, Enzyme Microb. Technol. 43 (2008) 544–549.

[138] A. Fassler, G. Bold, H.G. Capraro, R. Cozens, J. Mestan, B. Poncioni, J. Rosel, M. Tintelnot-Blomley, M. Lang, Aza-peptide analogs as potent human immunodeficiency virus type-1 protease inhibitors with oral bioavailability, J. Med. Chem. 39 (1996) 3203–3216.

[139] G. Bold, A. Faessler, H.-G. Capraro, R. Cozens, T. Klimkait, J. Lazdins, J. JMestan, B. Poncioni, J. Roesel, D. Stover, M. Tintelnot-Blomley, F. Acemoglu, W. Beck, E. Boss, M. Eschbach, T. Huerlimann, E. Masso, S. Roussel, K. Ucci-Stoll, D. Wyss, M. Lang, New aza-dipeptide analogs as potent and orally absorbed HIV-1 protease inhibitors: candidates for clinical development, J. Med. Chem. 41 (8) (1998) 3387–3401.

[140] B.S. Robinson, K.A. Riccardi, Y.F. Gong, Q. Guo, D.A. Stock, W.S. Blair, B.J. Terry, C.A. Deminie, F. Djang, R.J. Colonno, P.-F. Lin, BMS-232632, a highly potent human immunodeficiency virus protease inhibitor that can be used in combination with other available antiretroviral agents, Antimicrob. Agents Chemother. 44 (8) (2000) 2093–2099.

[141] R.N. Patel, L. Chu, R.H. Mueller, Diastereoselective microbial reduction of (S)- [3-chloro-2-oxo-1-(phenylmethyl)propyl]carbamic acid, 1,1-dimethylethyl ester, Tetrahedron Asymmetry 14 (20) (2003) 3105–3109.

[142] N.I. Bowers, P.M. Skonezny, G.L. Stein, T. Franceschini, S-J. Chiang, W.L. Anderson, L. You, Z. Xing, US Patent 2009/0,286,303.

[143] Z. Xu, J. Singh, M.D. Schwinden, B. Zheng, T.P. Kissick, B. Patel, M.J. Humora, F. Quiroz, L. Dong, D.-M. Hsieh, J.E. Heikes, M. Pudipeddi, M.D. Lindrud, S.K. Srivastava, D.R. Kronenthal, R.H. Mueller, Process research and development for an efficient synthesis of the HIV protease inhibitor BMS-232632, Org. Process Res. Dev. 6 (3) (2002) 323–328.

[144] Y.K. Bong, M. Vogel, S.J. Collier, V. Mitchell, J. Mavinahalli, Patent Application WO/2011/005527.

[145] T.S. Haque, W.R. Ewing, C. Mapelli, V.G. Lee, R.B. Sulsky, D.J. Riexinger, R.L. Martinez, Y.Z. Zhu, Human glucagon-like-peptide-1 modulators and their use in the treatment of diabetes and related conditions, WO2007,082,264.

[146] C. Mapelli, S.I. Natarajan, J.-P. Meyer, M.M. Bastod, M.S. Bernatowicz, V.G. Lee, J. Pluscec, D.J. Riexinger, E.S. Sieber-McMaster, K.L. Constantine, C.A. Smith-Monroy, R. Golla, Z. Ma, D.A. Longhi, D. Shi, L. Xin, J.R. Taylor, B. Koplowitz, C.L. Chi, A. Khanna, G.W. Robinson, S. Seethala, H.A. Antal-Zimanyi, R.H. Stoffel, S. Han, J.M. Whaley, C.S. Huang, J. Krupinski, W.R. Ewing, Eleven amino acid galucagon-like peptide-1 receptor agonists with antidiabetic activity, J. Med. Chem. 52 (23) (2009) 7788–7799.

[147] Y. Chen, S.L. Goldberg, R.L. Hanson, W.L. Parker, I. Gill, T.P. Tully, M. Montana, A. Goswami, R.N. Patel, Enzymatic Preparation of an (*S*)-amino acid from a racemic amino acid, Org. Process Res. Dev. 15 (1) (2011) 241–248.

[148] J.A. Robl, C. Sun, J. Stevenson, D.E. Ryono, L.M. Simpkins, M.P. Cimarusti, T. Dejneka, W.A. Slusarchyk, S. Chao, L. Stratton, R.N. Misra, M.S. Bednarz, M.M. Asaad, H.S. Cheung, B.E. Aboa-Offei, P.J. Smith, P.D. Mathers, M. Fox, T.R. Schaeffer, A.A. Seymour, N.C. Trippodo, Dual metalloprotease inhibitors: mercaptoacetyl-based fused heterocyclic dipeptide mimetics as inhibitors of angiotensin- converting enzyme and neutral endopeptidase, J. Med. Chem. 40 (1997) 1570–1577.

[149] A.A. Seymour, J.N. Swerdel, B.E. Abboa-Offei, Antihypertensive activity during inhibition of neutral endopeptidase and angiotensin converting enzyme, J. Cardiovasc. Pharmacol. 17 (1991) 456–465.

[150] R.N. Patel, A. Banerjee, V. Nanduri, S. Goldberg, R. Johnston, R. Hanson, C. McNamee, D. Brzozowski, T. Tully, R. Ko, T. LaPorte, D. Cazzulino, S. Swaminathan, L. Parker, J. Venit, Biocatalytic preparation of a chiral synthon for a vasopeptidase inhibitor: Enzymatic conversion of N2-{N-[(phenylmethoxy) carbonyl] L-homocysteinyl]-L-lysine (1 > 1′)-disulfide to [4S-(4a, 7a, 10ab)]1-octahydro-5- oxo-4- [(phenylmethoxy) carbonyl]amino]-7H-pyrido-[2,1-b][1,3]thiazepin-7-carboxylic acid methyl ester by a novel L-lysine ε − aminotransferase, Enzyme Microb. Technol. 27 (2000) 376–389.

[151] R.N. Patel, Enzymatic synthesis of chiral intermediates for Omapatrilat, an antihypertensive drug, Biomol. Eng. 17 (2001) 167–182.

[152] P.V. Chaturvedula, L. Chen, R. Civiello, A.P. Degnan, G.M. Dubowchik, X. Han, J.J. Jiang, J.E. Macor, G.S. Poindexter, G.O. Tora, G. Luo, Anti-migraine spirocycles, US Patent Application. Publ.2007/0149503 A1.

[153] X. Han, R.L. Civiello, C.M. Conway, D.A. Cook, C.D. Davis, R. Macci, S.S. Pin, S.X. Ren, R. Schartman, L.J. Signor, G. Thalody, K.A. Widmann, C. Xu, P.V. Chaturvedula, J.E. Macor, G.M. Dubowchik, The synthesis and SAR of calcitonin gene-related peptide (CGRP) receptor antagonists derived from tyrosine surrogates. Part1, Bioorg. Med. Chem. Lett. 22 (14) (2012) 4723–4727.

[154] P.V. Chaturvedula, S.E. Mercer, S.S. Pin, G. Thalody, C. Xu, C.M. Conway, D. Keavy, L. Signor, G.H. Cantor, N. Mathias, P. Moench, R. Denton, R. Macci, R. Schartman, V. Whiterock, C. Davis,

J.E. Macor, G.M. Dubowchik, Discovery of (R)-N- (3-(7-methyl-1H-indazol-5-yl)-1-(4-(1-methylpiperidin-4-yl)-1-oxopropan-2-yl)-4-(2- oxo-1,2-dihydroquinolin-3-yl)piperidine-1-carboxamide (BMS-742413): a potent human CGRP antagonist with superior safety profile for the treatment of migraine through intranasal delivery, Bioorg. Med. Chem. Lett. 23 (11) (2013) 3157–3161.

[155] R.L. Hanson, B.L. Davis, S.L. Goldberg, R.M. Johnston, W.L. Parker, T.P. Tully, M.A. Montana, R.N. Patel, Enzymatic preparation of a D-amino acid from a racemic amino acid or keto acid, Org. Process Res. Dev. 12 (6) (2008) 1119–1129.

[156] D. Kim, L. Wang, M. Beconi, G.J. Eiermann, M.H. Fisher, H. He, G.J. Hickey, J.E. Kowalchick, B. Bleating, K. Lyons, F. Marsilio, M.E. McCann, R.A. Patel, A. Petrov, G. Scapin, S.B. Patel, R.S. Roy, J.K. Wu, M.J. Wyvratt, B.B. Zhang, L. Zhu, N.A. Thornberry, A.E. Weber, (2*R*)-4-Oxo-4-[3-(Trifluoromethyl)-5,6- dihydro[1,2,4]triazolo[4,3-*a*]pyrazin-7(8*H*)-yl]-1-(2,4,5-trifluorophenyl)butan-2-amine: A potent, orally active dipeptidyl peptidase IV inhibitor for the treatment of type 2 diabetes, J. Med. Chem. 48 (2005) 141–151.

[157] K.B. Hansen, Y. Hsiao, F. Xu, N. Rivera, A. Clausen, M. Kubryk, S. Krska, T. Rosner, B. Simmons, J. Balsells, N. Ikemoto, Y. Sun, F. Spindler, C. Malan, E.J.J. Grabowski, J.D. Armstrong, Highly efficient asymmetric synthesis of sitagliptin, J. Am. Chem. Soc. 131 (2009) 8798–8804.

[158] N.B. Johnson, I.C. Lennon, P.H. Moran, J.A. Ramsden, Industrial- scale synthesis and applications of asymmetric hydrogenation catalysts, Acc. Chem. Res. 40 (2007) 1291–1297.

[159] J.-S. Shin, B.-G. Kim, Exploring the active site of amine:pyruvate aminotransferase on the basis of the substrate structure-reactivity relationship: how the enzyme controls substrate specificity and stereoselectivity, J. Org. Chem. 67 (2002) 2848–2853.

[160] B.-K. Cho, H.-Y. Park, J.-H. Seo, J. Kim, T.-J. Kang, B.-S. Lee, B.-G. Kim, Redesigning the substrate specificity of omega-aminotransferase for the kinetic resolution of aliphatic chiral amines, Biotechnol. Bioeng. 99 (2008) 275–284.

[161] M. Höhne, S. Kühl, K. Robins, U.T. Bornscheuer, Efficient asymmetric synthesis of chiral amines by combining transaminase and pyruvate decarboxylase, ChemBioChem 9 (2008) 363–365.

[162] M.D. Truppo, J.D. Rozzell, J.C. Moore, N.J. Turner, Rapid screening and scale- up of transaminase catalysed reactions, Org. Biomol. Chem. 7 (2009) 395–398.

[163] C.K. Savile, J.M. Janey, E.C. Mundorff, J.C. Moore, S. Tam, W.R. Jarvis, J.C. Colbeck, A. Krebber, F.J. Fleitz, J. Brands, P.N. Devine, G.W. Huisman, G.J. Hughes, Biocatalytic asymmetric synthesis of chiral amines from ketones applied to sitagliptin manufacture, Science (2010) 305–309 July.

[164] D.H. Overstreet, G. Griebel, Antidepressant-like effects of CRF1 receptor antagonist SSR;1; 125543 in an animal model of depression, Eur. J. Pharmacol. 497 (1) (2004) 49–53.

[165] W.-L. Yu, F. Lawrence, H. Wong, S. Lelas, G. Zhang, M.D. Lindner, T. Wallace, J. McElroy, N.J. Lodge, P. Gilligan, R. Zaczek, The Pharmacology of DMP696 and DMP904, Non-Peptidergic CRF1 Receptor Antagonists, CNS Drug Rev. 11 (1) (2006) 21–52.

[166] P.J. Gilligan, T. Clarke, L. He, S. Lelas, Y.-W. Li, K. Heman, L. Fitzgerald, K. Miller, G. Zhang, A. Marshall, C. Krause, J.F. McElroy, K. Ward, K. Zeller, H. Wong, S. Bai, J. Saye, S. Grossman, R. Zaczek, S.P. Arneric, P. Hartig, D.W. Robertson, G. Trainor, 8-(4-Methoxyphenyl) pyrazolo[1,5-a]-1,3,5-triazines: Selective and centrally active corticotropin-releasing factor receptor-1 (CRF1) antagonists, J. Med. Chem. 52 (9) (2009) 3084–3092.

[167] R.L. Hanson, B.L. Davis, Y. Chen, S.L. Goldberg, W.L. Parker, T.P. Tully, M.A. Montana, R.N. Patel, Preparation of (R)-Amines from racemic amines with an (*S*)-amine transaminase from *Bacillus megaterium*, Adv. Synth. Catal. 350 (9) (2008) 1367–1375.

[168] W. Leuchtenberger, K. Huthmacher, K. Karlheinz Drauz, Biotechnological production of amino acids and derivatives: current status and prospects, Appl. Microbiol. Biotechnol. 69 (2005) 1–8.

[169] A.S. Bommarius, M. Schwarm, K. Drauz, Biocatalysis to amino acid-based chiral pharmaceuticals: examples and perspective, J. Mol. Catal. B Enzym. 5 (1–4) (1998) 1–11.

[170] R.L. Hanson, Preparation of chiral amino acid intermediates for synthesis of pharmaceutical compounds using amino acid dehydrogenases. Asymmetric Synthesis and Application of (-Amino Acids, Chapter 19, pp. 306–321 ACS Symposium Series, vol. 1009.

[171] R.N. Patel, Chemo-enzymatic synthesis of pharmaceutical intermediates, Expert Opin. Drug Dis. Dev. 3 (2) (2008) 187–245.

[172] T. Ohshima, K. Soda, in: R.N. Patel (Ed.), Amino acid dehydrogenases and their applications in Stereoselective Biocatalysis, Marcel and Dekker, New York, 2000, pp. 877–903 page.

[173] E.M. Gordon, M.A. Ondetti, J. Pluscec, C.M. Cimarusti, D.P. Bonner, R.B. Sykes, O-Sulfated β – lactam hydroxamic acids (monosulfactams). Novel monocyclic β-lactam antibiotics of synthetic origin, J Am. Chem. Soc. 104 (1982) 6053–6060.

[174] J.D. Godfrey, R.H. Mueller, D.J. Van Langen, β-Lactam synthesis: cyclization versus 1,2-acyl migration-cyclization. The mechanism of the 1,2-acyl migration-cyclization, Tetrahedron Lett. 27 (1986) 2793–2796.

[175] R.L. Hanson, J. Singh, T.P. Kissick, R.N. Patel, L. Szarka, R. Mueller, Synthesis of L-β-hydroxyvaline from α-keto-β-hydroxyisovalerate using leucine dehydrogenase from *Bacillus* species, Bioorg. Chem. 18 (2) (1990) 116–130.

[176] G. Bold, A. Faessler, H.-G. Capraro, R. Cozens, T. Klimkait, J. Lazdins, J. JMestan, B. Poncioni, J. Roesel, D. Stover, M. Tintelnot-Blomley, F. Acemoglu, W. Beck, E. Boss, M. Eschbach, T. Huerlimann, E. Masso, S. Roussel, K. Ucci-Stoll, D. Wyss, M. Lang, New aza-dipeptide analogues as potent and orally absorbed HIV-1 protease inhibitors: candidates for clinical development, J. Med. Chem. 41 (8) (1998) 3387–3401.

[177] B.S. Robinson, K.A. Riccardi, Y.F. Gong, Q. Guo, D.A. Stock, W.S. Blair, B.J. Terry, C.A. Deminie, F. Djang, R.J. Colonno, P.-F. Lin, BMS-232632, a highly potent human immunodeficiency virus protease inhibitor that can be used in combination with other available antiretroviral agents, Antimicrob. Agents Chemother. 44 (8) (2000) 2093–2099.

[178] B. Degertekin, A.S. Lok, Update on viral hepatitis, Curr. Opin. Gastroenterol. 24 (3) (2008) 306–311.

[179] F.G. Njoroge, K.X. Chen, N.Y. Shih, J.J. Piwinski, Challenges in modern drug discovery: A case study of Boceprevir, an HCV protease inhibitor for the treatment of hepatitis C virus infection, Acc. Chem. Res. 41 (1) (2008) 50–59.

[180] U. Kragl, D. Nasic-Racki, C. Wandrey, Continuous production of L-*tert*-leucine in series of two enzyme membrane reactors, Bioprocess Eng. 14 (1996) 291–297.

[181] A.S.G. Krix, A.S. Bommarius, K. Kottenhahn, M. Schwarm, M.R. Kula, Enzymatic reduction of α-keto acids leading to L-amino acids, D- or L-hydroxy acids *J*, Biotechnol. 53 (1997) 29–39.

[182] A. Galkin, L. Kulakova, T. Yoshimura, K. Soda, N. Esaki, Synthesis of optically active amino acids from alpha-keto acids with *Escherichia coli* cells expressing heterologous genes, Appl. Environ. Microbiol. 63 (1997) 4651–4656.

[183] A. Menzel, H. Werner, J. Altenbuchner, H. Gröger, From enzymes to "designer bugs" in reductive amination: A new process for the synthesis of L-tert-leucine using a whole cell-catalyst, Eng.Life Sci. 4 (6) (2004) 573–576.

[184] J. Robl, C. Sun, J. Stevenson, D. Ryono, L. Simpkins, M. Cimarusti, T. Dejneka, W. Slusarchyk, S. Chao, L. Stratton, R. Misra, M. Bednarz, M. Asaad, H. Cheung, B. Aboa-Offei, P. Smith, P. Mathers, M. Fox, T. Schaeffer, A. Seymour, N. Trippodo, Dual metalloprotease inhibitors: mercaptoacetyl-based fused heterocyclic dipeptide mimetics as inhibitors of angiotensin-converting enzyme and neutral endopeptidase, J. Med. Chem. 40 (1997) 1570–1577.

[185] R.L. Hanson, J. Howell, T. LaPorte, M. Donovan, D. Cazzulino, V. Zannella, M. Montana, V. Nanduri, S. Schwarz, R. Eiring, S. Durand, J. Wasylyk, L. Parker, M. Liu, F. Okuniewicz, B.-C.

Chen, J. Harris, K. Natalie, K. Ramig, S. Swaminathan, V. Rosso, S. Pack, B. Lotz, P. Bernot, A. Rusowicz, D. Lust, K. Tse, J. Venit, L. Szarka, R.N. Patel, Synthesis of allylsine ethylene acetal using phenylalanine dehydrogenase from *Thermoactinomyces intermedius*, Enzyme Microb. Technol. 26 (2000) 348–358.

[186] R. Patel, Enzymatic synthesis of chiral intermediates for Omapatrilat, an antihypertensive drug, Biomol. Eng. 17 (2001) 167–182.

[187] E.M. Sinclair, D.J. Drucker, Glucagon-like peptide 1 receptor agonists and dipeptidyl peptidase IV inhibitors: new therapeutic agents for the treatment of type 2 diabetes, Curr. Opin. Endocrinol. Diabet. 12 (2005) 146–151.

[188] D.J. Augeri, J.A. Robl, D.A. Betebenner, D.R. Magnin, A. Khanna, J.G. Robertson, A. Wang, L.M. Simpkins, P. Taunk, Q. Huang, S.-P. Han, B. Abboa-Offei, M. Cap, L. Xin, L. Tao, E. Tozzo, G.E. Welzel, D.M. Egan, J. Marcinkeviciene, S.Y. Chang, S.A. Biller, M.S. Kirby, R.A. Parker, L.G. Hamann, A highly potent, long- acting, orally active dipeptidyl peptidase IV inhibitor for the treatment of type 2 Diabetes, J. Med. Chem. 248 (2005) 5025–5037.

[189] T.C. Vu, D.B. Brzozowski, R. Fox, J.D. Godfrey Jr., R.L. Hanson, S.V. Kolotuchin, J.A. Mazzullo Jr., R.N. Patel, J. Wang, K. Wong, J. Yu, J. Zhu, R.D. Magnin, D.J. Augeri, L.G. Hamann, Preparation of cyclopropyl-fused pyrrolidine-based inhibitors of dipeptidyl peptidase IV. 2004, PCT Int. Appl. (2004) 101 pp. WO 2004052850.

[190] R.L. Hanson, S.L. Goldberg, D.B. Brzozowski, T.P. Tully, D. Cazzulino, W.L. Parker, O.K. Lyngberg, T.C. Vu, M.K. Wong, R.N. Patel, Preparation of an amino acid intermediate for the dipeptidyl peptidase IV inhibitor, saxagliptin, using a modified phenylalanine dehydro genase, Adv. Synth. Catal. 349 (8 + 9) (2007) 1369–1378.

[191] H. Groeger, O. May, H. Werner, A. Menzel, J. Altenbuchner, A "second-generation process" for the synthesis of L-neopentylglycine: Asymmetric reductive amination using a recombinant whole cell catalyst, Org. Process Res. Dev. 10 (3) (2006) 666–669.

[192] M.J. Emmanuel, L.L. Frye, E.R. Hickey, W. Liu, T.M. Morwick, D.M. Spero, S. Sun, D.S. Thomson, Y.D. Ward, E.R.R. Young, Novel spiroheterocyclic compounds [morpholine-4-carboxylic acid amides of heterocyclic cyclohexylalanine and neopentylglycine derivatives and their analogs], useful as reversible inhibitors of cysteine proteases such as cathepsin S, PCT Int. Appl. 2001 (2001) 361pp WO2001019816.

[193] D.T. Gibson, R.E. Parales, Aromatic hydrocarbon dioxygenases in environmental biotechnology, Curr.Opin.Biotechnol. 11 (2000) 236–243.

[194] R.E. Parales, S.M. Resnick, Application of aromatic hydrocarbon dioxygenases, in: R.N. Patel (Ed.), Biocatalysis in the Pharmaceutical and Biotechnology Industries, CRC Press, FL, 2007, pp. 299–332.

[195] C. Louise, A.E. Nolan, K.E. O'Connor, Dioxygenase- and monooxygenase-catalysed synthesis of cis-dihydrodiols, catechols, epoxides and other oxygenated products, Biotechnol Lett. 30 (2008) 1879–1891.

[196] K.F. Lee, R. Parales, J.V. Parales, D.T. Gibson, S. Ramaswamy, Purification, crystallization and preliminary x-ray diffraction studies of the three components of the toluene 2,3-dioxygenase enzyme system, Acta Crystallogr. Sect. F Struct. Biol. Cryst. Commun. F61 (7) (2005) 669–672.

[197] A. Karlsson, J.V. Parales, R.E. Parales, D.T. Gibson, H. Eklund, S. Ramaswamy, Crystal structure of naphthalene dioxygenase: side-on binding of dioxygen to iron, Science 299 (5609) (2003) 1039–1042.

[198] T. Hudlicky, D. Gonzalez, D.T. Gibson, Enzymatic dihydroxylation of aromatics in enantioselective synthesis: expanding asymmetric methodology, Aldrichimica Acta 32 (2) (1999) 35–62.

[199] D.R. Boyd, N.D. Sharma, N.I. Bowers, I.N. Brannigan, M.R. Groocock, J.F. Malone, G. McConville, C.C.R. Allen, Biocatalytic asymmetric dihydroxylation of conjugated mono-and

poly-alkenes to yield enantiopure cyclic cis-diols, Adv. Synth. Catal. 347 (7 + 8) (2005) 1081–1089.

[200] D.R. Boyd, T.D.H. Bugg, Arene cis-dihydrodiol formation: from biology to application, Org. Biomol. Chem. 4 (2) (2006) 181–192.

[201] M. Chartrain, B. Jackey, C. Taylor, V. Sandfor, K. Gbewonyo, L. Lister, L. Dimichele, C. Hirsch, B. Heimbuch, C. Maxwell, D. Pascoe, B. Buckland, R. Greasham, Bioconversion of indene to cis-(1*S*,2*R*)-indandiol and trans- (1*R*,2*R*)-indandiol by *Rhodococcus* sp, J. Ferment. Technol. 86 (1998) 550–558.

[202] B.C. Buckland, D.K. Robinson, M. Chartrain, Biocatalysis for pharmaceuticals-status and prospects for a key technology, Metab. Eng. 2 (2003) 42–48.

[203] M. Chartrain, P.M. Salmon, D.K. Robinson, B.C. Buckland, Metabolic engineering and directed evolution for the production of pharmaceuticals, Curr. Opin. Biotechnol. 11 (2000) 209–214.

[204] N. Zhang, B.G. Stewart, J.C. Moore, R.L. Greasham, D.K. Robinson, B. Buckland, C. Lee, Directed evolution of toluene dioxygense from *Pseudomonas putida* for improved selectivity toward cis-(1*S*,2*R*)-indanediol during indane, Metab. Eng. 2 (4) (2000) 339–348.

[205] A. Amanullah, C.J. Hewitt, A.W. Nienow, C. Lee, M. Chartrain, B.C. Buckland, S.W. Drew, J.M. Woodley, Measurement of strain-dependent toxicity in the indene bioconversion using multiparameter flow cytometry, Biotechnol. Bioeng. 81 (4) (2003) 405–420.

[206] K. Gerth, S. Pradella, O. Perlova, S. Beyer, R. Muller, Myxobacteria: proficient producers of novel natural products with various biological activities-past and future biotechnological aspects with the focus on the genus *Sorangium*, J. Biotechnol. 106 (2–3) (2003) 233–253.

[207] D. Benigni, R. Stankavage, S.-J. Chiang, H. Hou, B. Eagan, D. Gu, D. Hou, L. Mintzmyer, T.P. Tully, B.L. Davis, I. Hargro, M. Mascari, G. Galvin, G. Stein, C.W. McConlogue, F.T. Comezoglu, Methods for the preparation, isolation and purification of epothilone B, and X-ray crystal structures of epothilone B, PCT Int. Appl. (2004) WO2004026254A1.

[208] S. Goodin, M.P. Kane, E.H. Rubin, Epothilones: Mechanism of action and biologic activity, J. Clin. Oncol. 22 (10) (2004) 2015–2025.

[209] K.C. Nicolaou, F. Roschangar, D. Vourloumis, Chemical biology of epothilones, Angew. Chem. Int. Ed. 37 (15) (1998) 2014–2045.

[210] K.-H. Altmann, The merger of natural product synthesis and medicinal chemistry: on the chemistry and chemical biology of epothilones, Org. Biomol. Chem. 2 (15) (2004) 2137–2152.

[211] C.N. Boddy, K. Hotta, M.L. Tse, R.E. Watts, C. Khosla, Precursor-directed biosynthesis of epothilone in *Escherichia coli*, J. Am. Chem. Soc. 126 (24) (2002) 7436–7437.

[212] N. Lin, K. Brakora, M. Seiden, BMS-247550 (Bristol-Myers Squibb/GBF), Curr. Opin. Inv. Drug. 4 (6) (2003) 746–756.

[213] J.A. Low, S.B. Wedam, J.J. Lee, A.W. Berman, A. Brufsky, S.X. Yang, M.S. Poruchynsky, S.M. Steinberg, N. Mannan, T. Fojo, S.M. Swain, Phase II clinical trial of ixabepilone (BMS-247550), an epothilone B analog, in metastatic and locally advanced breast cancer, J. Clin. Oncol. 23 (12) (2005) 2726–2734.

[214] J.D. Basch, S-J Chiang, S.W. Liu, A. Nayeem, Y-L Sun, The epothilone B hydroxylase and ferredoxin genes of Amycolatopsis and their use in the development of strains for the manufacture of hydroxylated epothilones, WO 2004,078,978 A1.

[215] J. Basch, S.J. Chiang, Cloning and expression of a cytochrome P450 hydroxylase gene from *Amycolatopsis orientalis*: hydroxylation of epothilone B for the production of epothilone F, J Ind. Microbiol. Biotechnol. 34 (2) (2007) 171–176.

[216] C.S. Miles, T.W.B. Ost, M.A. Noble, A.W. Munro, S.K. Chapman, Protein engineering of cytochromes P-450, Biochim. Biophys. Acta Protein Struct. Molec. Enzym. 1543 (2000) 383–407.

[217] P.C. Cirino, H. Frances, F.H. Arnold, Protein engineering of oxygenases for biocatalysis, Curr. Opin. Chem. Biol. 6 (2002) 130–135.

[218] R.T. Ruettinger, L.P. Wen, A.J. Fulco, Coding nucleotide, 5′ regulatory, and deduced amino acid sequences of P-450BM-3, a single peptide cytochrome P450: NADPH-P-450 reductase from *Bacillus megaterium*, J. Biol. Chem. 264 (1989) 10987–10995.

[219] J.H. Capdevila, S. Wei, C. Helvig, J.R. Falck, Y. Belosludtsev, G. Truan, S.E. Graham- Lorence, J.A. Peterson, The highly stereoselective oxidation of polyunsaturated fatty acids by cytochrome P450 BM-3, J. Biol. Chem. 271 (1996) 22663–22671.

[220] U. Schwaneberg, C. Otey, P.C. Cirino, E. Farinas, F.H. Arnold, Cost-effective whole-cell assay for laboratory evolution of hydroxylases in *Escherichia coli*, J. Biomol. Screen 6 (2001) 111–117.

[221] E.T. Farinas, U. Schwaneberg, A. Glieder, F.H. Arnold, Directed evolution of a cytochrome P450 monooxygenase for alkane oxidation, Adv. Synth. Catal. 343 (2001) 601–606.

[222] D.F. Muenzer, P. Meinhold, M.W. Peters, S. Feichtenhofer, H. Griengl, F.H. Arnold, A. Glieder, A. de Raadt, Stereoselective hydroxylation of an achiral cyclopentane carboxylic acid derivative using engineered P450s BM-3, Chem. Commun. 20 (2005) 2597–2599.

[223] A. de Raadt, H. Griengl, M. Petsch, P. Plachota, N. Schoo, H. Weber, G. Braunegg, I. Kopper, M. Kreiner, A. Zeiser, Microbial hydroxylation of 2-cycloalkyl benzoxazoles. Part III. Determination of product enantiomeric excess and cleavage of benzoxazoles, Tetrahedron Asymmetry 7 (2) (1996) 491–496.

[224] V.B. Urlacher, S. Lutz-Wahl, R.D. Schmid, Microbial P450 enzymes in biotechnology, Appl. Mivrobiol. Biotechnol. 64 (3) (2004) 317–325.

[225] Q.-S. Li, U. Schwaneberg, M. Fischer, J. Schmitt, J. Pleiss, S. Lutz-Wahl, R.D. Schmid, Rational evolution of a medium chain-specific cytochrome P-450 BM-3 variant, Biochimica et Biophysica Acta 1545 (1–2) (2001) 114–121.

[226] F.H. Arnold, C.R. Otey, Libraries of optimized cytochrome P 450 enzymes and the optimized P 450 enzymes exhibiting higher activities and new substrate specificities, WO2005,017,106 A2.

[227] C. Chassin, A biotechnological process for the production of nicotinamide, Chim. 14 (1996) 9–12.

[228] N.M. Shaw, K.T. Robins, A. Kiener, Lonza: 20 years of biotransformations, Adv. Synth. Catal. 345 (2003) 425–435.

[229] J. Heveling, E. Armbruster, L. Utiger, M. Rohner, H.-R. Dettwiler, R.J. Chuck, Process for preparing nicotinamide. US Patent 5,719, 045, 1998.

[230] K.T. Robins, T. Nagasawa, Enzymes for preparation of amides. U.S. Patent 6,444,451 B1, 2002.

[231] K.T. Robins, T.U.S. Nagasawa, Process for preparation of amides. Patent Appl. 20030148478 A1, 2003.

[232] K.T. Robins, T. Nagasawa, Microbiological method for producing amides. U.S. Patent Appl. 20040142447 A1, 2004.

[233] F. Watanabe, Jpn. Kokai Tokkyo Koho JP 2004215513 A2, 2004.

[234] F. Watanabe, F., Jpn. Kokai Tokkyo Koho JP 2004222538 A2, 2004.

[235] Ramey, S. and Tichenor, G., (2002) Enzymatically produced acrylamide: improving the environment, an additional benefit of building a better product, WEFTEC.02 Conference Proceedings, Chicago, September 28.

[236] M. Wieser, P. Pommares, PCT Int. Appl. WO 2003066800 A3, 2004.

[237] K.L. Petrillo, S. Wu, E.C. Hann, F.B. Cooling, A. Ben-Bassat, J.E. Gavagan, R. DiCosimo, M.S. Payne, Over-expression in *Escherichia coli* of a thermostable nitrile hydratase from *Comamonas testosteroni* 5-MGAM-4D, App. Microbiol. Biotechnol. 67 (2005) 664–670.

[238] L.J. Mersinger, E.C. Hann, F.B. Cooling, A. Ben-Bassat, J.E. Gavagan, S. Wu, K.L. Petrillo, M.S. Payne, R. DiCosimo, Production of acrylamide using alginate-immobilized *E. coli* expressing *Comamonas testosteroni* 5-MGAM-4D nitrile hydratase, Adv. Synth. Catal. 347 (2005) 1125–1131.

[239] T. Nagasawa, H. Shimizu, H. Yamada, The superiority of the third- generation catalyst, *Rhodococcus rhodochrous* J1 nitrile hydratase, for industrial production of acrylamide, Appl. Microbiol. Biotechnol. 40 (1993) 189–195.

[240] R. Padmakumar, P. Oriel, Bioconversion of acrylonitrile to acrylamide using a thermostable nitrile hydratase, Appl. Biochem. Biotechnol. 671 (1999) 77–79.

[241] D. Graham, R. Pereira, D. Barfield, D. Cowan, Nitrile biotransformations using free and immobilized cells of a thermophilic *Bacillus* spp, Enz. Microb. Technol. 26 (2000) 368–373.

[242] J. Han, Y. Zhang, J. Xue, J. Huang, X. Chen, Research on manufacture of acrylamide on industrial scale by microbial method, Shanghai Huagong 20 (1995) 7–11.

[243] M. Liu, C. Li, Y. Gao, Y. Wang, Z. Cao, Optimization of glucose-Co2 + coupling fed batch fermentation for production of nitrile hydratase of high activity by *Nocardia* sp. RS, Guocheng Gongcheng Xuebao 3 (2003) 555–560.

[244] M. Liu, C. Li, Y. Gao, Y. Huang, Z. Cao, Improved process for high acrylamide accumulation with acclimated microorganism *Nocardia* sp, Huagong Xuebao 55 (2004) 1678–1685.

[245] M. Liu, C. Li, Y. Huang, Y. Gao, Z. Cao, Improving the acrylamide-tolerance of nitrile hydratase in *Nocardia* sp. by extreme cultivation, Guocheng Gongcheng Xuebao 4 (2004) 250–256.

[246] R. Shapiro, R. DiCosimo, S.M. Hennessey, B. Stieglitz, O. Campopiano, G.C. Chiang, Discovery and development of a manufacturing-scale synthesis of azafenidin, Org. Process Res. Dev. 5 (2001) 593–598.

[247] O. Andresen, S.E. Godtfredsen, European Patent EP 178106 B1, 1993.

[248] T. Nagamune, H. Kurata, M. Hirata, J. Honda, A. Hirata, I. Endo, Photosensitive phenomena of nitrile hydratase of *Rhodococcus* sp. N-771, Photochem. Photobiol. 51 (1990) 87–90.

[249] T. Nagasawa, H. Nanba, K. Ryuno, K. Takeuchi, H. Yamada, Nitrile hydratase of *Pseudomonas chlororaphis* B23. Purification and characterization, Eur. J. Biochem. 162 (1987) 691–698.

[250] Y. Asano, T. Yasuda, Y. Tani, H. Yamada, Microbial degradation of nitrile compounds. Part VII. A new enzymic method of acrylamide production, Agric. Biol. Chem. 46 (1982) 1183–1189.

[251] M. Kobayashi, T. Nagasawa, H. Yamada, Enzymic synthesis of acrylamide: a success story not yet over, Trends Biotechnol. 10 (1992) 402–408.

[252] R. DiCosimo, E.C. Hann, A. Eisenberg, S.K. Fager, N.E. Perkins, F.G. Gallagher, S.M. Cooper, J.E. Gavagan, B. Stieglitz, S.M. Hennessey, Biocatalytic production of 5-cyanovaleramide from adiponitrile, ACS Symp. Ser. 767 (2000) 114–125.

[253] E.C. Hann, A. Eisenberg, S.K. Fager, N.E. Perkins, F.G. Gallagher, S.M. Cooper, J.E. Gavagan, B. Stieglitz, S.M. Hennessey, R. DiCosimo, 5-Cyanovaleramide production using immobilized *Pseudomonas chlororaphis* B23, Bioorg. Med. Chem. 7 (1999) 2239–2245.

[254] M. MacDonald, G.B. D'Cunha, A modern view on phenylalanine ammonia lyase, Biochem. Cell Biol. 85 (3) (2007) 273–282.

[255] F. Effenberger, S. Foerster, C. Kobler, in: R.N. Patel (Ed.), Biocatalysis in the Pharmaceutical and Biotechnology Industries, CRC Press, FL, 2007, pp. 677.

[256] V. Gotor, Lipases and (*R*)-oxynitrilases: useful tools in organic synthesis, J. Biotechnol. 96 (1) (2002) 35–42.

[257] M. Breuer, B. Hauer, Carbon-carbon coupling in biotransformation, Curr. Opin. Biotechnol. 14 (2003) 570–576.

[258] H. Kumagai, Microbial production of amino acids in Japan, in: T. Schepers (Ed.), Adv. Biochem. Eng./Biotechnol, 69, Springer Verlag, Berlin, Heidelberg, 2000, pp. 71–85.

[259] T. Tsuchida, Y. Nishimoto, T. Kotani, K. Iiizumi, Production of L-3,4- dihydroxy phenylalanine. Ajinomoto Co., Ltd., JP 5123177A, 1993.

[260] H. Yamada, Screening of novel enzymes for the production of useful compounds, in: K. Kieslich, C.P. van der Beek, J.A.M. deBont, W.J.J. van den Tweel (Eds.), New Frontiers in Screening for Microbial Biocatalysis, 53, Studies in Organic Chemistry, Elsevier, Amsterdam, 1998, pp. 13–17.

[261] R.A. Sheldon, Chirotechnology, Industrial Synthesis of Optically Active Compounds, Marcel Dekker-Verlag, New York, (1993).

[262] P.S.J. Cheetham, Bioprocess for the manufacture of ingredients for food and cosmetics, Adv. Biochem. Engin/Biotechnol 86 (2004) 83–158.

[263] J.P. Montgomery, J.J. Schruber, P.J. Vollmer, H-H Yang, Method for stabilizing the enzymatic activity of phenylalanine ammonia lyase during L-phenylalanine production. Genex Corporation, US 4,584,269 A, 1985.

[264] J. Crosby, Synthesis of optically active compounds: a large scale perspective, Tetrahedron 47 (1991) 4789–4846.

[265] J.E. Gavagan, R. DiCosimo, A. Eisenberg, S.K. Fager, P.W. Folsom, E.C. Hann, K.J. Schneider, R.D. Fallon, AGram negative bacterium producing a heat-stable nitrilase highly active on aliphatic dinitriles, Appl. Microbiol. Biotechnol. 52 (1999) 654–659.

[266] F.B. Cooling, S.K. Fager, R.D. Fallon, P.W. Folsom, F.G. Gallagher, J.E. Gavagen, E.C. Hann, F.S. Herkes, R.L. Phillips, A. Sigmund, L.W. Wagner, W. Wu, R. DiCosimo, Chemoenzymatic production of 1,5-dimethyl-2-piperidone, J. Mol. Catal. B Enzym. 11 (2001) 295–306.

[267] J.E. Gavagan, S.K. Fager, R.D. Fallon, P.W. Folsom, F.E. Herkes, A. Eisenberg, E.C. Hann, R. DiCosimo, Chemoenzymatic production of lactams from aliphatic- dinitriles, J. Org. Chem. 63 (1998) 4792–4801.

[268] H. Biebl, K. Menzel, A.-P. Zeng, W.-D. Deckwer, Microbial production of 1,3- propanodiol, Appl. Microbiol. Biotechnol. 52 (1999) 289–297.

[269] A.-P. Zeng, H. Biebl, Bulk Chemicals from Biotechnology Production: The case of 1,3-propanodiol production and the new trends, Advances in Biochemical Engineering and Biotechnology 74 (2002) 239–259.

[270] S. Papanikolaou, P. Ruiz-Sanchez, B. Pariset, F. Blanchard, M. Fick, High production of 1,3-propanediol from industrial glycerol by a newly isolated *Clostridium butyricum* strain, J. Biotechnol. 77 (2000) 191–208.

[271] M. González-Pajuelo, J.C. Andrade, I. Vasconcelos, Production of 1,3- propanediol by *Clostridium butyricum* VPI 3266 using a synthetic medium and raw glycerol, J. Ind. Microbiol. Biotechnol. 31 (2004) 442–446.

[272] J.C. Thompson, B.B. He, Characterization of crude glycerol from biodiesel production from multiple feedstocks, Appl. Eng. Agric. 22 (2) (2008) 261–265.

[273] D. Dietz, A.P. Zeng, Efficient production of 1,3-propanediol from fermentation of crude glycerol with mixed cultures in a simple medium, Bioprocess Biosyst. Eng. (2013) 2013 Jun 9.

[274] X. Tang, Y. Tan, H. Zhu, K. Zhao, W. Shen, Microbial Conversion of Glycerol to 1,3-Propanediol by an Engineered Strain of *Escherichia coli*, Appl. Environ. Microbiol. 75 (6) (2009) 1628–1634.

[275] L.A. Laffend, V. Nagarajan, C.E. Nakamura, Bioconversion of a fermentable carbon source to 1,3-propanediol by a single microorganism, US Patent No. 5,688,276, 1997.

[276] I.T. Tong, H.H. Liao, D.C. Cameron, 1,3 propanediol production by *Escherichia coli* expressing genes from the *Klebsiella pneumoniae* dha regulon, Appl. Environ. Microbiol. 57 (1991) 3541–3546.

[277] M.J. Syu, Biological production of 2,3-butanediol, Appl. Microbiol. Biotechnol. 55 (2001) 10–18.

[278] E. Celińska, W. Grajek, Biotechnological production of 2,3-butanediol–current state and prospects, Biotechnol. Adv. 27 (2009) 715–725 (2009).

[279] X.J. Ji, H. Huang, P.K. Ouyang, Microbial 2,3-butanediol production: a state- of-the- art review, Biotechnol. Adv. 29 (2011) 351–364.

[280] R. Liu, H.E. Högberg, Chemoenzymatic preparation of (2*S*,3*S*)- and (2*R*,3*R*)-2,3-butanediols and their esters from mixtures of *d*,*l*- and *meso*-diols, TetrahedronAsymmetry 12 (2001) 771–778.

[281] Y. Yan, C.C. Lee, J.C. Liao, Enantioselective synthesis of pure (*R*,*R*)-2,3-butanediol in *Escherichia coli* with stereospecific secondary alcohol dehydrogenases, Org. Biomol. Chem. 7 (2009) 3914–3917.

[282] S. Ui, Y. Takusagawa, T. Sato, T. Ohtsuki, A. Mimura, M. Ohkuma, T. Kudo, Production of l-2,3-butanediol by a new pathway constructed in *Escherichia coli*, Lett. Appl. Microbiol. 39 (2004) 533–537.

[283] L. Li, L. Zhang, K. Li, Y. Wang, Chao, C. Gao, B. Binbin Han, C. Cuiqing Ma, P. Xu, Biocatalytic production of (2*S*,3*S*)-2,3-butanediol from diacetyl using whole cells of engineered *Escherichia coli*, Bioresour. Technol. 115 (2012) 111–116.

[284] Y. Wang, L. Li, C. Ma, C. Gao, F. Tao, P. Xu, Engineering of cofactor regeneration enhances (2S,3S)-2,3-butanediol production from diacetyl Scientific Reports 3, Volume:Article number: 2643DOI:, doi:10.1038/srep02643, September 12, 2013.

[285] O. May, S. Verseck, A. Bommarius, K. Drauz, Development of dynamic kinetic resolution processes for biocatalytic production of natural and nonnatural L-amino acids, Org. Proc. Res. Dev. 6 (2002) 452–457.

[286] H. Trauthwein, O. May, U. Dingerdissen, S. Buchholz, K. Drauz, A new biocatalytic route to enantiopure *N*-carbamoyl amino acids by fast enzyme screening, Tetrahedron Lett. 44 (2003) 3737–3739.

[287] T.H.P. Maier, Semisynthetic production of unnatural L-α-amino acids by metabolic engineering of the cysteine-biosynthetic pathway, Nat. Biotechnol. 21 (2003) 422–427.

CHAPTER 12

Future of Biocatalysis in the Synthesis of Organic Compounds

Animesh Goswami*, Jon D. Stewart†
*Chemical Development, Bristol-Myers Squibb, New Brunswick, NJ, USA
†Department of Chemistry, University of Florida, Gainesville, FL, USA

More than a century has passed since Dakin first discovered that racemic ethyl mandelate could be resolved by enzyme-catalyzed hydrolysis [1]. The field of biocatalysis has grown tremendously since that modest beginning and it is now firmly established as an important aspect of modern organic synthesis. The high selectivities of enzymes – chemo-, regio-, and stereo- – are now widely appreciated and biocatalytic approaches are routinely evaluated when planning synthetic routes [2–4]. In particular, enzymatic resolutions to produce chiral alcohols, acids, and esters are highly mature and widely used technologies, including some examples at industrial scales [5]. Chiral alcohol synthesis by enzyme-catalyzed ketone reductions is likewise a well-established method that is now often considered the "first-option" approach in many cases [6]. Amino acids – both natural and unnatural – are another synthetic realm in which biocatalysis is a dominant contributor in the form of amino acid dehydrogenases and/or transaminases [7,8]. As detailed elsewhere in this volume, a transaminase-catalyzed conversion of a ketone to a primary amine has also been used industrially for production of a pharmaceutical [9], which augurs well for continued growth in this area.

In addition to catalytic efficiencies and selectivities, biocatalysis can also play a key role in increasing the sustainability of chemical processes. Most enzymes used for preparative synthesis are overexpressed in recombinant microorganisms that can be grown in fermentors. The host microbes are usually GRAS organisms and require only biosafety level I precautions. Growth-medium components are typically simple inorganic salts such as NaCl, $MgCl_2$, or $K_mH_nPO_4$, along with natural products or their derivatives, for example, yeast extract or milk protein hydrolysates. Reactions involving enzymes are normally carried out in aqueous mixtures with small quantities of inorganic salts for pH buffering. Even when an organic solvent is used, they are safe and environmentally favorable in most cases, for example, heptane, isopropanol, etc. Solvent concentrations are also minimized

Organic Synthesis Using Biocatalysis. http://dx.doi.org/10.1016/B978-0-12-411518-7.00012-3

in order to maintain enzyme activity. The "green" nature of biocatalysis has been recognized as a strong benefit [10] and several biocatalytic processes have received US Presidential Green Chemistry awards [4].

Until recently, the largest impediment to growth in biocatalysis was the difficulty in obtaining synthetically useful enzymes, particularly for laboratories unable to carry out their own culturing of microorganisms and enzyme purification. Even as recently as the end of the last century, nearly all of the commercially available enzymes were primarily hydrolases along with a handful of oxidoreductases. Poor access to additional enzymes significantly hampered progress. At the time, most commercial enzymes were produced for the food, beverage, detergent, paper, and leather industries; applications to chemical synthesis were adventitious additions to their primary markets. Other classes of enzymes such ketone reductases or amino acid dehydrogenases were prepared by fermentation on an as-needed basis, often by individual laboratories. Happily, the enzyme-supply problem has been largely solved in recent years with the emergence of several companies that specialize in enzymes targeted primarily at the organic synthesis market. These biocatalysts are generally available off-the-shelf on small scales. Although they are usually more expensive than the earlier generation of hydrolases, their ready availability dramatically simplifies catalyst screening and this crucial benefit more than makes up for their higher costs for laboratory scale application. Moreover, some enzymes later "graduated" to production on much larger scales when a specific market opportunity arose, for example, for the synthesis of Atorvastatin and Sitagliptin. As with all commercial products, once a large market arises, the motivation to optimize catalyst production becomes much stronger and unit costs drop correspondingly.

Significant advances in protein engineering have been a second driver for increasing the impact of biocatalysis on chemical synthesis. It is rare that a naturally occurring enzyme possesses all the desired characteristics for chemical synthesis; usually one or more parameters (rate, stereoselectivity, resistance to product inhibition, etc.) are suboptimal. In the early days, chemical processes were designed "around" the inherent enzyme limitations and in favorable cases, an acceptable final solution could be identified. Although the technology of site-directed mutagenesis was reported more than 35 years ago [11], efficient methodologies for evolving enzyme properties critical for synthetic applications have become routine only within the past decade. Now, one can evolve an enzyme by a number of "directed evolution" methodologies rather rapidly, much faster than otherwise achievable by traditional natural selection. Properties amenable to improvement include enzyme activity, thermal stability, solvent stability, and stereoselectivity [12–15]. Although many examples have been reported, perhaps the best known is the highly efficient process for Sitagliptin based on an evolved transaminase [9].

What does the future hold for biocatalysis? We foresee not only continued growth and increasing impact overall but also certain specific areas poised for strong growth. Progress will depend on both scientific advances and technological improvements.

1. *New types of biocatalytic reactions.* When compared to the myriad reaction types that have been developed for synthetic organic chemistry, the current range of biocatalytic transformations is relatively small. In some cases, this is a consequence of a particular reaction type not being utilized in nature. For example, olefin metathesis has had an enormous impact in chemical synthesis; however, it seems highly unlikely that a direct enzymatic counterpart exists in nature since this reaction relies on rare transition metal ions that are not accessible to most living organisms. For this reason, nature has had to "invent around" restrictions such as these. On the other hand, only a small subset of known enzymes has been applied to organic synthesis. The field would benefit from additional applications of known, but previously unexplored enzymes for *in vitro* biocatalysis.
2. *More systematic studies of new biocatalytic reaction types.* Several examples exist where potentially useful reaction types have been reported, but have not been explored sufficiently to define their value in chemical synthesis. For example, enzyme-catalyzed flourination [16–18] and a polymer immobilized fluorinase enzyme [19] have been reported, but as yet there are no broader application of the reaction. Similarly, though an enzyme-catalyzed Diels–Alder reaction was reported two decades ago [20,21], questions still remain and there are technical challenges for wider applications of these enzymes for chemical synthesis [22].
3. *Unusual reactions by already-known enzymes.* Both lipases [23–28] and proteases [29–31] were reported to catalyze aldol condensation reactions. A hydrolytic enzyme was also reported to carry out a Michael addition reaction [32]. It was shown that model-guided engineering of Cal B with a specific mutation (S105L) improved the activity of aldol condensation reaction [23], though the improved activity was still low similar to those of catalytic antibodies. An initial report that Cal B lipase catalyzed a Knoevenagel reaction [33] was later shown to be erroneous (the actual reaction involved enzyme-catalyzed ester hydrolysis by Cal B followed by a spontaneous condensation reaction) [34]. This example points out the importance of careful studies to ensure that "new" reactions of already-known catalysts actually follow the proposed catalytic pathway. Such mechanistic studies can also lead to new insights for predicting novel catalytic activities. For example, a thorough mechanistic understanding and molecular engineering of cytochrome P450 led to the identification of a novel catalytic activity (olefin cyclopropanation via carbene transfer reaction) [35].
4. *Computational design of new biocatalysts.* Enzymes have been designed *de novo* to catalyze reactions with no known natural counterpart [36–38]. Although the activities of these first-generation *de novo* enzymes are relatively low, the field is in its infancy and its future appears bright.
5. *"Hybrid" biocatalysts.* As noted previously, nature is constrained in its choice of reactions by the availability of certain elements, particularly transition metals. One way to increase the range of accessible chemistry is to couple synthetically prepared organometallic catalysts to enzymes.

For example, asymmetric C—H activation has been achieved by combining a biotinylated rhodium complex with engineered streptavidin [39].

There are also some technical and commercial issues that must be overcome in order to facilitate the wider use of biocatalysis in organic synthesis:

1. *Increased commercial availability of a larger number of enzymes.* Although many more synthetically useful enzymes are available today than in the past, there is still a need to increase this number further. Consistent supply and quality are *important* for small-scale laboratory synthesis, but these traits are *absolutely essential* if enzymes are to be used for manufacturing.
2. *Better supplies of key enzymes.* Once a suitable enzyme has been identified (either by screening or developed by protein engineering), it must become available in large scale rapidly. In nearly every case, this will involve production in a recombinant host. Decreasing the time required for this step and ensuring a consistent supply at a reasonable cost is critical. The bottom line is that an enzyme that cannot be purchased from a commercial supplier has almost no chance of being tested by a practicing synthetic chemist.
3. *Shorter pathways to enzyme improvement.* Many exciting reports have come from both academic and industrial labs in which molecular biology techniques such as directed evolution and random mutation have been used to improve enzymes for specific applications. Although scientifically interesting, it is also necessary to consider whether these approaches to improving enzyme-catalyzed chemical synthesis are the only, or even the best answer. Molecular biology-based approaches can clearly create enzymes with very high activities and great selectivities; however, each enzyme–substrate pair requires extensive, expensive, and time-consuming work. For this reason, protein-engineering strategies may be favored by molecular biologists, but may often be impractical for organic chemists.

   Rather than modifying the enzyme, one can also modify the reaction conditions or add modifiers such as solvents and/or salts to improve the synthetic outcome. Such strategies can be deployed much more rapidly and could allow better, quicker, and more widespread use of biocatalysis by organic chemists. A second approach is to develop enzymes that perform well on whole groups of substrates, rather than working very well on a tiny subset. It is worth noting that widespread use of the automobile by the general public did not come from the luxury of a Rolls-Royce, the thrill of a Lamborghini or the superb engineering of Mercedes; instead, it was the reliable and inexpensive Model T that fulfilled the critical need for people to go places at a reasonable cost. Even today, our roads are still largely travelled by Toyotas, Hondas, Fords, Chevrolets, and Volkswagens. Translated to the world of biocatalysis, one often does not need to develop the "perfect" enzyme for every case when a pretty good solution will do. Readily available low-cost enzymes are needed that accept a variety of substrates and perform well for the vast majority of them. Such enzymes, combined with literature guidance on adjusting reaction conditions, would make it much simpler (and therefore cheaper) to employ biocatalytic strategies in chemical synthesis.

4. *Process and engineering data for large-scale applications*. Compared to conventional chemical processes, where large volumes of data are available, the engineering and technological information for enzyme production and biocatalytic reactions is significantly sparser. This is expected for any relatively young field and more data will be accumulated as additional biocatalytic reactions are studied, developed, and commercialized. In other cases, however, the problem lies in the language used to describe the systems. Many scientists working in the field of biocatalysis, especially in the area of enzyme creation and production, speak the language of biologists and develop biologically focused data. Though these are valuable, the application of enzymes in organic synthesis requires the mindset of chemists and chemical engineers, particularly as scales increase. For this reason, appropriate process data must be generated for enzyme-catalyzed reactions, just as they are for any other type of chemical conversion.

We conclude with a hopeful message: The future of biocatalysis is more promising now than ever before. Our greater understanding of the fundamentals of enzyme catalysis combined with an expanding commercial availability of proteins and greater experience in their applications has provided a roadmap to future developments. Enzymes can enable the synthesis of many classes of compounds, often with yields, selectivities, and productivities that are unrivalled by other methodologies. What is needed is a willingness to understand past experience and use it as the basis for future discoveries. We hope that this book has fulfilled the first task and look forward to its readers writing the future chapters of the field with their own creativity.

References

[1] H.D. Dakin, The hydrolysis of ethyl mandelate by lipase, Proc. Chem. Soc. London 19 (1903) 161.

[2] M. Breuer, K. Ditrich, T. Habicher, B. Hauer, M. Keßeler, R. Stürmer, T. Zelinski, Industrial methods for the production of optically active intermediates, Angew. Chem. Int. Ed. 43 (2004) 788–824.

[3] B.M. Nestl, B.A. Nebel, B. Hauer, Recent progress in industrial biocatalysis, Curr. Opin. Chem. Biol. 15 (2011) 187–193.

[4] U.T. Bornscheuer, G.W. Huisman, R.J. Kazlauskas, S. Lutz, J.C. Moore, K. Robins, Engineering the third wave of biocatalysis, Nature 485 (2012) 185–194.

[5] C.A. Martinez, S. Hu, Y. Dumond, J. Tao, P. Kelleher, L. Tully, Development of a chemoenzymatic manufacturing process for pregabalin, Org. Proc. Res. Dev. 12 (2008) 392–398.

[6] J.C. Moore, C.K. Savile, S. Panuri, B. Kosjek, J.M. Janey, , in: D.L. Hughes (Ed.), Industrial application of asymmetric synthesis, vol. 9, Elsevier, Amsterdam, 2012, pp. 318–341.

[7] R.L. Hanson, S.L. Goldberg, D.B. Brzozowski, T.P.D.C. Tully, W.L. Parker, O.K. Lyngberg, T.C. Vu, M.K. Wong, R.N. Patel, Preparation of an amino acid intermediate for the dipeptidyl peptidase IV inhibitor, Saxagliptin, using a modified phenylalanine dehydrogenase, Adv. Synth. Catal. 349 (2007) 1369–1378.

[8] R.L. Hanson, R.M. Johnston, S.L. Goldberg, W.L. Parker, A. Goswami, Enzymatic preparation of an *R*-amino acid intermediate for a *gamma*-Secretase inhibitor, Org. Proc. Res. Dev. 17 (2013) 693–700.

[9] C.K. Savile, J.M. Janey, E.C. Mundorff, J.C. Moore, S. Tam, W.R. Jarvis, J.C. Colbeck, A. Krebber, F.J. Fleitz, J. Brands, P.N. Devine, G.W. Huisman, G.J. Hughes, Biocatalytic asymmetric synthesis of chiral amines from ketones applied to Sitagliptin manufacture, Science 329 (2010) 305–309.

[10] P.J. Dunn, The importance of green chemistry in process research and development, Chem. Soc. Rev. 41 (2012) 1452–1461.

[11] C.A. Hutchison, S. Phillips, M.H. Edgell, S. Gillam, P. Jahnke, M. Smith, Mutagenesis at a specific position in a DNA-sequence, J. Biol. Chem. 253 (1978) 6551–6560.

[12] S. Lutz, Beyond directed evolution – semirational protein engineering and design, Curr. Opin. Biotechnol. 21 (2010) 734–743.

[13] C. Jäckel, D. Hilvert, Biocatalysts by evolution, Curr. Opin. Biotechnol. 21 (2010) 753–759.

[14] A.S. Bommarius, J.K. Blum, M.J. Abrahamson, Status of protein engineering for biocatalysts: how to design an industrially useful biocatalyst, Curr. Opin. Chem. Biol. 15 (2011) 194–200.

[15] E.M. Brustad, F.H. Arnold, Optimizing non-natural protein function with directed evolution, Curr. Opin. Chem. Biol. 15 (2011) 201–210.

[16] M. Sanada, t. Miyano, S. Iwadare, J.M. Williamson, B.H. Arison, J.L. Smith, A.W. Douglas, J.M. Liesch, E. Inamine, Biosynthesis of fluorothreonine and fluoroacetic acid by the Thienamycin producer, *Streptomyces-cattleya*, J. Antibiot. 39 (1986) 259–265.

[17] T. Tamura, Y. Sawamoto, T. Kuriyama, K. Oba, H. Tanaka, K. Inagaki, Cosynthesis of monofluoracetate and 4-fluorothreonine by resting cells of blocked mutants of *Streptomyces cattleya* NRRL8057, J. Mol. Catal. B Enzym. 23 (2003) 257–263.

[18] C.D. Cadicamo, J. Cortieu, H. Deng, A. Meddour, D. O'Hagan, Enzymatic fluorination in *Streptomyces cattleya* takes place with an inversion of configuration consistent with an S_N2 reaction mechanism, ChemBioChem 5 (2004) 685–690.

[19] M.E. Sergeev, F. Morgia, M.R. Javed, M. Doi, P.Y. Keng, Polymer-immobilized fluorinase: recyclable catalyst for fluorination reactions, J. Mol. Catal. B Enzym. 92 (2013) 51–56.

[20] H. Oikawa, K. Katayama, Y. Suzuki, A. Ichihara, Enzymatic-activity catalyzing *exo*-selective Diels-Alder reaction in solanapyrone biosynthesis, J. Chem. Soc. Chem. Commun. (1995) 1321–1322.

[21] K. Katayama, T. Kobayashi, H. Oikawa, M. Honma, A. Ichihara, Enzymatic activity and partial purification of solanapyrone synthase: first enzyme catalyzing Diels-Alder reaction, Biochim. Biophys. Acta 1384 (1998) 387–395.

[22] H.J. Kim, M.W. Ruszczycky, H.W. Liu, Current developments and challenges in the search for a naturally selected Diels-Alderase, Curr. Opin. Chem. Biol. 16 (2012) 124–131.

[23] C. Branneby, P. Carlqvist, A. Magnusson, K. Hult, T. Brinck, P. Berglund, Carbon–carbon bonds by hydrolytic enzymes, J. Am. Chem. Soc. 125 (2003) 874–875.

[24] C. Li, X.-W. Feng, N. Wang, Y.-J. Zhou, X.-Q. Yu, Biocatalytic promiscuity: the first lipase-catalysed asymmetric Aldol reaction, Green Chem. 10 (2008) 616–618.

[25] M. Kapoor, A.B. Majumder, J. Mukherjee, M.N. Gupta, Decarboxylative aldol reaction catalysed by lipases and a protease in organic co-solvent mixtures and nearly anhydrous organic solvent media, Biocatal. Biotrans. 30 (2012) 399–408.

[26] Z. Guan, J.-P. Fu, Y.-H. He, Biocatalytic promiscuity: lipase-catalyzed asymmetric Aldol reaction of heterocyclic ketones with aldehydes, Tetrahedron Lett. 53 (2012) 4959–4961.

[27] Z.-B. Xie, N. Wang, G.-F. Jiang, X.-Q. Yu, Biocatalytic asymmetric Aldol reaction in buffer solution, Tetrahedron Lett. 54 (2013) 945–948.

[28] Z.-B. Xie, N. Wang, L.-H. Zhou, F. Wan, T. He, Z.-G. Le, X.-Q. Yu, Lipase-catalyzed stereoselective cross-Aldol reaction promoted by water, ChemCatChem 5 (2013) 1935–1940.

[29] C. Li, Y.-J. Zhou, N. Wang, X.-W. Feng, K. Li, X.-Q. Yu, Promiscuous protease-catalyzed Aldol reactions: a facile biocatalytic protocol for carbon–carbon bond formation in aqueous media, J. Biotechnol. 150 (2010) 539–545.

[30] B.-H. Xie, W. Li, Y. Liu, H.-H. Li, Z. Guan, Y.-H. He, The enzymatic asymmetric Aldol reaction using acidic protease from *Aspergillus usamii*, Tetrahedron 68 (2012) 3160–3164.

[31] Y.-L. Chen, W. Li, Y. Liu, Z. Guan, Y.-H. He, Trypsin-catalyzed direct asymmetric Aldol reaction, J. Mol. Catal. B Enzym. 87 (2013) 83–87.

[32] J.-F. Cai, Z. Guan, Y.-H. He, The lipase-catalyzed asymmetric C—C Michael addition, J. Mol. Catal. B Enzym. 68 (2011) 240–244.

[33] X.-W. Feng, C. Li, N. Wang, K. Li, W.-W. Zhang, Z. Wang, X.-Q. Yu, Lipase-catalyzed decarboxylative Aldol reaction and decarboxylative knoevenagel reaction, Green Chem. 11 (2009) 1933–1936.

[34] A.S. Evitt, U.T. Bornscheuer, Lipase CAL-B does not catalyze a promiscuous decarboxylative Aldol addition or knoevenagel reaction, Green Chem. 13 (2011) 1141–1142.

[35] P.S. Coelho, E.M. Brustad, A. Kannan, F.H. Arnold, Olefin cyclopropanation via carbene transfer aatalyzed by engineered cytochrome P450 enzymes, Science 339 (2013) 307–310.

[36] L. Jiang, E.A. Althoff, F.R. Clemente, L. Doyle, D. Rothlisberger, A. Zanghellini, J.L. Gallaher, J.L. Betker, F. Tanaka, C.F. Barbas III, D. Hilvert, K.N. Houk, B.L. Stoddard, D. Baker, *De novo* computational design of retro-Aldol enzymes, Science 319 (2008) 1387–1391.

[37] D. Rothlisberger, O. Khersonsky, A.M. Wollacott, L. Jiang, J. DeChancie, J. Betker, J.L. Gallaher, E.A. Althoff, A. Zanghellini, O. Dym, S. Albeck, K.N. Houk, D.S. Tawfik, D. Baker, Kemp elimination catalysts by computational enzyme design, Nature 453 (2008) 190–195.

[38] J.B. Siegel, A. Zhanghellini, H.M. Lovick, G. Kiss, A.R. Lambert, J.L. St. Clair, J.L. Gallaher, D. Hilvert, M.H. Gelb, B.L. Stoddard, K.N. Houk, F.E. Michael, D. Baker, Computational design of an enzyme catalyst for a stereoselective bimolecular Diels-Alder reaction, Science 329 (2010) 309–313.

[39] T.K. Hyster, L. Knorr, T.R. Ward, T. Rovis, Biotinylated Rh(III) complexes in engineered Streptavidin for accelerated asymmetric C—H activation, Science 338 (2012) 500–503.

Subject Index

E

F

G

H

I

K